Fundamentals of Optics

SECOND EDITION

DEVRAJ SINGH

Assistant Professor and Head
Department of Applied Physics
Amity School of Engineering and Technology
Bijwasan, New Delhi

PHI Learning Private Limited

Delhi-110092
2015

₹595.00

FUNDAMENTALS OF OPTICS, Second Edition
Devraj Singh

© 2015 by PHI Learning Private Limited, Delhi. All rights reserved. No part of this book may be reproduced in any form, by mimeograph or any other means, without permission in writing from the publisher.

ISBN-978-81-203-5146-2

The export rights of this book are vested solely with the publisher.

Second Printing (Second Edition)   ···   ···   **August, 2015**

Published by Asoke K. Ghosh, PHI Learning Private Limited, Rimjhim House, 111, Patparganj Industrial Estate, Delhi-110092 and Printed by Syndicate Binders, A-20, Hosiery Complex, Noida, Phase-II Extension, Noida-201305 (N.C.R. Delhi).

यन्मण्डलं दीप्तिकरं विशालं रत्नप्रभं तीव्रमनादि रूपम्।
दारिद्रय दुःख क्षयकारणं च पुनातु मां तत्सवितुर्वरेण्यम्।।

—श्री आदित्य हृदय स्तोत्रम्
श्लोक 142

The source of light sparkling like a gem,
who is eternal and destroyer of poverty;
and replica of the Sun, we pray to Him to make us purified.

Contents

Preface xv
Acknowledgements xvii

Part I   GEOMETRICAL OPTICS

1. Fermat's Principle 3–15
    1.1   Introduction 4
    1.2   Optical Path 4
    1.3   Fermat's Principle of Extreme Path 5
    1.4   Laws of Reflection by Fermat's Principle 6
          1.4.1   First Law 6
          1.4.2   Second Law 7
    1.5   Laws of Refraction by Fermat's Principle 8
          1.5.1   First Law 8
          1.5.2   Second Law 8
    1.6   Case of Maximum Path or Extremum Time 10
    Formulae at a Glance 11
    Solved Numerical Problems 11
    Conceptual Questions 13
    Exercises 13
          Theoretical Questions 13
          Numerical Problems 14
          Multiple Choice Questions 15

v

2. Geometrical Optics 16–42

- 2.1 Introduction 17
- 2.2 Lenses: Thin and Thick Lenses 17
- 2.3 Lens Equation 17
- 2.4 Lens Maker's Formula 19
  - 2.4.1 Lens Maker's Formula for a Double Convex Lens 20
  - 2.4.2 Lens Maker's Formula for a Double Concave Lens 22
- 2.5 Cardinal Points of a Coaxially Optical System 23
  - 2.5.1 Principal Points 23
  - 2.5.2 Focal Points 24
  - 2.5.3 Nodal Points 25
- 2.6 Combination of Two Thin Lenses (Equivalent Lenses): Coaxial Lens System 26
- 2.7 Matrix Method in Paraxial Optical System 30
  - 2.7.1 Translation Matrix 30
  - 2.7.2 The Refraction Matrix 31
  - 2.7.3 The Reflection Matrix 32

Formulae at a Glance 33
Solved Numerical Problems 35
Conceptual Questions 38
Exercises 39

- *Theoretical Questions* 39
- *Numerical Problems* 40
- *Multiple Choice Questions* 41

3. Dispersion of Light 43–67

- 3.1 Introduction 44
  - 3.1.1 Newton's Classic Experiment on Dispersion of White Light 44
  - 3.1.2 Cause of Dispersion (Cauchy's Relation) 45
- 3.2 Angular Dispersion and Dispersive Power 45
- 3.3 Combination of Prisms 47
  - 3.3.1 Deviation without Dispersion 47
  - 3.3.2 Dispersion without Deviation 49
- 3.4 Direct Vision Spectroscope 51
- 3.5 Rainbow 51
  - 3.5.1 Primary Rainbow 52
  - 3.5.2 Secondary Rainbow 56
- 3.6 Sundogs and Halos 57

Formulae at a Glance 59
Solved Numerical Problems 60
Conceptual Questions 62

Contents **vii**

Exercises 63
    Theoretical Questions 63
    Numerical Problems 64
    Multiple Choice Questions 65

4. Lens Aberrations 68–103

4.1   Introduction 69
4.2   Types of Aberrations 69
    4.2.1   Chromatic Aberration 69
    4.2.2   Monochromatic Aberration 69
4.3   Types of Monochromatic Aberration and their Reduction 70
    4.3.1   Spherical Aberration 70
    4.3.2   Coma 75
    4.3.3   Astigmatism 77
    4.3.4   Curvature of Field 78
    4.3.5   Distortion 80
4.4   Chromatic Aberration 81
    4.4.1   Types of Chromatic Aberration 82
    4.4.2   Achromatism of Lenses 85
4.5   Aplanatism and Aplanatic Points 90
    4.5.1   Aplanatic Points for a Spherical Refracting Surface 91
    4.5.2   Oil-immersion Objective of High Power Microscope 92
Formulae at a Glance 93
Solved Numerical Problems 95
Conceptual Questions 99
Exercises 100
    Theoretical Questions 100
    Numerical Problems 101
    Multiple Choice Questions 102

5. Optical Instruments 104–142

5.1   Introduction 105
5.2   Microscopes 105
    5.2.1   Simple Microscope 106
    5.2.2   Compound Microscope 108
5.3   Telescopes 112
    5.3.1   Astronomical Telescope 112
    5.3.2   Terrestrial Telescope 115
    5.3.3   Reflecting Telescopes 116
5.4   Eyepieces or Oculars 117
    5.4.1   Huygen's Eyepiece 119

viii Contents

      5.4.2   Ramsden's Eyepiece 122
      5.4.3   Relative Merits and Demerits of Huygen's and Ramsden's Eyepieces 125
      5.4.4   Gauss Eyepiece 125
  5.5   Spectrometer 126
      5.5.1   Measurement of Refractive Index by a Spectrometer 128
  5.6   Electron Microscope 129
Formulae at a Glance *131*
Solved Numerical Problems *133*
Conceptual Questions *137*
Exercises *138*
      Theoretical Questions *138*
      Numerical Problems *139*
      Multiple Choice Questions *140*

Part II   VIBRATIONS AND WAVES

6. Fundamentals of Vibrations 145–196

  6.1   Introduction 146
  6.2   Simple Harmonic Motion and Harmonic Oscillator 147
  6.3   Energy of Harmonic Oscillator 150
  6.4   Average Values of Kinetic and Potential Energy 151
  6.5   Example of Harmonic Oscillator: Simple Pendulum 152
  6.6   Damping Force/Damping Motion 155
  6.7   Damped Harmonic Oscillator 157
  6.8   Power Dissipation in Damped Harmonic Oscillator 162
  6.9   Quality Factor (Q) of Damped Harmonic Oscillator 163
  6.10 Forced Harmonic Oscillator 163
  6.11 Resonance 166
  6.12 Amplitude Resonance 167
  6.13 Sharpness of Resonance 169
  6.14 Half Width of Resonance Curve 169
  6.15 Velocity Resonance 170
  6.16 Power Absorption of Forced Harmonic Oscillator 171
  6.17 Quality Factor (Q) of Forced Harmonic Oscillator 173
Formulae at a Glance *175*
Solved Numerical Problems *179*
Conceptual Questions *184*
Exercises *186*
      Theoretical Questions *186*
      Numerical Problems *187*
      Multiple Choice Questions *191*

7. Wave Motion 197–236

- 7.1 Introduction 198
- 7.2 What Propagates in Wave Motion? 198
- 7.3 Characteristics of Wave Motion 198
- 7.4 Types of Wave Motion 199
  - 7.4.1 Transverse Wave Motion 199
  - 7.4.2 Longitudinal Wave Motion 200
- 7.5 Important Definitions 201
- 7.6 Differential Equation of Wave Motion 202
- 7.7 Plane Progressive Wave in Fluid Media 207
  - 7.7.1 Wave Equation of Propagation of Sound Waves in Fluid Media 208
- 7.8 Pressure Variation in Longitudinal Wave (Acoustic Pressure) 211
  - 7.8.1 Relation between ψ and (ΔP) 212
- 7.9 Energy of Plane Progressive Wave 212
  - 7.9.1 Intensity 214
- 7.10 Reflection and Transmission of Transverse Waves in a String 215
- 7.11 Reflection and Transmission of Longitudinal Waves at Discontinuity 217
- 7.12 Superposition of Progressive Waves: Stationary Wave 219
  - 7.12.1 Analytical Treatment of Stationary Waves and their Properties 219
  - 7.12.2 Energy of Stationary Waves 221
  - 7.12.3 Characteristics of Stationary Wave 223
- 7.13 Comparison between Progressive Waves and Stationary Waves 224
- *Formulae at a Glance* 225
- *Solved Numerical Problems* 228
- *Conceptual Questions* 231
- *Exercises* 232
  - *Theoretical Problems* 232
  - *Numerical Problems* 233
  - *Multiple Choice Questions* 234

Part III   PHYSICAL OPTICS

8. Interference of Light Waves 239–354

- 8.1 Introduction 240
- 8.2 Wavefront and Rays 240
  - 8.2.1 Wavefront 240
  - 8.2.2 Ray of Light 242
- 8.3 Huygen's Principle of Secondary Wavelets 243
- 8.4 Principle of Superposition of Light Waves 244

8.5	Groups of Interference		245
8.6	Young's Double Slit Experiment (YDSE)		246
8.7	Coherence		247
8.8	Phase Difference and Path Difference		249
8.9	Conditions for Constructive and Destructive Interference		249
	8.9.1	Expression for Constructive and Destructive Interference Pattern	249
	8.9.2	Constructive Interference	250
	8.9.3	Destructive Interference	251
	8.9.4	Conservation of Energy in Interference: Average Intensity	251
	8.9.5	Comparison of Intensities at Maxima and Minima	252
	8.9.6	Visibility of Fringes	252
	8.9.7	Intensity Variation in Interference	252
8.10	Theory of Interference Fringes		255
	8.10.1	Expression for Fringe Width	255
	8.10.2	Positions of Bright Fringes	257
	8.10.3	Positions of Dark Fringes	257
	8.10.4	Fringe Width	258
	8.10.5	Measurement of Wavelength	258
	8.10.6	Interference Pattern with White Light	258
	8.10.7	Shape of the Interference Fringes	259
	8.10.8	Angular Fringe Width	259
	8.10.9	Displacement of Fringes	260
8.11	Conditions for Interference of Light Waves		263
	8.11.1	Conditions for Sustained Interference	263
	8.11.2	Conditions for Good Visibility	264
	8.11.3	Conditions for Good Contrast	264
8.12	Interference Fringes with Fresnel's Biprism		264
	8.12.1	Interference Fringes with White Light	267
	8.12.2	Effect of Increasing the Slit Width on Fresnel's Fringes	267
	8.12.3	Effect of Increasing the Angle of Biprism on Fringes	267
	8.12.4	Location of Zero Order Fringe in Biprism Experiment	267
8.13	Lloyd's Single Mirror		269
8.14	Fresnel's Double Mirror		270
8.15	Stokes' Law		271
8.16	Interference from Parallel Thin Films or Colour of Thin films		272
	8.16.1	Interference due to Reflected Light	272
	8.16.2	Interference due to Transmitted Light	274
	8.16.3	Colours in Reflected and Transmitted Light be Complementary	278
	8.16.4	Origins of Colours in Thin Film	278
	8.16.5	Colour in Thick Films	278
	8.16.6	Necessity of an Extended Source	278
	8.16.7	How Thin Must be a Thin Film?	279
	8.16.8	Classification of Fringes Exhibited by Thin Films	280

	8.17	Interference in Non-uniform Thick Film: Wedge-shaped Film	280
		8.17.1 Spacing between Two Consecutive Dark Bands	282
		8.17.2 If White Light is Substituted for a Sodium Light	283
		8.17.3 Testing of Optical Flatness of Surfaces	284
	8.18	Newton's Rings	285
		8.18.1 Experimental Arrangement for Newton's Rings	286
		8.18.2 Formation of Newton's Rings	286
		8.18.3 Production of Coherent Sources in Newton's Rings Experiment	287
		8.18.4 Theory of Newton's Rings	288
		8.18.5 Determination of Wavelength of Monochromatic Light using Newton's Rings Experiment	290
		8.18.6 Determination of Refractive Index of Transparent Liquid using Newton's Rings Experiment	291
		8.18.7 Newton's Rings by Contact of Concave and Convex Surfaces	291
		8.18.8 Newton's Rings by Contact of Two Convex Surfaces	293
		8.18.9 The Perfect Blackness of the Central Spot in Newton's Rings System	296
		8.18.10 Newton's Rings are Circular but Air-wedge Fringes are Straight	296
		8.18.11 Newton's Rings with White Light	297
	8.19	Michelson's Interferometer	297
		8.19.1 Working: Formation of Interference Fringes	298
		8.19.2 Measurements with Michelson's Interferometer	303
	8.20	Multiple Beam Interferometry	306
		8.20.1 Fabry–Perot Interferometer	307
		8.20.2 Lummer–Gehrcke Plate	313
	8.21	Interference Refractometers	316
		8.21.1 Jamin's Refractometer	317
		8.21.2 Rayleigh's Refractometer	318
		8.21.3 Mach–Zehnder's Refractometer	319
	8.22	Interference in Optical Technology	320
		8.22.1 Interference Filters	320
		8.22.2 Antireflection Coatings	321
	Formulae at a Glance		324
	Solved Numerical Problems		329
	Conceptual Questions		337
	Exercises		343
		Theoretical Questions	343
		Numerical Problems	346
		Multiple Choice Questions	349

9. Diffraction of Light Waves 355–451

	9.1	Introduction	356
	9.2	Diffraction and Huygen's Principle	357

9.3	Distinction between Interference and Diffraction	358
9.4	Fresnel's Explanation of Rectilinear Propagation of Light	358
	9.4.1 Fresnel's Half-period Zones	359
	9.4.2 Governing Factors of Amplitude	359
	9.4.3 Resultant Amplitude Due to Wavefront	361
9.5	Zone Plate	363
	9.5.1 Theory of the Zone Plate	363
9.6	Comparison between the Action of a Zone Plate and That of a Convex Lens	367
9.7	Fresnel's Diffraction Due to a Straight Edge	368
	9.7.1 Mathematical Treatment	369
	9.7.2 Wavelength of Monochromatic Light	371
9.8	Fresnel's Diffraction at a Circular Disc	372
9.9	Fresnel's Diffraction at a Circular Aperture	373
9.10	Fraunhofer's Diffraction at a Single Slit	374
9.11	Fraunhofer's Diffraction at Double Slit	384
9.12	Plane Transmission Diffraction Grating: Fraunhofer's Diffraction at N Parallel Slits	390
	9.12.1 Intensity Distribution	391
	9.12.2 Grating Spectra	396
	9.12.3 Angular Half Width of Principal Maxima	397
	9.12.4 Absent Spectra or Missing Orders in N-slits Diffraction Pattern	398
	9.12.5 Dispersive Power of Grating	399
	9.12.6 Grating at Oblique Incidence	400
	9.12.7 Wavelength of Incident Light by Means of Diffraction Grating	401
9.13	Resolving Power of an Optical Instrument	406
9.14	Rayleigh Criterion for the Limit of Resolution	406
9.15	Resolving Power of a Plane Diffraction Grating	407
9.16	Resolving power of a Prism	410
9.17	Theory of Concave Grating	412
9.18	Mountings of Concave Grating	414
9.19	Echelon Grating	418
9.20	Difference between Prism Spectrum and Grating Spectrum	420
9.21	Difference between Dispersive Power and Resolving Power of a Grating	420

Formulae at a Glance ... 420
Solved Numerical Problems ... 424
Conceptual Questions ... 437
Exercises ... 440
    Theoretical Questions ... 440
    Numerical Problems ... 445
    Multiple Choice Questions ... 449

10. Polarization of Light Waves 452–518

    10.1 Introduction 453
    10.2 Polarization of Light by Tourmaline Crystals Experiments 454
    10.3 Plane of Vibration and Plane of Polarization 454
    10.4 Pictorial Representation of Light Vibrations 455
    10.5 Methods of Producing Plane Polarized Light 455

        10.5.1 Polarization by Reflection 456
        10.5.2 Polarization by Refraction (A Pile of Plates) 458
        10.5.3 Polarization by Double Refraction (Birefringence) 461
        10.5.4 Polarization by Scattering 475
        10.5.5 Polarization by Selective Absorption (Dichroism) 476

    10.6 Matrix Representation of Plane-Polarized Light Waves 478
    10.7 Optical Activity 481

        10.7.1 Biot's Laws of Optical Activity: Specific Rotation 481
        10.7.2 Fresnel's Explanation of Optical Rotation 484

    10.8 Polarimeter 486

        10.8.1 Laurent's Half Shade Polarimeter 487
        10.8.2 Biquartz Polarimeter 489
        10.8.3 Lippich's Polarimeter 490
        10.8.4 Soleil's Compensated Biquartz Polarimeter 491

    10.9 Photoelasticity 491

        10.9.1 Birefringence or Double Refraction 492
        10.9.2 Stress Optic Law 493
        10.9.3 Theory of Photoelasticity 493
        10.9.4 Applications of Photoelasticity 495

    Formulae at a Glance 496
    Solved Numerical Problems 498
    Conceptual Questions 503
    Exercises 506

        Theoretical Questions 506
        Numerical Problems 511
        Multiple Choice Questions 514

Part IV   ELECTROMAGNETIC WAVES

11. Electromagnetic Waves 521–574

    11.1 Introduction 522
    11.2 Production of Electromagnetic Waves by an Antenna 523
    11.3 Wave Equation for Waves in Space 525

	11.4	Wave Propagation in Lossy Dielectric Medium	534
	11.5	Conductors and Dielectrics	537
	11.6	Wave Propagation in Good Dielectrics	537
	11.7	Wave Propagation in a Good Conductor	538
	11.8	Depth of Penetration: Skin Depth	539
	11.9	Poynting Vector and Poynting Theorem	542
	11.10	Polarization	545
	11.11	Reflection of Uniform Plane Waves by Perfect Dielectric—Normal Incidence	546

Formulae at a Glance 551
Solved Numerical Problems 554
Conceptual Questions 563
Exercises 566
    Theoretical Questions 566
    Numerical Problems 567
    Multiple Choice Questions 570

Appendices **575–578**

Bibliography **579–580**

Index **581–589**

Preface

This book on *Fundamentals of Optics* is in its second edition, intended to serve as a textbook for B.Sc. (Physics) and Engineering students of Indian Universities. This revised edition is designed primarily rectifying remaining faults and checking the imminent inadequacies in the first edition of the book. The endeavour has been made to eradicate without changing the technique of presentation, whose characteristics are lucidity and simplicity. This book is an outcome of my teaching experience and discussions with the students. The need for a book offering a comprehensive background to those students of Physics who wish to specialize in Optics as well as to those who are anxious to have a general knowledge has been indeed long felt. The subject matter of the book has been selected and developed to bridge the gap between the introductory and the advanced level courses in Optics.

The new materials included, pertains to the following topics:

- Matrix method in paraxial optical system in Geometrical Optics
- Matrix representation of plane polarized waves in Polarization of Light Waves
- Reflection of uniform plane waves by perfect dielectrics in Electromagnetic Waves
- Two new chapters, 'Fundamentals of Vibrations' and 'Wave Motion'

Furthermore, with a view to facilitate the self-study on the part of the students, following pedagogical features of the book have been provided:

- Simplified development of each topic to improve subject understanding
- Presentation of fundamental concepts in simplified way
- Methodical and systematic treatment of all the chapters
- Concise and simplified derivations of important formulae
- Well-illustrated diagrams
- Many universities examination questions
- *Formulae at a Glance* to make simple numerical problems
- Chapter-wise conceptual questions with answers

- Topic-wise solved examples to understand the concepts clearly
- Miscellaneous solved numerical problems
- Numerous theoretical questions, numerical problems and multiple choice questions to understand the concepts

The book is divided into 11 chapters.

Chapter 1 deals with Fermat's principle in terms of basic phenomena of light, like reflection and refraction laws using simple geometry. In Chapter 2, dispersion of light is described giving emphasis on rainbows, sundogs, halos, etc. A simple analysis of dispersion of prism is also given. Geometrical optics is the subject matter of Chapter 3. It describes the basic geometry of lenses, lensmaker's formula, cardinal points, etc.

Chapter 4 discusses lens aberrations. The definitions and explanations of spherical and chromatic aberrations have been provided with mathematical approach. The optical instruments, like eyepieces, electron microscope, compound microscope, etc. are described in Chapter 5.

Chapter 6 introduces the concepts of vibrations, i.e. simple harmonic motion, damped harmonic motion and forced harmonic motion. Chapter 7 is devoted on characteristics, types, differential equations of wave motion.

In Chapter 8, the phenomenon of interference of light waves is discussed. Interference due to division of wavefront and amplitude has been described with simple and clear illustrations. Interferometry is also discussed in detail. Chapter 9 deals with diffraction of light waves of Fresnel and Fraunhofer classes. It also focuses on resolving power, dispersive power of grating and prism.

On the basis of the phenomena of interference and diffraction, it was established that the light is a form of wave motion. But those phenomena do not reveal the characteristics of wave motion, i.e. whether it is longitudinal or transverse. When the phenomenon of polarization of light waves was discovered, it was established beyond the doubt that the light waves are of transverse nature. This phenomenon of light waves is described in Chapter 10 and the concluding Chapter 11 presents the basic concepts of electromagnetic waves.

The text is supported by a large number of illustrative examples and review questions to reinforce the students' understanding of the subject matter.

In spite of my best efforts, some errors and omissions might have crept in such a volume. I shall feel obliged if they are brought in my notice and will be corrected in next edition. Any suggestions for improvement of the book will be thankfully acknowledged. Please mail me at dsingh13@amity.edu

DEVRAJ SINGH

Acknowledgements

First of all, I take this opportunity to place on record my indebtedness to a large number of relevant books and journals those I have freely consulted during the course of new edition of the book.

I am greatly indebted to my teachers especially Prof. S.K. Kor and Prof. R.R. Yadav, Department of Physics, University of Allahabad, who created in me an immense interest in this field.

I must owe special debt of gratitude to Dr. Ashok K. Chauhan, Founder President, Ritnand Balved Education Foundation for his encouragement during the course of the manuscript.

I wish to thank Prof. B.P. Singh, Senior Director and Prof. Rekha Agarwal, Director, ASET, Bijwasan, New Delhi for excellent and congenial academic environment of the college that inspired me to accept the challenge.

I am thankful to my friends particularly Dr. Rajesh Kumar, Department of Physics, Guru Gobind Singh Indraprastha University, New Delhi for stimulating discussion on each and every topic of the book.

I am especially grateful to my colleagues particularly Mr. Sudhanshu Tripathi, Dr. Pramod Yadav, Dr. Giridhar Mishra and Mrs. Vyoma Bhalla for fruitful suggestions during preparation of the book.

I am so grateful for the love and support of my sons, Shobhit and Parth for their beaming faces in the morning and eager inquisitiveness in the evening.

I wish to express my special thanks to my wife, Mamta, for her help and encouragement. Without her support, this work would not have been possible.

I express my special gratitude to PHI Learning for enthusiastic handling of the publication process and bringing out the book in record time. It was indeed a charm to work with the entire team, particularly Ms. Shivani Garg, Ms. Lakshmi S. Kumar, Mr. Suman Kumar, Mr. Sarvendra Kumar, Mr. Ajai Kumar Lal Das and Ms. Mini Uthaman.

DEVRAJ SINGH

PART I

Geometrical Optics

Chapter 1   Fermat's Principle
Chapter 2   Geometrical Optics
Chapter 3   Dispersion of Light
Chapter 4   Lens Aberrations
Chapter 5   Optical Instruments

CHAPTER 1

Fermat's Principle

"Optics, developing in us through study, teach us to see."

—Paul Cezanne

IN THIS CHAPTER

- Optical Path
- Fermat's Principle of Extreme Path
- Laws of Reflection by Fermat's Principle
- Laws of Refraction by Fermat's Principle
- Case of Maximum Path or Extremum Time

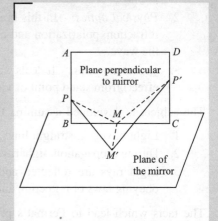

4 Fundamentals of Optics

1.1 INTRODUCTION

Light is a form of energy which evokes visual sensation of objects. However, nowadays radiations like ultraviolet, infrared are also included in this form of energy; which do not excite visual sensations but show similar effects.

Optics

Optics is the science of light and vision. It is also defined as the study of the phenomenon associated with generation, transmission and detection of electromagnetic radiations in the spectral range extending from the long wave edge of radio region. This range is often called optical region or the optical spectrum extended in wavelength from about 1 nanometre to about 1 millimetre.

Branches of optics: The following are the branches of optics:

1. *Geometrical optics:* In this branch, the rectilinear propagation of light, reflection, refraction and dispersion phenomena are studied assuming corpuscular nature of light.

    or

    It refers to the study of the image formation through the spherical surfaces, lenses and prisms.

2. *Physical optics:* In this branch, reflection, refraction, interference, diffraction, double refraction, polarization and other phenomena are studied assuming the nature of light as the wave.

3. *Quantum optics:* It deals with the interaction of radiation with matter (e.g. photoelectric effect) from stand point of view of quantum nature of light.

The subject is developed assuming the following properties of light:

1. Light travels in straight lines in a homogeneous medium.
2. During propagation, light rays can intersect each other without affecting their future path.
3. Light rays are reflected and refracted from optical surfaces and media respectively obeying laws of reflection and refraction.

The facts which lead to Fermat's principle of least time forms a basis of geometrical optics. The fundamental laws which form the basis of geometrical optics are:

(i) The rectilinear propagation of light
(ii) The laws of reflection of light
(iii) The laws of refraction of light

A single general principle which covers all these laws is known as Fermat's principle.

1.2 OPTICAL PATH

Optical path is the path taken by light ray through an optical system. It is also known as, the product of refractive index of the medium, *i.e.*, n and the distance travelled S by the light ray in medium *i.e.*,

$$nS = \text{Optical path, or equivalent air path}$$

If v is the velocity of light and t is the time taken for covering the distance S, then
$$S = vt \tag{1.1}$$
But, we know that the refractive index
$$n = \frac{c}{v} = \frac{\text{Velocity of light in vacuum}}{\text{Velocity of light in medium}}$$
so that
$$v = \frac{c}{n} \tag{1.2}$$
Substituting Eq. (1.2) in Eq. (1.1), we get
$$S = \frac{c}{n} t \tag{1.3}$$
or $\quad nS = ct = \Delta =$ optical path

When a ray travels S_1, S_2, S_3, S_4 ... distances in n_1, n_2, n_3, n_4, ... refractive indices media, then:
The optical path $\Delta = n_1 S_1 + n_2 S_2 + n_3 S_3 + n_4 S_4 + \cdots$
$$= \sum_i n_i S_i \tag{1.4}$$

For a medium of continuously varying optical density, the optical path of a ray PQ is expressed as an integral
$$\Delta = \int_P^Q n \, dS \tag{1.5}$$

EXAMPLE 1.1 Find the refractive index of water, if the optical ray of a monochromatic light is same for 2.25 cm of water or 2.0 cm of glass. The refractive of glass is 1.50.

Solution *Given:* $n_g = 1.50$, $s_g = 2.0$ cm, $s_w = 2.25$ cm.
Optical path through 2.25 cm water = Optical path through 2.0 cm glass
i.e. $\qquad n_w s_w = n_g s_g$
or $\qquad n_w \times 2.25 = 1.50 \times 2.0$
$$n_w = \frac{1.50 \times 2.0}{2.25}$$
$$= 1.33$$

1.3 FERMAT'S PRINCIPLE OF EXTREME PATH

Pierre de Fermat, a French mathematician, enunciated the principle of least time in the following way:

A ray of light is travelling from one point to another by any number of reflections and refractions follows that particular path for which the time taken is the least.

However, there are a number of cases in which the real path of light is the one for which time taken is maximum rather than minimum. Hence Fermat's principle in modified form states:

6 Fundamentals of Optics

A ray of light passing from one point to another through a set of media by a number of reflections and refractions selects a path for which the time taken is either a minimum or maximum or stationary.

This is known as "Fermat's principle of stationary time" or "Fermat's principle of extremum path".

Let dS be the small distance covered by light between two points P and Q in a medium and v is the velocity in that particular medium, then mathematical form of Fermat's principle is as:

$$\int_P^Q \frac{dS}{v} = \text{maximum or minimum or stationary}$$

or

$$\int_P^Q \frac{n\,dS}{c} = \text{maximum or minimum or stationary} \quad \left[\text{Here } n = \frac{c}{v}\right]$$

Since velocity of light (c) is constant, therefore, Fermat's principle of extremum path is:

$$\int_P^Q n\,dS = \text{maximum or minimum or stationary} \qquad (1.6)$$

where ndS is the optical path in a medium of refractive index n.

1.4 LAWS OF REFLECTION BY FERMAT'S PRINCIPLE

1.4.1 First Law

Statement: The incident ray, reflected ray and the normal ray at the point of incidence are in one plane.

Proof: Let us consider plane ABCD be normal to the plane mirror as shown in Figure 1.1. P is point object imaged by mirror as P'. Consider a point M' on the plane mirror; but not on plane *ABCD*. Let a ray PM' be reflected as $M'P'$. Drop a perpendicular $M'M$ on plane *ABCD*. M is the foot of perpendicular on *ABCD*.

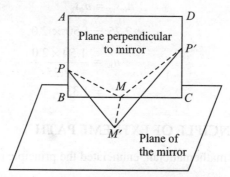

Figure 1.1 First law of reflection by Fermat's principle.

Now, PMM' and $P'MM'$ are right angle triangles. PM' and $P'M'$ are respective hypotenuses. Therefore, we have

$$PM' > PM \quad \text{and} \quad P'M' > P'M$$

But Fermat's principle demands that the path followed must be the shortest *i.e.*, the light would not travel along $PM'P'$. Now, shift M' towards M. It is seen that the shortest possible path is PMP', where the point of incidence M lies on plane $ABCD$. PM and MP' are the incident and reflected rays. This proves the first law of reflection.

1.4.2 Second Law

Let a ray of light from a point A (Figure 1.2) be reflected by a plane mirror MM' to another point B.

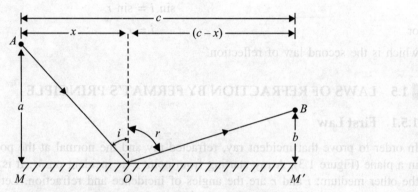

Figure 1.2 Second law of reflection by Fermat's principle.

Let i be the angle of incidence and r the angle of reflection. Let a and b be the lengths of the perpendiculars drawn from A to B on the mirror. Let $MM' = c$ and $MO = x$; so that $OM' = (c - x)$.

As the entire path AOB is travelled in air, the optical path between A and B is given by

$$l = AO + OB$$

$$= \sqrt{(AM)^2 + (MO)^2} + \sqrt{(BM')^2 + (OM')^2}$$

$$= \sqrt{a^2 + x^2} + \sqrt{b^2 + (c-x)^2} \qquad (1.7)$$

According to Fermat's principle, O will have a position such that the optical path l (or time of travel) is minimum or maximum. Therefore, first differential coefficient of l, with respect to x must be zero. Hence differentiating Eq. (1.7) with respect to x and equating the result to zero, we get

$$\frac{\partial l}{\partial x} = \frac{1}{2}(a^2 + x^2)^{-1/2}(2x) + \frac{1}{2}[b^2 + (c-x)^2]^{-1/2}(2)(c-x)(-1) = 0$$

This can be rearranged as

$$\frac{x}{\sqrt{a^2 + x^2}} = \frac{(c-x)}{\sqrt{b^2 + (c-x)^2}} \qquad (1.8)$$

8 Fundamentals of Optics

But from Figure 1.2, we have

$$\frac{x}{\sqrt{a^2 + x^2}} = \frac{MO}{AO} = \sin i$$

and

$$\frac{(c - x)}{\sqrt{b^2 + (c - x)^2}} = \frac{OM'}{OB} = \sin r$$

From Eq. (1.8), we have

$$\sin i = \sin r$$

or
$$i = r \qquad (1.9)$$

which is the second law of reflection.

1.5 LAWS OF REFRACTION BY FERMAT'S PRINCIPLE

1.5.1 First Law

In order to prove that incident ray, refracted ray and the normal at the point of incidence are in a plane (Figure 1.3). A ray starting from point P is incident on M. It is refracted as MP' in the other medium; i and r are the angles of incidence and refraction. Let us assume that the ray follows path $PM'P'$ instead of PMP'. It is evident that

$$PM' > PM \quad \text{and} \quad M'P' > MP'$$

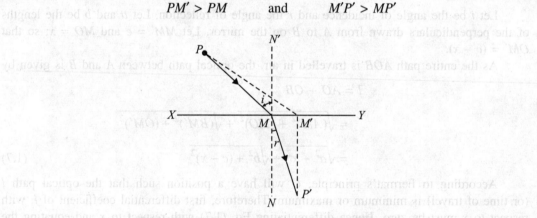

Figure 1.3 First law of refraction by Fermat's principle.

Therefore, path $PM'P'$ is not possible. Shift M' towards M, the path from P to P' becomes short. It is shortest when M' is coincident with M. It is inaccordance with Fermat's principle and proves the first law.

1.5.2 Second Law

In order to prove that the ratio of the $\sin i$ to $\sin r$ is equal to the refractive index of second medium with respect to the first (Figure 1.4).

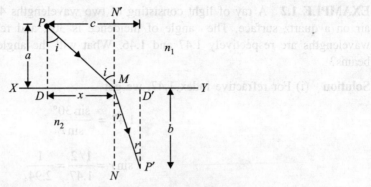

Figure 1.4 Second law of refraction by Fermat's principle.

XY be a plane surface dividing two media of refractive indices n_1 and n_2. Consider an object point P in the first medium, PM and MP' are the incident and refracted rays; i and r are the angle of incidence and refraction.

Let the distance be given as

$$PD = a,\ DM = x,\ DD' = c,\ D'P' = b$$

If a ray of light travels a distance l in a medium of refractive index n, then product nS is called the 'optical path' in the medium.

The optical path from P to P' is given by

$$l = PMP' = n_1 PM + n_2 MP' = n_1\sqrt{PD^2 + DM^2} + n_2\sqrt{D'M^2 + D'P^2}$$

$$= n_1\sqrt{a^2 + x^2} + n_2\sqrt{(c^2 - x^2) + b^2}$$

Now, for l to be minimum $\dfrac{dl}{dx}$ must be zero and $\dfrac{d^2 l}{dx}$ positive.

Differentiating S with respect to x, we get

$$\frac{dl}{dx} = \frac{n_1}{2} \times \frac{2x}{\sqrt{a^2 + x^2}} - \frac{n_2(c - x)}{\sqrt{(c - x)^2 + b^2}} = 0$$

or

$$\frac{n_1 x}{\sqrt{a^2 + x^2}} = \frac{n_2(c - x)}{\sqrt{(c - x)^2 + b^2}}$$

Using triangles PMD and $P'MD'$, the above relation is written as

$$n_1 \sin i = n_2 \sin r \quad \left[\text{Here } \sin i = \frac{x}{\sqrt{a^2 + x^2}} \text{ and } \sin r = \frac{(c - x)}{\sqrt{(c - x)^2 + b^2}}\right]$$

or

$$\frac{\sin i}{\sin r} = \frac{n_2}{n_1} = {}_1 n_2 \tag{1.10}$$

where, ${}_1 n_2$ is the refractive index of the second medium with respect to the first. This is the *Snell's law of refraction*.

The second differential coefficients of l may be shown to be positive. This proves that the second law of refraction is in accordance with Fermat's principle.

EXAMPLE 1.2 A ray of light consisting of two wavelengths 4000Å and 5000Å falls from air on a quartz surface. The angle of incidence is 30°, and refractive indices for the two wavelengths are respectively 1.47 and 1.46. What will the angle between the two refracted beams?

Solution (i) For refractive index 1.47, we get

$$1.47 = \frac{\sin 30°}{\sin r}$$

$$\sin r = \frac{1/2}{1.47} = \frac{1}{2.94}$$

$$\Rightarrow \quad r = 19.88°$$

(ii) For refractive index 1.46, we get

$$1.46 = \frac{\sin 30°}{\sin r'}$$

$$\sin r' = \frac{1/2}{1.46} = \frac{1}{2.92}$$

$$\Rightarrow \quad r' = 20.02°$$

Then $(r' - r) = 20.02° - 19.88° = 0.14°$

1.6 CASE OF MAXIMUM PATH OR EXTREMUM TIME

Consider a spherical mirror PQ as shown in Figure 1.5. A and B are the two points and AOB is the actual path of the ray. Now, draw an ellipse passing through O with A and B as its foci.

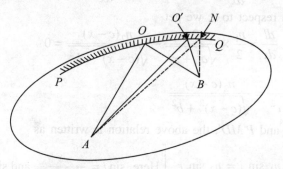

Figure 1.5 Arrangement of spherical mirror for the case of maximum path.

According to the property of ellipse, whatever may be the position of O, $(AO + OB)$ is constant. Let O' be the another point on ellipse, then due to this property of ellipse

$$AO + OB = AO' + O'B = \text{constant} \tag{1.11}$$

Suppose ANB is another path of the ray of light. The path difference (Δ) between AOB and ANB

$$\Delta = AOB - ANB = (AO + OB) - (AN + NB)$$
$$= (AO + OB) - [(AO' - NO') + NB]$$
$$= (AO' + O'B) - [(AO' - NO') + NB] \qquad [\text{Here } AO = AO']$$
$$\Delta = (O'B + NO') - NB \qquad (1.12)$$

Since $O'B$ and NO' are the two sides of the triangle $AO'B$ whose third side in NB. Therefore, path difference in Eq. (1.12) will be positive.

It means that any neighbouring path of ray of light (*e.g.* ANB) is smaller than actual path (AOB), *i.e.*, actual path in this case is maximum than any other path. Therefore, Fermat's principle is not the principle of least time but it is the principle of stationary time or extremum (either maximum or minimum). Thus, in general, the principle may be stated as the path taken by a ray of light in passing from one point to the other is that of minimum or maximum time.

FORMULAE AT A GLANCE

1.1 Optical path = nS
where n = refractive index of medium
and S = distance travelled by the ray.

1.2 Refractive index

$$n = \frac{c}{v} = \frac{\text{Velocity of light in vacuum}}{\text{Velocity of light in medium}}$$

1.3 Fermat's principle

(a) $\int_P^Q \frac{dS}{v}$ = maximum, minimum or stationary

or $\int_P^Q \frac{n\, dS}{v}$ = maximum, minimum or stationary

(b) $\int_P^Q n\, dS$ = maximum, minimum or stationary

1.4 Second law of reflection $< i = < r$.

1.5 Second law of refraction

$$\frac{\sin i}{\sin r} = \frac{n_2}{n_1} = {}_1n_2$$

where n_1 and n_2 are refractive indices of two media.

SOLVED NUMERICAL PROBLEMS

PROBLEM 1.1 The optical path of a monochromatic light is the same if it goes through 2.00 cm of glass or 2.50 cm of water. If the refractive index of water is 1.33, what is the refractive index of glass?

Solution When light travels through a distance S in a medium of refractive index n, its optical path is nS. Thus, if n is the refractive index of glass,

$$n \times 2.00 = 1.33 \times 2.50$$

or
$$n = \frac{1.33 \times 2.50}{2.00} = 1.66$$

PROBLEM 1.2 In Figure 1.6, light starts from point P and after reflection from the inner surface of the sphere reaches the diametrically opposite point Q. Calculate the length of hypothetical path PSQ and using Fermat's principle, find the actual path of light. Is the path minimum?

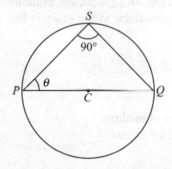

Figure 1.6 Illustration for Example 1.2.

Solution Refer to Figure 1.6. Let r be the radius of sphere and l be the length (hypothetical) of the path followed by light, i.e.

$$l = PS + SQ$$

If $\angle SPQ = \theta$, then the length l be

$$l = PQ \cos \theta + PQ \sin \theta \qquad [\because \angle PSQ = 90°]$$

or
$$l = 2r (\cos \theta + \sin \theta) \qquad [\because PQ = 2r] \quad \text{(i)}$$

According to Fermat's principle, the path length l should be maximum or minimum. But in both cases, we have

$$\frac{dl}{d\theta} = 0 = 2r(-\sin \theta + \cos \theta) \qquad \text{(ii)}$$

or
$$\sin \theta = \cos \theta$$

∴
$$\theta = 45° \text{ or } \frac{\pi}{4} \qquad \text{(iii)}$$

Substituting the value of θ in Eq. (i), the actual length which light shall follow:

$$l = 2r(\cos 45° + \sin 45°) = 2r\left(\frac{1}{\sqrt{2}} + \frac{1}{\sqrt{2}}\right)$$

$$= 2r \times \sqrt{2} = \sqrt{2} \times \text{diameter of the sphere}$$

Differentiating Eq. (ii) again to see whether the path is a maximum or minimum, we get

$$\frac{d^2 l}{d\theta^2} = 2r(-\cos\theta - \sin\theta) = -2r(\cos\theta + \sin\theta)$$

$$= -2r\left(\frac{1}{\sqrt{2}} + \frac{1}{\sqrt{2}}\right) = -\sqrt{2} \times 2r$$

which is a negative quantity, hence, actual path is maximum in this case.

CONCEPTUAL QUESTIONS

1.1 Define optical path.

Ans The optical path is the path taken by ray through an optical system. It is defined in optics as the length of the path multiplied by the index of refraction of the medium.

1.2 What is Fermat's principle?

Ans. In optics, Fermat's principle or the principle of least time is the principle in which the path taken between two points by a ray of light is the path that can be traversed in the least time.

1.3 Write down laws of reflection.

Ans There are following laws of reflection:
 (i) The incident ray, the refracted ray and the normal to the reflection surface at the point of the incidence lies in the same plane.
 (ii) The angle of the incident ray makes with the normal is equal to the angle of the reflected ray makes to the same normal.
 (iii) The reflected ray and the incident ray are on the opposite side of the normal.

1.4 Which law describes the refraction phenomenon? Describe it in brief.

Ans Snell's law describes the refraction phenomenon. Snell's law states that for a given pair of media and a wave with a single frequency, the ratio of the sines of the angle of incidence θ_1 and angle of refraction θ_2 is equivalent to the ratio of phase velocities of two media, i.e., v_1/v_2, or equivalent to the opposite ratio of incidices (n_2/n_1), i.e.,

$$\frac{\sin\theta_1}{\sin\theta_2} = \frac{v_1}{v_2} = \frac{n_2}{n_1}$$

EXERCISES

Theoretical Questions

1.1 What is meant by optical path?

1.2 State and explain Fermat's principle of extremum path and use it to deduce the laws of reflection and refraction of light.

1.3 Write a short note on optical path and Fermat's principle.

1.4 Fermat's principle is the principle of extremum path or time. Analyze this statement with example.

1.5 What is Fermat's principle? Prove that Snell's law follows the Fermat's principle.

1.6 State Fermat's principle use is to prove the laws of reflection.

1.7 Give examples to show that the path of reflected ray is
(a) Maximum in some cases, and
(b) Minimum in other

1.8 Discuss Fermat's principle in brief and prove laws of reflection and refraction with its help.

1.9 State Fermat's principle of stationary time. Derive the laws of reflection using this principle. Given an example where the path of light is maximum rather than minimum.

1.10 State Fermat's principle. What is its significance? [Agra, 2006]

Numerical Problems

1.1 A man walks on the hard ground with a speed of 1.55 m/s; but he has speed of 0.93 m/s. On the sandy ground. Suppose he is standing at the border of sandy and hard ground and wishes to reach a tree situated on the sandy ground. The man can reach the tree by walking 31 m along the border and 37.2 m on the sandy ground normal to the border. Find out the value of path, which requires minimum to reach the tree. [**Ans.** $x = 59$ m, 3 m]

1.2 In Figure 1.7, the point S is a source of light distant $0.8\,r$ from the centre of the sphere C of radius r. If the light starting from S is reflected at point P and reaches B shown by Fermat's principle.

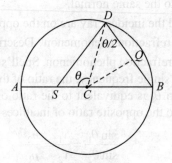

Figure 1.7

(a) Prove that $\cos\dfrac{\theta}{2} = \dfrac{3}{4}$.

(b) If distance $SC = 0.6\,r$, then prove that $\cos^2\dfrac{\theta}{2} = \dfrac{2}{3}$.

1.3 The optical path of a monochromatic light is the same, if it goes through 2.00 cm of glass or 2.25 cm of water. If the refractive index of water is 1.33, what is the refractive index of water is 1.33, what is the refractive index of glass? [**Ans.** $n = 1.50$]

Multiple Choice Questions

1.1 The product of refractive index of the medium and the distance travelled by the light in medium is known as
- (a) ray path
- (b) optical path
- (c) Fermat path
- (d) none of these

1.2 A ray of light travelling from one point to another by any number of reflections and refractions follows that particular path for which the time taken is
- (a) least
- (b) highest
- (c) maximum
- (d) none of these

1.3 Who enunciated the principle of least time
- (a) Pierre de Fermat
- (b) John Kepler
- (c) Albert Abraham Michelson
- (d) C.V. Raman

1.4 "The incident ray, reflected ray and the normal ray at the point of incidence are in one plane," is the statement for
- (a) first law of reflection
- (b) second law of reflection
- (c) first law of refraction
- (d) none of the above

1.5 The optical path of a monochromatic light is the same if it goes through 2.00 cm of glass or 2.25 cm of water. If the refractive index of water is 1.33, what is the refractive index of glass?
- (a) $n = 1.33$
- (b) 1.50
- (c) 1.23
- (d) None of these

1.6 If light travels 250 m distance in a material of refractive index 1.4 in a given time interval, then how much distance will it covers in a material of refractive index 1.75? [Agra, 2006]
- (a) 250 m
- (b) 312.5 m
- (c) 200 m
- (d) 212.5 m

Answers

1.1 (b) **1.2** (a) **1.3** (a) **1.4** (a) **1.5** (b) **1.6** (c)

CHAPTER 2

Geometrical Optics

"Music is the arithmetic of sounds as optics is the geometry of light."

—Claude Debussy

IN THIS CHAPTER

- Lenses: Thin and Thick Lenses
- Lens Equation
- Lens Maker's Formula
- Cardinal Points of a Coaxially Optical System
- Combination of Two Thin Lenses (Equivalent Lenses): Coaxial Lens System
- Matrix Method in Paraxial Optical System

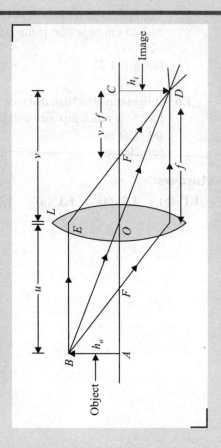

… Geometrical Optics 17

2.1 INTRODUCTION

Geometrical optics or ray optics describes the light propagation in terms of ray. The ray in geometrical optics is an abstraction or instrument, which can be used to approximate model how light will propagate? Light rays bends at the interface between two dissimilar media, and may curve in a medium where the refractive index changes. Geometrical optics provides rules for propagating these rays through an optical system. The path taken by the rays indicates how the actual wave will propagate. This is significant simplification of optics that fails to account for optical effects such as diffraction and polarization. It is a good approximation, however, when the wavelength is very small compared with size of structures with which the light interacts. Geometrical optics can be used to describe the geometrical aspect of imaging, including optical aberrations.

2.2 LENSES: THIN AND THICK LENSES

A lens is an optical device with perfect or approximate axial symmetry which transmits and refracts light, converging or diverging the beam. A simple lens consists of a single optical elements. A compound lens is an array of simple lenses (elements) with common axis; the use of multiple elements allows more optical aberrations to be corrected than is possible with a single element. Lenses are typically made of glass or transparent plastic. Elements which refract electromagnetic radiation outside the visual spectrum are also called *lenses*.

In optics, *a thin lens* is a lens with a thickness (distance along the optical axis between the two surfaces of the lens) that is a negligible compared to focal length of the lens. Lenses whose thickness is not negligible are sometimes called *thick lenses*.

2.3 LENS EQUATION

Let us consider an optical system as shown in Figure 2.1.

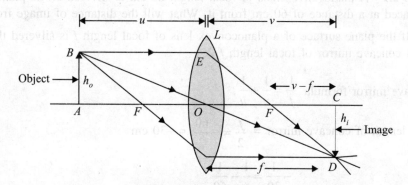

Figure 2.1 An optical system.

AB be the height of object, CD, the height image, F the focal points, u distance of the lens from object and v the distance between the lens from the image as shown in Figure 2.1.

Using the similar triangles *OEF* and *CDF*

$$\frac{CD}{EO} = \frac{CF}{OF} \tag{2.1}$$

or

$$\frac{v-f}{f} = \frac{v}{f} - 1 \tag{2.2}$$

and similar triangles *ABO* and *CDO*

$$\frac{CD}{AB} = \frac{OC}{AO} = \frac{v}{u} \tag{2.3}$$

$\therefore$ $\qquad AB = EO$

Then

$$\frac{CD}{AB} = \frac{CD}{EO} \tag{2.4}$$

We can combine Eqs. (2.1) and (2.3), then

$$\frac{v}{u} = \frac{v}{f} - 1$$

Dividing both sides by v

$$\frac{1}{u} = \frac{1}{f} - \frac{1}{v} \tag{2.5}$$

Rearrange Eq. (2.5), then we have the thin lens equation

$$\frac{1}{u} + \frac{1}{v} = \frac{1}{f} \tag{2.6}$$

Equation (2.6) is known as the *thin lens equation*.

EXAMPLE 2.1 A planoconvex lens of focal length 60 cm is silvered on its plane face. An object is placed at a distance of 60 cm from it. What will the distance of image from it?

Solution If the plane surface of a planoconvex lens of focal length f is silvered then it will behave as a concave mirror of focal length $f/2$.

We have mirror formula $\dfrac{1}{f} = \dfrac{1}{v} + \dfrac{1}{u}$.

Focal length of concave mirror $= \dfrac{f}{2} = -\dfrac{60}{2} = -30$ cm

$$-\frac{1}{30} = \frac{1}{v} - \frac{1}{60}$$

$\Rightarrow \qquad \dfrac{1}{v} = -\dfrac{1}{60}$ or $v = -60$ cm

Geometrical Optics

Magnification equation

We know that magnification $(m) = \dfrac{\text{Size of image}}{\text{Size of object}} = \dfrac{I}{O}$

Using similar triangles *ABO* and *CDO* again [see Figure 2.1]

$$\frac{CD}{AB} = \frac{OC}{AD} \quad \text{or} \quad \frac{I}{O} = \frac{v}{u}$$

By assigning a negative value to the inverted image, we have the magnification equation

$$m = \frac{h_i}{h_o} = \frac{-v}{u} \qquad (2.7)$$

EXAMPLE 2.2 A convex lens of focal length 20 cm produces a real image twice the size of the object, what will be the distance of the real object from the lens?

Solution Given $f = 20$ cm; $m = \dfrac{v}{u} = -2$

we know that
$$m = \frac{f}{f+u}$$

or
$$-2 = \frac{20}{20+u}$$

$$-40 - 2u = 20$$

or
$$u = -30 \text{ cm}$$

Sign convention

Sign convention for different distances are shown in Table 2.4.

Table 2.4 Sign Convention

Variable	Positive value	Negative value
u	Object is real	Impossible
v	Image is real	Image is virtual
f	Converging lens, mirror or optical system	Diverging lens, mirror or optical system
O	Object is upright	Object is inverted
I	Image is upright	Image is inverted

2.4 LENS MAKER'S FORMULA

The lens maker's formula relates the focal length of a lens to the refractive index of the lens material and radii of curvature of its two surfaces. It is used by manufacturers to design lenses of required focal length from a glass of given refractive index, therefore, it is called *lens maker's formula*.

Assumptions

1. The lens used is thin so that distances measured from its surfaces may be taken equal to those measured from its optical centre.

2. The object is a point object placed on the principal axis.
3. The aperture of the lens is small.
4. All the rays are paraxial *i.e.*, they make very small angles with the normals to the lens faces and with the principal axis.

2.4.1 Lens Maker's Formula for a Double Convex Lens

Let us consider a thin double convex lens of refractive index n_2 placed in a medium of refractive index n_1 as shown in Figure 2.2 ($n_1 < n_2$). Let B and D be the poles, C_1 and C_2 be the centres of curvature, and R_1 and R_2 be the radii of curvature of the two lens surfaces ABC and ADC respectively.

Suppose a point object O is placed on the principal axis in the rarer medium of refractive index n_1. The ray OM is incident on the first surface ABC. It is refracted along MN, bending towards the normal at this surface. If the second surface ADC were absent, the ray MN would have met the principal axis at I_1. So we can treat I_1 as the real image formed by first surface ABC in the medium of refractive index n_2.

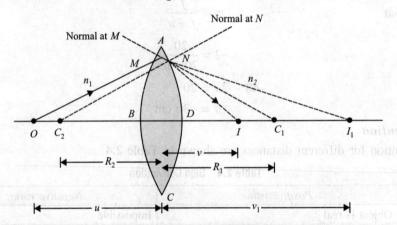

Figure 2.2 Refraction through a double convex lens.

For refraction at surface ABC, we can write the relation between the object distance u, image distance v_1 and the radius of curvature R_1 as

$$\frac{n_2}{v_1} - \frac{n_1}{u} = \frac{n_2 - n_1}{R_1} \tag{2.8}$$

But, actually the ray MN suffers another refraction at surface ADC, bending away from the normal at point N. The emergent ray meets the principal axis at point I, which is final image of O formed by the lens. For refraction at second surfaces I_1 acts as a virtual object placed in the medium of refractive index n_2 and I is the real image formed in the medium of refractive index n_1. Therefore, the relation between the object distance v_1, image distance v and radius of curvature R_2 can be written as

$$\frac{n_1}{v} - \frac{n_2}{v_1} = \frac{n_1 - n_2}{R_2} \tag{2.9}$$

Adding Eqs. (2.8) and (2.9), we have

$$\frac{n_1}{v} - \frac{n_1}{u} = (n_2 - n_1)\left(\frac{1}{R_1} - \frac{1}{R_2}\right)$$

or
$$\frac{1}{v} - \frac{1}{u} = \frac{n_2 - n_1}{n_1}\left(\frac{1}{R_1} - \frac{1}{R_2}\right) \tag{2.10}$$

If the object is placed at infinity ($u = \infty$), the image will be formed at the focus, i.e., $v = f$. Therefore,

$$\frac{1}{f} = \left(\frac{n_2 - n_1}{n_1}\right)\left(\frac{1}{R_1} - \frac{1}{R_2}\right) \tag{2.11}$$

Equation (2.11) is known as lens marker's formula.

When the lens is placed in air, $n_1 = 1$ and $n_1 = n$. Then, the lens maker's formula takes the form:

$$\frac{1}{f} = (n-1)\left(\frac{1}{R_1} - \frac{1}{R_2}\right) \tag{2.12}$$

From Eqs. (2.10) and (2.11), we have

$$\frac{1}{v} - \frac{1}{u} = \frac{1}{f} \tag{2.13}$$

This is the thin lens formula, which gives the relationship between u, v and f of a lens.

EXAMPLE 2.3 A convex lens of focal length 25 cm and refractive index 1.5 is immersed in water ($n = 1.33$). Find the change in the focal length of the lens.

Solution Given $n_2 = 1.5$ (for glass lens), $f = 25$ cm = focal length of the lens and $n_1 = 1.33$ (for water), f_1 = focal length of lens when immersed in water.

Then
$$\frac{1}{f} = (n-1)\left(\frac{1}{R_1} - \frac{1}{R_2}\right)$$

or
$$\frac{1}{f} = (1.5 - 1)\left(\frac{1}{R_1} - \frac{1}{R_2}\right)$$

or
$$\frac{1}{f} = 0.5 \times \left(\frac{1}{R_1} - \frac{1}{R_2}\right) \tag{i}$$

and
$$\frac{1}{f_1} = \left(\frac{n_2 - n_1}{n_1}\right)\left(\frac{1}{R_1} - \frac{1}{R_2}\right)$$

or
$$\frac{1}{f_1} = \frac{1.5 - 1.33}{1.33} \times \left(\frac{1}{R_1} - \frac{1}{R_2}\right)$$

or
$$\frac{1}{f_1} = \frac{0.17}{1.33} \times \left(\frac{1}{R_1} - \frac{1}{R_2}\right) \tag{ii}$$

Dividing Eq. (i) by Eq. (ii), we have

$$\frac{f_1}{f} = \frac{0.5 \times 1.33}{0.17}$$

or

$$f_1 = \frac{0.5 \times 1.33 \times 0.25}{0.17} = +0.978 \text{ m} \qquad [f = 0.25 \text{ m}]$$

The positive sign shows that the lens is converging. Increase in focal length = 0.978 − 0.25 = 0.728 m.

2.4.2 Lens Maker's Formula for a Double Concave Lens

Let us consider a thin double concave lens of refractive index n_2 placed in a medium of refractive index n_1 as shown in Figure 2.3 ($n_1 < n_2$). Let B and E be the poles and R_1 and R_2 be the radii of curvature of the two lens surfaces ABC and DEF respectively.

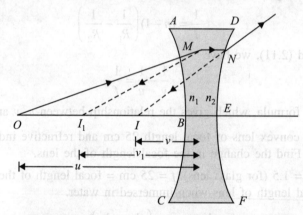

Figure 2.3 Refraction through a double concave lens.

Suppose a point object O is placed on the principal axis in the rarer medium of refractive index n_1. First, the spherical surface ABC forms its virtual image I_1. As refraction occurs from rarer to denser medium, so we can write the relation between object distance u, image distance v_1 and radius of curvature R_1 as

$$\frac{n_2}{v_1} - \frac{n_1}{u} = \frac{n_2 - n_1}{R_1} \qquad (2.14)$$

But the lens material is not continuous (A double convex lens is of n_1 and n_2 refractive indices). The ray MN suffers another refraction at N and emerges along NI. So I is the final image of the point object O. The image I_1 acts as an object for refraction at surface DEF from denser to rarer medium. So the relation between object distance v_1, image distance v and radius of curvature R_2 can be written as

$$\frac{n_1}{v} - \frac{n_2}{v_1} = \frac{n_1 - n_2}{R_2} \qquad (2.15)$$

Adding Eqs. (2.14) and (2.15), we get

$$\frac{n_1}{v} - \frac{n_1}{u} = (n_2 - n_1)\left(\frac{1}{R_1} - \frac{1}{R_2}\right)$$

or
$$\frac{1}{v} - \frac{1}{u} = \left(\frac{n_2 - n_1}{n_1}\right)\left(\frac{1}{R_1} - \frac{1}{R_2}\right) \tag{2.16}$$

If an object is placed at infinity, i.e., $u = \infty$, then the image is formed at the focus, i.e., $v = f$, so

$$\frac{1}{f} = \left(\frac{n_2 - n_1}{n_1}\right)\left(\frac{1}{R_1} - \frac{1}{R_2}\right) \tag{2.17}$$

Equation (2.17) is said to be lens maker's formula. When the lens is placed in air, $n_1 = 1$ and $n_2 = n$. Then the lens maker's formula takes the form

$$\frac{1}{f} = (n-1)\left(\frac{1}{R_1} - \frac{1}{R_2}\right) \tag{2.18}$$

2.5 CARDINAL POINTS OF A COAXIALLY OPTICAL SYSTEM

A coaxial optical system usually consists of a number of lenses placed apart and having a common principal axis. The position and size of the image of an object formed by such a system can be determined considering refraction at each lens separately. The process is, however, very tedious.

Gauss showed that if in an optical system the positions of certain specific points be known, the system may be treated as a 'single unit'. The position and size of the image of an object may then directly be obtained by same relations as used for thin lenses or single surface, however, complicated the system may be. These points are called *cardinal points* or *Gauss points of an optical system* or the traditional method of analyzing an optical system is to locate six characteristic points known as the *cardinal points*. These points are the front focal length (*FFL*), the back focal length (*BFL*), the two principal points, and the two nodal points. The effective focal length (*EFL*) and location of the vertices may be derived from these points.

2.5.1 Principal Points

The principal points (H_1 and H_2) in Figures 2.4 and 2.5 are a pair of conjugate points on the principal axis for which the linear transverse magnification is unity and positive.

If an object is placed at one principal point, then the image of same side is formed at other principal point.

Let L_1 and L_2 be the two convex lenses placed coaxially with foci F_1 and F_2.

Consider *AB* as an incident ray, which is parallel to principal axis. The emergent ray intersects the incident ray at a point *B*, ray BH_2 intersects the principal axis at H_2. This point H_2. This point H_2 is known as *second principal point*. The plane passing through H_2 and perpendicular to principal axis is known as *second pincipal plane*.

24 Fundamentals of Optics

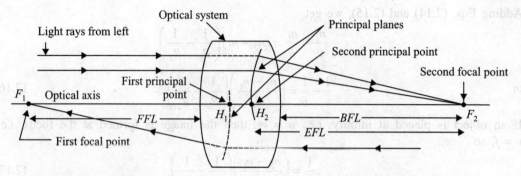

Figure 2.4 Location of the focal points and principal points of a generalized optical system.

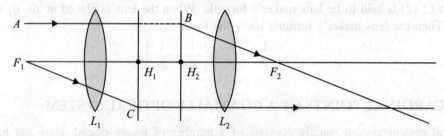

Figure 2.5 Principal points for convex lenses.

Now again consider ray F_1C passing through F_1. The emergent ray will be parallel to principal axis. The emergent ray and incident ray intersect at C. The perpendicular CH_1 intersects the principal axis at H_1. The point H_1 is known as *first principal point*. The plane passing through H_1 and perpendicular to principal axis is known as *first principal plane*.

2.5.2 Focal Points

The focal points are a pair of points lying on the principal axis and conjugate to points at infinity. They are of two types, *i.e.*, first focal point, and second focal point.

First focal point

The first focal point F_1 is an object point on the principal axis of an optical system for which the image points lie at infinity.

Case 1: *Convex or convergent lens:* The rays diverging from F_1 becomes parallel to the principal axis after refraction (Figure 2.6).

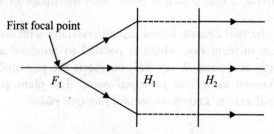

Figure 2.6 First focal point in case of convex lens.

Case 2: *Concave or divergent lens:* The rays directed towards F_1 become parallel to the principal axis after refraction (Figure 2.7).

A plane normal to the principal axis and passing through F_1 is called *first focal plane*.

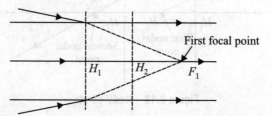

Figure 2.7 Second focal point in case of convex lens.

Second focal point

The second focal point F_2 is an image on the principal axis of an optical system for which object point lies at infinity.

Case 1: *Convex or convergent lens:* The rays parallel to the principal axis focus at this point after refraction through lens (Figure 2.8).

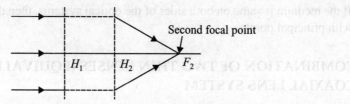

Figure 2.8 First focal point in case of concave lens.

Case 2: *Concave or divergent lens:* The parallel rays to the principal axis after refraction through the lens appear to diverge from F_2 (Figure 2.9).

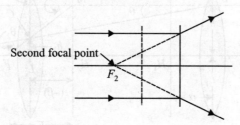

Figure 2.9 Second focal point in case of concave lens.

The plane normal to the principal axis and passing through F_2 is known as *second focal plane*.

2.5.3 Nodal Points

Nodal points are a pair of conjugate points having a unit positive angular magnification. These points lie on the principal axis and are represented by N_1 and N_2 (Figure 2.10).

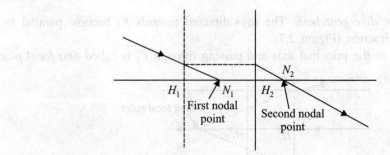

Figure 2.10 Nodal points.

The special feature of nodal points that a ray is directed towards the first nodal point N_1, the ray emerges from N_2 parallel to itself (parallel to its original direction).

Planes passing through the first and second nodal points and normal to the principal axis of the system are known as *first and second nodal planes*.

Note

1. The distance between two nodal points is equal to the distance between two principal points.
2. If the medium is same on both sides of the optical systems, then the nodal points coincide with principal points.

2.6 COMBINATION OF TWO THIN LENSES (EQUIVALENT LENSES): COAXIAL LENS SYSTEM

Consider the two thin convex lenses L_1 and L_2 separated at a distance d and placed in air. Let a ray AB parallel to the principal axis is incident on the first lens L_1 at height h_1 as shown in Figure 2.11.

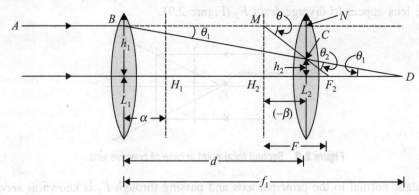

Figure 2.11 Coaxial lens system.

A ray AB is deviated through an angle θ_1 by the lens L_1. The deviated ray BC strikes the lens L_2 and meets the axis at F_2. If the ray would have not deviated by the lens L_2, it meets the axis at D.

Deviation θ_1 produced by the first lens, is

$$\theta_1 = \frac{h_1}{f_1} \qquad (2.19)$$

where f_1 is the focal length of lens L_1. Similarly deviation θ_2 produced by second lens is

$$\theta_2 = \frac{h_2}{f_2} \qquad (2.20)$$

where f_2 is the focal length of the second lens.

Now L_1 and L_2 may be supposed in the form of a combination in which AB is the incident ray and CF_2 is the emergent ray. In order to consider the deviation produced by this combination, we produced AB in the forward direction, while CF_2 in the backward direction. They meet at a point M. $\angle NMC = \theta$ is the deviation produced by combination. The value of θ is given by

$$\theta = \frac{h_1}{F} \qquad (2.21)$$

where h_1 is the height at which ray AB falls on the system and F is the combined focal length.

Focal length of the combination

From the geometry of Figure 2.11, $\theta = \theta_1 + \theta_2$. Substituting the values of θ, θ_1 and θ_2 from Eq. (2.19–2.21), we get

$$\frac{h_1}{F} = \frac{h_1}{f_1} + \frac{h_2}{f_2} \qquad (2.22)$$

Triangles CL_2D and BL_1D are similar, have

$$\frac{BL_1}{CL_2} = \frac{L_1D}{L_2D}$$

$\Rightarrow$
$$\frac{h_1}{h_2} = \frac{f_1}{f_1 - d} \qquad (2.23)$$

$\Rightarrow$
$$h_2 f_1 = h_1 f_1 - h_1 d$$

or
$$h_2 = h_1 - \frac{h_1 d}{f_1} \qquad (2.24)$$

From Eq. (2.24) substituting the value of h_2 in Eq. (2.22), we have

$$\frac{h_1}{F} = \frac{h_1}{f_1} + \frac{h_1}{f_2} - \frac{h_1 d}{f_1 f_2} \qquad (2.25)$$

Equation (2.25) can be put in a simplified form

$$F = \frac{f_1 f_2}{f_1 + f_2 - d} = \frac{-f_1 f_2}{\Delta} \qquad (2.26)$$

where $\Delta = (d - f_1 - f_2) =$ optical separation or optical interval.

Position of principal points

Suppose β be the distance between L_2 and second principal point and it is negative (according to sign convention). Triangles MH_2F_2 and CL_2F_2 are similar

$$\frac{MH_2}{CL_2} = \frac{H_2F_2}{L_2F_2}$$

or
$$\frac{h_1}{h_2} = \frac{F}{F-(-\beta)} \qquad (2.27)$$

Comparing Eqs. (2.23) and (2.27), we get

$$\frac{F}{F+\beta} = \frac{f_1}{f_1 - d}$$

or $\quad Ff_1 + \beta f_1 = Ff_1 - Fd$

or
$$\beta = -\frac{Fd}{f_1} \qquad (2.28)$$

Negative sign shows that *second principal point* lies towards left of the second lens.

Proceeding in a similar way, we can find the distance α between first lens and first principal point. The value of α is given by

$$\alpha = \frac{Fd}{f_2} \qquad (2.29)$$

Positive sign indicates the *first principal point* lies at the right of the lens.

Position of focal points

The distance L_2F_2 i.e., the distance between second lens and second focal point locate the position of second focal point.

$$L_2F_2 = H_2F_2 - H_2L_2 = F - (-\beta)$$

$\therefore$
$$L_2F_2 = F + \beta = F - \frac{Fd}{f_1} = F\left(1 - \frac{d}{f_1}\right) \qquad (2.30)$$

Similarly, we can locate the position of first focal point.

$$-L_1F_1 = H_1F_1 - L_1H_2 = F - \frac{Fd}{f_2}$$

$\therefore$
$$L_1F_1 = -F\left(1 - \frac{d}{f_2}\right) \qquad (2.31)$$

EXAMPLE 2.4 A lens combination consists of two lenses of focal length +20 cm and –10 cm separated by a distance of 8 cm. An object 2 cm high is placed at a distance of 40 cm from the convex lens. Find the position and size of the image.

Solution Given $f_1 = +20$ cm, $f_2 = -10$ cm and $d = 8$ cm.

Focal length of combination

$$F = \frac{f_1 f_2}{f_1 + f_2 - d} = \frac{20 \times (-10)}{20 - 10 - 8} \text{ cm}$$
$$= -100 \text{ cm}$$

Distance between first lens and first principal plane

$$\alpha = \frac{Fd}{f_2}$$
$$= \frac{(-100) \times 8}{(-10)} = 80 \text{ cm}$$

Distance between second lens and second principal plane

$$\beta = -\frac{Fd}{f_1}$$
$$= -\frac{(-100) \times 8}{20} = +40 \text{ cm}$$

Now, $u = -40$ cm, then

$$\frac{1}{v - \beta} - \frac{1}{u - \alpha} = \frac{1}{F}$$

or

$$\frac{1}{v - \beta} = \frac{1}{F} + \frac{1}{(u - \alpha)} = \frac{1}{-100} + \frac{1}{-40 - 80}$$
$$= -\frac{1}{100} - \frac{1}{120} = \frac{-11}{600}$$

or

$$v - \beta = -\frac{600}{11}$$

or

$$v = -\frac{600}{11} + 40 = \frac{-160}{11} \text{ cm}$$

So final image will be in left of the second lens of focal length $\frac{-160}{11}$ cm.

Now, magnification $m = \frac{\text{size of image}}{\text{size of object}} = \frac{I}{O} = \frac{v}{u} = \frac{\frac{-160}{11}}{-40} = \frac{4}{11}$

Now,

$$\frac{I}{O} = \frac{v}{u}$$

or

$$I = O \times \frac{v}{u} = 2 \times \frac{4}{11} \text{ cm}$$
$$= \frac{8}{11} \text{ cm}$$

2.7 MATRIX METHOD IN PARAXIAL OPTICAL SYSTEM

A coaxial optical system usually consists a large number of lenses placed apart and having a common principal axis. The position of the image of an object formed such a system can be determined considering refracting or reflecting at each lens separately. The cardinal points are discussed in Section 2.6. Now, we proceed to an analysis of large number of elements by application of multiplication of 2×2 matrices, which are representing the elementary refractions or reflections involved in this coaxial system. Hence, we achieved a system matrix for this optical system, which is related to same cardinal points, characterizing the thick lens.

Suppose any optical system consisting of large number of elements, *i.e.*, five or six lenses, these lenses constitute a pictorial lens. Then we need a systematic method for making complete analysis. We confined this analysis for paraxial rays, this systematic method is well handled by the matrix method. A treatment of image formation is presented, which employs matrices to describe changes in the height and angle of a ray as this ray makes its way by successive reflections and refractions through an optical system. The change in height and direction of ray are expressed by linear equations in the paraxial approximation for making this matrix method possible. A single matrix is obtained by combining matrices that represent individual refractions, reflections and translations through the given optical system. The essential properties of the composite optical system can be obtained by this single matrix.

A single ray through an arbitrary system is shown in Figure 2.12. The progress of ray can be described by changes in its elevation and direction.

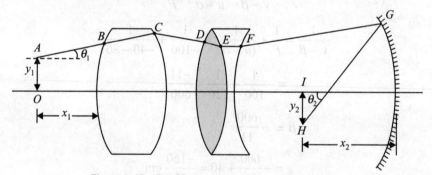

Figure 2.12 Tracing a ray through an optical system.

The ray is at a distance of x_1 from the first refracting surface in terms of height y_1 and angle θ_1 with respect to the optical axis. There is change in angle at each refraction, such as at point B through F and at each reflection, such as point G. The height of the ray changes during translations between these points. We find a method that will be useful to find out the height and slope angle of the ray at any point throughout the optical system. If the input data (x_1, y_1) is given at point A, then we can determine values (x_2, y_2) at point H as output data.

2.7.1 Translation Matrix

Let us consider a simple translation of the ray in a homogeneous medium as shown in Figure 2.13.

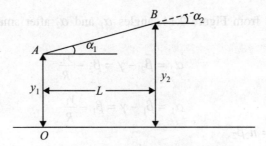

Figure 2.13 Simple translation of a ray.

Let the axial distance of the ray is L as shown in Figure 2.13, such that the coordinates of point B be (y_2, α_2). From the Figure 2.13,

$$\alpha_2 = \alpha_1 \text{ and } y_2 = y_1 + L \tan \alpha_1$$

This equation may be written in ordered form

$$y_2 = B(y_1) + L(\alpha_1) \tag{2.32}$$

and

$$\alpha_2 = 0(y_1) + B(\alpha_1) \tag{2.33}$$

Where the paraxial approximation is taken as $\tan \alpha_1 \cong \alpha_1$. In matrix notation, the two equations are written as

$$\begin{bmatrix} y_2 \\ \alpha_2 \end{bmatrix} = \begin{bmatrix} B & L \\ 0 & B \end{bmatrix} \begin{bmatrix} y_1 \\ \alpha_1 \end{bmatrix} \tag{2.34}$$

The effect of translation ray is represented by 2×2 ray – transfer matrix. The input data (y_1, α_1) is modified by this ray—transfer matrix to get output data (y_2, α_2).

2.7.2 The Refraction Matrix

Suppose the refraction of a ray at a spherical interface as shown in Figure 2.14. This spherical interface is separating two media of refractive indices n_1 and n_2. Our aim is to relate the ray coordinates (y_1, α_1) before refraction and those after refraction (y_2, α_2). Since refraction occurs at a point, there is no elevation change, so $y_1 = y_2$.

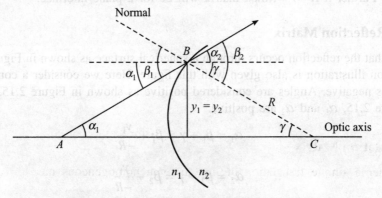

Figure 2.14 Refraction of a ray at spherical interface.

On the other hand from Figure 2.14, angles α_1 and α_2 after small angle approximation is given as

$$\alpha_2 = \beta_2 - \gamma = \beta_2 - \frac{y_1}{R}$$

and

$$\alpha_1 = \beta_1 - \gamma = \beta_1 - \frac{y_1}{R}$$

From Snell's law $n_1\beta_1 = n_2\beta_2$

Then, we have

$$\alpha_2 = \frac{n_1}{n_2}\beta_1 - \frac{y_1}{R}$$

$$\alpha_2 = \frac{n_1}{n_2}\left[\alpha_1 + \frac{y_1}{R}\right] - \frac{y_1}{R}$$

$$\alpha_2 = \frac{1}{R}\left(\frac{n_1}{n_2} - 1\right)y_1 + \frac{n_1}{n_2}\alpha_1$$

Then approximate linear equations may be written as

$$y_2 = (B)y_1 + (0)\alpha_1 \tag{2.35}$$

$$\alpha_2 = \frac{1}{R}\left(\frac{n_1}{n_2} - 1\right)y_1 + \left(\frac{n_1}{n_2}\right)\alpha_1 \tag{2.36}$$

Equations (2.35) and (2.36) may be written in matrix form as

$$\begin{bmatrix} y_2 \\ \alpha_2 \end{bmatrix} = \begin{bmatrix} B & 0 \\ \frac{1}{R}\left(\frac{n_1}{n_2} - 1\right) & \frac{n_1}{n_2} \end{bmatrix} \begin{bmatrix} y_1 \\ \alpha_1 \end{bmatrix} \tag{2.37}$$

Equation (2.37) shows the refraction matrix. If the surface is concave, R is positive, otherwise it is negative. Further if $R \to \infty$, then matrix will be for a plane interface.

2.7.3 The Reflection Matrix

We consider that the reflection occurs through a spherical surface as shown in Figure 2.15. The sign convention illustration is also given with this figure. Here we consider a concave mirror, for which R is negative. Angles are considered positive as shown in Figure 2.15.

In Figure 2.15, α_1 and α_2 are positive.

$$\alpha_1 = \beta_1 + \gamma = \beta_1 + \frac{y_1}{-R}$$

and

$$\alpha_2 = \beta_2 - \gamma = \beta_2 - \frac{y_1}{-R}$$

Geometrical Optics 33

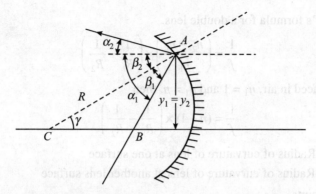

Figure 2.15 Reflection of a ray at a spherical surface.

The small angle approximation is considered here. Using law of reflection, we have $\beta_1 = \beta_2$, then we have

$$\alpha_2 = \beta_2 + \frac{y_1}{R} = \beta_1 + \frac{y_1}{R} = \alpha_1 + \frac{2y_1}{R}$$

So, the desired linear equation will be

$$y_2 = Ay_1 + C\alpha_1 \tag{2.38}$$

$$\alpha_2 = \frac{2}{R}y_1 + A\alpha_1 \tag{2.39}$$

Equations (2.38) and (2.39) may be written in matrix form as

$$\begin{bmatrix} y_2 \\ \alpha_2 \end{bmatrix} = \begin{bmatrix} A & C \\ 2/R & A \end{bmatrix} \begin{bmatrix} y_1 \\ \alpha_1 \end{bmatrix} \tag{2.40}$$

FORMULAE AT A GLANCE

2.1 Lens equation

$$\frac{1}{u} + \frac{1}{v} = \frac{1}{f}$$

where u = distance of lens from object
 v = distance of lens from image
 f = focal length of the lens

2.2 Magnification

$$m = \frac{\text{Size of image } (I)}{\text{Size of object } (O)} \quad \text{or} \quad \frac{I}{O} = \frac{v}{u}$$

2.3 Lens maker's formula for a double lens.

$$\frac{1}{f} = \left(\frac{n_2 - n_1}{n_1}\right) \times \left(\frac{1}{R_1} - \frac{1}{R_2}\right)$$

If lens is placed in air, $n_1 = 1$ and $n_1 = n$, then

$$\frac{1}{f} = (n - 1) \times \left(\frac{1}{R_1} - \frac{1}{R_2}\right)$$

where R_1 = Radius of curvature of lens at one surface
R_2 = Radius of curvature of lens at another lens surface

2.4 *Cardinal Points*

(i) Combined focal length $F = \dfrac{f_1 f_2}{f_1 + f_2 - d}$

where f_1 and f_2 are focal lengths of two lenses, which are at a distance of d.

(ii) Principal points

$$\alpha = +\frac{Fd}{f_2}$$

and

$$\beta = -\frac{Fd}{f_1}$$

(iii) Focal points

$$L_1 F_1 = -F\left(1 - \frac{d}{f_2}\right)$$

$$L_2 F_2 = F\left(1 - \frac{d}{f_1}\right)$$

2.5 Matrix analysis

(i) Translation matrix

$$\begin{bmatrix} y_2 \\ \alpha_2 \end{bmatrix} = \begin{bmatrix} B & L \\ 0 & B \end{bmatrix} \begin{bmatrix} y_1 \\ \alpha_1 \end{bmatrix}$$

(ii) Refraction matrix

$$\begin{bmatrix} y_2 \\ \alpha_2 \end{bmatrix} = \begin{bmatrix} B & 0 \\ \dfrac{1}{R}\left[\dfrac{n_1}{n_2} - 1\right] & \dfrac{n_1}{n_2} \end{bmatrix} \begin{bmatrix} y_1 \\ \alpha_1 \end{bmatrix}$$

(iii) Reflection matrix

$$\begin{bmatrix} y_2 \\ \alpha_2 \end{bmatrix} = \begin{bmatrix} A & C \\ 2/R & A \end{bmatrix} \begin{bmatrix} y_1 \\ \alpha_1 \end{bmatrix}$$

SOLVED NUMERICAL PROBLEMS

PROBLEM 2.1 The radii of curvature of the faces of a double convex lens are 10 cm and 15 cm. If focal length is 12 cm, what is the refractive index of glass?

Solution Given: $f = 12$ cm, $R_1 = +10$ cm and $R_2 = -15$ cm

We know that

$$\frac{1}{f} = (n-1)\left(\frac{1}{R_1} - \frac{1}{R_2}\right)$$

∴ $$\frac{1}{12} = (n-1)\left(\frac{1}{10} + \frac{1}{15}\right) = (n-1) \times \frac{5}{30}$$

or $$(n-1) = \frac{6}{12} = 0.5$$

∴ $$n = 1.5$$

PROBLEM 2.2 Find the radius of curvature of the convex surface of a plano-convex lens, whose focal length is 0.3 m and the refractive index of the material of the lens is 1.5.

Solution Given: $n = 1.5$, $f = +0.3$ m, $R_1 = \infty$ and $R_2 = -R$

Using lens maker's formula

$$\frac{1}{f} = (n-1)\left(\frac{1}{R_1} - \frac{1}{R_2}\right)$$

or $$\frac{1}{0.3} = (1.5 - 1)\left(\frac{1}{\infty} + \frac{1}{R}\right)$$

or $$\frac{1}{0.3} = 0.5 \times \frac{1}{R}$$

or $$R = 0.15 \text{ m}$$

PROBLEM 2.3 Two thin converging lenses of focal lengths 12 cm each are placed 8 cm apart. Calculate and plot the cardinal points of the optical system.

Solution Given: $f_1 = f_2 = 12$ cm and $d = 8$ cm

In usual notations the focal length of the combination is given by

$$F = \frac{f_1 f_2}{f_1 + f_2 - d} = \frac{12 \times 12}{12 + 12 - 8} = 9 \text{ cm}$$

Let L_1, L_2 be the positions of the lenses; H_1, H_2 be the principal points and F_1, F_2 the focal points, we have

$$L_1 H_1 = +\frac{Fd}{f_2}$$

$$= +\frac{9 \times 8}{12} + 6 \text{ cm } (H_1 \text{ is on the right of } L_1)$$

36 Fundamentals of Optics

$$L_2H_2 = -\frac{Fd}{f_1}$$

$$= -\frac{9 \times 8}{12} = -6 \text{ cm } (H_2 \text{ is on the left on } L_2)$$

$$L_1F_1 = -F\left(1 - \frac{d}{f_2}\right)$$

$$= -9\left(1 - \frac{8}{12}\right) = -3 \text{ cm } (F_1 \text{ is on the left } L_1)$$

$$L_2F_2 = +F\left(1 - \frac{d}{f_1}\right)$$

$$= +9\left(1 - \frac{8}{12}\right) = +3 \text{ cm } (F_2 \text{ is on the right of } L_2)$$

The positions of the cardinal points H_1, H_2, F_1, F_2 are plotted in Figure 2.16. The nodal points N_1, N_2 are same as the principal points. (Suppose the system is situated in air.)

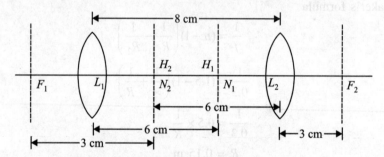

Figure 2.16 Location of cardinal points of a given system.

PROBLEM 2.4 Two thin convex lenses of focal lengths 6 cm and 18 cm are separated by a distance of 12 cm in air. Calculate the positions of principal, focal and nodal points.

Solution Given: $f_1 = 6$ cm, $f_2 = 18$ cm and $d = 12$ cm
The equivalent focal lengths of the combination

$$F = \frac{f_1 f_2}{f_1 + f_2 - d} = \frac{6 \times 18}{6 + 18 - 12} = 9 \text{ cm}$$

Cardinal points of the system

1. *Principal points:* The distance of the first principal point H_1 from the first lens L_1.

$$L_1H_1 = +\frac{Fd}{f_2} = \frac{9 \times 12}{18} = +6 \text{ cm}$$

The distance of second principal point H_2 from the lens L_2

$$L_2H_2 = -\frac{Fd}{f_1} = \frac{-9 \times 12}{6} = -18 \text{ cm}$$

2. *Focal points:* The distance of the first focal point F_1 from the first lens L_1

$$L_1F_1 = -F\left(1 - \frac{d}{f_2}\right) = -9\left(1 - \frac{12}{18}\right) = -3 \text{ cm}$$

The distance of second focal point F_2 from the second lens L_2

$$L_2F_2 = +F\left(1 - \frac{d}{f_1}\right) = +9\left(1 - \frac{12}{6}\right) = -9 \text{ cm}$$

3. *Nodal points:* Since the medium on both sides of the system (i.e. air) is same, the nodal points coincides with principal points, i.e.

$$L_1N_1 = L_1H_1 = +6 \text{ cm}$$
$$L_2N_2 = L_2H_2 = -18 \text{ cm}$$

PROBLEM 2.5 Two thin convex lenses of focal lengths 20 cm and 10 cm are separated by a distance 10 cm. If an object is placed at 25 cm from the first lens, deduce the location and magnification of the image.

Solution Given: $f_1 = 20$ cm, $f_2 = 10$ cm, $d = 10$ cm and $u = 25$ cm.

In usual notations, the focal length of the combination is given by

$$F = \frac{f_1 f_2}{f_1 + f_2 - d}$$

Here $f_1 = 20$ cm, $f_2 = 10$ cm, and $d = 10$ cm

$$\therefore \quad F = \frac{20 \times 10}{20 + 10 - 10} = 10 \text{ cm}$$

The distance of the first principal point H_1 from the first lens L_1

$$L_1H_1 = +\frac{Fd}{f_2} = \frac{10 \times 10}{10} = +10 \text{ cm (on the right of } L_1)$$

The distance of second principal point H_2 from the second lens L_2

$$L_2H_2 = -\frac{Fd}{f_1} = -\frac{10 \times 10}{20} = -5 \text{ cm (on the left of } L_2)$$

The object O is placed at a distance 25 cm from the first lens L_1, we know that if u and v be the distances of the object and the image measured from the first and second principal points respectively, then same formula holds for the system of lenses as for a single thin lens. Thus according to sign convention

$$u = -H_1O = -(H_1L_1 + L_1O)$$

$$= -(10 + 25) \text{ cm} = -35 \text{ cm}$$
$$F = 10 \text{ cm (calculated above)}$$
$$v = \text{Distance of image } I \text{ from } H_2 = ?$$

Putting these values in
$$\frac{1}{v} - \frac{1}{u} = \frac{1}{f}$$

We get
$$\frac{1}{v} + \frac{1}{35} = \frac{1}{10}$$
$$\frac{1}{v} = \frac{1}{10} - \frac{1}{35} = \frac{35-10}{35 \times 10} = \frac{25}{350} = \frac{1}{14}$$
$$v = +14 \text{ cm}$$

Thus image is at the right of the second principal; point H_2 at a distance of 14 cm

The magnification of the image
$$m = \frac{v}{u} = \frac{14}{-35} = -\frac{2}{5} = -0.4$$

The negative sign shows that the image is inverted.

CONCEPTUAL QUESTIONS

2.1 What is application of geometrical optics?

Ans Geometrical optics can be used to describe the geometrical aspect of imaging, including optical aberration.

2.2 What do you mean by thin lens?

Ans In optics, a thin lens is a lens with a thickness that is negligible in comparison to focal length of the lens.

2.3 What is lens maker's formula?

Ans The lens maker's formula relates the focal length of a lens to the refractive index of the lens material and radii of curvature of its two surfaces. Since it is used by manufacturer to design lenses of required focal length from a glass of given refractive index, so, it is known as lens maker's formula.

2.4 What do you mean by cardinal points of a coaxial optical system?

Ans A coaxial optical system usually consists of a number of lenses placed apart and having a common principal axis. The position and size of the image of an object formed by such a system can be determined considering refraction at each lens separately. The process is, however, very tedious. Gauss found that if in an optical system the positions of certain specific points be known, the system may be treated as a 'single unit'. The position and size of the image of an object may, then directly, be obtained by same relations as used for thin lenses or single surface, however, complicated the system may be. These points are called *cardinal points* or *Gauss points of the optical system*. There are six cardinal points, i.e., two focal points, two principal points and two nodal points.

2.5 Define the following:

(i) Principal points, (ii) focal points and (iii) nodal points.

Ans (i) *Principal points:* These are a pair of conjugate points on the principal axis for which linear transverse magnification is unity.

(ii) *Focal points:* The focal points are a pair of points lying on the principal axis and conjugate to points at infinity.

(iii) *Nodal points:* Nodal points are a pair of conjugate points having a unit positive angular magnification.

2.6 What is meant by optical separation of two lenses?

Ans If two lenses of focal lengths f_1 and f_2 are placed co-axially at a distance d from each other, then their equivalent focal length is given by

$$\frac{1}{F} = \frac{1}{f_1} + \frac{1}{f_2} - \frac{d}{f_1 f_2} = \frac{f_1 + f_2 - d}{f_1 f_2} = \frac{\Delta}{f_1 f_2}$$

The separation $\Delta = f_1 + f_2 - d$ is said to be optical separation of the lenses.

EXERCISES

Theoretical Questions

2.1 Derive the lens maker's formula for a double convex lens. State the assumptions made and the convention of sign used.

2.2 Prove that in the case of a thin convex lens

$$\frac{1}{f} = \frac{1}{v} - \frac{1}{u} = (n-1)\left(\frac{1}{R_1} - \frac{1}{R_2}\right)$$

2.3 Two thin convex lenses of focal lengths f_1 and f_2 cm are coaxial and separated by d. Show that the equivalent focal length F of the combination is given by the relation

$$F = \frac{f_1 f_2}{f_1 + f_2 - d}$$

2.4 Derive expression for the equivalent focal length and the positions of principal points, and focal points of a coaxial system of two thin lenses by a distance.

2.5 What are the cardinal points of a thick lens? Explain with the help of a neat ray diagram the positions of cardinal points for a thick lens in air.

2.6 What is an equivalent lens? In what respect it is called an equivalent lens?

2.7 Define cardinal points of a system of a coaxial lenses. Describe how you would determine experimentally the principal planes of a combination of two thin lenses separated by a distance.

2.8 Explain the significance of cardinal points of a coaxial lens system with suitable ray diagrams.

2.9 Obtain the system matrix for a thick lens and hence obtain the system matrix for a thin lens.

2.10 Obtain the reflection matrix and find its determinants.

2.11 Find the expression for refraction matrix. Hence calculate the value of determinant.

2.12 Show that minimum distance between an object and its real image in a convex lens is four times the focal length of the lens.

Numerical Problems

2.1 A biconvex lens has a focal length half the radius of curvature of either surface. What is the refractive index of lens material. [**Ans.** $n = 2$]

2.2 The radii of curvature of a double convex lens of glass ($n = 1.5$) are in the ratio 1:2. This lens renders the rays parallel coming from an illuminated filament at a distance of 6 cm. Calculate the radii of curvature of its surface.
[**Ans.** $R_1 = +4.5$ cm and $R_2 = -9.0$ cm]

2.3 Find the radius of curvature of the convex surface of a plano-convex lens, whose focal length is 0.3 m and the refractive index of the lens is 1.5. [**Ans.** $R = 0.15$ m]

2.4 The radii of curvature of a double convex lens are 10 cm and 20 cm respectively. Calculate its focal length when it is immersed in a liquid of refractive index 1.65. State the nature of the lens in the liquid. The refractive index of glass is 1.5. [**Ans.** $f = -73.33$ cm]

2.5 An equiconvex lens of focal length 15 cm is cut into two equal halves as shown in Figure 2.17.

Figure 2.17

What is the focal length of each half? [**Ans.** $f_1 = f_2 = 30$ cm]

2.6 Two thin convex lenses having focal lengths 5.0 cm and 2.0 cm are coaxial and separated by a distance of 3.0 cm. Find the equivalent focal length for the combination.
[**Ans.** $F = +2.5$ cm]

2.7 Two thin converging lenses of focal lengths 4 cm and 12 cm are placed 8 cm apart in air. Find the positions of the principal points and focal points of the lens system.
[**Ans.** $F = +6$ cm, $L_1H_1 = +4$ cm, $L_2H_2 = -12$ cm, $L_1F_1 = -2$ cm and $L_2F_2 = 6$ cm]

2.8 Two thin convergent lenses each of 20 cm focal length are set coaxillay 5 cm apart. An image of upright pole 200 m distant and 10 m high if formed by the combination. Find the position of the unit and focal planes and the image. Also find the size of the image.
[**Ans.** $F = +11.43$ cm, $L_1H_1 = 2.85$ cm, $L_2H_2 = -2.85$ cm, $v = 8.55$ cm and $I = -0.57$ cm]

2.9 A double convex lens of focal length 6 cm is made of glass of refractive index 1.5. Find the radius of curvature of one surface, which is double that of other surface. [**Ans.** 9 cm]

2.10 Two thin convex lenses of focal lengths 30 cm and 10 cm are separated by a distance of 25 cm in air. Calculate the positions of the cardinal points.
[**Ans.** $F = +20$ cm, $L_1H_1 = +50$ cm, $L_2H_2 = -16.7$ cm, $L_1F_1 = +30$ cm and $L_2F_2 = +3.3$ cm]

Multiple Choice Questions

2.1 Two thin lenses of focal lengths +60 cm and –20 cm are placed in contact. The focal length of the combination is
(a) –15 cm (b) +15 cm
(c) –30 cm (d) –30 cm

2.2 Two thin lenses of focal lengths +20 cm and +25 cm are placed in contact. The power of the combination is
(a) $6\,D$ (b) $45\,D$
(c) $1/9\,D$ (d) $9\,D$

2.3 The focal length of an equiconvex lens is equal to the radius of curvature of either face. What is the refractive index of the lens material?
(a) 1.32 (b) 1.25
(c) 1.50 (d) 1.23

2.4 A converging lens of refractive index 1.5 is kept in a liquid medium having same refractive index. What would be the focal length of the lens in this medium?
(a) 0 (b) ∞
(c) 10 (d) 100

2.5 How many cardinal points are an optical system?
(a) 2 (b) 4
(c) 6 (d) 8

2.6 The principal points are a pair of conjugate point on the principal axis for which linear transverse magnification is
(a) unity and positive (b) infinity and negative
(c) zero (d) infinity

2.7 The focal points are a pair of points lying on the principal axis and conjugate to points at
(a) finite distance (b) infinity
(c) first focal point (d) second focal point

2.8 The nodal points are a pair of conjugate points having a unit positive
(a) angular magnification (b) linear transverse magnification
(c) angular momentum (d) none of these

2.9 An equiconvex lens is cut into two halves by a plane *AB* as shown in Figure 2.17. The focal length of each half so obtained is
(a) *f* (b) *f*/2
(c) 2*f* (d) 3*f*/2

2.10 The pair of conjugate points having unit positive angular magnification is called
(a) principal points (b) focal points
(c) nodal points (d) axial points

Answers
2.1 (c) **2.2** (d) **2.3** (c) **2.4** (b) **2.5** (c) **2.6** (a) **2.7** (b) **2.8** (a)
2.9 (c) **2.10** (c)

CHAPTER 3

Dispersion of Light

"The multicoloured rays of the sun, being dispersed in a cloudy sky, are seen in the form of a bow, which is called the Rainbow."
—Bruhatsamhita

IN THIS CHAPTER

- Newton's Classic Experiment
- Cauchy's Relation
- Angular Dispersion and Dispersive Power
- Combination of Prisms
- Direct Vision Spectroscope
- Rainbow
- Sundogs and Halos

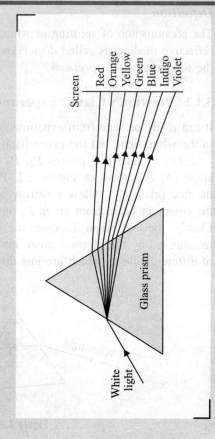

3.1 INTRODUCTION

When a narrow beam of sunlight is incident on a glass prism, the emergent light when made to fall on a screen shows seven coloured bands. Broadly, the component colours are in the sequence: violet, indigo, blue, green, yellow, orange and red (given by the acronym—VIBGYOR). The red colour bends the least and the violet colour bends the most as shown in Figure 3.1.

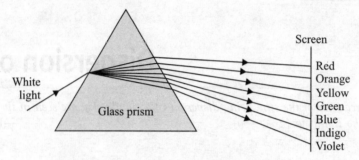

Figure 3.1 Dispersion of white light by a glass prism.

Definition

The phenomenon of splitting of white light into its components colours on passing through a refractive medium is called *dispersion of light*. The pattern of the coloured bands obtained on the screen is called *spectrum*.

3.1.1 Newton's Classic Experiment on Dispersion of White Light

It can easily be seen from experiments that a prism only separates the colours already present in the white light, but the prism itself does not create any colour.

Let us take two prisms P_1, P_2 of the same glass material and of the same refracting angle (A) as shown in Figure 3.2. Place the second prism P_2 upside dam with respect to the first prism P_1. Allow a narrow beam of white light to fall on the prism P_1 and observe the emergent beam from prism P_2 on a screen. A patch of white light is seen on the screen. Clearly, the first prism disperses the white light into its components colours, which are then recombined by the inverted prism into white light. This proves that white light itself consists of different colours which are just dispersed by the prism.

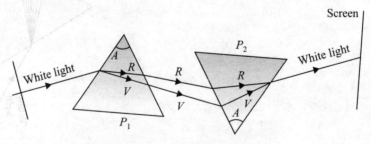

Figure 3.2 Recombination of white light.

3.1.2 Cause of Dispersion (Cauchy's Relation)

Each colour of light is associated with a definite wavelength. In the visible spectrum, red light is at the long wavelength end (~700 nm), while the violet light is at the short wavelengths end (~400 nm). *Dispersion takes place because the refractive index of the refractive medium is different for different wavelengths.* The refractive index n of a material for wavelength λ is given by the Cauchy's relation

$$n = a + \frac{b}{\lambda^2} + \frac{c}{\lambda^4} \tag{3.1}$$

where a, b, and c are constants, the values of which depend on the nature of the material. Also, for a small angled prism, the angle of deviation is given by

$$\delta = A(n - 1) \quad \text{[Here } A \text{ is angle of prism]}$$

Now, $\qquad\qquad\qquad\qquad \lambda_{\text{red}} > \lambda_{\text{violet}}$

Therefore, $n_{\text{red}} < n_{\text{violet}}$ and, hence, $\delta_{\text{red}} < \delta_{\text{violet}}$.

Thus the red colour is deviated the least and the violet is deviated the most. Other colours are deviated by angle between δ_{red} and δ_{violet}. So, different colours contained in white light emerge from the glass prism in different directions due to their different wavelengths, which is called *dispersion*.

3.2 ANGULAR DISPERSION AND DISPERSIVE POWER

When a beam of white light passes through a prism, it gets dispersed into its constituent colours. Let δ_V, δ_R, and δ_Y be the angles of deviation for violet, red and yellow (mean) colours respectively, as shown in Figure 3.3.

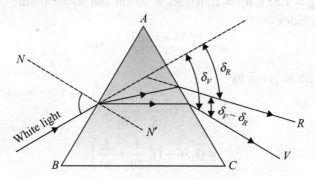

Figure 3.3 Angular dispersion.

Then, $\qquad\qquad\qquad\qquad \delta_V = (n_V - 1)A$

$\qquad\qquad\qquad\qquad\qquad \delta_R = (n_R - 1)A$

and $\qquad\qquad\qquad\qquad \delta_Y = (n_Y - 1)A$

where n_V, n_R, and n_Y be the refractive indices of the prism material for violet, red and yellow (mean) colours respectively.

The angular separation between the two extreme colours (violet and red) in the spectrum is called the angular dispersion.

$$\text{Angular dispersion} = \delta_V - \delta_R$$
$$= (n_V - 1)A - (n_R - 1)A$$
$$= (n_V - n_R)A$$

Clearly, the angular dispersion produced by a prism depends upon
 (i) the angle of the prism, and
 (ii) the nature of the material of the prism.

Dispersive power is the ability of the prism material to cause dispersion. It is defined as the ratio of the angular dispersion to the mean deviation. Therefore,

$$\text{Dispersive power } \omega = \frac{\text{Angular dispersion}}{\text{Mean deviation}} = \frac{\delta_V - \delta_R}{\delta_Y}$$

$$= \frac{(n_V - 1)A - (n_R - 1)A}{(n_Y - 1)A}$$

or
$$\omega = \frac{n_V - n_R}{(n_Y - 1)} \qquad (3.2)$$

EXAMPLE 3.1 The refractive indices of hard crown glass for red and violet light are 1.553 and 1.567 respectively. If a beam of white light falls on a convex lens of the same glass in a direction parallel to the axis and if the radii of curvature of the two surfaces of the lens are 30 cm and 20 cm, calculate dispersive power of the lens.

Solution *Given:* $n_R = 1.553$, $n_V = 1.567$, $R_1 = 20$ cm and $R_2 = 30$ cm
The mean refractive index

$$n = \frac{1.553 + 1.567}{2} = 1.560$$

The mean focal length is given by

$$\frac{1}{f} = (n-1)\left(\frac{1}{R_1} + \frac{1}{R_2}\right)$$

$$= (1.56 - 1)\left(\frac{1}{20} + \frac{1}{30}\right)$$

$$= 0.560 \times \frac{1}{12}$$

or
$$f = \frac{12}{0.560} = 21.43 \text{ cm}$$

∴ Dispersive power $\omega = \dfrac{n_V - n_R}{n - 1} = \dfrac{1.567 - 1.553}{1.56 - 1} = 0.025$

3.3 COMBINATION OF PRISMS

If a beam of white light be passed successively through two prisms similarly placed with respect to the incident light *i.e.*, bases of both the prisms are directed towards the same side, dispersion and deviation produced by the first prism is further increased by the second prism. On the other hand, if the bases of the two prisms are oppositely directed, deviation and dispersion produced by the first are reduced by the second prism. Since dispersive powers of different materials are different, two or more prisms of different materials and suitable angle can be combined so as to produce:

1. Deviation without dispersion, and
2. Dispersion without deviation

3.3.1 Deviation without Dispersion

Consider two prisms C and F of crown glass and flint glass of angles α and α' (Figure 3.4). Let the refractive indices of crown and flint glasses for violet, red and yellow light be n_V, n_R and n_Y and n'_V, n'_R and n'_Y respectively. The angular dispersion produced by crown glass is $(\delta_V - \delta_R)$ and that by flint glass is $(\delta'_V - \delta'_R)$. If there is no dispersion, the algebraic sum of the angular dispersions due to two prisms must be zero.

$$(\delta_V - \delta_R) + (\delta'_V - \delta'_R) = 0 \tag{3.3}$$

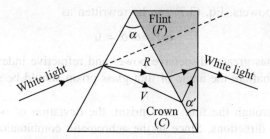

Figure 3.4 A chromatic prism combination.

If the prisms are of small angles, we know that
$$\delta = (n-1)\alpha$$
Then, we have
$$\left.\begin{aligned}
\delta_V &= (n_V - 1)\alpha \\
\delta_R &= (n_R - 1)\alpha \\
\delta_Y &= (n_Y - 1)\alpha, \text{ and} \\
\delta'_V &= (n'_V - 1)\alpha' \\
\delta'_R &= (n'_R - 1)\alpha' \\
\delta'_Y &= (n'_Y - 1)\alpha'
\end{aligned}\right\} \tag{3.4}$$

Combination of Eqs. (3.3) and (3.4) yields
$$(n_V - n_R)\alpha + (n'_V - n'_R)\alpha' = 0 \tag{3.5}$$

or
$$\alpha' = \frac{-(n_V - n_R)}{(n'_V - n'_R)}\alpha \qquad (3.6)$$

Equation (3.5) gives the relation between the angles of the two prisms in order that the combination does not produce any dispersion. The negative sign shows that the two prisms should be placed with thin angles opposite to each other as shown in Figure 3.4. Such a combination is known as an *achromatic combination*.

We may rewrite Eq. (3.5) as
$$(n_V - n_R)\alpha + (n'_V - n'_R)\alpha' = 0$$

or
$$\frac{n_V - n_R}{n_Y - 1}(n_Y - 1)\alpha - \frac{n'_V - n'_R}{n'_Y - 1}(n'_Y - 1)\alpha' = 0$$

or
$$\frac{n_V - n_R}{n_Y - 1}\delta_Y - \frac{n'_V - n'_R}{n'_Y - 1}\delta'_Y = 0 \qquad (3.7)$$

Let ω and ω' be the dispersive powers of the crown and flint glasses respectively. Then
$$\omega = \frac{n_V - n_R}{n_Y - 1} \qquad (3.8)$$

and
$$\omega' = \frac{n'_V - n'_R}{n'_Y - 1} \qquad (3.9)$$

In terms of dispersive powers, Eq. (3.7) can be rewritten as
$$\omega\delta - \omega'\delta' = 0 \qquad (3.10)$$

Now, since flint glass has greater dispersive power and refractive index than crown glass, for no dispersion by combination, the angle of flint glass prism should be smaller than that of the crown glass prism.

During passage through the flint glass prism, the deviation of both red and violet rays will be in the opposite directions. Since in the achromatic combination the difference in the deviation of red and violet is same in the two prisms, the emergent beam will be parallel. If a single ray of light is sent through the combination, the emergent beam will consist of the colours of the spectrum all travelling along parallel rays, but if a beam of finite width is used, these colours will overlap to produce white light except at the edges.

Although the above combination is achromatic, yet it does produce deviation, because the angle of the crown glass prism is greater than the angle of the flint glass prism. The mean deviation in crown glass prism C is greater than in the flint glass prism F, so that the difference in the two mean deviations is equal to the resultant mean deviation. Now, if Δ is the resultant mean deviation of the combination, then
$$\Delta = \delta_Y - \delta'_Y$$
$$= (n_Y - 1)\alpha - (n'_Y - 1)\alpha'$$

Substitute the value of α' from Eq. (3.6) in the above relation, so that
$$\Delta = (n_Y - 1)\alpha - (n'_Y - 1)\frac{n_V - n_R}{n'_V - n'_R}\alpha$$

$$= \frac{n_Y - 1}{n_V - n_R}(n_V - n_R)\alpha - \frac{n'_Y - 1}{n'_V - n'_R}(n_V - n_R)\alpha$$

$$= (n_V - n_R)\alpha \left(\frac{1}{\omega} - \frac{1}{\omega'} \right) \tag{3.11}$$

For two prisms of the same angle and of the same material, we have $n_V = n'_Y$, $\omega = \omega'$. This means that with such a combination, there will be a deviation, or the emergent beam will be parallel to the incident beam.

It must be emphasized that this combination is only achromatic for the two colours considered, i.e. violet and red. Other colours will emerge from the combination not quite parallel to these.

Achromatic prisms are not of much use in practice, but they lead to the formation of an achromatic combination of lenses.

EXAMPLE 3.2 Calculate the angle of a flint glass prisms which when combined with a crown glass prism of $10°$ produces deviation without dispersion. Find also the net deviation produced. Given

Crown glass $n_V = 1.523$, $n_R = 1.514$, $n = 1.517$
Flint glass $n'_V = 1.664$, $n'_R = 1.644$, $n' = 1.650$

Solution Let α and α' be the angles of crown and flint glass prism, we have $\alpha = 10°$. Substituting these values of various in parameter in following equation

$$\alpha' = -\frac{n_V - n_R}{n'_V - n'_R}\alpha = \frac{1.523 - 1.514}{1.664 - 1.644} \times 10°$$

$$= -45°$$

Let ω and ω' are dispersive powers of crown and flint glasses, then

$$\omega = \frac{n_V - n_R}{n - 1} = \frac{1.523 - 1.514}{1.517 - 1} = \frac{0.009}{0.517}$$

and

$$\omega' = \frac{n'_V - n'_R}{n' - 1} = \frac{1.664 - 1.644}{1.650 - 1} = \frac{0.02}{0.650}$$

The net deviation Δ is given as

$$\Delta = (n_V - n_R)\,\alpha \left(\frac{1}{\omega} - \frac{1}{\omega'} \right)$$

$$= (1.523 - 1.514) \times 10° \times \left(\frac{0.517}{0.009} - \frac{0.650}{0.02} \right)$$

$$= 2.245°$$

3.3.2 Dispersion without Deviation

If the combination of a crown glass prism C and flint glass prism F is so made that the mean deviation produced by the two prisms is equal and opposite, there will be no resultant mean deviation. But the angular dispersion produced by the two prisms will be different, so that there will be a finite amount of dispersion. Such a combination is said to produced dispersion without deviation (Figure 3.5).

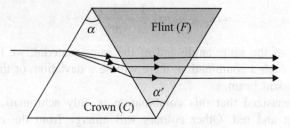

Figure 3.5 Dispersion without deviation.

Using the same notation as in previous case, the mean deviations δ_Y and δ'_Y produced by crown glass and flint glass prisms are given by

$$\delta_Y = (n_Y - 1)\alpha$$
$$\delta'_Y = (n'_Y - 1)\alpha'$$

In order that there is no resultant mean deviation

$$\delta_Y + \delta'_Y = 0$$

or
$$(n_Y - 1)\alpha + (n'_Y - 1)\alpha' = 0$$

or
$$\alpha = \frac{-(n'_Y - 1)}{(n_Y - 1)}\alpha' \qquad (3.12)$$

Equation (3.12) gives the relation between α and α', so that there is no deviation in the mean ray. The negative sign indicates that the two prisms should be oppositely placed with respect to one another.

The angular dispersion produced by the crown and flint glass prisms is $(\delta_V - \delta_R)$ and $(\delta'_V - \delta'_R)$ respectively. The total angular dispersion of the combination $(\Delta_V - \Delta_R)$ is the algebraic sum of the dispersion due to the two prisms.

Therefore,

$$\Delta_V - \Delta_R = (\delta_V - \delta_R) + (\delta'_V - \delta'_R) = (n_V - n_R)\alpha + (n'_V - n'_R)\alpha' \qquad (3.13)$$

Substituting the value of α' from Eq. (3.12), we get

$$\Delta_V - \Delta_R = (n_V - n_R)\alpha - \frac{(n'_V - n'_R)(n_Y - 1)}{(n'_Y - 1)}\alpha$$

$$= \frac{n_V - n_R}{n_Y - 1}(n_Y - 1)\alpha - \frac{n'_V - n'_R}{n'_Y - 1}(n_Y - 1)$$

$$= (\omega - \omega')(n_Y - 1)\alpha \qquad (3.14)$$

Since $\omega \neq \omega'$, $\Delta_V - \Delta_R$ is a finite quantity, the resultant dispersion depends upon the difference of the dispersive powers of the two prisms. Extreme red and violet rays lie on the opposite side of the undeviated mean ray. The principle is used in direct vision spectroscope.

3.4 DIRECT VISION SPECTROSCOPE

This is a compact form of a spectroscope with which we can see the spectrum of a source in the direction of the incident light. The instrument is constructed on the principle of producing dispersion without deviation. The instrument is very handly and can be small enough to be carried out in pocket.

The direct vision spectroscope consists of either two crown glass prisms and one flint glass prism or three crown glass prisms and two flint glass prisms of suitable angles α and α', defined by Eq. (3.12). The prisms are arranged with their refracting angles in opposite direction, so that the mid-part of the spectrum suffers no deviation. The prisms are cemented together by Canada balsam to avoid loss of intensity due to reflection at interfaces between the prisms. They are put in a tube which has an adjustable slit S and a collimating lens L_1 at one end. The other end consists of a telescope with an adjustable eyepiece E (Figure 3.6).

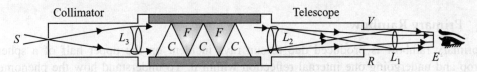

Figure 3.6   Direct vision spectroscope.

A narrow beam of white light passing through the slit is rendered parallel by the lens L_1. It then transverses the combination of the prisms, where it is dispersed without deviation. The paths of the blue and red rays are shown in Figure 3.7. It should be noted that these rays cross one another in the second crown glass prism. They emerge out as a parallel beam which falling on another converging lens L_2 is brought to focus in its second focal plane to form a spectrum VR. Thus spectrum is next examined by the eyepiece E, which acts as a simple magnifying glass. By directing the spectroscope towards a source of light, say a star, a qualitative study of the spectra can be made, but as the resolving power is small, lines close to each other cannot easily be separated. The resolving power of the instrument can be increased by using more prisms.

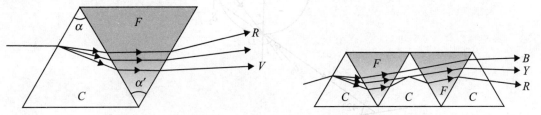

Figure 3.7   Paths of blue and red rays from direct vision spectroscope.

3.5 RAINBOW

The rainbow is nature's most spectral display of the spectrum of white light. The condition required for its appearance is that the sun is shining in one part of the sky and there be rain drops during or after shower in the other part. Under favourable conditions several bows may be seen. However, in all cases the observer's back must be towards the sun. The brightest rainbow is called the *primary rainbow*. Its radius subtends an angle of 40° at the observer's eye. The primary rainbow is violet at its inner edge and red at its outer. The arcs of intermediate colour occupy proper positions in spectral order between the two edges. In addition to the primary bow sometime another fainter bow, called the *secondary rainbow*, is also observed. The secondary rainbow subtends an angle of 52° at the observer's eye. It is violet at its outer edge and red at its inner edge, i.e. the order of spectral colours is reversed compared with that in the primary rainbow. In addition to the primary and the secondary bows sometimes other bows terms *supernumerary* or *spurious bows*, are also observed just near the inner edge of the primary and the outer edge of the secondary bow.

The elementary theory of rainbow was first given by *Antonian de Demini* in the year 1611 and later developed more exactly by *Descartes*. The general characteristics of the primary and secondary bows are explained by considering only the reflection and refraction of light by spherical raindrops.

3.5.1 Primary Rainbow

The primary rainbow is produced due to sun rays incident on the upper half of a spherical raindrop and undergoing one internal reflection within it. To understand how the phenomenon arises, we first confine our attention to an individual raindrop as shown in Figure 3.8. Let AB be a ray of light incident on the raindrop at the angle of incidence i. The radius OB of the drop is normal at B. This ray is refracted into the raindrop according to ordinary laws of refraction, giving rise to refracted ray BC inclined at angle r to the normal at B. The clockwise deviation δ_1 of the incident ray AB due to refraction at B is given by

$$\delta_1 = i - r \qquad (3.15)$$

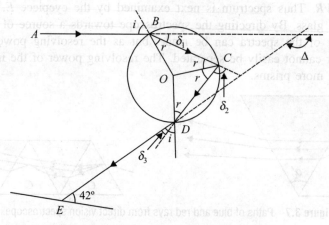

Figure 3.8 Ray diagram for primary rainbow.

The refracted ray BC is reflected at C on the back surface of the sphere. Since $OB = OC$, the angle of incidence at C is r and, consequently, the angle of reflection at C is also r. The clockwise deviation δ_2 of the ray BC due to reflection at C, is given by

$$\delta_2 = \pi - 2r \qquad (3.16)$$

Finally, the reflected ray *CD* suffers refraction at *D* into air. Since $OC = OD$, $\angle ODC = \angle OCD = r$. As a consequence of emergent ray *DE* makes an angle *i* with the radius *OD*, which is normal at *D*. The clockwise deviation δ_3 produced due to refraction at *D* is given by

$$\delta_3 = i - r \qquad (3.17)$$

Thus the resultant deviation Δ which is equal to the sum of deviations δ_1, δ_2, and δ_3 is given by

$$\Delta = (i - r) + (\pi - 2r) + (i - r)$$
$$= \pi + 2i - 4r \qquad (3.18)$$

The resultant deviation depends on the angle of incidence. The graph showing this, is given in Figure 3.9.

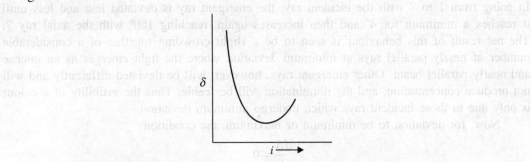

Figure 3.9 Graph between angle of incidence and angle of deviation in a primary rainbow.

It is found that corresponding to certain angle of incidence the deviation is minimum. It is the ray which undergoes minimum deviation that is responsible to produce the rainbow. To understand why these particular rays are responsible for the rainbow, and why other rays which enter the drop can be neglected (Figure 3.10).

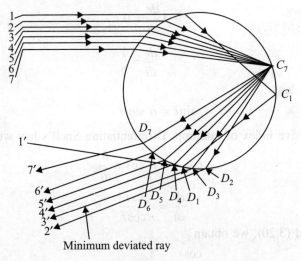

Figure 3.10 Responsible rays for primary rainbow.

Here, in the parallel beam of sunlight entering a single drop, our attention is confined to the rays of only one colour. By this simplification, the *phenomenon of dispersion* is dispensed with. We consider only the reflection and refraction of red light. As each of the parallel red rays 1, 2, 3, ..., 7 from the sun enter the drop, they are deviated according to the law of refraction. At the opposite side of the drop, in the region C_1 to C_7, these rays are partially reflected to the lower boundary at $D_1, D_2, ..., D_7$, where by refraction, they pass out into air in the direction 1', 2', 3', ..., 7' respectively.

From Figure 3.10 it will be noted that a ray like 1, entering at every edge of the drop, is deviated through an angle to 180°. It emerges almost in the opposite direction to that in which it entered. The same is true of ray 6 nearer the centre of the drop. Ray 7 is exactly reversed in direction and, therefore, deviated exactly through 180°. Ray 4, on the other hand, is seen to be deviated least of all and is, therefore, referred to as the minimum deviated ray. In going from 1 to 7 with the incident ray, the emergent ray is deviated less and less until it reaches a minimum for 4' and then increases again, reaching 180° with the axial ray 7. The net result of this behaviour is seen to be a slight crowding together of a considerable number of nearly parallel rays at minimum deviation where the light emerges as an intense and nearly parallel beam. Other emergent rays, however, will be deviated differently and will not produce concentration, and the illumination will be feeble. Thus the visibility of a colour is only due to those incident rays which undergo minimum deviation.

Now, for deviation to be minimum or maximum, the condition

$$\frac{d\Delta}{di} = 0$$

From Eq. (3.18), on differentiation, we obtain

$$\frac{d\Delta}{di} = 2 - 4\frac{dr}{di}$$

Thus the condition for maximum or minimum deviation is

$$2 - 4\frac{dr}{di} = 0$$

or

$$\frac{dr}{di} = \frac{1}{2} \qquad (3.19)$$

According to Snell's law

$$\sin i = n \sin r$$

where n is the refractive index of raindrop. Differentiating Snell's law with respect to i

$$\cos i = n \cos r \frac{dr}{di}$$

or

$$\frac{dr}{di} = \frac{\cos i}{n \cos r} \qquad (3.20)$$

From Eqs. (3.19) and (3.20), we obtain

$$\frac{\cos i}{n \cos r} = \frac{1}{2}$$

or
$$2\cos i = n \cos r$$
$$4\cos^2 i = n^2 \cos^2 r$$
$$= n^2(1 - \sin^2 r)$$
$$= n^2\left(1 - \frac{\sin^2 i}{n^2}\right)$$
$$= n^2 - \sin^2 i$$
or
$$3\cos^2 i + (\cos^2 i + \sin^2 i) = n^2$$
or
$$3\cos^2 i = n^2 - 1$$
$$\therefore \quad \cos i = \left(\frac{n^2 - 1}{3}\right)^{1/2} \tag{3.21}$$

Equation (3.21) gives the value of i, the angle of incidence, for which the deviation is minimum. Since the value of i is a function of n, the minimum deviation is different for different colours.

The refractive index of water for red colour $n_R = 1.329$. If we substitute this value of n in Eq. (3.21), we find $i = 59.6°$ and $r = 40.5°$ from Snell's law. Therefore, the angle of minimum deviation for red colour is given by

$$\Delta_R = \pi + 2i - 4r$$
$$= 180 + 2 \times 59.6 - 4 \times 40.5 = 137.2°$$

Therefore, the emergent red rays in the primary rainbow will be inclined to the horizon at an angle $(\pi - \Delta_R) = 180° - 137.2° = 42.6°$. Similarly, since for water $n_V = 1.343$, hence $i = 58.8°$, $r = 39.6°$, inclination of violet rays to the horizon will be 40.8°. Obviously, for intermediate spectral colours the inclination with the horizon will be between 40.8° and 42.8°.

Now consider an observer at P (Figure 3.11). The XY plane is horizontal and suppose the rays from the sun are incident on water drops in the sky from the left, parallel to X-axis.

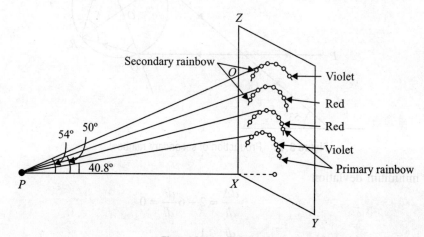

Figure 3.11 Rainbow.

The point O may be considered as shadow of observer's eye on the YZ plane. We have seen that the emergent red ray obtained after minimum deviation is inclined at 42.8° to the horizon. Hence all drops which lie on a circle, with centre O and radius subtending an angle 42.8° at P, will reflect red light strongly to P. Similarly, all drops, which lie on a circle with angular radius 40.8° at P, will reflect violet light strongly, while those situated at intermediate position will reflect intermediate colours to P. The primary bow is, therefore, violet in colour on the lower side and red on the upper, other colours falling at intermediate positions. This explains the formation of the primary rainbow.

3.5.2 Secondary Rainbow

The secondary rainbow is produced by the rays of sunlight incident on the lower half of the raindrop. Each ray is refracted and dispersed upon entering the drop. The ray undergoes two internal reflections before being refracted finally into air. This inverts the order of colours as compared with that of the primary bow. The path of the ray PQ incident on the raindrop is shown in Figure 3.12.

The deviation δ_1 produced due to refraction on entering the drop is $(i - r)$ in clockwise direction. One the first and second interval reflections, the deviations δ_2 and δ_3 are equal to $(\pi - 2r)$ and, finally on emergence from the drop, the deviation δ_4 is $(i - r)$ also in anticlockwise direction. Thus the total deviation Δ is

$$\Delta = \delta_1 + \delta_2 + \delta_3 + \delta_4$$
$$= (i - r) + (\pi - 2r) + (\pi - 2r) + (i - r)$$
$$= 2\pi + 2i - 6r$$

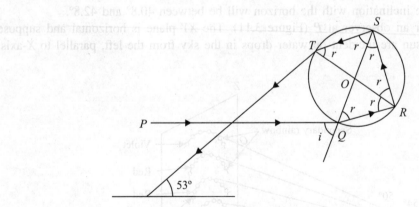

Figure 3.12 Production of secondary rainbow.

Hence the minimum deviation

$$\frac{d\Delta}{di} = 2 - 6\frac{dr}{di} = 0$$

or

$$\frac{dr}{dt} = \frac{1}{3}$$

But $\dfrac{\sin i}{\sin r} = n$

$$\Rightarrow \qquad \cos i = n \cos r \dfrac{dr}{di}$$

Thus for the minimum deviation

$$\dfrac{\cos i}{n \cos r} = \dfrac{1}{3}$$

Squarring

$$9 \cos^2 i = n^2 \cos^2 r$$
$$= n^2 (1 - \sin^2 r)$$

$$\Rightarrow \qquad 8 \cos^2 i = n^2 - (\sin^2 i + \cos^2 i) \qquad \left[\text{using Snell's law } n = \dfrac{\sin i}{\sin r} \right]$$

$$\Rightarrow \qquad \cos i = \sqrt{\dfrac{n^2 - 1}{8}} \qquad (3.22)$$

It can easily be shown, as in the case of primary rainbow, that at minimum deviation the red ray is inclined at an angle of about 54°. As in the primary rainbow, here also the light which is reflected in any given direction consists chiefly of the colour for which that direction is the angle of minimum deviation.

Referring again to Figure 3.11, all water drops lying on a circle with centre at O and angular raidus 54° at P, will reflect violet light strongly at P. The red light reflected strongly at P due to all drops which lie on the circle with angular radius 51° at P. The secondary bow is, therefore, situated above the primary and since violet is deviated more than red, it is red in colour on the lower edge and violet on the upper. Since rays producing secondary bow have undergone two internal reflections, the secondary rainbow is fainter than the primary. This explains the formation of the secondary bow.

3.6 SUNDOGS AND HALOS

Sundogs appear frequently at McMurdo. The most bright ones occur on a cold sunny morning or evening, when the sun is near the horizon and the air is loaded with ice crystals. The ice crystals refract the sunlight causing an image of the sun to appear on either or both sides of the sun as shown in Figure 3.13.

We also see rings of light surrounding to the sun or moon. These rings are called *halos*. Most halos appear as bright white rings, but in some instances, the dispersion of light as it passes through ice crystals can cause a halo to have colour, as shown in Figure 3.14.

Halos form when sunlight or moonlight is refracted by ice crystals associated with thin, high-level clouds (like cirrostratus clouds). A 22° halo is a ring of light 22° from the sun (or moon) and is the most common type of halo observed.

Light undergoes two refractions as it passes through an ice crystals and the amount of bending that occurs depends upon the diameter of the crystal. In the case of a 22° halo, we will focus on hexagonal ice crystals with diameter less than 0.5 μm (Figure 3.15).

58 Fundamentals of Optics

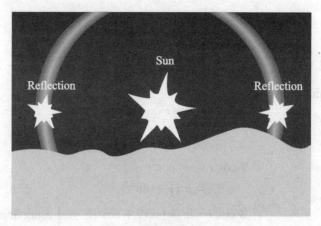

Figure 3.13 Sundogs.

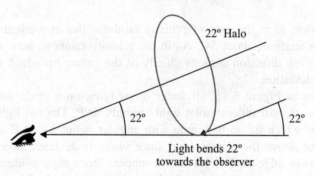

Figure 3.14 Halos.

A 22° halo develops when light enters one side of a columnar ice crystal and exists through another side. The light is refracted when it enters the ice crystal and once again when it leave the ice crystal as shown in Figure 3.15.

The two refractions bend the light by 22° from its original direction, producing a ring of light observed at 22° from the sun or moon.

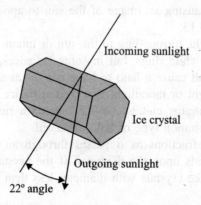

Figure 3.15 22° halo with hexagonal ice crystal.

The tangent arc is a patch of bright light that is occasionally observed along a halo. This occurs when sunlight is refracted by falling hexagonal "pencil-shaped" ice crystals whose long axes are oriented horizontally.

So, it is a technical explanation of something quite simple and beautiful to be hold. One some days the halos are quite vivid, but usually they are faint and difficult to see. With everything basically white in colour in the Antarctic, looking up at the sun's brilliance is not an easy thing to do.

FORMULAE AT A GLANCE

3.1 *Cauchy's relation*

The refractive index (n) of a material for wavelength λ.

$$n = a + \frac{b}{\lambda^2} + \frac{c}{\lambda^4}$$

where a, b and c are Cauchy's constant.

3.2 Angular dispersion

$$= \delta_V - \delta_R = (n_V - n_R)A$$

where A = angle of prism

δ_V and δ_R = angles of deviation for violet and red respectively and n_V and n_R = refractive indices for violet and red respectively.

3.3 Combination of prisms

(a) *Deviation without dispersion*

(i) $\alpha_F = \dfrac{-(n_V - n_R)}{(n'_V - n'_R)} \alpha_C$

where α_F = angle of prism of flint glass.

α_C = angle of prism of crown glass

n_V and n_R = refractive indices for violet and red colours respectively for crown glass.

n'_V and n'_R = refractive indices for violet and red colours respectively for flint glass.

(ii) *Dispersive power of crown glass*

$$\omega_C = \frac{n_V - n_R}{n - 1}$$

where n is the mean refractive index.

(iii) *Dispersive power of flint glass*

$$\omega_C = \frac{n'_V - n'_R}{n' - 1}$$

where n' = mean refractive index of flint glass.

(iv) The resultant *mean deviation*

$$\Delta = \delta_Y - \delta'_Y$$
$$= (n-1)\alpha_C - (n'-1)\alpha_F$$
$$= (n_V - n_R)\alpha_C\left(\frac{1}{\omega_C} - \frac{1}{\omega_F}\right)$$

(b) *Dispersion without deviation*

(i) $\alpha_C = \dfrac{-(n'-1)}{(n-1)}\alpha_F$

(ii) *The resultant mean deviation*

$$\Delta = (\omega_C - \omega_F)(n-1)\alpha_C$$

SOLVED NUMERICAL PROBLEMS

PROBLEM 3.1 Find the angle of dispersion between red and violet colours produced by a flint glass prism of refracting angle 60°. Refractive indices of prism for red and violet colours are 1.622 and 1.663 respectively.

Solution Given: $A = 60°$, $n_R = 1.622$, and $n_V = 1.663$
Angular dispersion between red and violet colours is:
$$\delta_V - \delta_R = A(n_V - n_R)$$
$$= 60°(1.663 - 1.622)$$
$$= 2.46°$$

PROBLEM 3.2 A thin prism of refracting angle 2° deviates an incident ray through an angle of 1°. Find the value of refractive index of the material of the prism.

Solution Here $A = 2°$, $\delta = 1°$, and $n = 1$
Deviation through a prism of small angle is given by
$$\delta = (n-1)A$$
$$\therefore \quad n = \frac{\delta}{A} + 1 = \frac{1}{2} + 1 = 1.5$$

PROBLEM 3.3 A prism with refracting angle of 60° gives angle of minimum deviation 53°, 51° and 52° for blue, red and yellow light respectively. What is the dispersive power of the material of the prism?

Solution Given: $\delta_B = 53°$, $\delta_R = 51°$, and $\delta_Y = 52°$
$$\text{Dispersive power } \omega = \frac{\delta_B - \delta_R}{\delta_Y} = \frac{53 - 51}{52} = 0.04$$

PROBLEM 3.4 The refractive indices for a material for red, violet and yellow lights are 1.52, 1.62, and 1.59 respectively. Calculate the dispersive power of the material. If the mean deviation of 40°, then what will be the angular dispersion produced by a prism of this material?

Solution Given: $n_R = 1.52$, $n_V = 1.62$, $n_Y = 1.59$ and $\delta_Y = 40°$

$$\text{Dispersive power } \omega = \frac{n_V - n_R}{n_Y - 1} = \frac{1.62 - 1.52}{1.59 - 1} = 0.169$$

Also $\omega = \dfrac{\delta_V - \delta_R}{\delta_Y}$

$\therefore$ Angular dispersion $= \delta_V - \delta_R = \omega \delta_Y = 0.169° \times 40° = 6.76°$.

PROBLEM 3.5 Find the angle of the flint glass prism required to combine with two crown glass prisms, each of refracting angle 20°, so as to form a direct vision spectroscope. Calculate the total dispersion from the following data:

crown glass $n_R = 1.514$, $n_V = 1.523$ and $n = 1.517$
flint glass $n'_R = 1.645$, $n'_V = 1.644$ and $n' = 1.650$

Solution Let α' be the desired angle of the flint glass prism. From equation

$$\alpha' = -\frac{n' - 1}{n - 1}\alpha$$

$$= -\frac{1.650 - 1}{1.517 - 1} \times 20°$$

$$= -\frac{0.650}{0.517} \times 20° = -31.8°$$

The minus sign indicates that flint glass prism should be placed with its angle opposite to the crown glass.

The dispersion produced by two crown glass prisms is

$$\delta_V - \delta_R = 2(n_V - n_R)\alpha$$

$$= 2 \times (1.523 - 1.514) \times 20°$$

$$= 0.36°$$

and the dispersion produced by the flint glass prism is

$$\delta'_V - \delta'_R = (n'_V - n'_R)\alpha'$$

$$= (1.664 - 1.645) \times (-31.8°)$$

$$= -0.6°$$

Therefore, the resultant dispersion is

$$\Delta_V - \Delta_R = (\delta_V - \delta_R) + (\delta'_V - \delta'_R)$$

$$= 0.36° - 0.6°$$

$$= -0.24°$$

CONCEPTUAL QUESTIONS

3.1 What is a prism?

Ans. A prism is a portion of refracting medium bounded by two plane faces inclined to one another.

3.2 Write the expression of refractive index in terms of deviation angle.

Ans. In passing through a prism, a monochromatic ray of light undergoes deviation δ given by the relation

$$\delta = i_1 + i_2 - A$$

where i_1 and i_2 stand for the angle of incidence on two faces of the prism and A is the refractive angle of the prism. The angle of deviation (δ_m) is minimum, when say passes symmetrically through the prism. The refractive index n of the prism is

$$n = \frac{\sin\dfrac{A+\delta_m}{2}}{\sin\dfrac{A}{2}}$$

3.3 What do you mean by dispersion?

Ans. When white (or polycromatic) light passes through a prism, it (light) breaks up into its constituent colours. This is known as *dispersion*. In general, the emergent rays are not in the direction of incident ray, *i.e.*, there is a deviation also. Refraction through a prism, therefore results in:

(i) dispersion

(ii) deviation

3.4 What is Cauchy's relation for a prism?

Ans. Dispersion is brought about by a prism because refractive index of prism material is different for different wavelengths. The refractive index, n for wavelength λ is given by Cauchy's relation

$$n = a + \frac{b}{\lambda^2} + \frac{c}{\lambda^4}$$

where a, b and c are constants, are known as Cauchy's constants.

3.5 What is dispersive power of a prism?

Ans. The dispersive power of a prism is defined as dispersion per unit deviation. It is constant characteristic of the prism material for small angled prism. If δ_V, δ_Y and δ_R stand for deviation of violet, yellow and red coloured rays, the dispersive power (ω) is

$$\omega = \frac{\delta_R - \delta_V}{\delta_Y} = \frac{n_R - n_V}{n_Y - 1}$$

3.6 What is irrotational dispersion?

Ans. When light undergoes dispersion in prisms of different materials and angles, the ratio of the partial dispersion of any two colours to that of the extreme colours is different. This feature is known as irrotational dispersion.

3.7 What is rainbow?

Ans. Rainbow is nature's spectral display of the phenomenon of dispersion. It is observed when after rain, the sun shines. A person standing back towards sun observes concentric circular arcs of different colours in the horizon. Rainbow is due to *refraction in water drops*. The rainbow formed due to one *internal reflection* in the water drop is known as the *primary rainbow*.

Here the innermost circular arc is violet in colour subtending the semicone angle of 40.8°. The outermost arc is red in colour subtending an angle of 54°.

If the sun rays undergo two internal reflections inside the water drop before emerging out, the rainbow produced is known as *secondary rainbow*. Due to two internal reflections, the sequence of colours in secondary rainbow is opposite to what it is in primary rainbow.

EXERCISES

Theoretical Questions

3.1 What is dispersion of light? Explain it with a ray diagram. Also explain the cause of dispersion of light.

3.2 Define angular dispersion and dispersive power. Write expression for these quantities in terms of refractive index.

3.3 What is dispersion? Explain the phenomenon of dispersion and define dispersive power.

3.4 Define dispersive power. Is dispersive power constant for a material, in general?

3.5 Show how prisms should be combined to obtain

(a) dispersion without deviation, and

(b) deviation without dispersion

3.6 Describe in detail the construction, theory and working of a direct vision spectroscope.

3.7 What is rainbow? Give a theory of the formation of primary rainbow. In what order use the colours arranged and why? In what part of the sky rainbow is seen in the morning?

3.8 Discuss the formation of primary and secondary rainbow. Will two observers see the same rainbow? Explain your answer.

3.9 How is secondary rainbow obtained? Why sequence of colours in secondary rainbow is opposite to that in primary rainbow?

3.10 What is rainbow? Explain the formation of primary and secondary rainbow.

Numerical Problems

3.1 The refractive indices of flint glass for blue and red colours are 1.664 and 1.644. Calculate its dispersive power. [**Ans.** $\omega = 0.0305$]

3.2 A glass prism deviates the red and blue rays through 10° and 12° respectively. A second prism of equal angle deviates than through 8° and 10° respectively. Find the ratio of their dispersive powers. [**Ans.** $\omega/\omega' = 9:11$]

3.3 White light is passed through a prism of 5°. If the refractive indices for red and blue rays are 1.641 and 1.659 respectively, calculate the angle of dispersion between them. [**Ans.** 0.09°]

3.4 The refractive indices for lights of violet, yellow and red colours for a flint glass prism are 1.632, 1.620 and 1.613 respectively. Find the dispersive power of the prism material. [**Ans.** 0.0306]

3.5 Calculate the dispersive power of crown and flint glass prisms from the following data:
for crown glass $n_B = 1.522$, $n_R = 1.514$
for flint glass $n_B = 1.662$, $n_R = 1.644$ [**Ans.** $\omega_{crown} = 0.015$ and $\omega_{flint} = 0.028$]

3.6 The refractive indices of crown glass for violet and red colours are 1.523 and 1.513 respectively. Determine the dispersive power of this glass. If a crown glass prism produces a mean deviation of 40°, what will be the angular dispersion? [**Ans.** $\omega = 0.019$, 0.77°]

3.7 A ray of light is incident on the surface of a transparent sphere ($n = 1.45$). After refraction, it is internally reflected (not total) and then refracted out of the sphere. Find the deviation suffered by the ray and the angle of incidence to the nearest degree so that the deviation may be minimum. [**Ans.** 152°42′]

3.8 A crown glass convex lens radii of curvature 10 cm and 20 cm. If n_B and n_R are 1.523 and 1.513 respectively. Find the dispersive power of the lens. [**Ans.** $\omega = 0.0193$]

3.9 The angle of minimum deviation for a prism of refractive angle A is found to be 51°. The angle of deviation is 62°48′ for angles of incidence 40°6′ and 82°42′. Calculate the refractive index of the prism. [**Ans.** $n = 1.648$]

3.10 Light undergoes refraction through a prism XYZ as shown in Figure 3.16.

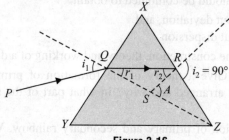

Figure 3.16

Show that the refractive index n of the prism is

$$n = \sqrt{1 + \left[\frac{\sin i_1 + \cos A}{\sin A}\right]^2}$$

3.11 Find the angle of the flint glass prism required to combine with two crown glass prisms, each of refractive angle 20°, so as to form a direct vision spectroscope. Calculate the total dispersion from the following data:

Crown glass: $n_R = 1.514$ $n_V = 1.523$ $n = 1.517$
Flint glass: $n'_R = 1.645$ $n'_V = 1.664$ $n' = 1.650$

[**Ans.** $\alpha' = -31.8°$, net dispersion $\Delta = -0.2442°$]

Multiple Choice Questions

3.1 A ray incident at 15° on one refracting surface of a prism of angle 60°, suffers a deviation of 55°, what is the angle of emergence?
(a) 95°   (b) 45°
(c) 30°   (d) none of these

3.2 In Figure 3.17, what is the angle of prism?
(a) A   (b) B
(c) C   (d) D

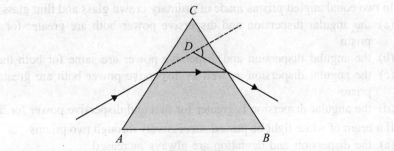

Figure 3.17

3.3 When white light enters a prism, it gets split into its constituent colours. This is due to
(a) high density of prism material
(b) because n is different for different wavelengths
(c) diffraction of light
(d) velocity changes for different frequency

3.4 Dispersion of light is caused due to
(a) wavelength   (b) intensity of light
(c) density of medium   (d) none of these

3.5 A thin prism P_1 with angle 4° made from a glass of refractive index 1.54 is combined with another thin prism P_2 made from glass of refractive index 1.72 to produce dispersion without deviation. The angle of the prism P_2 is
(a) 5.33°   (b) 4.33°
(c) 3.33°   (d) 2.66°

3.6 White light is incident on one of the refracting surfaces of a prism of angle 5°. If the refractive indices for red and blue colours are 1.641 and 1.659 respectively, the angular separation between these two colours when they emerge out of the prism is
(a) 0.9°   (b) 0.09°
(c) 1.8°   (d) 1.2°

3.7 In the formation of a rainbow, the light from the sun on water droplets undergoes
  (a) dispersion only
  (b) only total internal reflections
  (c) dispersion and total internal reflection
  (d) none of the above

3.8 The phenomenon of dispersion arises because of
  (a) the decomposition of a white light beam by a prism
  (b) the refraction of light
  (c) the refractive index of the prism material being different for different wavelengths
  (d) the corpuscular nature of light

3.9 When a beam of white light is dispersed with the help of a prism, the dispersion is greatest for
  (a) orange
  (b) green
  (c) red
  (d) indigo

3.10 Among the following, the dispersive power of which material is greatest
  (a) flint glass
  (b) crown glass
  (c) carbon disulphide
  (d) none of these

3.11 In two equal angled prisms made of ordinary crown glass and flint glass
  (a) the angular dispersion and dispersive power both are greater for the crown glass prism
  (b) the angular dispersion and dispersive power are same for both the prisms
  (c) the angular dispersion as well as dispersive power both are greater for flint glass prism
  (d) the angular dispersion is greater for first and dispersive power for the second prism

3.12 If a beam of white light be passed successively through two prisms
  (a) the dispersion and deviation are always increased
  (b) the dispersion and deviation are always decreased
  (c) the dispersion is always increased and the deviation always decreased or vice versa
  (d) the dispersion as well as deviation are increased when the prisms are similarly placed and when the prisms are oppositely placed with respect to the incident lights

3.13 It is observed that if a large number of transparent glass beads are hung from a window curtain and sun is shining brightly the beads under favourable conditions exhibit various colours. The colour effect is due to
  (a) dispersion of light
  (b) reflection of light
  (c) interference of light
  (d) diffraction of light

3.14 In a room, artificial rain is produced in one end and a strong source of white light is switched ON at the other end. To observe the rainbow, an observer must
  (a) look anywhere in the room
  (b) look towards the source
  (c) look towards raindrops
  (d) look in a direction equally inclined to the source of raindrops

3.15 If sun is shining brightly in one part of the sky after rain, how many rainbows are observed?
  (a) one
  (b) two
  (c) four
  (d) infinite

3.16 The rays responsible for primary rainbow undergo
(a) multiple reflection and refraction
(b) one reflection and two refractions
(c) one refraction and two reflections
(d) two reflections and two refractions

3.17 The rays producing secondary rainbow undergo
(a) two reflections and two refractions
(b) multiple reflection and refraction
(c) one reflection and two refractions
(d) one refraction and one reflection

3.18 Soap bubbles appear to be coloured due to the phenomenon of
(a) polarisation
(b) interference
(c) diffraction
(d) dispersion

Answers

3.1 (d)   **3.2** (c)   **3.3** (b)   **3.4** (a)   **3.5** (a)   **3.6** (b)   **3.7** (c)   **3.8** (c)
3.9 (d)   **3.10** (c)   **3.11** (c)   **3.12** (d)   **3.13** (a)   **3.14** (c)   **3.15** (b)   **3.16** (b)
3.17 (a)   **3.18** (d)

CHAPTER 4

Lens Aberrations

"There are two kinds of light—the glow that illuminates, and the glare that obscures."
—James Thurber

IN THIS CHAPTER

- Types of Aberrations
- Types of Monochromatic Aberration and Their Reduction
- Chromatic Aberration
- Aplantism and Aplanatic Points

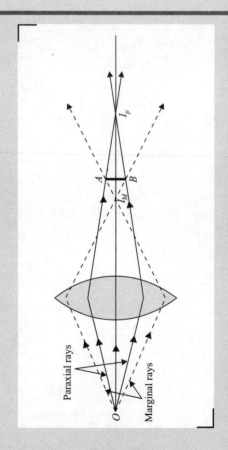

4.1 INTRODUCTION

In various image formulae, between object and image distances, radii of curvature, refractive indices, focal lengths, etc. were based on following assumptions:
1. All the incident rays made small angle with principal axis.
2. The aperture of the lens was small.

In practice, however, lenses are used to form images of points situated off the axis also. Moreover, due to finite size of the lens, the cone of the light rays which forms the image of the point object is of finite size.

We know that deviation produced in any ray depends on the height of point of incidence, on the lens, different rays come to focus at different points. Thus, the image is not sharp. Also, the refractive indices and the focal lengths of a lens are different for different wavelength of light. When the incident light is not monochromatic, different colours are focused at different points. Thus, the lens forms a number of images of different colours at different positions. Hence, the *deviations from the actual size, shape and positions of an image as calculated by simple equations are called aberrations produced by a lens.*

4.2 TYPES OF ABERRATIONS

The aberrations are of two types.

4.2.1 Chromatic Aberration

The aberration in image due to variation of refractive index with wavelength (due to colour) is called *chromatic aberration.*

4.2.2 Monochromatic Aberration

For monochromatic light, due to single wavelength or no variation of refractive index, chromatic aberration can be reduced but another type of aberration arises. The monochromatic light rays incident on the optical surface after refraction from image of different sizes and different positions. Such aberration is known as *monochromatic aberration*, which is classified as:
  (i) Spherical aberration
  (ii) Coma
  (iii) Astigmatism
  (iv) Curvature of field, and
  (v) Distortion

The aberration can also be understood mathematically. From Abbe's sine condition

$$n_1 h_1 \sin \theta_1 = n_2 h_2 \sin \theta_2 \tag{4.1}$$

where n_1, h_1, θ_1 are the refractive index, lateral size of the object and angle subtended by object with principal axis in object space, while n_2, h_2, θ_2 are respective values in image space. Expanding $\sin \theta$ as

$$\sin\theta = \theta - \frac{\theta^3}{3!} + \frac{\theta^5}{5!} - \frac{\theta^7}{7!} + \cdots \qquad (4.2)$$

If θ is small, $\sin\theta = \theta$, which is the case of paraxial rays. It is known as *first order theory*. In such case, the principal aberration (monochromatic aberration) is reduced, but chromatic aberration arises.

Furthermore, if θ is not small, then angle of sine (i.e. θ) may be replaced by $\left(\theta - \frac{\theta^3}{3!}\right)$. The results obtained due to $\left(\theta - \frac{\theta^3}{3!}\right)$ angle represent *third order theory*. This theory was developed by Ludwig Von Seidal in 1855 which gives the various departures from ideal image. These departures are known as *monochromatic* or *Seidal aberrations*.

4.3 TYPES OF MONOCHROMATIC ABERRATION AND THEIR REDUCTION

4.3.1 Spherical Aberration

Consider a point source O of monochromatic light placed on the axis of a large aperture convex lens as shown in Figure 4.1.

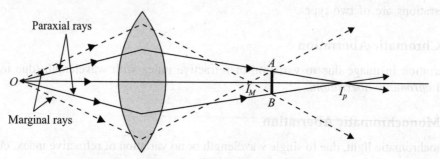

Figure 4.1 Spherical aberration.

The rays, which are incident near the axis (called paraxial rays) come to focus at point I_P, while the rays incident near the rim of the lens (called marginal or peripheral rays) come to focus at I_M. The intermediate rays are brought to focus between I_M and I_p. It is clear from Figure 4.1, that the paraxial rays from the image focus at a point distance than marginal rays. Thus, the image is not sharp at any point on the axis. The failure or inability of the lens to form a point image of an axial point object is called *spherical aberration*.

Reason

The lens can be supposed to be divided into circular zones. It can be proved mathematically that the focal lengths slightly vary with the radius of the zone, i.e. different zones have different focal lengths. The focal length of marginal rays is lesser than the paraxial zone, hence the marginal rays are focused first. The spherical aberration can also be explained by saying that the marginal rays suffer greater deviation than the paraxial rays, because they are incident at a greater height than the latter. The distance between I_M and I_p is the measure of longitudinal or axial spherical aberration.

Spherical aberration due to spherical surface

Let AB be the spherical surface of radius R. Let an incident ray parallel to principal axis meet the spherical surface at a height h from the principal axis. The refracted ray intersects the axis at a point F_h (Figure 4.2). Here OF_h is equal to the focal length f_h for rays in zone of height h.

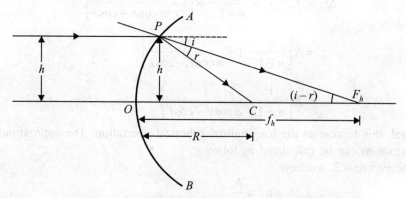

Figure 4.2 Spherical aberration due to spherical surface.

From Figure 4.2

$$f_h = R + CF_h \qquad (4.3)$$

Now, in $\triangle CPF_h$, we have

$$\frac{CF_h}{\sin r} = \frac{R}{\sin(i-r)}$$

$$\Rightarrow \quad CF_h = \frac{R \sin r}{\sin(i-r)} = \frac{R \sin r}{\sin i \cos r - \cos i \sin r}$$

$$= \frac{R \sin r}{n \sin r \cos r - \cos i \sin r} \quad \left[\because n = \frac{\sin i}{\sin r}\right]$$

$$= \frac{R}{n \cos r - \cos i} \qquad (4.4)$$

Substituting the value of CF_h from Eq. (4.4) in Eq. (4.3), we get

$$f_h = R + \frac{R}{n \cos r - \cos i}$$

$$= R \left[1 + \frac{1}{n \cos r - \cos i} \right] \qquad (4.5)$$

For paraxial rays

In limiting case, when $h \to 0$, $\cos i$ and $\cos r$ tend to 1.

In this case

$$f_h = \frac{nR}{n-1} \tag{4.6}$$

which is equal to f_h, i.e. the focal length of paraxial rays.

The change in focal length for h zone as compared to axial zone is given as

$$\Delta f_h = f_p - f_h = \frac{nR}{n-1} - R\left[1 + \frac{1}{n\cos r - \cos i}\right]$$

$$= R\left[\frac{n}{n-1} - 1 - \frac{1}{n\cos r - \cos i}\right]$$

$$= R\left[\frac{1}{n-1} - \frac{1}{n\cos r - \cos i}\right] \tag{4.7}$$

In general, this represents the longitudinal spherical aberration. The approximate value of spherical aberration can be calculated as follows:

From the Figure 4.2, we have

$$\sin i = \frac{h}{R}$$

or

$$\sin r = \frac{h}{nR} \quad \left[\because \frac{\sin i}{\sin r} = n\right]$$

Now,

$$\cos i = \sqrt{1 - \sin^2 i} = \sqrt{1 - \frac{h^2}{R^2}} = 1 - \frac{h^2}{2R^2} \quad \text{(By binomial expansion)}$$

and

$$\cos r = \sqrt{1 - \sin^2 r} = \sqrt{1 - \frac{h^2}{n^2 R^2}} = 1 - \frac{h^2}{2n^2 R^2} \quad \text{(By binomial expansion)}$$

Substituting these values in Eq. (4.7), we get

$$\Delta f_h = R\left[\frac{1}{n-1} - \frac{1}{n\left(1 - \frac{h^2}{2n^2 R^2}\right) - \left(1 - \frac{h^2}{2R^2}\right)}\right]$$

solving it, we get

$$\Delta f_h = \frac{h^2}{2(n-1)^2 f_p} \tag{4.8}$$

where

$$f_p = \frac{nR}{n-1}$$

Equation (4.8) gives the approximate value of the spherical aberration due to spherical surface.

Minimization of spherical aberration

The following are the methods for minimizing the spherical aberration.

(i) By means of stops: We know that the spherical aberration is due to different focal lengths of different zones. The spherical aberration can be minimized by using stops (Figure 4.3). The stops used may be of such a nature that they permit only the axial rays and stop the marginal rays or permits the marginal rays and stop the axial rays. The method is not generally used, since the intensity of the image is very much reduced.

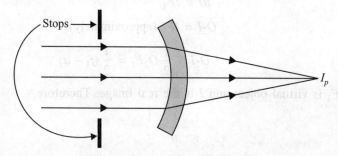

Figure 4.3 Use of stops.

(ii) By using two suitable lenses in contact: It has been observed that in case of a convex lens, the marginal image I_M lies towards the left of paraxial image I_p, while in case of a concave lens, the marginal image I_M lies towards the right of paraxial image I_p. Thus, by a suitable combination of two lenses, the spherical aberration may be minimized. The difficultly with this combination is that it works only for a particular pair of object and image for which it is designed.

(iii) Using planoconvex lens: Spherical aberration can also be made minimum by using two planoconvex lenses separated by a distance equal to the difference in their focal lengths. In this case, the total deviation produced, is equally divided among the deviations, which are produced by two lenses. This *condition is for minimum spherical aberration.*

Let f_1 and f_2 be the focal lengths of two lenses and d, the separation between two lenses as shown in Figure 4.4.

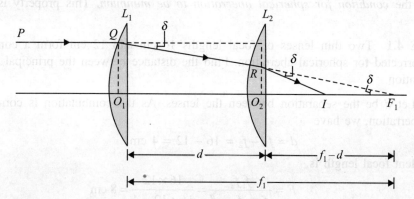

Figure 4.4 Illustration for minimum spherical aberration by using planoconvex lens.

The deviations produced by the two lenses are equal to δ. A ray of light PQ parallel to the principal axis is incident on the first lens L_1 and after refraction, it deviates through an angle δ. The refracted ray from L_1 is incident on second lens L_2 and after refraction, it deviates through same angle δ and intersects the axis at I. From Figure 4.4,

$$\angle RF_1 I = \delta = \angle IRF_1$$

and from ΔRIF_1

$$RI = IF_1$$

or

$$O_2 I = IF_1 \text{ (approximately)}$$

$\therefore$
$$O_2 I = \frac{1}{2} O_2 F_1 = \frac{1}{2}(f_1 - d) \qquad (4.9)$$

For second lens, F_1 is virtual object and I is the real image. Therefore,

$$\frac{1}{v} - \frac{1}{u} = \frac{1}{f}$$

or

$$\frac{1}{O_2 I} - \frac{1}{O_2 F_1} = \frac{1}{f_2} \qquad (4.10)$$

Substituting values of Eq. (4.9) in Eq. (4.10), we get

$$\frac{1}{(f_1 - d)/2} - \frac{1}{f_1 - d} = \frac{1}{f_2}$$

or

$$\frac{2}{f_1 - d} - \frac{1}{f_1 - d} = \frac{1}{f_2}$$

or

$$\frac{1}{f_1 - d} = \frac{1}{f_2}$$

or

$$f_1 - f_2 = d \qquad (4.11)$$

Thus, it is the *condition for spherical aberration to be minimum*. This property is used in eyepieces.

EXAMPLE 4.1 Two thin lenses of focal lengths 16 cm and 12 cm form a combination which is corrected for spherical aberration. Find the distance between the principal points of the combination.

Solution Let d be the separation between the lenses. As the combination is corrected for spherical aberration, we have

$$d = f_1 - f_2 = 16 - 12 = 4 \text{ cm}$$

The equivalent focal length is

$$F = \frac{f_1 f_2}{f_1 + f_2 - d} = \frac{16 \times 12}{16 + 12 - 4} = 8 \text{ cm}$$

The distance of the first principal point H_1 from the first lens L_1 is

$$L_1 H_1 = +\frac{Fd}{f_2} = +\frac{8 \times 4}{12} = +\frac{8}{3} \text{ cm} \quad \text{(on right of } L_1\text{)}$$

and that of the second principal point H_2 from the second lens L_2 is

$$L_2 H_2 = -\frac{Fd}{f_1} = -\frac{8 \times 4}{16} = -2 \text{ cm} \quad \text{(on left of } L_2\text{)}$$

The distance between the principal points is

$$\begin{aligned} H_2 H_1 &= H_2 L_2 - H_1 L_2 \\ &= H_2 L_2 - (L_1 L_2 - L_1 H_1) \\ &= 2 - \left(4 - \frac{8}{3}\right) = \frac{2}{3} \text{ cm} \end{aligned}$$

4.3.2 Coma

As mentioned earlier, the spherical aberration arises due to point object placed on the axis. If the object be situated off the axis, the image of such object suffers from aberration known as *coma*. This aberration arises due to different zones of lens. These zones produce different lateral magnifications in images, which are responsible for coma. The coma is illustrated in Figure 4.5.

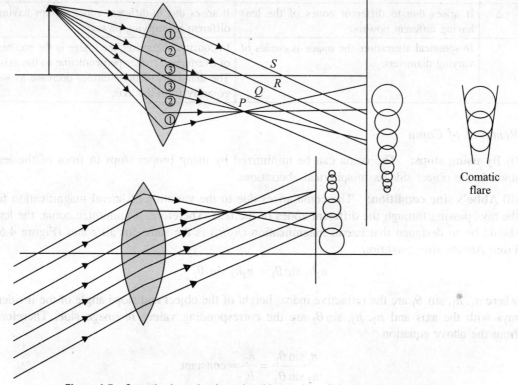

Figure 4.5 Comatic aberration for point object and parallel rays and comatic flare.

The different zones of the lens are shown as (1, 1), (2, 2), (3, 3), ..., etc. Light rays incident on the lens after refraction through the zones meet at the screen, off the axis, at different points. For example, the rays passing through (1, 1), (2, 2), (3, 3), ... zones focus at points P, Q, R respectively. It is also observed that the rays incident on outer zones after refraction focus nearer to the lens than those of inner zones.

Therefore, the image as the overlapping circular paths of increasing diameter is obtained off the axis. The resultant of these paths is comet-shaped and, hence, named as *comatic aberration* or *coma*.

If the magnification of image produced due to outer zones is smaller than the inner zones $\left(m = \dfrac{h_2}{h_1} = \dfrac{I}{O} \right)$, the coma is known as *negative*. If the magnification produced due to outer zones is greater than the inner zones, the coma is said to be *positive*. It is also seen that a lens corrected for coma will not be completely free from spherical aberration and that corrected for spherical aberration will not be free from coma.

Difference between Spherical Aberration and Coma

S.No.	Spherical aberration	Coma
1.	Spherical aberration arises due to the object situated on the axis.	Coma arises due to object situated off the axis.
2.	It arises due to different zones of the lens having different powers.	It arises due to different zones of lens having different magnifying powers.
3.	In spherical aberration, the image is circles of varying diameters.	In comatic aberration, the image is the patches of overlapped circles perpendicular to the axis. The diameters of these circles decrease as we go away from the axis.

Removal of Coma

(i) By using stops: The coma can be minimized by using proper stops in front of the lens towards the object side as in spherical aberrations.

(ii) Abbe's sine condition: The coma arises due to the variation of lateral magnification for the rays passing through the different zones of the lens. Therefore, to minimize coma, the lens should be so designed that lateral magnification (h_2/h_1) is the same for all zones (Figure 4.6). From Abbe's sine condition

$$n_1 h_1 \sin \theta_1 = n_2 h_2 \sin \theta_2$$

where n_1, h_1, $\sin \theta_1$ are the refractive index, height of the object and slope angle of the incident rays with the axis and n_2, h_2, $\sin \theta_2$ are the corresponding values in image side. Therefore, from the above equation

$$\frac{n_1 \sin \theta_1}{n_2 \sin \theta_2} = \frac{h_2}{h_1} = \text{constant}$$

If the system (lens) is situated in air, then ($n_1 = n_2$)

$$\frac{\sin \theta_1}{\sin \theta_2} = \text{constant}$$

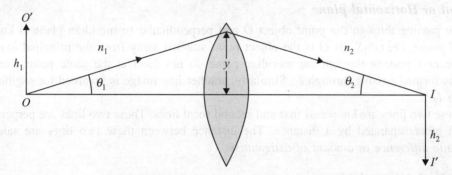

Figure 4.6 Illustration of Abbe's sine condition.

For distant object ($u \to \infty$) and $\sin \theta_1 \propto y$ (where y is the height of the ray), then the condition for no coma

$$\frac{y}{\sin \theta_2} = \text{constant}$$

This condition may be satisfied for a pair of conjugate points only. The lens that satisfies above condition is called *planatic lens*.

4.3.3 Astigmatism

In astigmation, a lens or mirror has different focal lengths across different diameters and focuses off axis, rays on two times instead of a single point (Figure 4.7).

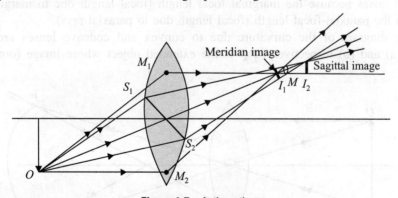

Figure 4.7 Astigmatism.

When a point object is situated far-off the axis of a lens, the image formed by lens is not in a perfect focus. The image consists of two mutually perpendicular lines separated by a finite distance. Moreover, the two lines lie in perpendicular planes. The defect of the image is called *astigmatism*.

Meridian plane, Vertical plane or Tangential plane

A plane passing through point object O and principal axis is known as *meridian plane*, i.e. OM_1M_2.

Sagittal or Horizontal plane

A plane passing through the point object O and perpendicular to meridian plane is known as *sagittal plane*, i.e. OS_1S_2, if O is the object point situated away from the principal axis.

The rays passing through the meridian plane do not meet at the same point, but a line image is formed passing through I_1. Similarly, another line image is obtained for sagittal plane through I_2.

These two lines are known as first and second focal lines. These two lines are perpendicular to each other separated by a distance. The distance between these two lines are said to be *astigmatic difference* or *amount of astigmatism*.

Removal of astigmatism

(i) Using stops: The astigmatism can be minimized by using stops in case of single lens. Stops are placed such that only less oblique rays pass through the lens to form image.

(ii) Anastigmat: Since astigmatism may be positive or negative depending on the nature of the lens, therefore, using a convex and concave lenses of suitable focal length and separated by a distance, the astigmatism can be minimized. Such a lens combination is called *anastigmat*.

4.3.4 Curvature of Field

It is observed that if a single lens is free from spherical aberration—coma and astigmatism, even then the image of extended plane object obtained is not a flat one, but curved. The central portion of the image near the axis is in focus, while outer regions of the image away from the axis are blurred (curved). This defect in image is called *curvature of field* or *curvature*. It arises because the marginal focal length (focal length due to marginal rays) is smaller than the paraxial focal length (focal length due to paraxial rays).

The ray diagram of the curvature due to convex and concave lenses are shown in Figures 4.8(a) and (b) respectively. PQ is the extended object whose image formed $P'Q'$ is curved.

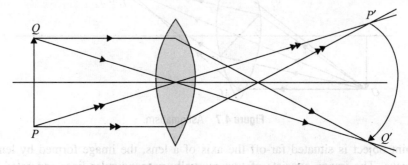

Figure 4.8(a) Ray diagram for curvature due to convex lens.

Lens Aberrations 79

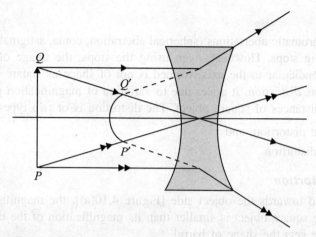

Figure 4.8(b) Ray diagram for curvature due to concave lens.

Removal of curvature

(i) Using stops: By using proper stops in front of the lens, the curvature of the field can be minimized (Figure 4.9).

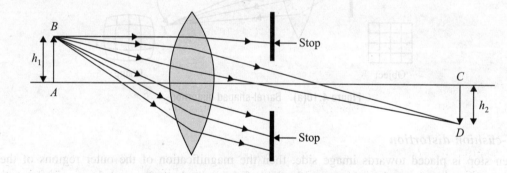

Figure 4.9 Removal of curvature by using stops.

(ii) Using Petzval condition: According to Petzval, the curvature of field will be zero for two equicurved lenses in contact or separated by a distance if they satisfy the condition

$$n_1 f_1 + n_2 f_2 = 0$$

or
$$\frac{n_1}{n_2} = -\frac{f_2}{f_1} \qquad (4.12)$$

where n_1, f_1 are the refractive index and focal length of one lens and n_2, f_2, the refractive index and focal length of second lens. Since n_1, n_2 are positive, therefore, f_1 and f_2 must be of opposite sign. Thus, if the combination of a convex lens and a concave lens of different materials satisfy the above *Petzval condition*, the curvature of image can be minimized.

4.3.5 Distortion

The various monochromatic aberrations (spherical aberration, coma, astigmatism and curvature) are reduced by using stops. However, even using the stops, the image of plane square-like object (placed perpendicular to the axis) formed is not of shape as square object. This defect in image is known as *distortion*. It arises due to variation of magnification produced by a lens for different axial distances of square object. The distortion is of two types:

1. Barrel-shaped distortion, and
2. Pin-cushion distortion

Barrel-shaped distortion

When stop is placed towards the object side [Figure 4.10(a)], the magnification of the outer regions of the plane square object is smaller than its magnification of the central portion and, consequently, image gets the shape of barrel.

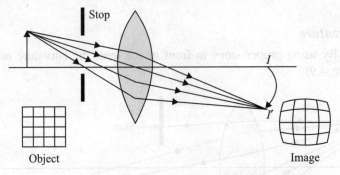

Figure 4.10(a) Barrel-shaped distortion.

Pin-cushion distortion

When stop is placed towards image side, then the magnification of the outer regions of the square object is greater than the magnification of the central portion and, hence, image gets the shape of pin-cushion as shown in Figure 4.10(b).

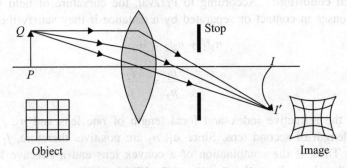

Figure 4.10(b) Pin-cushion distortion.

Removal of distortion

To remove distortion, the two meniscus lenses are placed such that the concave surfaces of two lenses face each other. Now place an aperture stop between the two lenses as shown in Figure 4.11.

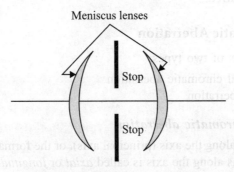

Figure 4.11 Elimination of distortion.

In this situation, the pin-cushion distortion produced by the first lens is compensated by the barrel-shaped distortion produced by second lens. This aberration is important in camera.

4.4 CHROMATIC ABERRATION

When the rays of white light (light having waves of different wavelengths) parallel to the principal axis are incident on a lens, the different colours are refracted by different amounts or are dispersed into various colours, and are focused at different distances from the lens as shown in Figures 4.12 and 4.13.

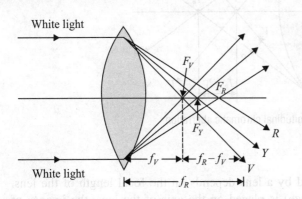

Figure 4.12 Chromatic aberration in case of convex lens.

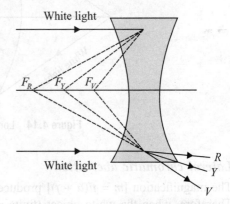

Figure 4.13 Chromatic aberration in case of concave lens.

This is due to the dependance of the refractive index of the material of the lens on the wavelength of light. The refractive index of the material of the lens (glass) is greater for violet rays than that of the red rays, i.e. $n_V > n_R$ and $f_V > f_R$ (as the focal length of the lens is inversely proportional to the refractive index). Since violet light gets refracted more than red light, the point at which the violet light would focus is nearer the lens than the point at which the red light would focus. Thus, the inability of a lens to form a single image of the white object is called *chromatic aberration*. The image of the white object formed by the lens is usually coloured and blurred.

4.4.1 Types of Chromatic Aberration

The chromatic aberration is of two types:
1. Longitudinal or axial chromatic aberration
2. Lateral chromatic aberration

Longitudinal or axial chromatic aberration

The spreading of an image along the axis (principal axis), or the formation of images of different colours at different positions along the axis is called *axial* or *longitudinal chromatic aberration*. The axial distance between the positions of red and violet images is a measure of longitudinal or axial chromatic aberration as shown in Figure 4.14. The quantity $(I_R - I_V)$ is the measure of this aberration. If an object is situated at inifinity, then longitudinal chromatic aberration becomes equal to the difference in the focal lengths for red and violet colours, i.e. equal to $(f_R - f_V)$ as shown in Figures 4.12 and 4.13. The longitudinal chromatic aberration is positive for a convex lens and is negative for a concave lens because of sign convention used.

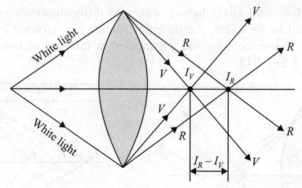

Figure 4.14 Longitudinal chromatic aberration.

Lateral chromatic aberration

The magnification $[m = f/(u + f)]$ produced by a lens depends on the focal length of the lens. Therefore, when the white object (finite size) is placed on the axis of the lens, the images of different colours do not focus at different sizes as shown in Figure 4.15.

Thus, the formation of images of different sizes for different wavelengths (colours) due to a variation of the lateral magnification with the wavelength is called *lateral chromatic aberration*. The sizes of violet and red images are $A'B'$ and $A''B''$ respectively as shown in Figure 4.15. The distance $y = (A''B'' - A'B')$ is the measure of the lateral chromatic aberration.

Calculation of longitudinal chromatic aberration of a thin lens: Now we shall calculate the longitudinal chromatic aberration in two situations, which are as follows:

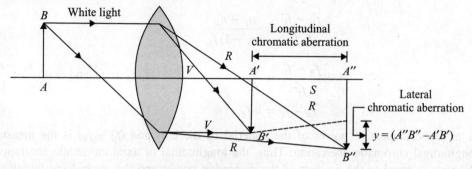

Figure 4.15 Lateral chromatic aberration.

(i) *For an object at infinity:* If the white object is situated at infinity, the images of different colours are focused at different focal points on the principal axis of the lens. The focal length of a thin lens is given by

$$\frac{1}{f} = (n-1)\left(\frac{1}{R_1} - \frac{1}{R_2}\right) \tag{4.13}$$

where n is the refractive index of the material of the lens and R_1, R_2 are the radii of the curvature of the two surfaces of the lens.

If f_V, f_R and f_Y are the focal length of the lens for violet, red and yellow colours respectively and n_V, n_R and n_Y are their respective refractive indices for the material of the lens, then

$$\frac{1}{f_V} = (n_V - 1)\left(\frac{1}{R_1} - \frac{1}{R_2}\right) \tag{4.14}$$

$$\frac{1}{f_R} = (n_R - 1)\left(\frac{1}{R_1} - \frac{1}{R_2}\right) \tag{4.15}$$

and

$$\frac{1}{f_Y} = (n_Y - 1)\left(\frac{1}{R_1} - \frac{1}{R_2}\right) \tag{4.16}$$

Subtracting Eq. (4.15) from Eq. (4.14), we get

$$\frac{1}{f_V} - \frac{1}{f_R} = (n_V - n_R)\left(\frac{1}{R_1} - \frac{1}{R_2}\right)$$

or

$$\frac{f_R - f_V}{f_V f_R} = (n_V - n_R)\left(\frac{1}{R_1} - \frac{1}{R_2}\right) \tag{4.17}$$

From Eq. (4.16),

$$\left(\frac{1}{R_1} - \frac{1}{R_2}\right) = \frac{1}{(n_Y - 1)f_Y}$$

Substituting this value of $\left(\dfrac{1}{R_1} - \dfrac{1}{R_2}\right)$ in Eq. (4.17), we get

$$\frac{f_R - f_V}{f_R f_V} = \frac{n_V - n_R}{(n_Y - 1)f_Y}$$

or
$$\frac{f_R - f_V}{f_Y^2} = \frac{\omega}{f_Y} \qquad \left(\because\ f_V f_R = f_Y^2 \text{ and } \omega = \frac{n_V - n_R}{(n_Y - 1)f_Y}\right)$$

or
$$f_R - f_V = \omega f_Y \qquad (4.18)$$

where ω is the dispersive power of the material of the lens and $(f_R - f_V)$ is the measure of the longitudinal chromatic aberration. Thus, the longitudinal or axial chromatic aberration for parallel rays is equal to the product of the dispersive power and the mean focal length of the lens.

(ii) *For an object an infinite distance:* If an object is placed on the principal axis of the lens and is illuminated by a polychromatic source of light (like white light), the coloured images are formed along the axis. If u is the distance of the object and v is the mean distance of the image from the lens as shown in Figure 4.16, then the mean focal length of the lens is given by

$$\frac{1}{v} - \frac{1}{u} = \frac{1}{f} \qquad (4.19)$$

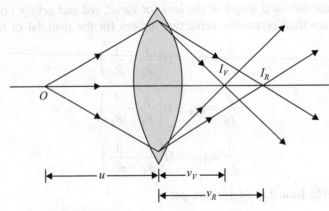

Figure 4.16 Formation of coloured image by a lens.

If f_V and f_R are the focal lengths and v_V and v_R are the distances of images for violet and red colours respectively, then

$$\frac{1}{f_V} = \frac{1}{v_V} - \frac{1}{u} \qquad (4.20)$$

and
$$\frac{1}{f_R} = \frac{1}{v_R} - \frac{1}{u} \qquad (4.21)$$

Subtracting Eq. (4.21) from Eq. (4.20)

$$\frac{1}{f_V} - \frac{1}{f_R} = \frac{1}{v_V} - \frac{1}{v_R}$$

$$\frac{f_R - f_V}{f_R f_V} = \frac{v_R - v_V}{v_V v_R}$$

or

$$v_R - v_V = \frac{v_V v_R}{f_R f_V}(f_R - f_V)$$

or

$$v_R - v_V = \frac{v_Y^2}{f_Y^2}(f_R - f_V) \qquad (4.22)$$

$$[\because v_R v_V = v_Y^2 \text{ and } f_R f_Y = f_Y^2]$$

where v_Y is the distance of the yellow (mean) image and f_Y is the focal length of the lens for the yellow colour.
But,
$$f_R - f_V = \omega f_Y$$
Therefore, Eq. (4.22) becomes

$$v_R - v_V = \frac{\omega v_Y^2}{f_Y}$$

Therefore, axial chromatic aberration

$$v_R - v_V = v_Y^2 \frac{\omega}{f_Y} \qquad (4.23)$$

Thus, for an object at finite distance, the axial chromatic aberration depends upon the mean distance of the image from the lens and, hence, the distance of the object from the lens.

EXAMPLE 4.2 A lens made of crown glass has focal length 50 cm. Calculate longitudinal chromatic aberration. Given: $n_C = 1.5565$, $n_F = 1.5200$.

Solution We have longitudinal chromatic aberration

$$f_R - f_V = \omega f_Y$$

But $\omega = \dfrac{n_C - n_F}{n - 1}$ where $n = \dfrac{n_C + n_F}{2}$

$$\therefore \quad \omega = \frac{1.5565 - 1.5200}{\left(\dfrac{1.5565 + 1.5200}{2}\right) - 1} = \frac{0.0365}{0.5382} = 0.067$$

Then longitudinal chromatic aberration

$$f_R - f_V = 0.067 \times 50 \text{ cm}$$
$$= 3.35 \text{ cm}$$

4.4.2 Achromatism of Lenses

It is well known that when a white object is placed in front of a lens, a single lens will given a continuous sequence of images of different colours. The coloured images are neither at the same place nor of the same size. This defect of the lens is called *chromatic aberration*. If two or more lenses are combined together in such a way that the lens combination produces images of different colours at the same place or position and of the same size, then the combination is called achromatic combination of lenses and the property is called *achromatism*. The minimization or removal of chromatic aberration is known as *achromatization*. The acromatic combination of lenses is called *ideal* if all the coloured images are of the same size and are formed at one point. Practically, ideal achromatism cannot be achieved or chromatic aberration cannot be totally removed. It is possible that a combination of a crown glass. Convex lens of low focal length and a flint glass concave lens of greater focal length may be free from chromatic aberration for two colours. Such a combination of lenses is called an *achromatic doublet*.

Achromatism of two lenses in contact

For an achromatic doublet, consider two lenses of different materials placed in contact with each other. For the combination of two lenses to be achromatic, we will find the condition for this lens combination to have the focal length independent of the colours.

The focal length f of a thin lens is given by

$$\frac{1}{f} = (n-1)\left(\frac{1}{R_1} - \frac{1}{R_2}\right) \tag{4.24}$$

or

$$\left(\frac{1}{R_1} - \frac{1}{R_2}\right) = \frac{1}{(n-1)f} \tag{4.25}$$

where n is the refractive index of the lens and R_1 and R_2 are the radii of the curvature of the two surfaces of the lens.

If δf is the change in the focal length f corresponding to a change (δn) in the refractive index n, then differentiating Eq. (4.24) with respect to f, we get

$$-\frac{\delta f}{f^2} = \delta n \left(\frac{1}{R_1} - \frac{1}{R_2}\right)$$

From Eq. (4.25), we have

$$-\frac{\delta f}{f^2} = \frac{\delta n}{n-1} \times \frac{1}{f} \tag{4.26}$$

But $\dfrac{\delta n}{n-1} = \omega$ the dispersive power of the lens between the two colours for which the difference in refractive index is δn and mean refractive index is n.

$$-\frac{\delta f}{f^2} = \frac{\omega}{f} \tag{4.27}$$

The negative sign in the left-hand side indicates that the focal length of the lens decreases as the refractive index increases. If f_1 and f_2 are the mean focal lengths of the two thin lenses of combination ω_1 and ω_2 are the dispersive powers between the two colours for which the combination is to be achromatized.

As the lenses are in contact, the focal length of the combination is given by

$$\frac{1}{F} = \frac{1}{f_1} + \frac{1}{f_2} \qquad (4.28)$$

Differentiating Eq. (4.28) partially, we get

$$-\frac{\delta F}{F^2} = -\frac{\delta f_1}{f_1^2} - \frac{\delta f_2}{f_2^2}$$

But from Eq. (4.27), we have

$$-\frac{\delta f_1}{f_1^2} = \frac{\omega_1}{f_1} \quad \text{and} \quad -\frac{\delta f_1}{f_2^2} = \frac{\omega_2}{f_2}$$

$$\therefore \qquad -\frac{\delta F}{F^2} = \frac{\omega_1}{f_1} + \frac{\omega_2}{f_2} \qquad (4.29)$$

For an achromatic combination $\delta F = 0$, i.e. F should not change with colour

$$\therefore \qquad \frac{\omega_1}{f_1} + \frac{\omega_2}{f_2} = 0$$

or

$$\frac{f_1}{f_2} = -\frac{\omega_1}{\omega_2} \qquad (4.30)$$

This is the required condition for an achromatic doublet. According to this condition,
  (i) the ratio of the focal lengths of the two lenses is numerically equal to the ratio of the dispersive powers of their materials, and
  (ii) since ω_1 and ω_2 are both positive, the focal lengths of the two lenses must be opposite sign, i.e. if one lens is convex, the other should be concave as shown in Figure 4.17.

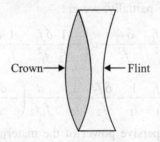

Crown → ← Flint

Figure 4.17 An achromatic doublet.

Discussion of the condition: If we put $\omega_1 = \omega_2$ in Eq. (4.30), we get

$$\frac{1}{f_1} + \frac{1}{f_2} = 0$$

or

$$\frac{1}{F} = 0$$

or

$$F = \infty$$

i.e. the combination behaves like a plane glass plate and not a lens. Hence $\omega_1 = \omega_2$.

For the combination to act as convex lens, it is an essential that $\omega_1 > \omega_2$. Hence for a converging lens system, the convex lens of a material of low dispersive power should be taken. The dispersive power of crown glass is smaller than the flint glass. Hence in an achromatic doublet, the convex lens of crown glass and the concave lens of flint glass is used. The function of an achromatic doublet is shown in Figure 4.18.

Achromatism for two lenses separated by a distance

It is possible to minimise achromatic aberration by using two convex lenses of the same material provided they are separated by a finite distance. To find the condition of the minimum chromatic aberration, consider two convex lenses of focal lengths f_1 and f_2 placed coaxially at a distance d apart. The equivalent focal length F of the combination is given by

$$\frac{1}{F} = \frac{1}{f_1} + \frac{1}{f_2} - \frac{d}{f_1 f_2} \tag{4.31}$$

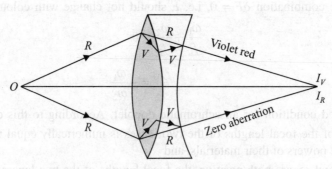

Figure 4.18 Function of an achromatism doublet.

Differentiating the above relation partially, we get

$$-\frac{\delta F}{F^2} = -\frac{\delta f_1}{f_1^2} - \frac{\delta f_2}{f_2^2} - d\left(-\frac{1}{f_2}\frac{\delta f_1}{f_1^2} - \frac{1}{f_1}\frac{\delta f_2}{f_2^2}\right)$$

$$= -\frac{\delta f_1}{f_1} \cdot \frac{1}{f_1} - \frac{\delta f_2}{f_2} \cdot \frac{1}{f_2} - \frac{d}{f_1 f_2}\left(-\frac{\delta f_1}{f_1} - \frac{\delta f_2}{f_2}\right)$$

We know that $-\delta f/f = \omega$, the dispersive power of the material of the lens. Therefore,

$$-\frac{\delta f_1}{f_1} = \omega_1$$

and

$$-\frac{\delta f_2}{f_2} = \omega_2$$

where ω_1 and ω_2 are the dispersive powers of the first and second lenses respectively.

Thus, $$-\frac{\delta F}{F_2} = \frac{\omega_1}{f_1} + \frac{\omega_2}{f_2} - \frac{d}{f^2} \times \frac{\omega_1}{f_1} - \frac{d}{f_1} \times \frac{\omega_2}{f_2} = 0$$

For an achromatic combination of lenses $\delta F = 0$, the focal length of the combinazion for all colours must be the same.

$$\therefore \quad \frac{\omega_1}{f_1} + \frac{\omega_2}{f_2} - \frac{d}{f_2} \times \frac{\omega_1}{f_1} - \frac{d}{f_1} \times \frac{\omega_2}{f_2} = 0$$

or
$$\frac{\omega_1 f_2 + \omega_2 f_1 - d(\omega_1 + \omega_2)}{f_1 f_2} = 0$$

or
$$\omega_1 f_2 + \omega_2 f_1 - d(\omega_1 + \omega_2) = 0$$

or
$$d = \frac{\omega_1 f_2 + \omega_2 f_1}{\omega_1 + \omega_2} \tag{4.32}$$

If both the lenses are of the same material, i.e. $\omega_1 = \omega_2$, then from Eq. (4.32), we get

$$d = \frac{f_1 + f_2}{2}$$

Thus for a minimum chromatic aberration, the distance between the two lenses of the same material must be equal to half the sum of their focal lengths. Here we find that (in the result) the separation between the lenses is independent of ω. It mean that the focal length of an achromatic combination is the same for all colours.

Nature of achromatism: The achromatic combination has the same focal length of all colours. Actually, coloured images are of the same size. It means that the rays of all the colours are initially parallel to the axis and are deviated through the same angle, so that each coloured image subtends at the same angle at the eye causing the sensation of whiteness due to overlapping. In fact, the lateral chromatic aberration minimizes in this combination though the actual positions of the images of various colours are different. An achromatic contact doublet is widely used in the objectives of telescope, collimators and cameras. The separated lens achromate is used in eyepiece.

EXAMPLE 4.3 An achromatic doublet of a convex lens and a concave lens of focal length 0.30 m is to be formed out of flint and crown glass whose dispersive powers are 0.03 and 0.02 respectively. Calculate the focal length of the two lenses.

Solution *Given:* $F = 0.30$ m $= 30$ cm, $\omega_1 = 0.03$, $\omega_2 = 0.02$

If f_1 and f_2 are the focal lengths and ω_1 and ω_2 are the dispersive powers of the two lenses, then condition for achromatism,

$$\frac{\omega_1}{f_1} + \frac{\omega_2}{f_2} = 0$$

or
$$\frac{\omega_1}{f_1} = -\frac{\omega_2}{f_2}$$

or
$$\frac{1}{f_1} = -\frac{\omega_2}{\omega_1} \times \frac{1}{f_2} = -\frac{0.02}{0.03} \times \frac{1}{f_2} = -\frac{2}{3 f_2}$$

Also the combined focal lengths

$$\frac{1}{F} = \frac{1}{f_1} + \frac{1}{f_2}$$

$$\frac{1}{30} = \frac{-2}{3f_2} + \frac{1}{f_2} = \frac{1}{f_2}\left[1 - \frac{2}{3}\right] = \frac{1}{3f_2}$$

or $\qquad 3f_2 = 30$

or $\qquad f_2 = 10 \text{ cm} = 0.10 \text{ m}$

Again $\qquad \dfrac{1}{f_1} = -\dfrac{2}{3f_2} = \dfrac{-2}{3 \times 10} = -\dfrac{1}{15}$

$$f_1 = -15 \text{ cm} = -0.15 \text{ m}$$

EXAMPLE 4.4 The focal length of an achromatic combination of two lenses is 1.5 m. If the dispersive powers of the materials of two lenses are 0.018 and 0.027, calculate the focal length of two lenses.

Solution *Given:* $F = 1.5$ m $= 150$ cm, $\omega_1 = 0.018$ and $\omega_2 = 0.027$
Now, condition for achromatism

$$\frac{\omega_1}{f_1} + \frac{\omega_2}{f_2} = 0$$

$$\Rightarrow \frac{1}{f_1} = -\frac{\omega_2}{\omega_1} \times \frac{1}{f_2}$$

$$= -\frac{0.027}{0.018} \times \frac{1}{f_2} = \frac{-3}{2} \times \frac{1}{f_2}$$

$$\frac{1}{F} = \frac{1}{f_1} + \frac{1}{f_2}$$

or $\qquad \dfrac{1}{150} = -\dfrac{3}{2f_2} + \dfrac{1}{f_2}$

or $\qquad -\dfrac{3}{2f_2} = \dfrac{1}{150}$

or $\qquad f_2 = -75$ cm

$$\frac{1}{f_1} = +\frac{3}{2 \times 75 \text{ cm}}$$

or $\qquad f_1 = 50$ cm

4.5 APLANATISM AND APLANATIC POINTS

The *aplanatism* is the property of a surface by virtue of which all the rays starting from a point object on its axis, after reflection or refraction at the surface, converge to or appear to diverge from a single point image. The particular object and image points are called 'aplanatic points', and the surface is said to be aplanatic with respect to these two points'.

4.5.1 Aplanatic Points for a Spherical Refracting Surface

Let SS' (Figure 4.19) be a spherical refracting surface with axis PI and centre of curvature C. Let R be the radius of curvature of the surface. Let the refractive index of the medium on the left side of the surface be n, the medium on the right side being air (or vacuum).

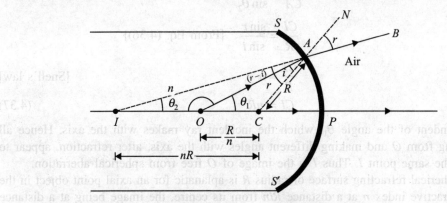

Figure 4.19 Illustration for aplanatic points.

Let O be a point object in the medium placed on the axis of the surface at a distance R/n from the centre of curvature C. An incident ray OA making an angle θ_1 with the axis, after refraction, bends away from the normal CAN, and follows the path AB. Another ray OP coincident with the axis meets the surface normally and goes straight. The refracted rays produced back meet at I, which is the virtual image of O. Let $\angle AIO = \theta_2$.

Let i and r be the angles of incidence and refraction.

Then
$$\angle OAC = i$$
and
$$\angle BAN = \angle CAI = r.$$

As the light is going from denser to rarer medium, we have

$$\frac{\sin i}{\sin r} = \frac{1}{n} \qquad \text{(Snell's law)}$$

or
$$n \sin i = \sin r \tag{4.33}$$

Now in $\triangle CAO$, we have

$$\frac{CA}{CO} = \frac{R}{R/n} = \frac{\sin \theta_1}{\sin i}$$

or
$$n \sin i = \sin \theta_1 \tag{4.34}$$

Comparing Eqs. (4.33) and (4.34), we have

$$\sin \theta_1 = \sin r$$

or
$$\theta_1 = r \tag{4.35}$$

Again in $\triangle OAI$, we have

External angle $COA = \angle AIO + \angle IAO$

$$\theta_1 = \theta_2 + (r - i)$$

or $\qquad \theta_2 = \theta_1 - r + i = r - r + i$

or $\qquad \theta_2 = i$ (4.36)

Now, in ΔCIA, we have

$$\frac{CI}{CA} = \frac{\sin r}{\sin \theta_2}$$

or $\qquad \dfrac{CI}{R} = \dfrac{\sin r}{\sin i}$ [From Eq. (4.36)]

or $\qquad \dfrac{CI}{R} = n$ [Snell's law]

or $\qquad CI = nR$ (4.37)

This is independent of the angle θ_1, which the incident ray makes with the axis. Hence all the rays starting from O and making different angles with the axis, after refraction, appear to diverge from the same point I. Thus I is the image of O free from spherical aberration.

Thus, a spherical refracting surface of radius R is aplanatic for an axial point object in the medium of refractive index n at a distance R/n from its centre, the image being at a distance nR from its centre.

The sine condition $\dfrac{\sin \theta_1}{\sin \theta_2}$ = constant, for elimination of coma, is satisfied in the case. Here

$\qquad \theta_1 = r$ [From Eq. (4.35)]

and $\qquad \theta_2 = i$ [From Eq. (4.36)]

$\therefore \qquad \dfrac{\sin \theta_1}{\sin \theta_2} = \dfrac{\sin r}{\sin i} = n$ = constant

Hence the refracting surface is free from spherical aberration as well as coma for a particular position of the object at a distance R/n from the centre.

4.5.2 Oil-immersion Objective of High Power Microscope

The focal length of a high power microscope objective is very small and the object is placed very close to the objective. Hence the objective receives a wide-angled pencil of rays diverging from the object, which is likely to produce large spherical aberration and coma.

In order to minimise these defects, oil-immersion objectives based on the principle of aplantism are used.

According to this principle, *if an object is placed at one aplanatic point of a surface, an image free from spherical aberration and coma is formed at the second aplanatic point. In case of a single spherical surface, these points are at distances R/n and nR from the centre of curvature.*

An oil-immersion objective is made of a combination of lenses, as shown in Figure 4.20. The lens L_1 which is closest to the object O is hemisphere with its flat surface in contact with cedarwood oil whose refractive index is the same as that of L_1. The object O is placed in the oil. The depth of oil is so adjusted that the distance of O is R/n from the centre of curvature C

of the lens L_1. The object O is, thus, at the first aplanatic point for L_1. Hence all rays starting from it after refraction through the lens appear to come from the corresponding aplanatic point I_1, where

$$CI_1 = nR$$

The image I_1 is free from spherical aberration.

L_2 is a meniscus lens so placed that I_1 lies at the centre of curvature of the first surface of L_2. Therefore, rays enter L_2 undeviated. I_1 also lies at the first aplanatic point of the second surface of L_2 which forms a virtual image I_2, free from spherical aberration, at its second aplanatic point. Thus a wide-angled pencil starting from O is changed into a narrow pencil diverging from O is changed into a narrow pencil diverging from I_2. The lenses L_3 and L_4 (not shown), therefore, form the final image almost free from spherical aberration. These lenses also correct for the chromatic aberration introduce by L_1 and L_2.

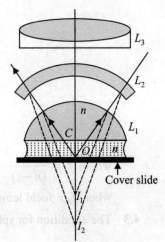

Figure 4.20 Oil-immersion objective of high (power microscope).

Function of the first component L_1

The semiangle θ_1 of the widest possible cone of incident rays from O is given by

$$\tan\theta_1 = \frac{R}{CO} = \frac{R}{R/n} = n$$

while the semiangle θ_2 of the cone of the corresponding emergent rays is given by

$$\tan\theta_2 = \frac{R}{CI_1} = \frac{R}{nR} = \frac{1}{n}$$

Thus, the lens L_1 reduces the semiangle of the cone of light from $\tan^{-1}(n)$ to $\tan^{-1}(1/n)$.

It is clear from the above description that oil-immersion objection admits a wider cone of rays than a dry objective of the same diameter would be.* Hence numerical aperture of oil-immersion objective is much greater than that of a dry objective.

FORMULAE AT A GLANCE

4.1 Abbe's sine condition

$$n_1 h_1 \sin\theta_1 = n_2 h_2 \sin\theta_2$$

where n_1, h_1, θ_1 are the refractive index, lateral size of the object and angle subtended by the object with principal axis in object space. n_2, h_2, θ_2 are the respective values in image space.

---

* In a dry objective, the rays emerging from the cover slide (which protects the object) pass into air so that they refract outwards. Hence their divergence is increased.

4.2 Change in focal length for h zone as compared to axial zone is

$$\Delta f_h = f_p - f_h$$

$$= R\left[\frac{1}{n-1} - \frac{1}{n\left(1 - \frac{h^2}{2n^2 R^2}\right) - \left(1 - \frac{h^2}{2R^2}\right)}\right]$$

$$= \frac{h^2}{2(n-1)^2 f_p}$$

where $f_p = \dfrac{nR}{(n-1)}$

where f_h = focal length for ray in zone of height h.

4.3 The condition for spherical aberration to be minimum

$$f_1 - f_2 = d$$

where f_1 and f_2 be the focal length of two lenses and d is the separation between two lenses.

4.4 *Petzval conditoin for removal of curvature*

$$n_1 f_1 + n_2 f_2 = 0$$

or

$$\frac{n_1}{n_2} = -\frac{f_2}{f_1}$$

4.5 *Longitudinal chromatic aberration of a thin lens*

(a) For an object at infinity

$$f_R - f_V = \omega f_Y$$

$f_R - f_V$ = measure of longitudinal chromatic aberration

ω = dispersive power

(b) For an object, an infinite distance

$$v_R - v_V = \frac{\omega v_Y^2}{f_Y}$$

Axial chromatic aberration

$$v_R - v_V = v_Y^2 \frac{\omega}{f_Y}$$

4.6 The required condition for an achromatic doublet

$$\frac{f_1}{f_2} = -\frac{\omega_1}{\omega_2}$$

4.7 Condition for minimum chromatic aberration

$$d = \frac{f_1 + f_2}{2}$$

4.8 Aplantism and aplanatic points condition

$$\frac{\sin \theta_1}{\sin \theta_2} = \frac{\sin r}{\sin i} = n = \text{constant}$$

SOLVED NUMERICAL PROBLEMS

PROBLEM 4.1 Two lenses separated by a distance 2 cm and having equivalent focal length 10 cm is free from a. Find the focal length of the second lens such that the curvature of field may be removed. The focal length of the first lens is 33 cm.

Solution Given: $d = 2$ cm, $f_1 = 33$ cm, $F = 10$ cm

We know that condition for minimum spherical aberration is

$$f_1 - f_2 = d \quad \text{or} \quad f_1 - f_2 = 2 \Rightarrow f_1 = 2 + f_2$$

The equivalent focal length (F) is given by

$$F = \frac{f_1 f_2}{f_1 + f_2 - d} = 10 \text{ cm}$$

Substituting

$$F = \frac{(f_2 + 2)f_2}{(f_2 + 2) + (f_2 - 2)} = 10 \text{ cm}$$

or

$$f_2 = 18 \text{ cm}$$

Therefore, $f_2 = 18$ cm and $f_1 = 20$ cm.

PROBLEM 4.2 Make a combination of two lenses such that the spherical aberration be minimum and chromatic aberration is zero. The focal length of combination of the two lenses is 5.0 cm.

Solution Given: $F = 5$ cm

We know the condition of minimum spherical aberration is

$$d = f_1 - f_2 \qquad \text{(i)}$$

and condition for achromatism,

$$2d = f_1 + f_2 \qquad \text{(ii)}$$

From these two equations

$$f_1 = 3f_2 \qquad \text{(iii)}$$

Substituting Eq. (iii) in Eq. (i), we get

$$d = 3f_2 - f_2 = 2f_2 \qquad \text{(iv)}$$

But,

$$F = \frac{f_1 f_2}{f_1 + f_2 - d} = 5$$

or

$$\frac{3f_2 \times f_2}{3f_2 + f_2 - 2f_2} = 5$$

or

$$f_2 = 3.3 \text{ cm}$$

substituting $f_2 = 3.3$ cm in Eqs. (iii) and (iv), we get

$$f_1 = 10.0 \text{ cm}$$
and
$$d = 6.7 \text{ cm}$$

PROBLEM 4.3 A equiconvex lens of crown glass and an equiconcave lens of flint glass make an achromatic combination. The radius of curvature of the convex lens is 0.54 m. If the focal length of the combination for the mean colour is 1.50 m, $n_V = 1.55$ and $n_R = 1.53$, then find the dispersive power of the flint glass.

Solution Given: $R = 0.54$ m, $F = 1.5$ m, $n_V = 1.55$ and $n_R = 1.53$

The condition of an achromatic combination,

$$\frac{\omega_1}{f_1} + \frac{\omega_2}{f_2} = 0$$

If ω_1 is the dispersive power of crown glass, then

$$\omega_1 = \frac{n_V - n_R}{n_Y - 1}$$

where n_Y is the refractive index of mean (yellow) colour

Hence, $\quad n_Y = \frac{n_V + n_R}{2} = \frac{1.55 + 1.53}{2} = 1.54 \qquad [\because \ n_V = 1.55 \text{ and } n_R = 1.53]$

and $\quad \omega_1 = \frac{1.55 - 1.53}{(1.54 - 1)} = \frac{0.02}{0.54} = \frac{1}{27}$

If f_1 is the focal length of a convex lens, then from the lens formula,

$$\frac{1}{f_1} = (n_Y - 1)\left(\frac{1}{R_1} - \frac{1}{R_2}\right)$$

and for equiconvex lens,

$$R_1 = +R$$
and
$$R_2 = -R$$

$\therefore \quad \dfrac{1}{f_1} = (n_Y - 1)\left(\dfrac{1}{R} - \dfrac{1}{-R}\right)$

or $\quad \dfrac{1}{f_1} = \dfrac{2(n_Y - 1)}{R}$

$\therefore \quad f_1 = \dfrac{R}{2(n_Y - 1)} = \dfrac{0.54}{2(1.54 - 1)} = \dfrac{1}{2} = 0.5$ m

If F is the focal length of the combination for the mean colours and f_2 that of the second lens, then

$$\frac{1}{F} = \frac{1}{f_1} + \frac{1}{f_2}$$

or $\quad \dfrac{1}{f_2} = \dfrac{1}{F} - \dfrac{1}{f_1}$

Here

$$F = 1.50 \text{ m}$$

$$\frac{1}{f_2} = \frac{1}{1.5} - \frac{1}{0.5} = -\frac{2}{1.5}$$

or
$$f_2 = -\frac{1.5}{2} = -0.75 \text{ m}$$

Hence, the second lens is concave.

From the condition of achromatism

$$\omega_2 = -\frac{f_2}{f_1} \times \omega_1 = -\frac{0.75}{0.5} \times \frac{0.02}{0.54} = \frac{1}{18} = 0.055$$

Thus, the dispersive power of flint glass = 0.055.

PROBLEM 4.4 A convex lens of crown glass is perfectly connected to a planoconvex lens of flint glass to form an achromatic combination of power +5D. Calculate the radii of curvature of the convex lens from the following data:

	n	ω
Crown glass	1.50	0.01
Flint glass	1.60	0.02

Solution According to condition of achromatism

$$\frac{\omega_1}{f_1} + \frac{\omega_2}{f_2} = 0$$

or
$$f_2 = -\frac{\omega_2}{\omega_1} \times f_1$$

From the given data,
$$f_2 = -\frac{0.02}{0.01} f_1 = -2f_1$$

The focal length of combination

$$F = \frac{100 \text{ cm}}{\text{Power}} = \frac{100}{5}$$

Thus,
$$F = 20 \text{ cm}$$

From the formula for the focal length of combination

$$\frac{1}{F} = \frac{1}{f_1} + \frac{1}{f_2}$$

$$\frac{1}{20} = \frac{1}{f_1} + \frac{1}{-2f_1} = \frac{1}{2f_1}$$

or
$$f_1 = \frac{20}{2} = 10 \text{ cm} \quad \text{(Convex lens)}$$

and
$$f_2 = -2f_1 = -20 \text{ cm} \quad \text{(Planoconvex lens)}$$

For radii of curvature of the convex lens we have

$$\frac{1}{f} = (n-1)\left(\frac{1}{R_1} - \frac{1}{R_2}\right) \quad \text{(i)}$$

For the planoconvex lens $R_2 = \infty$, $n = 1.60$ and $f = -20$ cm

$$\therefore \quad \frac{1}{(-20)} = (1.60 - 1)\left(\frac{1}{R_1} - \frac{1}{\infty}\right) = \frac{0.60}{R_1}$$

$$\therefore \quad R_1 = -0.60 \times 20 = -12 \text{ cm}$$

The radius of curvature of the concave part of the planoconcave lens is the same as the radius of curvature of the second surface of the convex lens because the two surfaces are in perfect contact (Figure 4.21). Therefore, for the convex lens of crown glass.

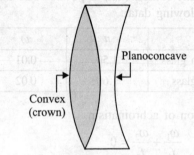

Figure 4.21 Perfect contact of two surfaces.

$R_2 = -12$ cm, $n = 1.50$ and $f = +10$ cm, Thus from Eq. (i)

$$\frac{1}{10} = (1.50 - 1)\left(\frac{1}{R_1} + \frac{1}{12}\right) = \frac{0.50}{R_1} + \frac{0.50}{12}$$

or

$$\frac{1}{R_1} = \frac{1}{5} - \frac{1}{12} = \frac{7}{60}$$

or

$$R_1 = +8.6 \text{ cm}$$

Thus, two radii of curvature of the convex lens are +8.6 cm and −12 cm.

PROBLEM 4.5 n_B and n_R are the refractive indices respectively for blue and red light in crown glass, n'_B and n'_R are the corresponding indices in flint glass. An equiconvex crown glass lens of radius of face R when placed in contact with a double concave flint glass lens of radii r_1 and r_2 respectively forms an achromatic lens and shows that

$$\frac{n_B - n_R}{n'_B - n'_R} = \frac{r_1 + r_2}{2r_2}$$

Solution Let f_1, f_2 be the focal lengths of the crown and flint glass lenses respectively; and n_1, n_2 be their mean refractive indices.

For a thin lens $\dfrac{1}{f} = (n-1)\left(\dfrac{1}{R_1} - \dfrac{1}{R_2}\right)$

For crown glass lens, $R_1 = +r_1$, $R_2 = -r_1$

$$\therefore \quad \frac{1}{f} = (n_1 - 1)\left(\frac{1}{r_1} - \frac{1}{r_2}\right) = \frac{2(n_1 - 1)}{r_1}$$

For flint glass lens, $R_1 = -r_1$, $R_2 = -r_2$

$$\therefore \quad \frac{1}{f_2} = (n_2 - 1)\left(\frac{1}{-r_1} - \frac{1}{r_2}\right) = -\frac{(n_2 - 1)(r_1 + r_2)}{r_1 r_2}$$

$$\therefore \quad \frac{f_1}{f_2} = -\frac{(n_2 - 1)(r_1 + r_2)}{r_1 r_2} \cdot \frac{r_1}{2(n_1 - 1)}$$

$$= -\frac{(n_2 - 1)(r_1 + r_2)}{2r_2(n_1 - 1)} \qquad (i)$$

But from the condition of achromatism

$$\frac{f_1}{f_2} = -\frac{\omega_1}{\omega_2} = -\frac{(n_B - n_R)/(n_1 - 1)}{(n'_B - n'_R)/(n_2 - 1)}$$

or $$\frac{f_1}{f_2} = -\frac{(n_B - n_R)(n_2 - 1)}{(n_1 - 1)(n'_B - n'_R)} \qquad (ii)$$

From Eqs. (i) and (ii)

$$-\frac{(n_2 - 1)(r_1 + r_2)}{2r_2(n_1 - 1)} = -\frac{(n_B - n_R)(n_2 - 1)}{(n_1 - 1)(n'_B - n'_R)}$$

$$\therefore \quad -\frac{n_B - n_R}{n'_B - n'_R} = \frac{r_1 + r_2}{2r_2}$$

PROBLEM 4.6 The focal lengths of two lenses are 8 cm and 4 cm. What should be separation between them to minimise chromatic aberration and spherical aberration. **[Bundelkhand, 2009]**

Solution (i) To minimize chromatic aberration, separation between two lenses, i.e.,

$$d = \frac{f_1 + f_2}{2} = \frac{(8 + 4)}{2} \text{ cm} = 6 \text{ cm}$$

(ii) To minimize spherical aberratoin, the separation between two lenses

$$d = f_1 - f_2 = (8 - 4) \text{ cm} = 4 \text{ cm}$$

CONCEPTUAL QUESTIONS

4.1 What do you understand by an achromatic combination of prisms?

Ans The dispersive powers of prisms of different materials are different. The dispersive power of flint glass is greater than that of crown glass. It is, therefore, possible to have a combination of two prisms, such that a ray of white light, after passing through it may suffer deviation without dispersion. Such a combination of prisms is called an *achromatic combination* of prisms.

4.2 Explain comatic error.

Ans When a lens is corrected for spherical aberration, it produces a point-image of a point-object situated on its principal axis. It, however, still fails to give a point image of a point-object situated slightly off the axis. The image of such a point-object is found to have a comet-like shape. This defect in the image is formed by a lens is called "coma" or "comatic error". It is particularly serious when an optical instrument carrying lenses is being used to observe highly resolved image.

4.3 What are the conditions for minimum spherical and chromatic aberrations for two thin lenses of focal lengths f_1 and f_2 separated by a distance d?

Ans (i) Condition for minimum spherical aberration $d = f_1 - f_2$
(ii) Condition for minimum chromatic aberration $d = (f_1 + f_2)/2$

4.4 What is Abbe's sine condition?

Ans For a lens or lens system, free four spherical aberration, coma is completely removed if Abbe's sine condition

$$n_1 h_1 \sin \theta_1 = n_2 h_2 \sin \theta_2$$

is satisfied for all the zones, where θ_1 and θ_2 are the angles which the conjugate rays make with the axis, n_1 and n_2 are the refractive indices of the object and image spaces, and h_1 and h_2 are the lengths of the object and of the image for the zone.

4.5 What is astigmatism?

Ans When a point-object is situated *far off the axis* of a lens, its image consists of two mutually perpendicular lines at a distance from each other and lying in perpendicular planes. The defect is known as "astigmatism".

4.6 What are aplanatic points of a refracting surface?

Ans The aplanatism is the property of a surface by virtue of which all the rays starting from a point-object on its axis, after reflection or refraction at the surface, converge to or appear to diverge from a single point-image. The particular object and image points are called "aplanatic points" and the surface is said to be "aplanatic with respect to these points".

EXERCISES

Theoretical Questions

4.1 Discuss the monochromatic aberrations present in the image formed by a lens.
4.2 What is spherical aberration? Explain its causes.
4.3 What is coma?
4.4 Distinguish between spherical aberration and coma.
4.5 What is meant by chromatic aberration of a lens?
4.6 What is achromatism?
4.7 Discuss the condition for achromatism of two lenses separated by a distance.

4.8 Write a short notes on:
(a) Coma
(b) Astigmatism
(c) Curvature
(d) Distortion

4.9 What are defects in images? What are common defects in images produced in a lens and how are they removed?

4.10 Explain chromatic and spherical aberrations. Deduce the condition for achromatism of two lenses separated by a distance.

4.11 Explain chromatic aberration? Show that axial chromatic aberration is equal to $f_R - f_V = \omega f_Y$.

4.12 Derive the condition for minimum spherical aberration exhibited by two lenses separated coaxially from each other.

4.13 Prove that the condition for minimum spherical aberration of two thin lenses of focal lengths f_1 and f_2 separated by a distance d is given by
$$d = f_1 - f_2$$

4.14 Write any two applications of aplanatic points. **[Bundelkhand, 2005, 2008]**

4.15 What is chromatic aberration? Deduce the condition for achromatism of two lenses separated by distance d. **[Bundelkhand, 2008]**

Numerical Problems

4.1 The dispersive powers of crown and flint glasses are 0.02 and 0.04 respectively. What will the focal length of the crown glass convex lens, which forms an achromatic doublet with a flint glass concave lens of focal length 80 cm? **[Ans. 40 cm]**

4.2 A double convex lens has a radius of curvature of 40 cm and 10 cm. Find the longitudinal chromatic aberration for an object at infinity. Given $n_V = 1.5230$ and $n_R = 1.5145$.
[Ans. $f_R - f_V = 0.253$ cm]

4.3 A lens made of crown glass has a focal length 40 cm. Calculate the longitudinal chromatic aberration if $n_C = 1.5164$ and $n_F = 1.5249$. Here C and F are used for crown and flint glasses respectively. **[Ans. $f_R - f_V = 0.0656$ cm]**

4.4 A converging achromatic doublet of 40 cm focal length is to be constructed out of a thin crown glass and a thin flint glass lens, and the radius of the crown surface is 25 cm. Calculate the radius of curvature of the second surface of each lens, given that the dispersive power and mean refractive index are respectively 0.017 and 1.5 for crown glass and 0.034 and 1.7 for flint glass. **[Ans. $R_1 = 16.66$ cm (for convex) and $R_2 = 233.25$ cm (for concave)]**

4.5 Two convex lenses of focal lengths 20 cm and 30 cm of crown and flint glasses respectively are placed at a certain distance apart for achromatism. The dispersive powers of crown and flint glasses are 0.01 and 0.02 respectively
(a) Find the distance between the two lenses.
(b) If both the lenses are of the same material, what will be the distance between them?
[Ans. (a) 23.33 cm, (b) 25 cm]

4.6 A convergent doublet of separated lenses, corrected for spherical aberration, has an equivalent focal length of 10 cm. The lenses of the doublet are separated by 2 cm. What are the focal lengths of its component lenses. [**Ans.** $f_1 = 20$ cm and $f_2 = 18$ cm]

4.7 From the condition of no chromatic aberration and minimum spherical aberration of the combination for two separated thin lenses, design a combination of equivalent focal length 5.0 cm. [**Ans.** $d = 6.67$ cm, $f_1 = 10$ cm, $f_2 = 3.33$ cm]

4.8 Two thin lenses of focal lengths f_1 and f_2 separated by a distance d have an equivalent focal length 50 cm. There is no chromatic and spherical aberration for the combination. If the lenses are made of same material, find focal lengths f_1, f_2 and d.
[**Ans.** $f_1 = 100$ cm, $f_2 = 33.33$ cm, $d = 66.6$ cm]

4.9 Two convex lenses made from same material and of focal lengths 0.20 m and 0.30 m are to be placed at a certain distance apart so that they form an image free from chromatic aberration. Find the distance between them. [**Ans.** $d = 0.25$ m]

4.10 It is desired to make an achromatic lens by placing a thin convex lens in contact with a thin concave lens. The convex and concave lenses are made up of materials of dispersive power in the ratio of 1:2. If the focal length of the convex lens is 20 cm, determine the focal length of the achromatic combination. How does the combination behave?
[**Ans.** $F = +40$ cm, convergent lens]

Multiple Choice Questions

4.1 Spherical aberrations occur because the focal points of rays far from the principal axis of a spherical lens are different from
(a) the focal points of rays of the same wavelength passing near to axis
(b) the focal planes of rays of different wavelength passing near the axis
(c) the nodal points of the rays of different wavelengths passing far from the axis
(d) none of the above

4.2 The aberration in image due to variation of refractive index with wavelength (due to colour) is called
(a) spherical aberration (b) chromatic aberration
(c) coma (d) astigmatism

4.3 Spherical aberration can be removed by using
(a) convex lens (b) concave lens
(c) planoconvex lens (d) cylindrical lens

4.4 The condition for achromatic combination of lenses in contact is
(a) $\dfrac{\omega_1}{f_1} = \dfrac{\omega_2}{f_2}$ (b) $\dfrac{f_1}{\omega_1} = \dfrac{f_2}{\omega_2}$
(c) $\dfrac{\omega_1}{f_1} + \dfrac{\omega_2}{f_2} = 0$ (d) $\dfrac{\omega_1}{\omega_2} = \dfrac{f_1}{f_2}$

4.5 An achromatic combination of lens produces
(a) image in black and white (b) coloured images
(c) image unaffected by variation of refractive index with wavelength
(d) highly enlarged images

4.6 Two thin lenses of focal lengths f_1 and f_2 are made of materials of dispersive powers ω_1 and ω_2 respectively. The lenses are kept in contact. The combination will show no chromatic aberration if
  (a) $\omega_1 = \omega_2$
  (b) $f_1/f_2 = -(\omega_1/\omega_2)$
  (c) $f_2\omega_1 = f_2\omega_2$
  (d) $f_1\omega_1 \times f_2\omega_2 = 0$

4.7 A lens made from crown glass has focal length 50 cm. The longitudinal chromatic aberration will be
  (a) 3.35 cm
  (b) 4.35 cm
  (c) 9.35 cm
  (d) 2.35 cm

4.8 Condition for achromatism for two lenses of same material separated by a finite distance is
  (a) $d = \dfrac{f_1 + f_2}{2}$
  (b) $d = \dfrac{f_1 - f_2}{2}$
  (c) $d = f_1 - f_2$
  (d) $d = f_1 + f_2$

4.9 If a screen is place through these point of intersection, an image of least aberration is formed which is known as
  (a) circle of high confusion
  (b) circle of least confusion
  (c) circle of spherical aberration
  (d) circle of coma

4.10 The condition for spherical aberration to be minimum is
  (a) $d = f_1 - f_2$
  (b) $d = f_1 + f_2$
  (c) $d = \dfrac{f_1 + f_2}{2}$
  (d) $d = \dfrac{f_1 - f_2}{2}$

4.11 The focal length of field achromatic combination of a telescope is 90 cm. The dispersive powers of lenses are 0.024 and 0.036 respectively. Their focal lengths will be

[Bundelkhand, 2005]
  (a) 30 cm and 60 cm
  (b) 45 cm and 90 cm
  (c) 15 cm and 45 cm
  (d) 30 cm and 45 cm

4.12 Condition of achromatism for two lenses in contact is [Bundelkhand, 2006, 2009]
  (a) $\omega_1 f_1 + \omega_2 f_2 = 0$
  (b) $\omega_1 f_2 + \omega_2 f_1 = 0$
  (c) $f_1 f_2 = \omega_1 \omega_2$
  (d) $\omega_1 f_1 - \omega_2 f_2 = 0$

Answers

4.1 (a)   **4.2** (b)   **4.3** (c)   **4.4** (c)   **4.5** (c)   **4.6** (b)   **4.7** (a)   **4.8** (a)
4.9 (b)   **4.10** (a)   **4.11** (d)   **4.12** (b)

CHAPTER 5

Optical Instruments

"The camera introduces us to unconscious optics as does psychoanalysis to unconscious impulses."
—Walter Benjamin

IN THIS CHAPTER

- Microscopes
- Telescopes
- Eyepieces
- Spectrometer
- Electron microscope

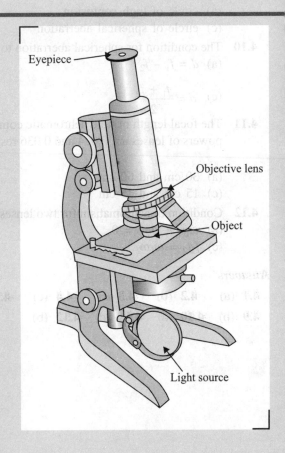

5.1 INTRODUCTION

Optical instruments are the devices which make use of mirrors, lenses and prisms, and are primarily used to extend the range of vision of human eye. For example, microscopes are used for viewing tiny objects clearly, while telescopes are used to see distant objects clearly. The design of an optical instrument must meet the following two requirements:

1. High magnification: Magnification is the ratio of the size of the final image to the size of the object. An optical instrument with high magnification makes viewing more clear and comfortable by increasing the size of image.

2. Adequate resolution: The resolution of an optical instrument is its ability to resolve the images of two closely spaced objects so that they can be seen separately. An optical instrument with high resolution reveals the finer details of the objects.

In this chapter, we will discuss simple microscope, compound microscopes, telescopes, eyepieces, spectrometer and electron microscopes.

5.2 MICROSCOPES

When we need greater magnification then we can get with a simple magnifier, the instrument that we usually use is the *microscope*, sometimes called a *compound microscope*. The essential elements of a microscope is shown in Figure 5.1.

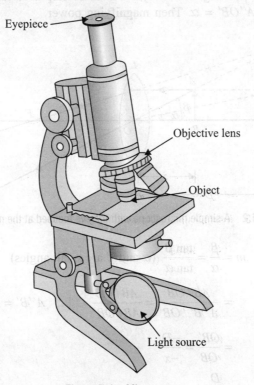

Figure 5.1 Microscope.

5.2.1 Simple Microscope

A simple microscope or a magnifying glass is just a convex lens of short focal length, held close to the eye.

Working principle

1. When the final image is formed at the least distance of the distinct vision: When an object AB is placed between the focus F and optical centre O of a convex lens; a virtual, erect and magnified image $A'B'$ is formed on the same side of the lens as the object. Since a normal eye can see an object clearly at the least distance of the distinct vision $D (= 25$ cm), the position of the lens is so adjusted that the final image is formed at the distance D from the lens, as shown in Figure 5.2.

Magnifying power: The magnifying power of a simple microscope is defined as the ratio of the angles subtended by the image and the object at the eye when both are at the least distance of distinct vision from the eye. Thus,

$$\text{Magnifying power} = \frac{\text{Angle subtended by the image at the least distance of distinct vision}}{\text{Angle subtended by the object at the least distance of distinct vision}}$$

As the eye is held close to the lens, the angles subtended at the lens may be taken to be the angles subtended at the eye. The image $A'B'$ is formed at the least distance of the distinct object D. Let $\angle A'OB' = \beta$. Imagine the object AB to be displaced to position $A''B'$ at distance D from the lens. Let $\angle A''OB' = \alpha$. Then magnifying power

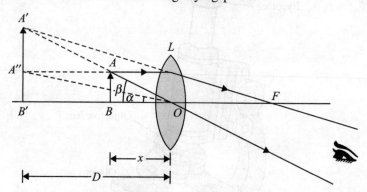

Figure 5.2 A simple microscope with the eye focused at the near point.

$$m = \frac{\beta}{\alpha} = \frac{\tan \beta}{\tan \alpha} \quad (\alpha \text{ and } \beta \text{ are small angles})$$

$$= \frac{AB/OB}{A''B'/OB'} = \frac{AB/OB}{AB/OB'} \quad [\because A''B' = AB]$$

$$= \frac{OB'}{OB} = \frac{-D}{-x}$$

or

$$m = \frac{D}{x}$$

Let f be the focal length of the lens. As the image is formed at the least distance of the distinct vision from the lens, so
$$v = -D$$
Using thin lens formula
$$\frac{1}{v} - \frac{1}{u} = \frac{1}{f}$$
We get
$$\frac{1}{-D} - \frac{1}{-x} = \frac{1}{f}$$
or
$$\frac{1}{x} = \frac{1}{D} + \frac{1}{f}$$
or
$$\frac{D}{x} = 1 + \frac{D}{f}$$
$\therefore$
$$m = 1 + \frac{D}{f} \qquad (5.1)$$

Thus, shorter the focal length of the convex lens, the greater is its magnifying power.

2. When the final image is formed at infinity: When we see an image at the near point [Figure 5.3(a)], it causes some strain in eye. Often the object is placed at the focus of the convex lens, so that parallel rays enter the eye, as shown in Figure 5.3(b). The image is formed at infinity, which is more suitable and comfortable for viewing by the relaxed eye.

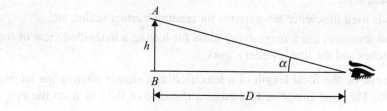

Figure 5.3(a) Object at the near point.

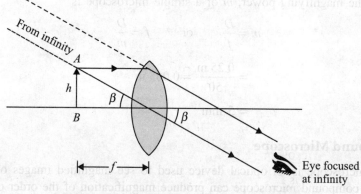

Figure 5.3(b) With object at F, image is at infinity.

Magnifying power: It is defined as the ratio of the angle formed by the image (when situated at infinity) at the eye to the angle formed by the object at the eye when situated at the least distance of the distinct vision.

$$m = \frac{\beta}{\alpha} = \frac{\tan \beta}{\tan \alpha} \quad [\alpha \text{ and } \beta \text{ are small}]$$

From Figure 5.3(a)

$$\tan \alpha = \frac{h}{D}$$

From Figure 5.3(b)

$$\tan \beta = \frac{h}{f}$$

$\therefore$

$$m = \frac{h/f}{h/D}$$

or

$$m = \frac{D}{f} \tag{5.2}$$

This magnification is one less than the magnification when the image is formed at the near point. But viewing is more comfortable when the eye is focused at infinity.

Applications of simple microscope

1. In magnifying the printed letters in a book, textures of fibres or thread of cloth, engravings, details of stamps, etc.
2. Magnifying glass is used in science laboratories for reading Vernier scales, etc.
3. Watch makers and jewellers use a magnifying glass for having a magnified view of the small parts of watches and the fine jewellery work.

EXAMPLE 5.1 What must be the focal length of a lens used as a simple microscope having a magnifying power of 50. The final image is formed at a distance of 0.25 m from the eye.

Solution Given: $m = 50$, $D = 0.25$ m

We know that the magnifying power, m of a simple microscope is

$$m = \frac{D}{f} \quad \text{or} \quad f = \frac{D}{m}$$

$$= \frac{0.25 \text{ m}}{50} = 0.005 \text{ m}$$

$$= 5 \text{ mm}$$

5.2.2 Compound Microscope

A compound microscope is an optical device used to see magnified images of tiny objects. A good quality compound microscope can produce magnification of the order of 1000.

Construction

It consists of two convex lenses of short focal length, arranged coaxially at the ends of two sliding metal tubes.

1. Objective: It is a convex lens of very short focal length f_0 and small aperture. It is positioned near the object to be magnified.

2. Eyepiece or ocular: It is convex lens of comparatively larger focal length f_e and larger aperture than the objective ($f_e > f_o$). It is positioned near the eye for viewing the final image.

The distance between the two lenses can be varied by using rack and pinion arrangement.

Working

1. When the final image is formed at the least distance of the distinct vision: The object AB to be viewed is placed at a distance u_o, slightly larger than the focal length f_o of the objective O. The objective forms a real, inverted and magnified image $A'B'$ of the object AB on the other side of the lens O, as shown in Figure 5.4. The separation between the objective O and the eyepiece E is so adjusted that the image $A'B'$ lies within the focal length f_e of the eyepiece. The image $A'B'$ acts as an object for the eyepiece which essentially acts like a simple microscope. The eyepiece E forms a virtual and magnified final image $A''B''$ of the object AB. Clearly, the final image $A''B''$ is inverted with respect to the object AB.

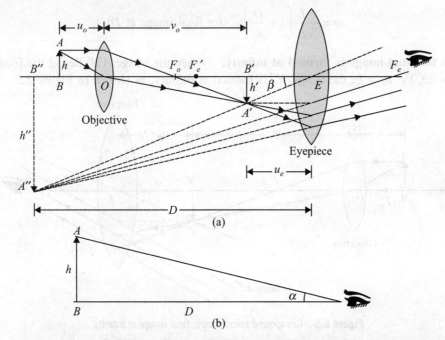

Figure 5.4 Compound microscope, final image at D.

Magnifying power: The magnifying power of a compound microscope is defined as the ratio of the angle subtended at the eye by the final virtual image to the angle subtended at the eye by the object when both are at the least distance of the distinct vision from the eye.

$$m = \frac{\beta}{\alpha} = \frac{\tan\beta}{\tan\alpha} = \frac{h'/u_e}{h/D} = \frac{h'}{h} \times \frac{D}{u_e} = m_o m_e$$

Here
$$m_o = \frac{h'}{h} = \frac{v_o}{u_o}$$

As the eyepiece acts as a simple microscope, so

$$m_e = \frac{D}{u_e} = 1 + \frac{D}{f_e}$$

$\therefore$
$$m = \frac{v_o}{u_o}\left(1 + \frac{d}{f_e}\right)$$

As the object AB is placed close to the focus F_o of the objective, therefore,
$$u_o \cong -f_o$$

Also the final image $A'B'$ is formed close to the eye lens whose focal length is short, therefore, $v_o \cong L$ = the length of the microscope tube or the distance between the two lenses.

$\therefore$
$$m_0 = \frac{v_o}{u_o} = \frac{L}{-f_o}$$

$\therefore$
$$m = -\frac{L}{f_o}\left(1 + \frac{D}{f_e}\right) \quad \text{(for final image at } D\text{)} \tag{5.3}$$

2. When the final image is formed at infinity: When the image $A'B'$ lies at the focus F'_e of the eyepiece, i.e. $u_e = f_e$, the image h'' is formed at infinity, as shown in Figure 5.5.

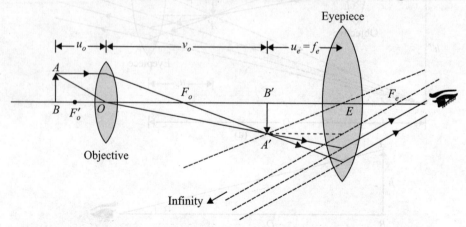

Figure 5.5 Compound microscope, final image at infinity.

When the final image is formed at infinity

$$m_e = \frac{D}{f_e}$$

$$\therefore \qquad m = -\frac{L}{f_o} \times \frac{D}{f_e} \quad \text{(for final image at infinity)} \qquad (5.4)$$

Obviously, magnifying power of the compound microscope is large when both f_o and f_e are small.

EXAMPLE 5.2 A compound microscope is made using a lens of focal length 10 mm as objective and another lens of focal length 15 mm as eyepiece. An object is placed at a distance of 11 mm from the objective and microscope so adjusted that final image is obtained at infinity. Calculate (i) the magnfying power; and (ii) distance between the objective and eye lens.

Solution Let the image formed by objective be at a distance b_1 from it.
Given: $u_1 = -11$ mm, $v_1 = +b_1$, $f_o = 10$ mm

(i) Therefore,
$$\frac{1}{+b_1} - \frac{1}{-11} = \frac{1}{10}$$

or
$$\frac{1}{b_1} = \frac{1}{10} - \frac{1}{11} = \frac{1}{110}$$

$$\therefore \qquad b_1 = 110 \text{ mm}$$

Since the final image is formed at infinity, this image must be at the focus of the eye lens. This is shown in Figure 5.6. The magnifying power m, by definition is

$$m = \frac{\beta}{\alpha}$$

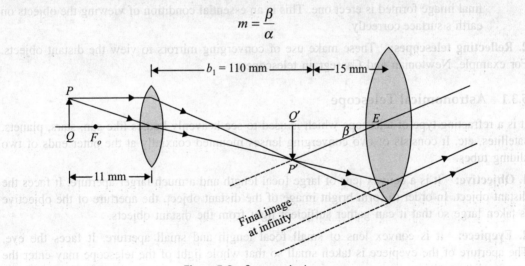

Figure 5.6 Compared microscope.

From Figure 5.6, we have

$$\beta = \frac{P'Q'}{EQ'} \quad \text{and} \quad \alpha = +\frac{PQ}{250} \quad [\because \; D = 250 \text{ mm}]$$

$$m = \frac{P'Q' \times 250}{EQ' \times PQ}$$

From lens formula

$$\frac{P'Q'}{PQ} = -\frac{v_1}{u_1} = -\frac{110}{(-11)} = 10$$

Also $EQ' = 15$ mm (given)

$$\therefore \quad m = \frac{10 \times 250}{15} = 166.67$$

(ii) It is obvious from Figure 5.6, the distance between objective and eye lens, D_1 is

$$D_1 = 110 + 15 = 125 \text{ mm}$$

5.3 TELESCOPES

A telescope is an optical device which enables us to see the distant objects clearly. It provides angular magnification of the distant objects.

1. Refracting telescopes: These make use of lenses to view distant objects. These are of two types:

(i) *Astronomical telescope:* It is used to see heavenly objects like the sun, stars, planets, etc. The final image formed is inverted one which is immaterial in the case of heavenly bodies because of their round shape.

(ii) *Terrestrial telescope:* It is used to see distant objects on the surface of the earth. The final image formed is erect one. This is an essential condition of viewing the objects on earth's surface correctly.

2. Reflecting telescopes: These make use of converging mirrors to view the distant objects. For example, Newtonian and Cassegrain telescopes.

5.3.1 Astronomical Telescope

It is a refracting type of telescope which is used to see heavenly bodies like sun, stars, planets, satellites, etc. It consists of two converging lenses mounted coaxially at the outer ends of two sliding tubes.

1. Objective: It is a convex lens of large focal length and a much larger aperture. It faces the distant object. In order to form bright image of the distant object, the aperture of the objective is taken large so that it can gather sufficient light from the distant objects.

2. Eyepiece: It is convex lens of small focal length and small aperture. It faces the eye. The aperture of the eyepiece is taken small so that whole light of the telescope may enter the eye for distant vision.

Working

1. When the final image if formed at the least distance of the distinct vision: The parallel beam of light coming from the distant object falls on the objective at some angle α as shown in Figure 5.7. The objective focuses the beam in its focal plane and forms a real, inverted and diminished image $A'B'$. This image $A'B'$ acts as an object for the eyepiece. The distance of the eyepiece is so adjusted that the image $A'B'$ lies within its focal length.

The eyepiece magnifies this image so that final image $A''B''$ is magnified and inverted with respect to the object. The final image is seen distinctly by the eye at the least distance of the distinct vision.

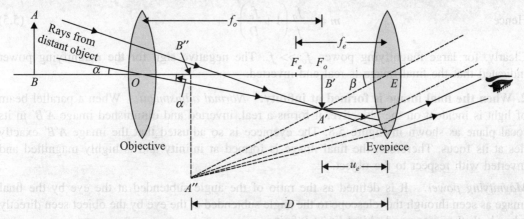

Figure 5.7 Astronomic telescope focused for least distance of distinct vision.

Magnifying power: The magnifying power of a telescope is defined as the ratio of the angle subtended at the eye by the final image formed at the least distance of the distinct vision to the angle subtended at the eye by the object at infinity, when seen directly.

As the object is very far off, the angle subtended by it at the eye is practically equal to the angle α subtended by it at the object. Thus,

$$\angle A'OB' = \alpha$$

Also, let $\angle A''EB'' = \beta$

$\therefore$ Magnifying power $m = \dfrac{\beta}{\alpha}$

$$= \frac{\tan \beta}{\tan \alpha} \quad (\because \alpha \text{ and } \beta \text{ are small angles})$$

According to the Cartesian sign convention

$OB' = +f_o =$ focal length of the objective

$B'E = -u_e =$ distance of $A'B'$ from the eyepiece acting as an object for it

$\therefore \qquad m = -\dfrac{f_o}{u_e}$

Again, for the eyepiece

$$u = -u_e \text{ and } v = -D$$

As
$$\frac{1}{v} - \frac{1}{u} = \frac{1}{f}$$

$\therefore$
$$\frac{1}{-D} - \frac{1}{-u_e} = \frac{1}{f_e}$$

or
$$\frac{1}{u_e} = \frac{1}{f_e} + \frac{1}{D} = \frac{1}{f_e}\left(1 + \frac{f_e}{D}\right)$$

Hence
$$m = -\frac{f_o}{f_e}\left(1 + \frac{f_e}{D}\right) \tag{5.5}$$

Clearly, for large magnifying power, $f_o \gg f_e$. The negative sign for the magnifying power indicated that the final image is real and inverted.

2. When the final image is formed at infinity: *Normal adjustment:* When a parallel beam of light is incident on the objective, it forms a real, inverted and diminished image $A'B'$ in its focal plane as shown in Figure 5.8. The eyepiece is so adjusted that the image $A'B'$ exactly lies at its focus. Therefore, the final image is formed at infinity and is highly magnified and inverted with respect to the object.

Magnifying power: It is defined as the ratio of the angle subtended at the eye by the final image as seen through the telescope to the angle subtended at the eye by the object seen directly when both the image and object lie at infinity.

As the object is very far off, the angle subtended by it at the objective. Thus,
$$\angle A'OB' = \alpha$$
Also, let $\angle A'EB' = \beta$

∴ Magnifying power $m = \dfrac{\beta}{\alpha}$

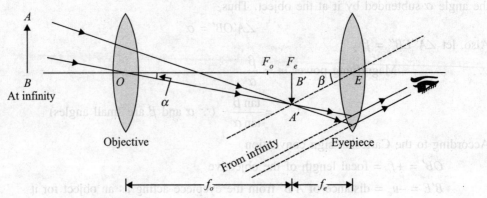

Figure 5.8 Astronomical telescope in normal adjustment.

$$= \frac{\tan \beta}{\tan \alpha} \quad (\because \alpha \text{ and } \beta \text{ are small angles})$$

$$= \frac{A'B'/B'E}{A'B'/OB'} = \frac{OB'}{B'E}$$

Applying the Cartesian sign convention.

$OB' = +f_o$ = Distance of $A'B'$ from the objective along the incident light

and $\quad B'E = -f_e$ = Distance of $A'B$ from the eyepiece against the incident light

$$m = -\frac{f_o}{f_e} \qquad (5.6)$$

Clearly, for large magnifying power, $f_o \gg f_e$. The negative sign for m indicates that the image is real and inverted.

EXAMPLE 5.3 A telescope consists of two convex lenses. The focal length of the objective and eye lens is 500 mm and 50 mm respectively. In normal adjustment of telescope, obtain the magnifying power.

Solution The magnifying power m of telescope is

$$m = \frac{f_o}{f_e} = \frac{500}{50} = 10$$

5.3.2 Terrestrial Telescope

It is a refracting type of telescope which is used to see erect images of distant earthy objects. It uses an additional convex lens between objective and eyepiece for obtaining an erect image.

As shown in Figure 5.9, the objective forms a real, inverted and diminished image $A'B'$ of the distant object in its focal length from the focal plane of the objective. This lens forms a real, inverted and equal size image $A''B''$ of $A'B'$. This image is now erect with respect to the distant object. The eyepiece is so adjusted that the image $A''B''$ lies at its principal focus. Hence the final image is formed at infinity and is highly magnified and erect with respect to the distant object.

As the erecting lens does not cause any magnification, the angular magnification of the terrestrial telescope is the same as that of the astronomical telescope

$$\therefore \quad m = \frac{f_o}{f_e} \qquad (5.7)$$

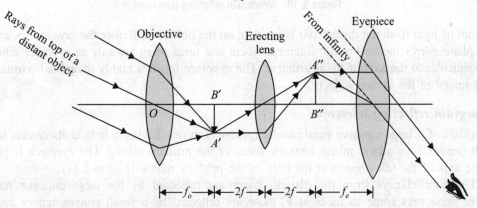

Figure 5.9 Terrestrial telescope.

When the image is formed at infinity,

$$m = \frac{f_o}{f_e}\left(1 + \frac{f_e}{D}\right) \qquad (5.8)$$

when the image is formed at the least distance of distinct vision.

Drawbacks

1. The length of the terrestrial telescope is much larger than the astronomical telescope. In normal adjustment, the length of a terrestrial telescope $= f_o + 4f + f_e$, where f is the focal length of the erecting lens.
2. Due to extra reflection at the surfaces of the erecting lens, the intensity of the final image decreases.

5.3.3 Reflecting Telescopes

These are of two types of reflecting telescopes.

Newtonian reflecting telescope

The first reflecting telescope was set up by Newton in 1668. It consists of a large concave mirror of large focal length as the objective, made of an alloy of copper and tin as shown in Figure 5.10.

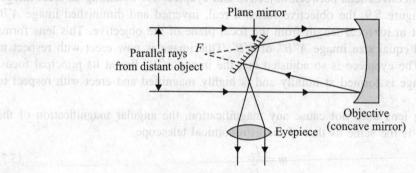

Figure 5.10 Newtonian reflecting telescope.

A beam of light from the distant star is incident on the objective. Before the rays are focused at F, a plane mirror inclined at 45° intercepts them and turns them towards an eyepiece adjusted perpendicular to the axis of the instrument. The eyepiece forms a highly magnified, virtual and erect image of the distant object.

Cassegrain reflecting telescope

It consists of a large concave paraboloidal (primary) mirror having a hole at its centre. It is a small convex (secondary) mirror near the focus of the primary mirror. The eyepiece is placed on the axis of the telescope near the hole of the primary mirror (Figure 5.11).

The parallel rays from the distant object are reflected by the large concave mirror. Before these rays come to focus at F, these are reflected by a small convex mirror and are converged to a point I, just outside the hole. The final image formed at I is viewed through the eyepiece. As the first image at F is inverted with respect to the distant object and the second image I is erect with respect to the first image F, hence the final image is inverted with respect to the object.

Let f_o be the focal length of the objective and f_e that of the eyepiece.

Optical Instruments 117

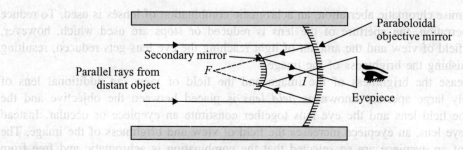

Figure 5.11 Cassegrain reflecting telescope.

For the final image formed at the least distance of distinct vision

$$m = \frac{f_o}{f_e}\left(1 + \frac{f_e}{D}\right)$$

For the final image formed at infinity

$$m = \frac{f_o}{f_e} = \frac{R/2}{f_e} \tag{5.9}$$

Advantages of a reflecting-type telescope

A reflecting-type telescope has the following advantages over a refracting-type telescope.
- A concave mirror of large aperture has high gathering capacity and absorbs very less amount of light than the lenses of large apertures. The final image formed in reflecting telescope is very bright. So even very distant or faint stars can be easily viewed.
- Due to the large aperture of the mirror used, the reflecting telescopes have high resolving power.
- As the objective is a mirror and not a lens, it is free from chromatic aberration (formation of coloured image of a white object).
- The use of paraboidal mirror reduces the spherical aberration (formation of non-point, blurred image of a point object).
- A mirror requires griding and polishing of one surface only. So it costs much less to construct a reflecting telescope than a refracting telescope of equivalent optical quality.
- A lens of large aperture tends to be very heavy, and therefore, difficult to make and support by its edges. On the other hand, a mirror of equivalent optical quality weighs less and can be supported over it entire back surface.

5.4 EYEPIECES OR OCULARS

In optical instruments such as telescope and microscope, the objective lens faces the object and forms an image of the object and eye lens receives the light from the objective lens, and forms the magnified image. An image formed by a lens suffers from two main defects:
1. Chromatic aberration, and
2. Spherical aberration

To minimise chromatic aberration, an achromatic combination of lenses is used. To reduce spherical aberration, the aperture of the lens is reduced or stops are used which, however, reduces the field of view and the amount of light reaching the eye lens gets reduced, resulting in the diminishing the brightness of the image.

To increase the brightness of the image and the field of view, an additional lens of comparatively large aperture known as *field lens* is placed between the objective and the eye lens. The field lens and the eye lens together constitute an eyepiece or occular. Instead of a single eye lens, an eyepiece increases the field of view and brightness of the image. The two lenses of an eyepiece are so selected that the combination is achromatic and free from spherical aberration.

L_1, L_2, and L_3 are the objective lens, field lens and eye lens respectively as shown in Figure 5.12.

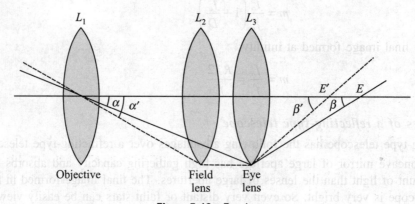

Figure 5.12 Eyepiece.

The lenses L_2 and L_3 together constitute an eyepiece. The extreme ray can be collected by eye lens, L_3, in the absence of the field lens L_2, makes an angular object field α and the angular image field β. The best position for the eye or the centre of the exit pupil in this case is E. When the field lens L_2 is placed at the location where the image of the object due to the objective lens is formed, the ray represented by dotted line which was earlier not collected by the eye lens, is now collected by it, i.e. the field of view and the brightness of the image increase on placing the field lens. Now, the centre of the exit pupil shifts to E' (nearer than earlier) and the angular object field is α', where $\alpha' > \alpha$ and the angular image field is β', where $\beta' > \beta$.

From the above discussion, we can conclude that the field lens:
  (i) Increases the angular object field,
  (ii) Brings the centre of the exit pupil near the eye lens, and
  (iii) Helps minimise aberrations.

Following three eyepieces are commonly used:
  (a) Huygen's eyepiece,
  (b) Ramsden's eyepiece, and
  (c) Gauss's eyepiece.

5.4.1 Huygen's Eyepiece

Construction

Huygen's eyepiece consists of two planoconvex lenses of focal lengths $3f$ and f. They are separated by a distance of $2f$. The convex surfaces of both the lenses are towards the object (light) as shown in Figure 5.13. To minimize spherical aberration, we have to use planoconvex lenses.

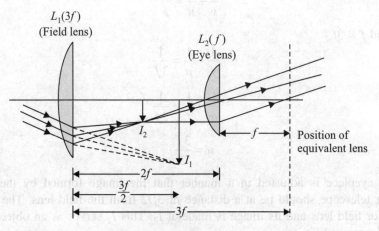

Figure 5.13 Huygen's eyepiece.

To obtain the condition for minimum spherical and chromatic aberrations, the distance between the two lenses is kept $2f$ as can be seen from the following description.

Condition of minimum chromatic aberration or achromatism: For achromatism, the distance between the two lenses should be equal to average of the focal lengths of eye lens and field lens. i.e.,

$$d = \frac{f_1 + f_2}{2}$$

Here $f_1 = 3f, f_2 = f$

$\therefore \qquad d = \dfrac{3f + f}{2} = 2f \qquad (5.10)$

Condition for minimum spherical aberration: For this, the separation (d) between the lenses should be equal to the difference of the focal lengths of eye lens and field lens. i.e.,

$$d = f_1 - f_2 = 3f - f = 2f \qquad (5.11)$$

Hence conditions for minimum chromatic aberration and minimum spherical aberration are satisfied by this eyepiece.

Working

Huygen's eyepiece is not used to see any objects directly, it is used either in telescope or in microscope. The image formed by the objective of microscope or telescope works as an object for Huygen's eyepiece. The eyepiece is adjusted in this manner that the image I_2 formed by the field lens lies in the first focal plane of the eyelens. Then this eyepiece forms the final image at infinity. Therefore, image I_2 formed by field lens should be at a distance f to the left of eye lens or at a distance f to the right of field lens. The image I_1 formed by the objective of microscope or telescope serves as an object for field lens. Suppose u be the distance of I_1 from field lens, then

$$\frac{1}{v} - \frac{1}{u} = \frac{1}{f}$$

Here $v = f$ and $f = 3f$

∴

$$\frac{1}{f} - \frac{1}{u} = \frac{1}{3f}$$

or

$$\frac{1}{u} = \frac{1}{f} - \frac{1}{3f}$$

or

$$u = \frac{3f}{2} \tag{5.12}$$

Thus the eyepiece is adjusted in a manner that the image formed by the objective of microscope or telescope should be at a distance of $3f/2$ from the field lens. The image serves as an object for field lens and its image is made at I_2. This I_2 serves as an object for eye lens and it forms image at infinity.

Negative eyepiece

Huygen's eyepiece is known as negative eyepiece because the real inverted image formed by the objective lies behind the field lens and this image acts as virtual object for eye lens. This eyepiece is used either in microscope or in telescope using white light only.

Position of crosswires

The cross-wires must be placed between the field lens and eye lens for the measurement of actual position of final image. But the cross-wires are viewed through eye lens only while the distant object is viewed by the rays refracted by both the lenses. Due to this reason, relative lengths of the cross-wires and image are disproportionate. Hence the cross-wires cannot be used in this eyepiece. Therefore, only distances and angles can be measured.

Position of cardinal points of Huygen's eyepiece

1. Equivalent focal length: The equivalent focal length (F) is given by

$$\frac{1}{F} = \frac{1}{f_1} + \frac{1}{f_2} - \frac{d}{f_1 f_2}$$

⇒

$$\frac{1}{F} = \frac{1}{3f} - \frac{1}{f} - \frac{2f}{(3f)(f)} = \frac{2}{3f}$$

or

$$F = \frac{3}{2}f \tag{5.13}$$

2. Principal points: Suppose H_1 and H_2 are the positions of first and second principal points from the lenses L_1 and L_2 respectively, then the first principal point is

$$L_1H_1 = \frac{Fd}{f_2} = \frac{(2f)(3f/2)}{f} = 3f \tag{5.14}$$

Hence the first principal point would be at a distance $3f$ to the right of the first lens.
The second principal point is

$$L_2H_2 = -\frac{Fd}{f_1} = -\frac{(2f)(3f/2)}{3f} = -f \tag{5.15}$$

Hence, the second principal point would be situated at a distance f to the left of second lens.

3. Focal points: Let F_1 and F_2 be the positions of the first and second focal points from the lenses L_1 and L_2, then first focal point is

$$L_1F_1 = -F\left(1 - \frac{d}{f_2}\right) = \frac{-3f}{2}\left(1 - \frac{2f}{f}\right) = \frac{3f}{2} \tag{5.16}$$

Hence the first focal point F_1 locates towards the right of first lens at a distance $(3f/2)$.
The second focal point is

$$L_2F_2 = F\left(1 - \frac{d}{f_1}\right) = \frac{3f}{2}\left(1 - \frac{2f}{3f}\right) = \frac{f}{2} \tag{5.17}$$

Thus the second focal point F_2 lies towards right of the second lens at a distance $f/2$.

4. Nodal points: Since medium for image and object is the same. Hence nodal points N_1 and N_2 coincide with the principal points H_1 and H_2.
The positions of cardinal points of Huygen's eyepiece are shown in Figure 5.14.

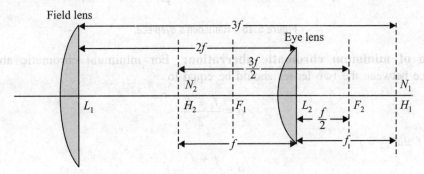

Figure 5.14 Positions of cardinal points of Huygen's eyepiece.

EXAMPLE 5.4 Two lenses of focal length 10 cm and 30 cm are separated by 20 cm in an eyepiece. Show that both spherical and chromatic aberrations are removed. Name the eyepiece.

Solution *Given:* $f_1 = 10$ cm, $f_2 = 30$ cm and $d = 20$ cm
(i) Condition for minimum spherical aberration

$$d = f_1 - f_2 = (30 - 10)\text{cm} = 20 \text{ cm}$$

(ii) Condition for minimum chromatic aberration

$$d = \frac{f_1 + f_2}{2} = \frac{30 + 10}{2} = 20 \text{ cm}$$

Both conditions are satisfied in this case. This eyepiece is Huygen eyepiece.

5.4.2 Ramsden's Eyepiece

Construction

Ramsden's eyepiece consists of two planconvex lenses of equal focal lengths (f), placed with their convex surfaces facing each other. These two planconvex lenses are placed at a distance of $(2/3)f$. The lens facing the object is known as field lens whose plane side faces the incident rays. Schematic diagram of Ramsden's eyepiece is shown in Figure 5.15.

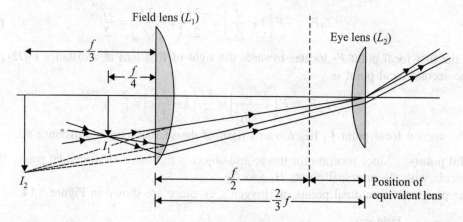

Figure 5.15 Ramsden's eyepiece.

Condition of minimum chromatic aberration: For minimum chromatic aberration, the distance between the two lenses should be equal to

$$d = \frac{f_1 + f_2}{2}$$

Here $f_1 = f$ and $f_2 = f$

$\therefore$
$$d = \frac{f + f}{2} = f \qquad (5.18)$$

But actual distance is $d = \frac{2}{3}f$, which is not much different from f. So chromatic aberration is not serious in this case.

Condition for minimum spherical aberration: This condition is not satisfied in Ramsden's eyepiece, but spherical aberration can be reduced by putting the lenses with their convex surfaces facing each other as shown in Figure 5.15.

Working

Although we may use this eyepiece to see the image of any object directly, but Ramsden's eyepiece is generally used in telescope or microscope. The image formed by the objective of telescope or microscope serves as object for eyepiece. The eyepiece is adjusted in the manner that the image I_2 is formed at a distance of $f/3$ from the field lens. This image I_2 serves as an object for eye lens and final image is formed at infinity. The image I_1 is formed by the objective of microscope or telescope serves as an object for field lens. Suppose u be the distance of I_1 from the field lens, then

$$\frac{1}{v} - \frac{1}{u} = \frac{1}{f}$$

Here $v = -\frac{f}{3}, f = f$

∴

$$\frac{1}{-f/3} - \frac{1}{u} = \frac{1}{f}$$

or

$$\frac{1}{u} = -\frac{3}{f} - \frac{1}{f} = -\frac{4}{f}$$

or

$$u = -\frac{f}{4} \qquad (5.19)$$

Here negative sign indicates that the image I_1 is formed by the objective lies to the left of the field lens. The rays coming from I_1 appears to come from I_2 after emerging from the emergent rays from eye lens becomes parallel and, thus, final image is formed at infinitely as shown in Figure 5.15.

Position of crosswires

The crosswires are placed in the position of real image I_1, i.e. at a distance $f/4$ towards the left of the field lens.

Cardinal points of Ramsden's eyepiece

1. Equivalent focal length: The equivalent focal length of Ramsden's eyepiece is given by

$$\frac{1}{F} = \frac{1}{f_1} + \frac{1}{f_2} - \frac{d}{f_1 f_2}$$

or

$$\frac{1}{F} = \frac{1}{f} + \frac{1}{f} - \frac{2f}{3ff} = \frac{1}{f} + \frac{1}{f} - \frac{2}{3f} = \frac{4}{3f}$$

or

$$F = \frac{3}{4}f \qquad (5.20)$$

2. Principal points: Let H_1 and H_2 be the positions of the first and second principal points from lenses L_1 and L_2 respectively, then for first principal point

$$L_1 H_1 = \frac{Fd}{f_2} = \frac{\frac{3}{4}f \cdot \frac{2}{3}f}{f} = \frac{f}{2} \qquad (5.21)$$

Thus H_1 lies to the right of L_1 at a distance $f/2$.

For the second principal point

$$L_2H_2 = \frac{-Fd}{f_1} = \frac{-\frac{3}{4}f \cdot \frac{2}{3}f}{f} = -\frac{f}{2} \qquad (5.22)$$

Thus H_2 lies to the left of L_2 at a distance $f/2$.

3. Focal points: Let F_1 and F_2 be the positions of the first and second focal points from the lenses L_1 and L_2 respectively, then for the first focal point

$$L_1H_1 = -F\left(1 - \frac{d}{f_2}\right) = -\frac{3}{4}f\left(1 - \frac{\frac{2}{3}f}{f}\right) = -f/4 \qquad (5.23)$$

Thus F_1 lies to the left of the field lens at a distance $f/4$.

For, the second focal point

$$L_2H_2 = -F\left(1 - \frac{d}{f_1}\right) = -\frac{3}{4}f\left(1 - \frac{\frac{2}{3}f}{f}\right) = +f/4 \qquad (5.24)$$

Thus F_2 lies to the right of the eye lens L_2 at a distance $f/4$.

4. Nodal points: Since the medium for image and object is the same. Therefore, nodal points N_1 and N_2 coincide with the principal points H_1 and H_2. All cardinal points are shown in Figure 5.16.

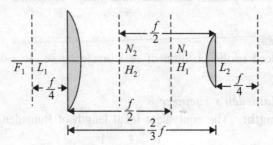

Figure 5.16 Cardinal points of Ramsden's eyepiece.

EXAMPLE 5.5 Light from the sun is falling upon the Ramsden's eyepiece. Locate the position of the image thus formed and also find the point from which the distance of the image is to be measured. The focal length of each lens of the eyepiece is 3.0 cm.

Solution Given: $f_1 = f_2 = f = 3.0$ cm, $d = \frac{2}{3}f = \frac{2}{3} \times 3 = 2$ cm

Equivalent focal length

$$F = \frac{f_1 f_2}{f_1 + f_2 - d} = \frac{3 \times 3}{3 + 3 - 2} = \frac{9}{4} = 2.25 \text{ cm}$$

∵ Sun is at infinity, the image is formed at second focal plane of the eyepiece, i.e. at a distance of 2.25 cm from second principal plane.

Position of second principal point from the eye lens

$$= \frac{-dF}{f_1} = \frac{-2 \times 2.25}{3} = -1.5 \text{ cm}$$

The negative sign means that the second principal plane lies behind the eye lens.

5.4.3 Relative Merits and Demerits of Huygen's and Ramsden's Eyepieces

S.No.	Huygen's eyepiece	Ramsden's eyepiece
1.	This eyepiece satisfies the condition of achromatism completely and the condition of minimum spherical aberration.	This eyepiece does not satisfy the condition of achromatism completely and the condition of minimum spherical aberration.
2.	The cross-wires, if used, are to be put midway between the two lenses. This involves mechanical difficulty.	The cross-wires are placed outside the eyepiece, hence no mechanical difficulty arises.
3.	As the cross-wires are fixed inside the eyepiece, the eyepiece is permanently adjusted for persons of normal vision only.	As the cross-wires lie outside the eyepiece, the eyepiece can be adjusted for different persons by altering the distance between the cross-wires and the eyepiece.
4.	Size measurements are reliable, because the final image undergoes refraction through both lenses, while the scale is refracted by one lens.	Size measurements are truthworthy, because the image and scale are equally magnified.
5.	It is generally used either in microscope for biological work or where no measurement is necessary.	It is used either in telescope or in microscope of various optical instruments.
6.	Final magnified image is convex towards the eye.	Final magnified image is almost flat.
7.	This eyepiece can examine only, the image.	It can be used as a simple magnifier to examine the real objects.

5.4.4 Gauss Eyepiece

Gauss eyepiece is a modified form of Ramsden's eyepiece. In Gauss eyepiece, the two lenses, i.e. field lens and eye lens are of equal focal length and is separated by a distance $(2/3)f$. Where f is focal lengths of field lens and eye lens. There have been made a small modification in Ramsden's eyepiece, a thin transparent glass plate is mounted at 45° to the axis of the tube. There is an opening in the side of the tube through which the light passes in the eyepiece. This eyepiece is used to set the optical axis of the spectrometer telescope. The schematic diagram is shown in Figure 5.17.

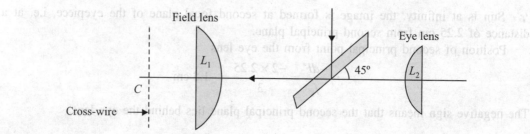

Figure 5.17 Gauss eyepiece.

5.5 SPECTROMETER

It is an optical device used for producing and studying the spectra of different light sources (Figure 5.18). Spectrometer, which generally used for experimental purposes, is shown in Figure 5.18.

Construction

A spectrometer (Figure 5.18) has three main parts:

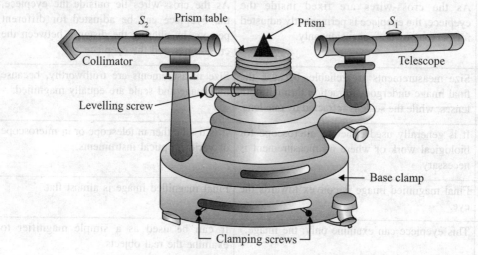

Figure 5.18 Spectrometer.

1. Collimator: The collimator is shown in Figure 5.18. It produces a parallel beam of light. It consists of two co-axial metal tubes. The outer tube is mounted horizontally and carries a convex lens L_1 at its free end. The inner tube has an adjustable vertical slit at the free end and can be slided inside the outer tube by a rack and pinion arrangement. The slit is adjusted in the focal plane of lens L_1. When the light source is kept in front of the slit, a parallel beam of light emerges from the collimator.

2. Prism table: It is shown in Figure 5.18. It is a circular horizontal plate on which the prism or grating is placed. It can be adjusted at a desired height with the help of a clamping screw.

It can be rotated about a vertical axis. Its position can be noted with the help of the verniers V_1 and V_2 attached to it and moving over a graduated circular scale carried by the telescope.

3. Telescope: It is shown in Figure 5.18. It is used for observing the spectrum. It is an astronomical telescope having a convex lens L_2 at one end and a Ramsden eyepiece at the other end. It is mounted horizontally on a vertical arm attached to the main circular scale. It can be rotated about the same vertical axis about which the prism table rotates. Its position can be noted on the circular scale by Vernier's V_1 and V_2. A cross-wire is fixed at the focus of the eyepiece.

Working

For getting a pure spectrum, the following adjustments are made in a spectrometer:

1. Focusing the eyepiece: The eyepiece of the telescope is moved in and out so that the cross-wires are clearly visible.

2. Focusing the telescope for parallel rays: The telescope is turned towards a distant object. The distance between the eyepiece and the object is so adjusted that object becomes clearly visible. This sets the telescopes for receiving the parallel rays.

3. Focusing the collimator for parallel rays: Illuminate the slit with a bright source and view it through the telescope. Adjust the distance between the slit and collimator lens till a clear image of the slit is seen. This sets the collimator to provide a parallel beam of light.

4. Setting the prism: The prism is placed at the centre of the prism table. The prism table is rotated so that light from the collimator falls on refracting face AB and after refraction emerges from the other face AC. The prism causes dispersion. The rays of a given colour emerge parallel to each other. They are received by the telescope. All red rays are focused at R, all violet rays at V and rays of other colours in between and a spectrum RV is formed in the focal plane of the objective, as shown in Figure 5.19. A magnified spectrum is viewed through the eyepiece.

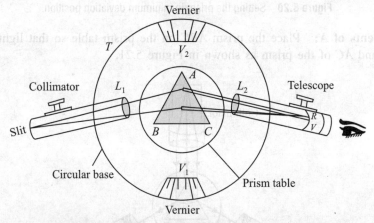

Figure 5.19 Setting a spectrometer to get pure spectrum.

5. To get rid of overlapping of colours: The prism is set in the minimum deviation position for same mean (yellow) colour. This gives a pure spectrum.

5.5.1 Measurement of Refractive Index by a Spectrometer

The refractive index n of the material of a prism is given by

$$n = \frac{\sin\dfrac{A+\delta_m}{2}}{\sin\dfrac{A}{2}} \qquad (5.25)$$

To determine n, we need to measure

1. Angle of minimum deviation (δ_m), and
2. The refracting angle of the prism (A).

1. Measurements of δ_m: The minimum deviation position is shown in Figure 5.20. Turn the telescope so that its crosswire coincides with mean (yellow) colour. This gives a pure spectrum.

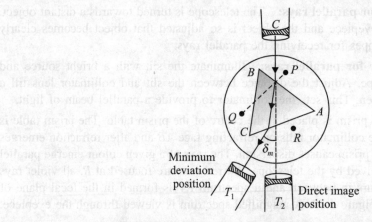

Figure 5.20 Setting the prism in minimum deviation position.

2. Measurements of A: Place the prism ABC on the prism table so that light falls directly on faces AB and AC of the prism as shown in Figure 5.21.

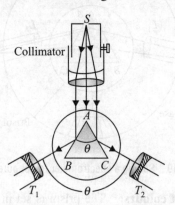

Figure 5.21 To measure angle of prism.

Look for the brightest image of the slit formed by reflection of light from faces AB and AC. Set the telescope in position T_1. So that cross-wire coincides with the image of the slit from the face AB. Turn the telescope to the position T_2 so as to focus the image of the slit from face AC. Let θ be the angle through which telescope turn. Then

$$A = \frac{\theta}{2}$$

5.6 ELECTRON MICROSCOPE

The human eye is a wonderful optical instrument having a wide field of view and a great range in distance. However, it has limitations too. We know that the limiting angle of resolution of a human eye is about 1', i.e. if the angle subtended by two objects is less than 1', it ceases to be distinguishable. In fact whether an object is visible clearly or not, depends both on its size and distance. The human eye cannot clearly see anything which is closer than 25 cm and has diameter less than 0.01 cm. Optical microscope is employed to see the details of the objects having dimensions less than what can be seen clearly by an unaided eye. It is well known that lower the wavelength of light used, higher is the resolving power of the microscope.

In 1924, de Broglie argued (later experimentally verified) that when electrons are scattered or diffracted, they behave as waves of wavelength $\lambda = h/mv$ (h = Planck's constant, m = mass of the electron, v = velocity of the electron). An electron accelerated through a potential difference of 60,000 volts ($v = \sqrt{2eV/m}$) the wavelength works out to be 5×10^{-10} cm (or 0.05 Å). This showed that if a microscope could be constructed with electron beam, it was possible to have much larger resolving power than that with an optical microscope. In 1927, Busch showed that a beam of electrons can be focused by suitable electric and magnetic fields as light is focused using glass lens is optical microscope. These developments paved the way for the construction of electron microscope and the first electron microscope was constructed in 1932 by German electrical engineer Ernst August Freiedrich Ruska. With continuous perfection of technique, an electron microscope can give a magnification up to 1,00,000 compared to the maximum magnification of an optical microscope up to 6000.

Thus, the working principle of an electron microscope can be stated as

(i) A beam of electrons exhibits wave nature similar to light rays, but of much shorter wavelength, and

(ii) A beam of electrons can be focussed by suitable electric and magnetic fields, very much like light rays are focused by glass lenses.

Construction

The schematic diagram of an electron microscope is shown in Figure 5.22(a). For comparison with an optical microscope, schematic diagram of an optical microscope is also shown in Figure 5.22(b). The essential parts of an electron microscope are given as follows.

130 Fundamentals of Optics

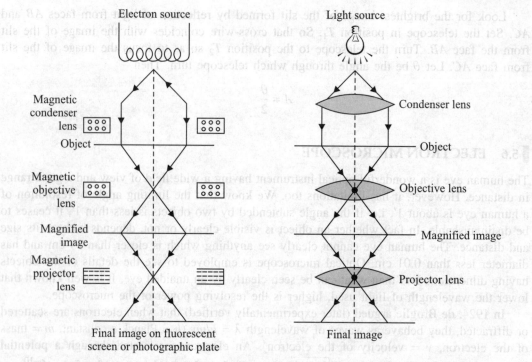

Figure 5.22 Schematic diagrams: (a) electron microscope and (b) optical microscope.

1. Electron gun: The electrons are produced by the electron gun. Electrons are emitted from the hot filament. The filament is surrounded by a metallic cylinder kept at a negative potential, which stops the electrons from spreading due to repulsion and a fine beam of electrons is produced. The beam is then accelerated through a potential of 60 kV. Now, with these accelerated electrons, the waves of very short wavelengths are associated.

There are five main types of electron microscope. These are:

(i) Transmission electron microscope (TEM)
(ii) Scanning electron microscope (SEM)
(iii) Scanning transmission electron microscope (STEM)
(iv) Reflection electron microscope (REM) and
(v) Scanning tunneling microscope (STM).

In first-two, electrons are generated in an electron gun to act upon the atomic nuclei of the specimen. While in field emission type, the specimen itself is a source of radiation.

2. Magnetic (or electrostatic) lens: The basic function of the condenser lens is to focus the electron beam from the electron gun on to the specimen to permit illumination and formation of image. The different parts of specimen absorb electrons differently and a corresponding image is formed. The objective lens forms an initial enlarged image of the illuminated specimen in a plane that is suitable for further enlargement by the projector lens. The projector lens, as the name indicates, is used to project the final magnified image on to the screen or photographic plate.

3. Fluorescent screen or photographic plate: The electron image is converted into visible light image on the fluorescent screen or photographic plate. To avoid the collisions of electrons with air molecules, the entire system is enclosed inside the metallic frame and a high degree of vacuum is maintained with a high speed diffusion pump.

Uses of electron microscope

Electron microscope is a very useful tool for researchers such as biologists for studying viruses, bacteria and other microorganisms, for material's scientist for studying crystal structure, textile technologist for the investigation of fibres, paints, etc.

FORMULAE AT A GLANCE

5.1 Simple microscope

(a) *Finite distance*

$$\text{Magnifying power} = \frac{\text{Angle subtended by the image at the least distance of distinct vision}}{\text{Angle subtended by the object at the least distance of distinct vision}}$$

$$m = \frac{\beta}{\alpha} = \frac{D}{x}$$

$$= \frac{\text{Distance between image and lens}}{\text{Distance between object and lens}}$$

$$= 1 + \frac{D}{f}$$

where f = focal length of the lens.

(b) *When the final image at infinity*

$$m = \frac{\beta}{\alpha} = \frac{D}{f}$$

5.2 Compound microscope

(a) When final image at the least distance of the distinct vision

$$m = \frac{\beta}{\alpha} = m_o m_e$$

Here

$$m_o = \frac{h'}{h} = \frac{v_o}{u_o} = -\frac{L}{f_o}$$

$$m_e = \frac{D}{u_e} = 1 + \frac{D}{f_o}$$

where u_o = distance between object and lens

f_o = focal length of the objective

(b) When image is infinity
$$m = \frac{D}{f_e} = -\frac{L}{f_o} \times \frac{D}{f_e}$$

5.3 Telescopes
(a) *Astronomical telescope*
  (i) Image at finite
$$m = \frac{\beta}{\alpha} = -\frac{f_o}{f_e}\left(1 + \frac{f_e}{D}\right)$$

where f_o = focal length of objective
f_e = focal length of eyepiece
D = distance between final image and eyepiece

  (ii) When image at infinity
$$m = -\frac{f_o}{f_e}$$

(b) *Terrestrial telescope*
$$m = \frac{f_o}{f_e}\left(1 + \frac{f_e}{D}\right)$$

(c) *Cassegrain reflecting telescope*
  (i) For finite (image)
$$m = \frac{f_o}{f_e}\left(1 + \frac{f_e}{D}\right)$$

  (ii) For final image (infinity)
$$m = \frac{f_o}{f_e} = \frac{R/2}{f_e}$$

5.4 Eyepiece
(a) (i) *Condition for minimum chromatic aberration or achromatism*
$$d = \frac{f_1 + f_2}{2}$$

where f_1 and f_2 are focal length of field lens and eye lens and d is the distace between them.

  (ii) Condition for minimum spherical aberration $d = f_1 - f_2$

(b) *Huygen's eyepiece*
  (i) Combined focal length $F = \dfrac{3f}{2}$
  (ii) Principal points $L_1H_1 = 3f$, $L_1H_1 = f$
  (iii) Focal points $L_1F_1 = \dfrac{3f}{2}$, $L_2F_2 = \dfrac{f}{2}$

(b) *Ramsden's eyepiece*

(i) $F = \dfrac{3f}{4}$

(ii) $L_1H_1 = \dfrac{f}{2}$, $L_2H_2 = -\dfrac{f}{2}$

(iii) $L_1F_1 = -\dfrac{f}{4}$, $L_2F_2 = \dfrac{f}{4}$

5.5 Spectrometer

The refractive index n of the material of a prism is

$$n = \frac{\sin\dfrac{A+\delta_m}{2}}{\sin\dfrac{A}{2}}$$

where A = refracting angle of prism and δ_m = angle of minimum deviation.

SOLVED NUMERICAL PROBLEMS

PROBLEM 5.1 A converging lens of focal length 6.25 cm is used as a magnifying glass. If the near point of the observer is 25 cm from the eye and the lens is held close to the eye, calculate:

(a) The distance of the object from the lens, and
(b) The angular magnification.

Also, find the angular magnification, when the final image is formed at infinity.

Solution Given: $f = 6.25$ cm, $v = -D = -25$ cm

(a) Using thin lens formula

$$\frac{1}{u} = \frac{1}{v} - \frac{1}{f} = \frac{1}{-25} - \frac{1}{6.25}$$

$$= -\frac{1}{25} - \frac{4}{25} = \frac{-5}{25} = -\frac{1}{5}$$

or $\qquad u = -5$ cm

(b) Angular magnification

$$m = 1 + \frac{D}{f} = 1 + \frac{25}{6.25} = 1 + 4 = 5$$

When the final image is formed at infinity, the angular magnification becomes

$$m = \frac{D}{f} = \frac{25}{6.25} = 4$$

PROBLEM 5.2 A man with normal near point (25 cm) reads a book with small print using a magnifying glass a thin convex lens of focal length 5 cm.
(a) What are the closest distance and farthest distance at which he can read the book when viewing through the magnifying glass?
(b) What are the maximum angular magnification and minimum angular magnification (magnifying power) possible using the above simple microscope?

Solution *Given:* $v = -25$ cm, $f = 5$ cm
(a) For the closest distance:

As
$$\frac{1}{v} - \frac{1}{u} = \frac{1}{f}$$

∴
$$\frac{1}{u} = \frac{1}{v} - \frac{1}{f} = \frac{1}{-25} - \frac{1}{5} = \frac{-1-5}{25} = \frac{-6}{25}$$

or
$$u = -\frac{25}{6} \text{ cm} = -4.2 \text{ cm}$$

This is closest distance at which the man can read the book
For the farthest image: $v = \infty$, $f = 5$ cm, $u = ?$

∴
$$\frac{1}{u} = \frac{1}{v} - \frac{1}{f} = \frac{1}{\infty} - \frac{1}{5} = 0 - \frac{1}{5} = -\frac{1}{5}$$

or
$$u = -5 \text{ cm}$$

This is the farthest distance at which the man can read the book.
(b) Maximum angular magnification

$$= \frac{D}{u_{\min}} = \frac{25}{25/6} = 6$$

Minimum angular magnification

$$= \frac{D}{u_{\max}} = \frac{25}{5} = 5$$

PROBLEM 5.3 A compound microscope with an objective of 1.0 cm focal length and an eyepiece of 2.0 cm focal length has a tube length of 20 cm. Calculate the magnifying power of the microscope, if the final image is formed at the near point of the eye.

Solution *Given:* $f_o = 1.0$ cm, $f_e = 2.0$ cm, $L = 20$ cm and $D = 25$ cm.
When the final image is formed at the near point of the eye, the magnifying power is

$$m = \frac{L}{f_o}\left(1 + \frac{D}{f_e}\right) = \frac{20}{1.0}\left(1 + \frac{25}{2}\right)$$

$$= 20 \times 13.5 = 270$$

PROBLEM 5.4 A compound microscope is used to enlarge an object which is kept at a distance of 0.30 m from its objective, which consists of several convex lenses and has focal length 0.02 m. If a lens of focal length 0.1 m is removed from the objective, find out the distance by which the eyepiece of the microscope must be moved to refocus the image.

Solution For the objective, *Given:* $u_o = -0.03$ m $= -3$ cm, $f_o = 0.02$ m $= 2$ cm

$$\therefore \quad \frac{1}{v_o} = \frac{1}{f_o} + \frac{1}{u_o} = \frac{1}{2} - \frac{1}{3} = \frac{1}{6}$$

or $\quad v_o = +6$ cm

The image is formed at 6 cm behind the objective. Let f'_o be the new focal length of the objective when a lens of focal length 0.1 m or 10 cm is removed from it, then

$$\frac{1}{f'_o} = \frac{1}{f_o} - \frac{1}{10} = \frac{1}{2} - \frac{1}{10} = \frac{2}{5}$$

or $\quad f'_o = \dfrac{+5}{2}$ cm

If v'_o is the new distance of the image formed by the objective, then

$$\frac{1}{v'_o} = \frac{1}{u_o} + \frac{1}{f_o} = -\frac{1}{3} + \frac{2}{5} = \frac{1}{15}$$

or $\quad v'_o = +15$ cm

Distance through which the eyepiece should be moved to refocus image $= v'_o - v_o = 15 - 6 = 9$ cm.

PROBLEM 5.5 Draw a labelled ray diagram of an astronomical telescope forming the image at infinity. An astronomical telescope uses two lenses of powers 10 diopter, and 1 diopter.
(a) State with reason, which lens is preferred as objective and eyepiece.
(b) Calculate the magnifying power of the telescope, if the final image is formed at the near point.
(c) How do the light gathering power and resolving power of a telescope change, if the aperture of the objective lens is doubled.

Solution For ray diagram [Refer to Figure 5.8].

(a) The lens of power 1 diopter should be used as objective because of its larger focal length and the lens of 10 diopter should be used is eyepiece because of its smaller focal length.

(b) Here, $\quad f_o = \dfrac{1}{1D} = 1$ m $= 100$ cm

$$f_e = \frac{1}{10D} = 0.1 \text{ m} = 10 \text{ cm}$$

$$\therefore \quad m = \frac{f_o}{f_e}\left(1 + \frac{f_e}{D}\right) = \frac{100}{10}\left(1 + \frac{10}{25}\right) = 14$$

(c) Light gathering capacity of telescope $\propto$ Area of the objective, i.e.

$$Q \propto \frac{\pi D^2}{4} \quad \text{or} \quad Q \propto D^2$$

when aperture (D) is doubled, light gathering capacity increases 4 times

$$\text{R.P. of a telescope} \propto D$$

When aperture (D) is doubled, resolving power also gets doubled.

PROBLEM 5.6 The focal length of eye lens of Huygen's eyepiece is 4 cm. The sun rays are incident on eyepiece. Calculate the position of image in this situation and find the second principal point of the eye lens.

Solution Given: $f_2 = 4$ cm $= f$, $f_1 = 3f = 12$ cm, and $d = f_1 - f_2 = 8$ cm

Thus, equivalent focal length (F) of Huygen's eyepiece is

$$F = \frac{f_1 f_2}{f_1 + f_2 - d} = \frac{12 \times 4}{12 + 4 - 8} = 6 \text{ cm}$$

∵ The sun is at infinity, therefore its image will be at F_2 (i.e. second focal point)

$$L_2 F_2 = F\left(1 - \frac{d}{f_1}\right) = 6\left(1 - \frac{8}{12}\right) = 2 \text{ cm}$$

Hence, the image will be formed at a distance of 2 cm from the right of eye lens.

The second principal point

$$L_2 H_2 = -\frac{Fd}{f_1} = -\frac{8 \times 6}{12} = -4 \text{ cm}$$

Therefore, the position of the second principal point is 4 cm left to the lens L_2, i.e. eye lens.

PROBLEM 5.7 Light from the sun is falling on Ramsden's eyepiece. Locate the position of the image thus formed and also find the point from which the distance is to be measured. The focal length of each lens of the eyepiece is 3.0 cm.

Solution For Ramsden's eyepiece, Given: $f_1 = f_2 = f = 3$ cm and $d = \frac{2}{3}f = \frac{2}{3} \times 3 = 2$ cm

Then equivalent focal length

$$F = \frac{f_1 f_2}{f_1 + f_2 - d} = \frac{3 \times 3}{3 + 3 - 2} = \frac{9}{4} = 2.25 \text{ cm}$$

∵ The sun (object) is at infinity, so its image will be at F_2, i.e.

$$L_2 F_2 = F\left(1 - \frac{d}{f_1}\right) = \frac{f}{4} = \frac{3}{4} \text{ cm} = 0.75 \text{ cm}$$

Hence, the image will be at a distance of 0.75 cm from the right of the eye lens.

The second principal point

$$L_2 H_2 = \frac{Fd}{f_1} = \frac{2 \times 2.25}{3} = 1.5 \text{ cm}$$

PROBLEM 5.8 The magnifying power of an astronomial telescope is 8 and the distance between two lenses is 54 cm. Obtain the focal lengths of eye lens and objective lens.

Solution

$$m = \frac{f_o}{f_e}$$

Let $f_o = x$, then $f_e = 54 - x$

∴

$$8 = \frac{x}{54 - x} \quad \Rightarrow \quad x = 48 \text{ cm}$$

Then $f_o = x = 48$ cm and $f_e = (54 - 48)$ cm $= 6$ cm.

CONCEPTUAL QUESTIONS

5.1 What is angular magnification of an optical instrument?

Ans The angular magnification of an optical instrument is the ratio of β and α i.e.,

$$m = \frac{\beta}{\alpha} = \frac{\text{Angle subtended by the image at eye}}{\text{Angle subtended by the object when placed at distance of distinct vision}}$$

5.2 What is a simple microscope? What is its magnifying power?

Ans A simple microscope is a short focal length convex lens. The magnified object is placed at or within the focus. The final image formed is virtual, erect and magnified. The magnifying power of a simple microscope is

$$m = 1 + \frac{D}{f}$$

5.3 What is a compound microscope?

Ans A compound microscope consists of two converging lenses. The object to be magnified is just beyond the focus of the objective lens which forms a real, inverted image. The image is either at a focus or within the focus of eye lens. The eye lens, acting as a simple microscope forms final image that is virtual, erect and magnified. In normal adjustment of microscope, the final image is formed at infinity.

5.4 What type of image is obtained in case of a terrestial telescope?

Ans In terrestrial telescope, the final image formed is erect instead of inverted as is there in an astronomical telescope. This is achieved by placing an inverting lens between the intermediate image and the eye lens.

5.5 Why do we call Huygen's eyepiece theoretically perfect and negative?

Ans It is called theoretically perfect as both the chromatic and spherical aberrations are minimized simultaneously by taking the focal length of field lens $3f$ and eye lens f. These two are separated by $2f$. It is called a negative eyepiece as image formed by the objective lies on the negative side i.e., behind the field lens.

5.6 What is an eyepiece?

Ans An eyepiece is primarily a magnifier, designed to give more perfect image than obtained by a single lens.

5.7 Are crosswires used in Huygen's eyepiece? Why?

Ans No, when the measurement of final image is required, the cross-wires should be placed between the field lens and the eye lens. But the crosswires are viewed through the eye lens only, while the distant object is viewed by ray refracted through the lens. Due to this reason, relative lengths of the crosswires and the image are disproportionate. Hence, the crosswires cannot be used in a Huygen's eyepiece, which is a disadvantage. Therefore, this eyepiece is not used in telescopes and other instruments which are used for distance and angle measurements.

5.8 What is the working principle of electron microscope?

Ans The working principle of an electron microscope can be stated as:

(i) a beam of electron exhibits wave nature similar to light rays but of much shorter wavelengths, and

(ii) a beam of electron can be focused by suitable electric and magnetic fields, very much like light rays are focused by glass lenses.

5.9 What are conditions of minimum spherical and chromatic aberrations?

Ans *For spherical aberration*, to be minimum the distance between the two lenses must be equal to the difference of their focal lengths i.e., $d = f_1 - f_2$, where f_1 and f_2 be the focal lengths of field lens and eye lens respectively.

For chromatic aberration, to be minimum the distance between the two lenses must be equal to mean of their focal lengths i.e.,

$$d = \frac{(f_1 + f_2)}{2}$$

5.10 What are the uses of electron microscope?

Ans An electron microscope has a variety of uses in industry, crystal structure analysis, medical industry, analysis of functions of microorganism, bacteria, fungi, etc. due to high order of magnification, great resolving power and increased depth of focus.

EXERCISES

Theoretical Questions

5.1 What is a simple microscope? Give its working principle. Write an expression for its magnifying power when it forms final image at the least distance of the distinct vision and at infinity.

5.2 With the help of a ray diagram, explain the construction and working of a compound microscope. Write an expression for its magnifying power.

5.3 What is a telescope? What are the different types of telescopes commonly used?

5.4 What is an astronomical telescope? Give its construction. With the help of ray diagrams, explain its working when it forms final image at the least distance of distinct vision and at infinity. Deduce expression for magnifying power in each case.

5.5 What is a terrestrial telescope? With the help of a ray diagram, explain its working. Write expression for its magnifying power. State its main drawbacks.

5.6 With the help of a ray diagram, explain the construction and working of a Newtonian reflecting telescope.

5.7 With the help of a labelled diagram, explain the construction and working of a Cassegrain reflecting telescope.

5.8 State some important advantages of a reflecting type telescope over a refracting type telescope.

5.9 Describe construction and working of a Ramsden eyepiece. Find the position and mention the utility of crosswires.

5.10 What is an eyepiece? Deduce the positions of cardinal points of Huygen's eyepiece and indicate them on diagram.

5.11 Mention the advantages of an eyepiece over a simple lens. Draw a neat diagram of Ramsden's eyepiece. Compare Huygen's eyepiece with Ramsden's eyepiece.

5.12 Give some of the advantages of electron microscope over the ordinary optical microscope.

5.13 What is an eyepiece and what are its advantages over a single lens?
[Bundelkhand, 2006]

5.14 What is the basic requirement of having multiple lens eyepiece? Describe the construction and working of Ramsden's eyepiece. [Bundelkhand, 2008; Agra, 2006]

5.15 With the help of neat diagram discuss the working of Huygen's eyepiece. [Agra, 2007]

Numerical Problems

5.1 Magnifying power of a simple microscope A is 1.25 less than that of a simple microscope B. If the power of the lens used in B is 25 D, find the power of lens used in A. Given that the distance of distinct vision is 25 cm. [**Ans.** 20 D]

5.2 A child has a near point at 10 cm. What is the maximum angular magnification, the child can have with a convex lens of focal length 10 cm? [**Ans.** 2]

5.3 The focal lengths of the objective and the eyepiece of a microscope are 2 cm and 5 cm respectively and the distance between them is 20 cm. Find the distance of the object from the objective when the final image seen by the eye is 25 cm from the eyepiece. Also find the magnifying power. [**Ans.** $u_o = 2.3$ cm, $m = 41.5$]

5.4 An astronomical telescope is designed to have a magnifying power of 50 in normal adjustment. If the length of tube is 102 cm, find the power of the objective and the eyepiece.
[**Ans.** 1 D, 50 D]

5.5 A refracting telescope has an objective of focal length 1 m and an eyepiece of focal length 20 cm. The final image of the sun 10 cm in diameter is formed at a distance of 24 cm from the eyepiece. What angle does the sun subtend at the objective? [**Ans.** 0.0455 rad]

5.6 Focal lengths of the lenses of Huygen's eyepiece are 4 cm and 12 cm respectively. Find the position of cardinal points and plot them in a diagram. If an object is situated 6 cm in front of the field lens, find the positions of the image formed by the eyepiece.
[**Ans.** $F = 6$ cm, $L_1H_1 = 12$ cm, $L_2H_2 = -4$ cm,
$L_1F_1 = 6$ cm, $L_2F_2 = 2$ cm, $L_1I = 5$ cm]

5.7 From the condition for 'no chromatic aberration' and 'minimum spherical aberration' of a combination of two separated thin lenses, design a combination of equivalent focal length 6 m.

5.8 A Ramsden's eyepiece is to have an effective focal length of 3 cm. Calculate the focal length of lens component and their distance of separation.
[**Ans.** $F = 3$ cm, $f_1 = f_2 = f = 4$ cm, and $d = 2.67$ cm]

5.9 The equivalent focal length of an Huygen's eyepiece is 5.0 cm. Calculate the focal length of the field lens. [**Ans.** 10 cm]

5.10 Two thin plano-convex lenses of same material in the Huygen's eyepiece are 10 cm apart. Find the focal lengths of two lenses and the equivalent focal length of the eyepiece. [**Ans.** 15 cm, 5 cm, 7.5 cm]

Multiple Choice Questions

5.1 In a compound microscope, the intermediate image is
(a) virtual, erect and magnified
(b) real, erect and magnified
(c) real, inverted and magnified
(d) virtual, erect and reduced

5.2 The focal lengths of the objective and the eyepiece of a compound microscope are 2.0 cm and 3.0 cm respectively. The distance between the objective and the eyepiece is 15.0 cm. The final image formed by the eyepiece is at infinity. The two lenses are thin. The distance in cm of the object and the image produced by the objective, measured from the objective lens, are respectively
(a) 2.4 and 12.0
(b) 2.4 and 15.0
(c) 2.0 and 12.0
(d) 2.0 and 3.0

5.3 An astronomical telescope has an angular magnification of magnitude 5 for distant objects. The separation between the objective and the eyepiece is 36 cm and the final image is formed at infinity. The focal length f_o of the objective and the focal length f_e of the eyepiece are:
(a) $f_o = 45$ cm and $f_e = -9$ cm
(b) $f_o = 50$ cm and $f_e = 10$ cm
(c) $f_o = 7.2$ cm and $f_e = 5$ cm
(d) $f_o = 30$ cm and $f_e = 6$ cm

5.4 A telescope has an objective lens of focal length 200 cm and an eyepiece with focal length 2 cm. If this telescope is used to see a 50 metre tall building at a distance of 2 km, what is the height of the image of the building formed by the objective lens?
(a) 5 cm
(b) 10 cm
(c) 1 cm
(d) 2 cm

5.5 If a scale is fitted in the eyepiece, it should not be
(a) Ramsden eyepiece
(b) Huygen's eyepiece
(c) Gauss's eyepiece
(d) Kellner eyepiece

5.6 In Ramsden eyepiece the condition of achromatism would be obtained if the separation between the eye lens and the field lens is kept (where f is the focal length of eye lens and the field lens)
(a) $\frac{2}{3}f$
(b) $\frac{3}{2}f$
(c) $\frac{1}{2}f$
(d) f

5.7 An electron microscope using an accelerating potential 50 keV has
   (a) worse resolving than an optical microscope
   (b) it is not possible from the data to ray whether an electron microscope or an optical microscope has better resolving power
   (c) better resolving power than the optical microscope
   (d) the same resolving power as the optical microscope

5.8 Resolving power of a microscope depends upon
   (a) the focal length and aperture of the eye lens
   (b) the focal lengths of the objective and the eye lens
   (c) the aperture of the objective and the eye lens
   (d) the wavelength of light illuminating the object

5.9 You are supplied with four convex lenses of focal lengths 100 cm, 25 cm, 3 cm and 2 cm. For astronomical telescope with maximum magnifying power, we will use lenses of focal lengths.
   (a) 100 cm and 2 cm
   (b) 100 cm and 25 cm
   (c) 100 cm and 3 cm
   (d) none of these

5.10 An eyepiece consists of two lenses of focal lengths f_1 and f_2. In order that the combination is achromatic, the lenses are to be separated by a distance
   (a) $\dfrac{(f_1 - f_2)}{2}$
   (b) $\dfrac{(f_1 + f_2)}{2}$
   (c) $(f_1 - f_2)$
   (d) $(f_1 + f_2)$

Answers

5.1 (c)   **5.2** (a)   **5.3** (d)   **5.4** (a)   **5.5** (b)   **5.6** (d)   **5.7** (b)   **5.8** (d)
5.9 (a)   **5.10** (b)

PART II

Vibrations and Waves

Chapter 6   Fundamentals of Vibrations
Chapter 7   Wave Motion

Part II

CHAPTER 6

Fundamentals of Vibrations

"If you sing alone in your, the vibrations return to you as a reaction. But in community singing what you have is not a reaction but a wave of vibrations. They enter into the atmosphere and purify the polluted air."

—Shri Sathya Sai Baba

IN THIS CHAPTER

- Simple Harmonic Motion and Harmonic Oscillator
- Energy of Harmonic Oscillator
- Average Values of Kinetic and Potential Energy
- Example of Harmonic Oscillator: Simple Pendulum
- Damping Force/Damping Motion
- Damped Harmonic Oscillator
- Power Dissipation in Damped Harmonic Oscillator
- Quality Factor (Q) of Damped Harmonic Oscillator
- Forced Harmonic Oscillator
- Resonance
- Amplitude Resonance
- Sharpness of Resonance
- Half Width of Resonance Curve
- Velocity Resonance
- Power Absorption of Forced Harmonic Oscillator
- Quality Factor (Q) of Forced Harmonic Oscillator

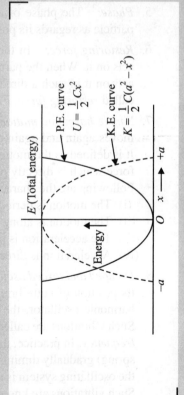

6.1 INTRODUCTION

A motion which repeats itself after equal intervals of time is called *periodic motion* or *harmonic motion*. Examples of periodic motion are the spin of earth, the motion of satellite around a planet, vibrations of atoms in molecules, etc. A body or a particle is said to possess oscillatory or vibratory motion if it moves back and forth repeatedly about the mean position. The pendulum of clock swings back and forth and is said to perform an oscillatory motion. Similarly, the motion of the prongs of a tuning fork, motion of simple pendulum, the vertical oscillations of a loaded spring, to and fro motion of the piston of an engine are the examples of oscillatory motion. Here, we shall define few terms regarding the oscillatory motion.

1. *Periodic time:* The periodic time, T of an oscillatory motion is defined as the time taken for one oscillation.
2. *Frequency:* The frequency (f or v) is defined as the number of oscillations in one second. It is reciprocal of periodic time, *i.e.*, $v = 1/T$ cycles per second.
3. *Displacement:* The distance of the particle in any direction from the equilibrium position at any instant is called the *displacement* of the particle at that instant.
4. *Amplitude:* The maximum displacement or the distance between the equilibrium position and the extreme position is known as *amplitude* (a) of the oscillation.
5. *Phase:* The phase of an oscillatory particle at any instant defines the states of the particle as regards its position and direction of motion at that instant.
6. *Restoring force:* In the equilibrium position of the oscillating particle, no net force acts on it. When the particle is displaced from its equilibrium position, a periodic force acts on it in such a direction as to bring the particle to its equilibrium position. This is called *restoring force*, F.
7. *Simple harmonic motion:* This is special case of periodic motion in which the body moves again and again over the same path about a fixed point (equilibrium position). It is defined as the motion of an oscillatory particle which is acted upon by a restoring force which is directly proportional to the displacement but opposite to it in direction. Following are the characteristics of simple harmonic motion:
   (i) The motion is periodic.
   (ii) The motion is along a straight line about the mean or equilibrium position.
   (iii) The acceleration is proportional to displacement.
   (iv) Acceleration is directed towards the mean or equilibrium position.
8. *Damped and forced oscillations:* When a body capable of vibrations is displaced from its position of equilibrium and then released, it begins to vibrate. In case of an ideal harmonic oscillator, the amplitude of vibration remains constant for an infinite time. Such vibrations are called *free vibrations* and the frequency of vibration is called *natural frequency*. In practice, the vibrations of a freely vibrating body (such as a pendulum or spring) gradually diminish in amplitude and ultimately die away. The reason being that the oscillating system is always subjected to frictional forces arising from air resistance. Such vibrations are known as *damped vibrations*. When a body is made to vibrate by an external periodic force (which may or may not have its frequency equal to the natural

frequency of the body), tries to vibrate with its own natural frequency but ultimately it vibrates with the frequency of applied force. Such vibrations are called *forced vibrations*. The forced vibrations, after the removal of external periodic force, become free and die out in due course of time. When a body is set into oscillations by an external periodic force of the same frequency as the natural frequency of the body, the amplitude of oscillation is very much increased. Such vibrations are called as *sympathetic vibrations* or *resonant vibrations*. This phenomenon is called as *resonance*. Thus, the resonance is a special case of forced oscillation. Radio tuning is an example of resonance. Now, we will discuss simple, damped and forced harmonic oscillators in this chapter.

6.2 SIMPLE HARMONIC MOTION AND HARMONIC OSCILLATOR

Simple harmonic motion is a particular type of periodic motion and is very common in nature (Figure 6.1).

"A particle is said to execute simple harmonic motion (SHM) when it vibrates periodically in such a way that at any instant the restoring force acting on it is proportional to its displacement from a fixed point in its path and is always directed towards that points."

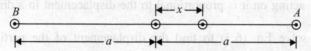

Figure 6.1 SHM of a particle.

A system executing SHM is called *simple harmonic oscillator*. Let us consider a particle of mass m, executing simple harmonic motion. If the displacement of the particle at any instant t be x, then its acceleration will be d^2x/dt^2.

Now, according to the definition of simple harmonic motion, we have restoring force directly proportional to displacement

i.e.,
$$m\frac{d^2x}{dt^2} \propto -x$$

or
$$m\frac{d^2x}{dt^2} = -Cx \tag{6.1}$$

where C is a constant.

Negative sign indicates that the force on the particle is directed opposite to x increasing.

From Eq. (6.1),
$$m\frac{d^2x}{dt^2} + Cx = 0 \tag{6.2}$$

If we put $\omega^2 = \dfrac{C}{m}$, then Eq. (6.2), becomes

$$\frac{d^2x}{dt^2} + \omega^2 x = 0 \tag{6.3}$$

Equations (6.2) and (6.3) are known as the differential equation of motion for a simple harmonic oscillator.

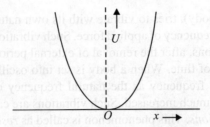

Figure 6.2 Potential energy vs. distance.

From Eq. (6.1) [$F = -Cx$], it is clear that harmonic oscillator is an example of conservative system. At any instant, its potential energy $U = \frac{1}{2}Cx^2$ [i.e., $\int_0^x Cx\,dx$ and total energy $E = \frac{1}{2}mv^2 + \frac{1}{2}Cx^2$ is constant.

If a graph is plotted between U and x, it will be a parabola as shown in Figure 6.2. Thus, for a harmonic oscillator, U-x curve or potential energy curve is a parabola and the force $\left[F = -\dfrac{dU}{dx}\right]$ acting on it is proportional to the displacement in a direction opposite to it.

Now, let us solve Eq. (6.3) to find the displacement of the particle at any instant t. Multiplying Eq. (6.3) by $2\dfrac{dx}{dt}$, we get

$$2\frac{dx}{dt} \times \frac{d^2x}{dt^2} + \omega^2 2x \frac{dx}{dt} = 0 \qquad (6.4)$$

Integrating it, we get

$$\left(\frac{dx}{dt}\right)^2 + \omega^2 x^2 = A \qquad (6.5)$$

where A is a constant of integration.

When the displacement is maximum

i.e.,
$$x = a$$

Velocity
$$\frac{dx}{dt} = 0$$

∴
$$0 + \omega^2 a^2 = A$$

or
$$A = \omega^2 a^2$$

∴
$$\left(\frac{dx}{dt}\right)^2 + \omega^2 x^2 = a^2 \omega^2$$

or
$$\frac{dx}{dt} = \omega\sqrt{(a^2 - x^2)} \qquad (6.6)$$

Equation (6.6) gives the velocity of the particle at any time t.

From Eq. (6.6), we get

$$\frac{dx}{\sqrt{a^2 - x^2}} = \omega dt$$

Integrating further, we get

$$\sin^{-1}\frac{x}{a} = \omega t + \phi$$

or
$$x = a \sin(\omega t + \phi) \tag{6.7}$$

where a is the maximum value of displacement, is called the amplitude of oscillation. ϕ is the constant, is known as *initial phase* or *phase constant*. The term $(\omega t + \phi)$ is phase of vibration.

Equation (6.7) is the solution of the equation of harmonic oscillator and provides us the displacement (x) of the particle at any instant t.

In Eq. (6.7), if we put $\phi = \varphi + \dfrac{\pi}{2}$

then,
$$x = a \cos(\omega t + \varphi) \tag{6.8}$$

Thus, a simple harmonic may be represented by either a sine or cosine function. But the phase constant has different values in two cases.

If the time is measured, when particle is at its mean position, i.e., at $x = 0$, $t = 0$, then we have,

$$\phi = 0$$

Hence, Eq. (6.7) reduces to

$$x = a \sin \omega t \tag{6.9}$$

Now, let us find the physical significance of the constant ω. If the time t in Eq. (6.7) is increased by $(2\pi/\omega)$, the function becomes

$$x = a \sin\left[\omega\left(t + \frac{2\pi}{\omega}\right) + \phi\right]$$

or
$$= a \sin(\omega t + 2\pi + \phi) = a \sin(\omega t + \phi) \tag{6.10}$$

That is, the displacement of the particle is the same after a time $(2\pi/\omega)$. Therefore, $2\pi/\omega$ is the period (T) of motion.

i.e.,
$$T = \frac{2\pi}{\omega} \times 2\pi\sqrt{\frac{m}{C}} \qquad \left(\because \omega^2 = \frac{C}{m}\right) \tag{6.11}$$

The number of vibration per second (ν) is called the *frequency* of the oscillator and is given by

$$\nu = \frac{1}{T} = \frac{\omega}{2\pi} = \frac{1}{2\pi}\sqrt{\frac{C}{m}} \tag{6.12}$$

$$\therefore \qquad \omega = 2\pi\nu = \frac{2\pi}{T} \tag{6.13}$$

The quantity ω is called the *angular frequency*.

6.3 ENERGY OF HARMONIC OSCILLATOR

In general, a harmonic oscillator possesses two types of energy:

(i) Potential energy, which is due to its displacement from the mean position.
(ii) Kinetic energy, which is due to its velocity.

Hence, at any instant the total energy of the oscillator will be sum of these two energies. As the system is conservative (i.e., no dissipative or frictional force are acting), the total mechanical energy $E(=K + U)$ must be conserved.

Let the displacement of the harmonic oscillator at any instant t is given by

$$x = a \sin(\omega t + \phi) \tag{6.14}$$

Its velocity,
$$v = \frac{dx}{dt} = a\omega \cos(\omega t + \phi) = \omega\sqrt{(a^2 - x^2)} \tag{6.15}$$

and acceleration $\frac{d^2x}{dt^2} = -\omega^2 a \sin(\omega t + \phi) = -\omega^2 x$

Then, the force $= -m\omega^2 x = -Cx \quad \left[\because \omega^2 = \frac{C}{m}\right] \tag{6.16}$

If the oscillator is displaced through dx distance, then workdone on the oscillator is

$$dW = C x\, dx$$

If it is displaced from $x = 0$ to $x = x$, then work done

$$W = \int_0^x Cx\, dx = \frac{1}{2}Cx^2 \tag{6.17}$$

This workdone on the oscillator becomes its potential energy U

i.e.,
$$U = \frac{1}{2}Cx^2 = \frac{1}{2}Ca^2 \sin^2(\omega t + \phi) \tag{6.18}$$

Kinetic energy of the oscillator at the displacement x is given as

$$K = \frac{1}{2}mv^2 = \frac{1}{2}m\omega^2(a^2 - x^2) = \frac{1}{2}C(a^2 - x^2) = \frac{1}{2}Ca^2 \cos^2(\omega t + \phi) \tag{6.19}$$

∴ Total energy,

$$E = U + K = \frac{1}{2}Cx^2 + \frac{1}{2}C(a^2 - x^2)$$

$$= \frac{1}{2}Ca^2 \tag{6.20}$$

$$= \frac{1}{2}m\omega^2 a^2 = \frac{1}{2}m\left(\frac{2\pi}{T}\right)^2 a^2 = \frac{2m\pi^2 a^2}{T^2}$$

$$E = 2m\pi^2 a^2 \nu^2 \tag{6.21}$$

Thus, we see that the total mechanical energy is constant, as we expect and has the value equal to $\frac{1}{2}Ca^2$ i.e., the total energy is proportional to the square of the amplitude. Hence, if the system is once oscillated the motion will continue for indefinite period without any decrease in amplitude, provided no damping (frictional) forces are acting on the system.

It is clear from the expression $U = 1/2 Cx^2$ that if we plot a graph between the potential energy U and displacement x, we have a parabola vertex at $x = 0$ (Figure 6.3).

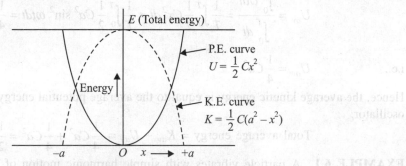

Figure 6.3 P.E. and K.E. with total energy (SHM).

The curve for total energy E (constant) is the horizontal line. The particle cannot go beyond the points, where this line intersects the potential energy curve, because U can never be larger than E. These points are called *turning points* of the motion and correspond to the maximum displacement. At these position (i.e., $x = \pm a$), the total energy of the oscillator is completely potential (i.e., $E = U = \dfrac{1}{2} Ca^2$) but the kinetic energy is zero and so the amplitude of motion is $a = \pm\sqrt{2E/C}$.

At the equilibrium position, the potential energy is zero, but the kinetic energy has maximum value $\left[K = \dfrac{1}{2} m v_{max}^2 = \dfrac{1}{2} Ca^2 = E \right]$ with maximum velocity $\left[v_{max} = \sqrt{\dfrac{2E}{m}} \right]$ while at other intermediate points the energy is partly potential but their sum (total energy) is always $\dfrac{1}{2} Ca^2$.

6.4 AVERAGE VALUES OF KINETIC AND POTENTIAL ENERGY

Average kinetic energy for a period T is given by

$$K_{av} = \dfrac{\int_0^T K\, dt}{\int_0^T dt} = \dfrac{1}{T} \int_0^T \dfrac{1}{2} m v^2\, dt$$

$$= \dfrac{1}{T} \int_0^T \dfrac{1}{2} m a^2 \omega^2 \cos^2 \omega t\, dt \qquad [\because v = a\omega \cos \omega t]$$

$$= ma^2 \dfrac{\omega^2}{2T} \int_0^{T=2\pi/\omega} \dfrac{1}{2}[1 + \cos 2\omega t]\, dt$$

$$= \dfrac{1}{4} \dfrac{m a^2 \omega^2}{2\pi/\omega} \times \dfrac{2\pi}{\omega} = \dfrac{1}{4} m a^2 \omega^2 = \dfrac{1}{4} Ca^2$$

$$K_{av} = \dfrac{1}{4} Ca^2 \qquad (6.22)$$

Also, the average value of potential energy for a period T is given by

$$U_{av} = \frac{\int_0^T U dt}{\int_0^T dt} = \frac{1}{T}\int_0^T \frac{1}{2}Cx^2 dt = \frac{1}{T}\int_0^T \frac{1}{2}Ca^2 \sin^2 \omega t\, dt = \frac{1}{4}Ca^2$$

i.e., $$U_{av} = \frac{1}{4}Ca^2 \tag{6.23}$$

Hence, the average kinetic energy is equal to the average potential energy of a simple harmonic oscillator.

$$\text{Total average energy} = K_{av} + U_{av} = \frac{1}{4}Ca^2 + \frac{1}{4}Ca^2 = \frac{1}{2}Ca^2$$

EXAMPLE 6.1 A particle vibrates with simple harmonic motion of the amplitude 0.09 m and time period 30 s. Calculate maximum velocity.

Solution Simple harmonic motion can be represented by

$$x = a \sin(\omega t + \phi) \tag{i}$$

Given: $a = 0.09$ m, $T = 30$ s

and $$\omega = 2\pi\nu = \frac{2\pi}{T} = \frac{2\pi}{30} \text{ rad/s}$$

Differentiating Eq. (i), we get

$$\text{velocity } (v) = \frac{dx}{dt} = a\omega \cos(\omega t + \phi) \tag{ii}$$

$$v_{max} = a\omega \quad [\because \cos(\omega t + \phi) = 1]$$

$$= \frac{0.09 \times 2\pi}{30} = 0.02 \text{ m/s}$$

6.5 EXAMPLE OF HARMONIC OSCILLATOR: SIMPLE PENDULUM

A simple pendulum consists of a heavy particle (or ideally point mass), suspended by an inextensible weightless and flexible string from a point in a right support, about which the pendulum oscillates without friction.

In practice, it is very difficult to achieve these ideal conditions of a simple pendulum. For experimental purposes, a small heavy metallic spherical bob, suspended by means of a thin and flexible cotton thread, constitutes a practical simple pendulum, closely approximate to the ideal one. The thread at its upper end is held between the two halves of a cork, supported rightly in a clamp. This arrangement provides a rigid support and avoids friction.

Let Figure 6.4 represents a simple pendulum with S as the point of suspension, O being the equilibrium position of the bob. If the bob is drawn to one side and then left free, it begins to oscillate about its mean position O.

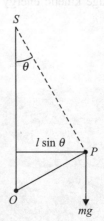

Figure 6.4 A simple pendulum.

Let θ be the angular displacement at any time t. If m be the mass of the bob, then due to its weight the moment of force about the point $S = -mgl \sin \theta$.

If the moment of inertia of the bob (assume it the point mass) about the point S is I, then torque or moment of force $= I \dfrac{d^2\theta}{dt^2}$, where $\dfrac{d^2\theta}{dt^2}$ is the angular acceleration at the displacement θ.

Therefore,
$$I \frac{d^2\theta}{dt^2} = -mgl \sin \theta$$

or
$$ml^2 \frac{d^2\theta}{dt^2} + mgl \sin \theta = 0 \qquad \text{(as } I = ml^2\text{)}$$

or
$$\frac{d^2\theta}{dt^2} + \frac{g}{l} \sin \theta = 0$$

If θ is small, $\sin \theta = \theta$ $\qquad \left(\because \sin \theta = \theta - \dfrac{\theta^3}{3!} + \dfrac{\theta^5}{5!} - \dfrac{\theta^7}{7!} + \cdots \right)$

$\therefore$
$$\frac{d^2\theta}{dt^2} + \frac{g}{l} \theta = 0 \qquad (6.24)$$

Equation (6.24) represents a simple harmonic motion, whose solution is given by
$$\theta = \theta_{\max} \sin(\omega t + \theta) \qquad (6.25)$$

where $\omega = \sqrt{\dfrac{g}{l}}$ and ϕ = phase constant.

Its period, $\qquad T = \dfrac{2\pi}{\omega} = 2\pi \sqrt{\dfrac{l}{g}} \qquad (6.26)$

In the above case, the displacement of the system is angular, hence such a motion is called *angular simple harmonic motion*.

For the motion to be simple harmonic, the value of $\theta_{\max}$ should be small. If we neglect the dissipative force due to friction, the law of conservation of mechanical energy should be hold.

At any instant, the kinetic energy of the bob is given by
$$K = \frac{1}{2} mv^2 = \frac{1}{2} ml^2 \omega^2$$

as the linear velocity $v = l\omega$

$\therefore \qquad K = \dfrac{1}{2} ml^2 \left(\dfrac{d\theta}{dt} \right)^2 \qquad \left(\text{as } \omega = \dfrac{d\theta}{dt} \right) \qquad (6.27)$

Again the potential energy of the bob
$$U = mgh = mg(l - l \cos \theta) \qquad (6.28)$$

The total energy of the bob

$$E = K + U = \frac{1}{2}ml^2\left(\frac{d\theta}{dt}\right)^2 + mgl(1-\cos\theta) \quad (6.29)$$

According to the law of conservation of energy,

$$\frac{d\theta}{dt} = \left[\frac{2}{ml^2}[E - mgl(1-\cos\theta)]\right]^{1/2} \quad (6.30)$$

The energy remains conserved during motion.

EXAMPLE 6.2 A bob of mass m is hung by an elastic string. If the bob stretches the string through 1 cm, calculate the time period of vertical oscillations.

Solution Suppose the original length of the string is L, and the area of cross-section is A.

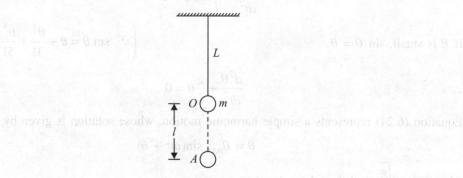

Figure 6.5

$$\text{Stress} = \frac{mg}{A} \text{ and Strain} = \frac{l}{L}$$

∴ Young's modulus

$$Y = \frac{mgL}{Al}$$

or

$$mg = \left(\frac{YA}{L}\right)l \quad \text{(i)}$$

The quantities inside the bracket are constant and can be put equal to k. Again in equilibrium position, the downward force mg is balanced by the tension of the string.

Hence tension,

$$t = mg = kl \quad \text{(ii)}$$

If the bob is pulled downward further through a small distance x, the new tension

$$t' = k(l + x)$$

∴ Resultant upward force $= t' - t = k(l + x) - kl = kx$ (iii)

But from Eq. (iii),
$$k = \frac{mg}{l}$$

$$\therefore \quad t' - t = \frac{mg}{l} x$$

Now, remaining that the force acts towards x decreasing the equation of motion is

$$\text{Force} = \text{mass} \times \text{acceleration}$$

or
$$-\frac{mg}{l} x = m \frac{d^2 x}{dt^2}$$

or
$$\frac{d^2 x}{dt^2} = -\frac{g}{l} x = -\omega^2 x$$

which shows simple harmonic motion, where time period is given by

$$T = \frac{2\pi}{\omega} = \frac{2\pi}{\sqrt{\frac{g}{l}}} = 2\pi \sqrt{\frac{l}{g}}$$

6.6 DAMPING FORCE/DAMPING MOTION

The frictional force, acting on a body opposite to the direction of its motion, is called *damping force*. Such a force reduces the velocity and kinetic energy of the moving body so this is known as *retarding* or *dissipative force*. These forces arise due to the viscosity or friction of the medium and are non-conservative in nature.

Where the velocity are not sufficiently high, the damping force is found to the proportional to the velocity (v) of the particle and may be expressed as

$$F = -\gamma v = -\gamma \frac{dx}{dt} \quad (6.31)$$

where γ is a positive constant and known as *damping coefficient*. If it is the only external force acting on the moving particle, then according to Newton's law, we have equation of motion as

$$m \frac{dv}{dt} = -\gamma v$$

or
$$\frac{dv}{dt} + \frac{\gamma}{m} v = 0 \quad (6.32)$$

It is sometimes useful to define a constant $\tau = \frac{m}{\gamma}$, called the *relaxation time*, so that the equation of motion becomes

$$\frac{dv}{dt} + \frac{v}{\tau} = 0 \quad (6.33)$$

This is very important differential equation and may be written as

$$\frac{dv}{v} = -\frac{dt}{\tau}$$

or
$$\int_{v_0}^{v} \frac{dv}{v} = -\frac{1}{\tau} \int_0^t dt \tag{6.34}$$

where v_0 is the velocity at $t = 0$.

On integrating Eq. (6.34), we have

$$\log v - \log v_0 = -\frac{t}{\tau}$$

or
$$\log_e \frac{v}{v_0} = -\frac{t}{\tau} \text{ so that } v = v_0 e^{-t/\tau} \tag{6.35}$$

Therefore, the velocity decreases exponentially with time (Figure 6.6) and we say that the velocity has been damped with time constant τ.

When $t = \tau, v = \dfrac{v_0}{e}$. Hence, the relaxation time or time constant may be defined as the time in which velocity becomes $1/e$ times the initial velocity. Sometime $\dfrac{1}{\tau}$ or $\dfrac{\gamma}{m}$ is put equal to $2k$, when k is called *damping constant*.

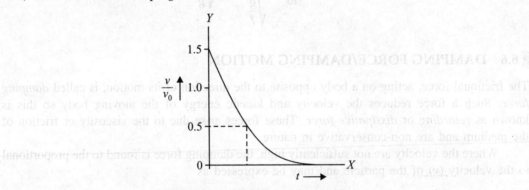

Figure 6.6 Exponential graph of velocity.

The instantaneous kinetic energy of the particle is

$$K = \frac{1}{2} mv^2 = \frac{1}{2} m v_0^2 e^{-2t/\tau}$$

$$\Rightarrow \qquad K = K_0 e^{-2t/\tau} \tag{6.36}$$

Thus, the kinetic energy also decreases exponentially with a relaxation time $\tau/2$.

Putting $v = \dfrac{dx}{dt}$ in Eq. (6.35), and then

$$x = \int_0^t v_0 e^{-t/\tau} dt, \text{ where at } t = 0, x = 0$$

$$\therefore \qquad x = v_0 \tau [e^{-t/\tau}]_0^t = v_0 \tau (1^{-t/\tau}) \tag{6.37}$$

The example, for representing Eq. (6.31), is furnished by an ohmic resistance, if i is the current at any instant, then voltage drop across resistor, R is iR and the induced e.m.f. across inductance

L is $-L\dfrac{dI}{dt}$ (Figure 6.7). There is no external e.m.f., so that

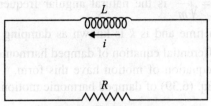

Figure 6.7 R-L circuit.

$$IR = -L\frac{dI}{dt}$$

or
$$L\frac{dI}{dt} + IR = 0$$

which is exactly of the form as Eqs. (6.32) and (6.33) with $\tau = L/R$.

The current will decrease in undriven L-R circuit as

$$I = I_0 e^{-(R/L)t}$$

$\Rightarrow \qquad I = I_0 e^{-t/\tau} \qquad (6.38)$

6.7 DAMPED HARMONIC OSCILLATOR

In discussing simple harmonic oscillator, we did not consider the effect of frictional forces on the oscillator. If no such forces are acting, a pendulum or a tuning fork, if vibrate once, will oscillate indefinitely with constant amplitude, i.e., without any loss of energy.

In actual practice, the amplitude of vibrations/ oscillations gradually decreases to zero as the result of frictional forces, arising due to viscosity of the medium in which the oscillator is moving. The motion of the oscillator is said to be damped by friction and vibrating system is called *damped harmonic oscillator*.

If the damping is to be taken into account, then a harmonic oscillator experiences:

(i) a restoring force $(-Cx)$
(ii) damping force (proportional to the velocity dx/dt but opposes to it, i.e., $-\gamma(dx/dt)$

where x is the displacement of the oscillating system and dx/dt is its velocity at this displacement.

Hence, the equation of motion of the damped harmonic oscillator is

$$m\frac{d^2x}{dt^2} = -\gamma\frac{dx}{dt} - Cx$$

or
$$\frac{d^2x}{dt^2} + \frac{\gamma}{m}\frac{dx}{dt} + \frac{C}{m}x = 0$$

or
$$\frac{d^2x}{dt^2} + 2k\frac{dx}{dt} + \omega_0^2 x = 0 \qquad (6.39)$$

Note: Only for small velocities, the damping force is proportional to the velocity.

where, $2k = \dfrac{\gamma}{m} = \dfrac{1}{\tau}$ and $\omega_0 = \sqrt{\dfrac{C}{m}}$ is the natural angular frequency in absence of damping force. This τ is the relaxation time and is k is known as damping constant.

Equation (6.39) is the differential equation of damped harmonic oscillator and is applicable to any system for which the equation of motion have this form.

Now, we have to solve Eq. (6.39) of damped harmonic motion. Let its possible solution be

$$x = Ae^{\alpha t} \qquad (6.40)$$

Then $\qquad \dfrac{dx}{dt} = A\alpha e^{\alpha t} \qquad$ and $\qquad \dfrac{d^2 x}{dt^2} = A\alpha^2 e^{\alpha t}$

Putting these values in Eq. (6.39)

$$A(\alpha^2 + 2k\alpha + \omega_0^2)e^{\alpha t} = 0, \qquad Ae^{\alpha t} \neq 0$$

$$\alpha^2 + 2k\alpha + \omega_0^2 = 0$$

$$\left[\because \text{ we know for } ax^2 + bx + c = 0, \text{ then } x = \dfrac{-b \pm \sqrt{b^2 - 4ac}}{2a} \right]$$

Then $\qquad \alpha = -k \pm \sqrt{k^2 - \omega_0^2}$

Hence, solutions of Eq. (6.39) are:

$$x_1 = A_1 e^{\left[-k+\sqrt{k^2-\omega_0^2}\right]t} \qquad \text{and} \qquad x_2 = A_2 e^{\left[-k-\sqrt{k^2-\omega_0^2}\right]t}$$

The most general solution of the Eq. (6.39) is

$$x = A_1 e^{\left[-k+\sqrt{k^2-\omega_0^2}\right]t} + A_2 e^{\left[-k-\sqrt{k^2-\omega_0^2}\right]t}$$

$$\Rightarrow \qquad x = e^{-kt}\left[A_1 e^{\left[\sqrt{k^2-\omega_0^2}\right]t} + A_2 e^{-\left[\sqrt{k^2-\omega_0^2}\right]t} \right] \qquad (6.41)$$

where A_1 and A_2 are constants, depending upon the initial conditions of motions.

The quality $\sqrt{k^2 - \omega_0^2}$ is either imaginary or real or zero, it depends on the relative values of k and ω_0.

(i) If $k < \omega_0$; then $\sqrt{k^2 - \omega_0^2}$ is imaginary and it is called *underdamped* case.

(ii) If $k > \omega_0$; then $\sqrt{k^2 - \omega_0^2}$ is real and it is called *overdamped* case, non-oscillatory motion, dead beat or aperiodic.

(iii) If $k = \omega_0$; then $\sqrt{k^2 - \omega_0^2}$ is zero and it is called *critically damped* case.

Underdamped case

If damping is so low that $k < \omega_0$, then $\sqrt{k^2 - \omega_0^2}$

$$= \sqrt{-(\omega_0^2 - k^2)} = i\sqrt{(\omega_0^2 - k^2)} = i\omega$$

where $i = \sqrt{-1}$ and $\sqrt{\omega_0^2 - k^2}$, which is a real quantity.

Then from Eq. (6.41), we have

$$x = e^{-kt}[A_1 e^{i\omega t} + A_2 e^{-i\omega t}]$$
$$= e^{-kt}[A_1(\cos \omega t + i \sin \omega t) + A_2(\cos \omega t - i \sin \omega t)]$$
$$= e^{-kt}[(A_1 + A_2)\cos \omega t + i(A_1 - A_2)\sin \omega t]$$

As x is a real quantity, $(A_1 + A_2)$ and $i(A_1 - A_2)$ must be real quantities. Clearly A_1 and A_2 are complex quantities.

If
$$A_1 + A_2 = a_0 \sin \phi$$
and
$$i(A_1 - A_2) = a_0 \cos \phi$$

$$x = a_0 e^{-kt} \sin(\omega t + \phi) \tag{6.42}$$

Equation (6.42), which represents a damped harmonic motion. This motion is *oscillatory* or *ballistic*, whose periodic time is given by

$$T = \frac{2\pi}{\omega} = \frac{2\pi}{(\omega_0^2 - k^2)^{1/2}} \tag{6.43}$$

Thus, the effect of damping is to increase the periodic time. If $k = 0$, i.e., no damping, then $T = \frac{2\pi}{\omega_0}$. But in actual case, the effect of damping on periodic time is usually negligible except few extreme cases.

The amplitude of motion [Eq. (6.42)] is given by

$$a = a_0 e^{-kt} = a_0 e^{-t/2\tau} \tag{6.44}$$

(at $x = a$, $\sin(\omega t + \phi) = 1$ or $\omega t + \phi = \pi/2$)

where a_0 = amplitude in the absence of damping.

In the presence of damping, the amplitude decreases exponential with time. In Figure 6.8, time displacement curve is shown for damped harmonic motion. As the maximum value of $\sin(\omega t + \phi)$ is alternately +1 and –1, obviously the time displacement curve of the vibrating body lies entirely between the curves $a = a_0 e^{-kt}$ and $a = -a_0 e^{-kt}$, shown by dotted lines in Figure 6.8.

Decrement and logarithmic decrement: Logarithmic decrement measures the rate at which the amplitude dies away. The amplitude of damped harmonic oscillator is given by $a = a_0 e^{-kt}$ at $t = 0$, $a = a_0$.

Let $a_1, a_2, a_3, a_4, \ldots$ be the amplitudes at time $t = T, 2T, 3T, 4T, \ldots$ respectively, where T is the period of oscillation, then

$$a_1 = a_0 e^{-kT}, \; a_2 = a_0 e^{-2kT}, \; a_3 = a_0 e^{-3kT}, \ldots$$

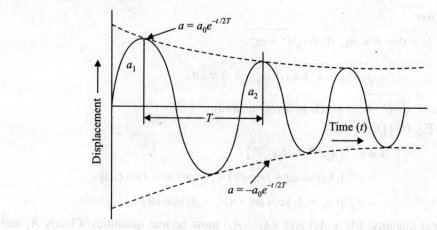

Figure 6.8 Time-displacement curve.

Now, $$\frac{a_0}{a_1} = \frac{a_1}{a_2} = \frac{a_2}{a_3} = e^{kT} = e^{\lambda} \quad \text{[where } kT = \lambda\text{]} \tag{6.45}$$

Taking logarithms of Eq. (6.45), we get

$$\lambda = \log_e \frac{a_0}{a_1} = \log_e \frac{a_1}{a_2} = \log_e \frac{a_2}{a_3} = \cdots \text{ (logarithmic decrement)} \tag{6.46}$$

Thus, *logarithmic decrement* is defined as the natural logarithmic of the ratio between two successive maximum amplitudes, which are separated by one period.

Overdamped case

If the damping is so high that $k > \omega_0$, then $\sqrt{k^2 - \omega_0^2} = \beta$ (say) is a real quantity and then from Eq. (6.41), we have

$$x = (A_1 e^{-(k-\beta)t} + A_2 e^{-(k+\beta)t}) \tag{6.47}$$

As $k > \beta$, both quantities of right hand side decrease exponentially with time and motion is non-oscillatory. Such a motion is called *dead-beat* or *aperiodic*. The time displacement curve for overdamped case is shown in Figure 6.9.

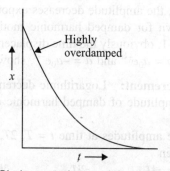

Figure 6.9 Displacement time graph for highly damped case.

Application: Dead-beat galvanometer.

Critically damped case

If $k = \omega_0$, then Eq. (6.41), becomes

$$x = (A_1 + A_2)e^{-kt} = Ce^{-kt}, \quad \text{where } C = A_1 + A_2$$

In this equation, there is only one constant, hence it does not provide us the solution of differential equation [Eq. (6.39)] of second order. Now, suppose $(k^2 - \omega_0^2)^{1/2} = h$, which is small quantity. Hence, from Eq. (6.41), we have

$$x = e^{-kt}[A_1 e^{ht} + A_2 e^{-ht}]$$

$$= e^{-kt}\left[A_1\left(1 + ht + \frac{h^2 t^2}{2!} + \cdots\right) + A_2\left(1 - ht + \frac{h^2 t^2}{2!} + \cdots\right)\right]$$

Neglecting the small terms, containing h^2 and higher power of h, we get

$$x = e^{-kt}[(A_1 + A_2) + h(A_1 - A_2)t]$$

$$= e^{-kt}[P + Qt] \tag{6.48}$$

where $P = (A_1 + A_2)$ and $Q = h(A_1 - A_2)$.

Also
$$\frac{dx}{dt} = e^{-kt}Q + e^{-kt}(-k)(P + Qt)$$

If initially at $t = 0$, the displacement of the particle is $x = x_0$ and there the velocity is v_0, then

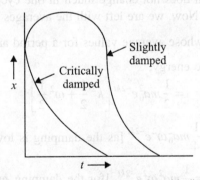

Figure 6.10 Displacement time graph for critically damped.

$$x_0 = P \quad \text{and} \quad v_0 = Q - kP = Q - kx_0$$

or
$$Q = v_0 + kx_0$$

∴
$$x = [x_0 + (v_0 + kx_0)t]e^{-kt} \tag{6.49}$$

This represents that initially the displacement increases due to the factor $[x_0 + (v_0 + kx_0)t]$ but as time elapses, the exponential term becomes relatively more important and the displacement returns continuously from the maximum value to zero and the oscillatory motion does not just occur. Such a motion is called *critically damped* or *just aperiodic* (Figure 6.10).

6.8 POWER DISSIPATION IN DAMPED HARMONIC OSCILLATOR

If a particle oscillates in a medium then due to the viscosity of the medium, the damping forces act on the particle in a direction opposite to its movement. In this process work is done by the particle in overcoming the resistance forces. Consequently, the mechanical energy of the vibrating particle continuously decreases so that the amplitude of oscillation becomes less and less. Here we want to find out a relation for power dissipation (i.e., rate of dissipation of energy).

At any instant t, the displacement of a damped harmonic oscillator is given by

$$x = a_0 e^{-kt} \sin(\omega t + \phi) \qquad (6.50)$$

∴ Velocity of the particle $\dfrac{dx}{dt} = a_0 [e^{-kt} \omega \cos(\omega t + \phi) - k e^{-kt} \sin(\omega t + \phi)]$

$$= a_0 e^{-kt} [\omega \cos(\omega t + \phi) - k \sin(\omega t + \phi)] \qquad (6.51)$$

∴ Kinetic energy of vibration

$$K = \frac{1}{2} m a_0^2 e^{-2kt} [k^2 \sin^2(\omega t + \phi) + \omega^2 \cos^2(\omega t + \phi) - 2k\omega \sin(\omega t + \phi)\cos(\omega t + \phi)] \qquad (6.52)$$

Potential energy $\quad U = \dfrac{1}{2} C x^2 = \dfrac{1}{2} m \omega^2 x^2 \qquad \left[\because U = \int_0^x C x\, dx \right]$

$$= \frac{1}{2} m a_0^2 e^{-2kt} \{\omega^2 \sin^2(\omega t + \phi)\} \qquad (6.53)$$

The average total energy for a period will be sum of time average of kinetic energy and potential energy.

If amplitude of oscillation does not change much in one cycle of motion, then the factor e^{-2kt} may be taken as constant. Now, we are left with the averages of $\sin^2(\omega t + \phi)$, $\cos^2(\omega t + \phi)$ and $2 \sin(\omega t + \phi) \cos(\omega t + \phi)$, whose average values for a period are $\dfrac{1}{2}$, $\dfrac{1}{2}$ and 0 respectively.

Hence the average kinetic energy,

$$= \frac{1}{2} m a_0^2 e^{-2kt} \left[k^2 \frac{1}{2} + \omega^2 \frac{1}{2} \right]$$

⇒ Average kinetic energy $= \dfrac{1}{4} m a_0^2 \omega^2 e^{-2kt}$ [as the damping is low, so $k^2 \ll \omega^2$]

and average potential energy $= \dfrac{1}{4} m a_0^2 \omega^2 e^{-2kt}$ [for the damping $\omega = \omega_0$]

∴ Average total energy,

$$<E> = \frac{1}{2} m a_0^2 \omega^2 e^{-2kt}$$

$$= E_0 e^{-2kt} = E_0 e^{-t/\tau} \qquad \left[\because E_0 = \frac{1}{2} m a_0^2 \omega^2 \right] \qquad (6.54)$$

Average power dissipated,

$$P = -\frac{dE}{dt} = k m a_0^2 \omega^2 e^{-2kt}$$

$$\Rightarrow \quad P = 2kE = \frac{E}{\tau} \tag{6.55}$$

Average rate of doing work $= <Fv^2> = <r\left(\frac{dx}{dt}\right)^2> = \frac{2kma_0^2 e^{-2kt}\omega_0^2}{2} = 2kE = \frac{E}{\tau}$

This loss of energy generally appears in the form of heat in oscillating system.

6.9 QUALITY FACTOR (Q) OF DAMPED HARMONIC OSCILLATOR

The quality factor is defined as 2π times the ratio of energy stored in the system to the energy lost per unit time.

$$Q = 2\pi \times \frac{\text{Energy stored in system}}{\text{Energy lost per unit time}} = 2\pi \times \frac{E}{PT} \tag{6.56}$$

where, P is the power dissipated and T is known as periodic time.

$$\therefore \quad Q = 2\pi \frac{E}{\left(\frac{E}{\tau}\right)T} = \frac{2\pi\tau}{T} = \omega\tau \tag{6.57}$$

where $\omega = \frac{2\pi}{T}$ = angular frequency.

So, it is clear that higher is the value of Q, higher would be the value of relaxation time τ, i.e., lower damping.

EXAMPLE 6.3 A simple pendulum has a period of 1s and an amplitude of 100. After 10 complete oscillations, its amplitude has been reduced to 50. What is the relaxation time of the pendulum? Calculate the quality factor.

Solution The amplitude at a time t is $a(t) = a_0 e^{-2kt} = a_0 e^{-t/\tau}$

Where, a_0 is the initial amplitude, k is the damping coefficient and $\tau = \frac{1}{2k}$ is called relaxation time.

Thus, $5^0 = 10^0 e^{-10/\tau}$

$$\frac{10}{\tau} = \log_e 2$$

$$\tau = \frac{10}{\log_e 2} = \frac{10}{2.303 \times 0.3010} = 14.49 \text{ s}$$

Thus, $Q = \omega\tau = \frac{2\pi\tau}{T} = \frac{2\pi}{1} \times 14.49 = 45.55$

6.10 FORCED HARMONIC OSCILLATOR

If an external periodic force is applied on a damped harmonic oscillator, then the oscillating system is called driver or *forced harmonic oscillator* and its oscillations are called *forced* (or driven) *vibrations*.

Let the periodic force be represented by $F_P = F_0 \sin pt$, where F_0 be the maximum value of the force frequency $p/2\pi$. The forces acted on the particle are:

(i) a restoring force, $-Cx$
(ii) a frictional force (damping force)
(iii) external periodic force, $F = F_0 \sin pt$

According to Newton's second law

$$F = ma = m\frac{d^2x}{dt^2} \qquad (6.58)$$

Forces acted on the particle must be equal to Eq. (6.58)

$$m\frac{d^2x}{dt^2} = -\gamma\frac{dx}{dt} - Cx + F_P \qquad (6.59)$$

or

$$\frac{d^2x}{dt^2} + \frac{\gamma}{m}\frac{dx}{dt} + \frac{C}{m}x = \frac{F_0}{m}\sin pt$$

or

$$\frac{d^2x}{dt^2} + 2k\frac{dx}{dt} + \omega_0^2 x = f_0 \sin pt \qquad (6.60)$$

where $2k = \dfrac{\gamma}{m}$ and $\omega_0 = \sqrt{\dfrac{C}{m}}$ is the natural frequency in absence of damping and driven forces, k is the damping constant and C is the force constant

$$f_0 = \frac{F_0}{m}$$

The natural frequency of the oscillating system may be different to the frequency ($p/2\pi$) of the applied force.

When a periodic force acts on a body, it delivers periodic impulses to the body so that the loss of energy in doing work against the dissipative forces is recovered. The result of this is that the body is thrown into continuous vibrations. In the initial stages, the body tends to vibrate with its natural frequency, while the impressed force tries to impose it own frequency upon it. But soon the free vibrations of the body die out and ultimately the body vibrates with a constant amplitude and with the same frequency as that of the impressed force.

Such vibrations of constant amplitude, performed by a body under the influence of an impressed periodic force, with a frequency equal to that of the force, are known as forced vibrations and the oscillating system is itself called driver or forced harmonic oscillator.

The impressed periodic force is called driver and the body, executing forced oscillations is called the driven. Now, let us suppose that steady state solution of Eq. (6.60) is

$$x = A \sin(pt - \theta) \qquad (6.61)$$

because the amplitude (A) of the forced oscillations remains constant and the frequency of vibrations is equal to the ($p/2\pi$) of the force. Here θ represents the phase difference between the force and the resultant displacement of the system.

Now, from Eq. (6.61), we have

$$\frac{dx}{dt} = pA\cos(pt - \theta)$$

$$\frac{d^2x}{dt^2} = -p^2 A\sin(pt - \theta)$$

Substituting these values in Eq. (6.60), we get

$$-p^2 - p^2 A \sin(pt - \theta) + 2kpA \cos(pt - \theta) + \omega_0^2 A \sin(pt - \theta) = f_0 \sin(pt - \theta + \theta)$$
[$f_0 \sin pt$ may be written as $\sin(pt - \theta + \theta)$]

or $\quad -p^2 - p^2 A \sin(pt - \theta) + 2kpA \cos(pt - \theta) + \omega_0^2 A \sin(pt - \theta)$
$$= f_0 \sin(pt - \theta)\cos\theta + f_0 \cos(pt - \theta)\sin\theta$$

or $\quad A(\omega_0^2 - p^2) \sin(pt - \theta) + 2kpA \cos(pt - \theta)$
$$= f_0 \cos\theta \sin(pt - \theta) + f_0 \sin\theta \cos(pt - \theta) \quad (6.62)$$

If this relation holds for all values of t, the coefficient of $\sin(pt - \theta)$ and $\cos(pt - \theta)$ terms on both sides of this equation must be equal, i.e., comparing the coefficients of $\sin(pt - \theta)$ and $\cos(pt - \theta)$ on both sides, we have

$$A(\omega_0^2 - p^2) = f_0 \cos\theta \quad (6.63)$$

$$2kpA = f_0 \sin\theta \quad (6.64)$$

Squaring and adding Eqs. (6.63) and (6.64), we get

$$A^2(\omega_0^2 - p^2)^2 + 4k^2 p^2 A^2 = f_0^2$$

or $\quad A^2[(\omega_0^2 - p^2)^2 + 4k^2 p^2] = f_0^2$

or $$A = \frac{f_0}{[(\omega_0^2 - p^2)^2 + 4k^2 p^2]^{1/2}} \quad (6.65)$$

If we divide Eq. (6.64) by Eq. (6.63), we get

$$\tan\theta = \frac{2kp}{(\omega_0^2 - p^2)}$$

$$\theta = \tan^{-1}\left(\frac{2kp}{\omega_0^2 - p^2}\right) \quad (6.66)$$

Substituting the value of θ and A in Eq. (6.61), we get

$$x = \frac{f_0}{\sqrt{(\omega_0^2 - p^2)^2 + 4k^2 p^2}} \sin\left(pt - \tan^{-1}\frac{2kp}{\omega_0^2 - p^2}\right) \quad (6.67)$$

Here, we will assume that the damping is low. Now, we will consider the different cases, when $\omega_0 \gg p$, $\omega_0 = p$ and $\omega_0 \ll p$.

Case 1: For low driving frequency, when $p \ll \omega_0$

In this case, the amplitude of vibration is given by

$$A = \frac{f_0}{[(\omega_0^2 - p^2)^2 + 4k^2 p^2]^{1/2}} \tag{6.68}$$

$$\cong \frac{f}{\omega_0^2} = \text{Constant}$$

and $\qquad \theta = \tan^{-1}\left(\frac{2kp}{\omega_0^2 - p^2}\right) \cong \tan^{-1}(0) \cong 0 \qquad (6.69)$

This shows that the amplitude of vibration is independent of frequency of force. The amplitude depends on the magnitude of the applied force and force constant. The force and displacement are always in phase.

Case 2: When $p = \omega_0$ i.e., the frequency of the force is equal to the frequency of the body. In this case the amplitude of vibration is given by

$$A = \frac{f_0}{2k\omega_0} = \frac{f_0 \tau}{\omega_0}$$

also $\qquad \theta = \tan^{-1}\left(\frac{2kp}{0}\right) \cong \tan^{-1}(\infty) = \frac{\pi}{2} \qquad (6.70)$

Thus, the amplitude of vibration is governed by damping and for small damping forces, the amplitude of vibration will be quite large. The displacement lags behind the force by a phase $\pi/2$.

Case 3: For high driving frequency, when $p \gg \omega_0$
i.e., the frequency of driving force is greater than the natural frequency ω_0 of the body.

In this case, $\qquad A = \dfrac{f_0}{[p^4 + 4k^2 p^2]^{1/2}} = \dfrac{f_0}{p^2} = \dfrac{F_0}{mp^2}$

and $\qquad \theta = \tan^{-1}\left(\dfrac{2kp}{\omega_0^2 - p^2}\right) \qquad (6.71)$

$$= \tan^{-1}\left(-\frac{2k}{p}\right) \approx \tan^{-1}(-0) = \pi$$

6.11 RESONANCE

If we bring a vibrating fork near another stationary tuning fork of the same natural frequency as that of vibrating tuning fork, we find that stationary tuning fork also starts vibrating. This phenomenon is known as *resonance*.

Let us consider three springs S_1, S_2 and S_3 suspended from a flexible rod *AB* such that S_1 and S_2 are identical in all respect and carry equal masses at the ends, while S_3 has different spring constant and carries different mass. Now, if S_1 spring is set in vibration by pulling down the attached mass and let go, we find that springs S_2 and S_3 also start vibrating. The vibrations in S_3 die out quickly while the vibrations set in spring (S_2 is identical to S_1) keeps on increasing in its amplitude till it is very nearly equal to the amplitude of the spring S_1. The vibrations of spring S_2 have the same frequency as that of S_1 and are called resonant vibrations and this phenomenon is called *resonance*.

Definition

The phenomenon of making a body vibrate with its natural frequency under the influence of another vibrating body with the same frequency is called *resonance*.

Examples of resonance

1. Tuning of a radio or transistor, when natural frequency is so adjusted, by moving the tuning knob of the receiver set that it equals the frequency of radio waves, the resonance takes place and the incoming sound waves can be listened after being amplified.
2. Musical instrument can be made to vibrate by bringing them in contact with vibrations which have the frequency equal to the natural frequency of the instruments.
3. When an atom or a molecule is struck by electromagnetic radiation having frequency exactly equal to the natural frequency of the atom or molecule, the radiation is absorbed more rapidly than at other frequencies. This is known as 'resonance absorption of radiation'. The phenomenon is known as 'optical resonance'.

6.12 AMPLITUDE RESONANCE

The amplitude of forced oscillations varies with the frequency of applied force and becomes maximum at a particular frequency. This phenomenon is known as *amplitude resonance*.

Condition of amplitude resonance

In case of forced vibrations, we have

$$A = \frac{f_0}{[(\omega_0^2 - p^2)^2 + 4k^2 p^2]^{1/2}} \tag{6.72}$$

and

$$\theta = \tan^{-1}\left[\frac{2kp}{\omega_0^2 - p^2}\right] \tag{6.73}$$

Equation (6.72) shows that the amplitude varies with the frequency of the force p. For a particular value of p, the amplitude becomes maximum. The phenomenon is known as amplitude resonance.

The amplitude is maximum when $[(\omega_0^2 - p^2)^2 + 4k^2 p^2]^{1/2}$ is minimum.

or

$$\frac{d}{dp}[(\omega_0^2 - p^2)^2 + 4k^2 p^2] = 0$$

or

$$2(\omega_0^2 - p^2)(-2p) + 4k^2(2p) = 0$$

or
$$\omega_0^2 - p^2 = 2k^2$$
or
$$p = \sqrt{\omega_0^2 - 2k^2} \qquad (6.74)$$

Thus, the amplitude is maximum when the frequency ($p/2\pi$) of the impressed force becomes ($\sqrt{\omega_0^2 - 2k^2}/2\pi$). This is the resonant frequency. This gives frequency of the system both in presence of damping i.e., ($\sqrt{\omega_0^2 - 2k^2}/2\pi$) and in the absence of damping i.e., ($\omega_0/2\pi$).

If the damping is small then it can be neglected and the condition amplitude reduced to
$$p = \omega_0$$

Putting the value of p from Eq. (6.74) in Eq. (6.72), we get

$$A_{max} = \frac{f}{\{(\omega_0^2 - \omega_0^2 - 2k^2)^2 + 4k^2(\omega_0^2 - 2k^2)\}^{1/2}}$$

$$= \frac{f}{\sqrt{4k^2\omega_0^2 - 4k^4}} = \frac{f}{2k\sqrt{\omega_0^2 - k^2}} \qquad (6.75)$$

and for low damping, it reduces to

$$A_{max} \cong \frac{f}{2kp}$$

showing that $A_{max} \to \infty$ as $p \to 0$.

Figure 6.11 shows the variation of amplitude with forcing frequency at different amounts of damping.

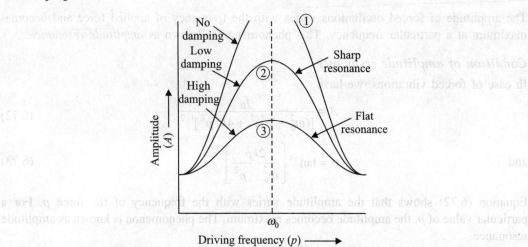

Figure 6.11 Amplitude resonance.

Curve (1) shows the amplitude when there is no damping i.e., $k = 0$. In this case the amplitude becomes infinite at $p = \omega_0$. This case is never attained in practice due to frictional resistance, as slight damping is always present. Curves (2) and (3) show the effect of damping on the amplitude. It is observed that the peak of the curve moves towards the left. It is also observed that the value A, which is different for different value of k (damping), diminishing as the value of p increases. For smaller value of k, the fall in the curve about $\omega_0 = p$ is steeper than for large values. This shows that the smaller is the value of damping, greater is the departure of amplitude of forced vibration from the maximum value and vice versa.

6.13 SHARPNESS OF RESONANCE

At resonance, the amplitude of the oscillating system becomes maximum. It decreases from the maximum value with the change (decrease or increase) of the frequency of the impressed force. Figure 6.11, for different values of damping, curves have been drawn between the driven frequency and the amplitude.

The term sharpness of resonance refers to the rate of fall in amplitude with the change of forcing frequency on each side of the resonance frequency.

When the damping is low, the response (amplitude) falls of very rapidly on other side of resonant frequency and we say that the response is sharp. Thus, the resonance is flat or sharp according to the damping for the oscillating system is large or small. Familiar examples of a flat and sharp resonance are the resonance of an air column and a sonometer wire with a tuning fork respectively. Due to its large damping, the air column responds with tuning fork over a fairly wide range in the neighbourhood of resonance. Thus in this case, it is actually difficult to get an exact point of resonance and resonance is said to be flat. But in the case of sonometer wire, the damping is small so that the resonance is sharp and hence to have the resonant points, the length of the wire is adjusted very precisely.

6.14 HALF WIDTH OF RESONANCE CURVE

If p_h is the value of angular frequency when the amplitude falls to half the value of resonance, then the change in Δp is called half width of the resonance curve i.e., half width

$$\Delta p = |\, p_h - p_r \,| \tag{6.76}$$

At resonance $A_{\max} = \dfrac{f_0}{2k\sqrt{\omega_0^2 - k^2}} = \dfrac{f_0}{\sqrt{4k^2\omega_0^2 - 4k^4}}$ = [From Eq. (6.75)] $\tag{6.77}$

At the frequency p_h, the amplitude is $\dfrac{A_{\max}}{2}$.

$\therefore$
$$\dfrac{A_{\max}}{2} = \dfrac{f_0}{\sqrt{(\omega_0^2 - 2k^2 - p_h^2)^2 + 4k^2\omega_0^2 - 4k^4}} \tag{6.78}$$

From Eqs. (6.77) and (6.78), we have

$$(\omega_0^2 - 2k^2 - p_h^2)^2 + 4k^2\omega_0^2 - 4k^4 = 4(4k^2\omega_0^2 - 4k^4)$$

But $\qquad p_r^2 = \omega_0^2 - 2k^2$ [From Eq. (6.77)]

$\Rightarrow \qquad (p_r^2 - p_h^2)^2 = 3(4k^2\omega_0^2 - 4k^4) = 3(4k^2 p_r^2 + 4k^4) \tag{6.79}$

If damping is small, $4k^4$ is negligible and then we have

$$(p_r^2 - p_h^2) = \pm\sqrt{3}(2kp_r)$$

or

$$p_h^2 = p_r^2 \mp \sqrt{3}(2kp_r)$$

or

$$p_h = p_r\left(1 \mp \sqrt{3}\frac{2k}{p_r}\right)^{1/2} = p_r\left(1 \mp \frac{\sqrt{3}k}{p_r}\right) = p_r \mp \sqrt{3}k$$

∴ Half width $\Delta p = |p_h - p_r| = \sqrt{3}k$ (6.80)

6.15 VELOCITY RESONANCE

The velocity of the driven oscillator is

$$v = \frac{dx}{dt} = \frac{f_0 p}{\sqrt{(\omega_0^2 - p^2)^2 + 4k^2p^2}} \cos(pt - \theta) = v_0 \sin\left(pt - \theta + \frac{\pi}{2}\right) \quad (6.81)$$

where,

$$v_0 = \frac{f_0 p}{\sqrt{(\omega_0^2 - p^2)^2 + 4k^2p^2}} = \frac{f_0}{\sqrt{\frac{\omega_0^4 + p^4 - 2\omega_0^2 p^2}{p^2} + 4k^2}}$$

$$= \frac{f_0}{\sqrt{\frac{\omega_0^4}{p^2} + p^2 - 2\omega_0^2 + 4k^2}} \text{ is velocity amplitude.} \quad (6.82)$$

and

$$\theta = \tan^{-1}\left(\frac{2kp}{\omega_0^2 - p^2}\right) \quad (6.83)$$

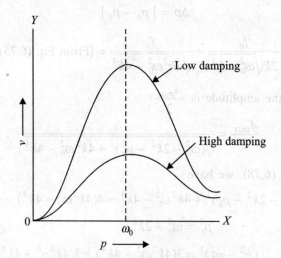

Figure 6.12 Velocity resonance curve.

We see that the velocity amplitude v_0 varies with driven frequency p (Figure 6.12). When $p = 0$, $v_0 = 0$ and $p = \omega_0$, v_0 attains the maximum value. Hence, at the frequency $p = \omega_0 = \sqrt{\dfrac{C}{m}}$ of the impressed force, the velocity has the maximum value and we call it *velocity resonance*. At velocity resonance,

$$\theta = \tan^{-1}\left(\frac{2kp}{0}\right) = \frac{\pi}{2}$$

Velocity phase constant $\quad \theta = -\theta + \dfrac{\pi}{2} = -\dfrac{\pi}{2} + \dfrac{\pi}{2} = 0$

This means that the velocity of the oscillator is in phase with applied force. This is most favourable situation for transfer of energy from the applied force to the oscillator, because the rate of work done on the oscillator by the impressed force is Fv. Which is always positive for F and v is phase.

When $p \leq \omega_0$ the velocity is smaller than at $p = \omega_0$.

6.16 POWER ABSORPTION OF FORCED HARMONIC OSCILLATOR

When an oscillator executes vibration in presence of damping force, it loses energy in doing work against these forces.

This loss of energy is supplied by the periodic impressed force so as to continue the oscillations. In a small time interval dt the energy supplied by the driving force $F_p(= F_0 \sin pt)$ will be equal to the work done on the oscillating system by this force,

i.e., $\qquad dE = F_p dx = F_p \dfrac{dx}{dt} dt \qquad \left[\because x = \dfrac{f_0}{\sqrt{[(\omega_0^2 - p^2)^2 + 4k^2 p^2]}} \sin(pt - \theta)\right]$

or $\qquad \dfrac{dE}{dt} = F_p \dfrac{dx}{dt} \qquad \qquad (6.84)$

where dx is the displacement in the time dt

$\therefore \qquad F_0 = m f_0 \sin pt \qquad \left[\because f_0 = \dfrac{F}{m} \sin pt\right]$

and $\qquad \dfrac{dx}{dt} = \dfrac{p f_0 \cos(pt - \theta)}{\sqrt{[(\omega_0^2 - p^2)^2 + 4k^2 p^2]}}$

Hence at any time, the energy absorbed by the oscillating system is

$$P = \frac{dE}{dt} = F_0 \frac{dx}{dt} = m f_0 \sin pt \frac{p f_0 \cos(pt - \theta)}{[(\omega_0^2 - p^2)^2 + 4k^2 p^2]^{1/2}} \qquad (6.85)$$

Hence, the average power absorbed is

$$P_{av} = \left(F_0 \frac{dx}{dt}\right)_{av} = \frac{m f_0^2}{[(\omega_0^2 - p^2)^2 + 4k^2 p^2]^{1/2}} [\sin pt \cos(pt - \theta)_{av}] \qquad (6.86)$$

Now, the average of $\sin pt \cos(pt - \theta)$ for one period T or $\frac{2\pi}{\omega}$

$$= \frac{1}{T}\int_0^T \sin pt \cos(pt-\theta)dt = \frac{1}{T}\int_0^T \frac{1}{2}\{\sin(2pt-\theta)-\sin\theta\}dt = \frac{\sin\theta}{2}$$

or

$$\frac{\sin\theta}{2} = \frac{1}{2}\frac{2kp}{\sqrt{(\omega_0^2-p^2)^2+4k^2p^2}}$$

$$\therefore \qquad P_{av} = \frac{1}{2}mf_0^2 \frac{2kp^2}{(\omega_0^2-p^2)^2+4k^2p^2} \qquad (6.87)$$

Replacing $\dfrac{f_0^2 p^2}{(\omega_0^2-p^2)^2+4k^2p^2}$ by v_0^2.

$$P_{av} = mkv_0^2 = \frac{mv_0^2}{2\tau} = \frac{\gamma v_0^2}{2} \qquad (6.88)$$

From Eq. (6.87), we see that when $p = \omega_0$, the absorbed average power is maximum and is equal to $\dfrac{1}{4}mf_0^2/k$ or $\dfrac{1}{2}mf_0^2\tau$. Thus, the power absorbed is maximum at the frequency of velocity resonance and not at the frequency of amplitude resonance as shown in Figure 6.13.

When steady state is attained, the average power dissipated should be same as the average power supplied by the driving force.

Instantaneous power dissipated = Force $\times \dfrac{dx}{dt}$

$$= 2km\left(\frac{dx}{dt}\right)^2 = 2km\frac{f_0^2 p^2}{(\omega_0^2-p^2)^2+4k^2p^2}\cos^2(pt-\theta)$$

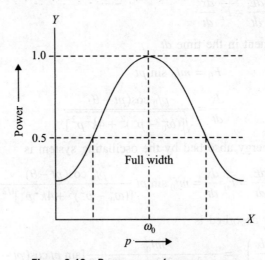

Figure 6.13 Power versus frequency curve.

Average power dissipated $= \frac{1}{2}mf_0^2 \frac{2kp^2}{(\omega_0^2 - p^2)^2 + 4k^2p^2}$ $\left[\because \cos^2(pt-\theta)_{av} = \frac{1}{2}\right]$

From Eq. (6.87), we see that the power drops to half its maximum value in the condition.

$$\frac{1}{2}(P_{av})_{res} = (P_{av})_{\text{half power frequency}}$$

$$\frac{1}{2}\frac{mf_0^2}{4k} = \frac{1}{2}mf_0^2 \times \frac{2kp^2}{(\omega_0^2 - p^2)^2 + 4k^2p^2}$$

or $\qquad 8k^2p^2 = (\omega_0^2 - p^2)^2 + 4k^2p^2$

or $\qquad (\omega_0^2 - p^2) = \pm 2kp$

or $\qquad (\omega_0 - p) = \pm \frac{2kp}{(\omega_0 - p)} = \pm \frac{2k}{\left(1 + \frac{\omega_0}{p}\right)}$

$$\cong \pm k \text{ or } \frac{1}{2\tau} \qquad \left[\text{talking } \frac{\omega_0}{p} = 1(\text{approx.})\right]$$

∴ Half width of the power versus frequency curve

$$\Delta p = \left|\omega_0^2 - p\right| = k = \frac{1}{2\tau} \qquad (6.89)$$

Hence, $\qquad Q = \omega_0 \tau = \frac{\omega_0}{2\Delta p} = \frac{\text{Frequency at resonance}}{\text{Full width at half maximum power}}$

6.17 QUALITY FACTOR (Q) OF FORCED HARMONIC OSCILLATOR

An expression for quality factor (Q) of a driven harmonic oscillator

$$Q = 2\pi \times \frac{\text{Average energy stored}}{\text{Power dissipated}}$$

But at any instant $x = A\sin(pt - \theta)$ and $\frac{dx}{dt} = Ap\cos(pt - \theta)$

(i) Stored energy of the oscillator

$$= \text{Kinetic energy} + \text{Potential energy} = \frac{1}{2}m\left(\frac{dx}{dt}\right)^2 + \frac{1}{2}Cx^2$$

(ii) Rate of dissipation of energy $= \left(\gamma\frac{dx}{dt}\right)\left(\frac{dx}{dt}\right)$

∴ Dissipative force at displacement $x = \gamma\left(\frac{dx}{dt}\right)$

$$\therefore \quad \frac{\text{Stored energy}}{(\text{Period})(\text{Rate of dissipation of energy})} = \frac{\frac{1}{2}m\left(\frac{dx}{dt}\right)^2 + \frac{1}{2}Cx^2}{T\left(\gamma\frac{dx}{dt}\right)\left(\frac{dx}{dt}\right)}$$

$$= \frac{\frac{1}{2}mp^2A^2\cos^2(pt-\theta) + \frac{1}{2}m\omega_0^2 A^2\sin^2(pt-\theta)}{\frac{2\pi}{p}\left(\frac{m}{\tau}\right)p^2 A^2\cos^2(pt-\theta)} \quad \left[\because \gamma = \frac{m}{\tau}\right]$$

As the average value of $\cos^2(pt-\theta)$ or $\sin^2(pt-\theta)$ for a period is $\frac{1}{2}$.

$$\therefore \quad Q = 2\pi \times \frac{\frac{1}{4}p^2 A^2 + \frac{1}{4}\omega_0^2 A^2}{\left(\frac{2\pi}{p}\right)\left(\frac{p^2}{\tau}\right)A^2 \times \frac{1}{2}} = \frac{1}{2}\left[1 + \left(\frac{\omega_0}{p}\right)^2\right]p\tau \quad (6.90)$$

Near resonance, when $p = \omega_0$, we see that

$$Q = \omega_0 \tau$$

For low damping, Q will be high. Hence, Q measures the sharpness of resonance.

EXAMPLE 6.4 In case of a forced harmonic oscillator, the amplitude of vibrations increase from 0.02 mm at very low frequencies to a value 5 mm at the frequency 100 Hz. Find
  (i) Q-factor of the system
  (ii) damping constant
  (iii) the relaxation time and
  (iv) half width of resonance curve.

Solution *Given*: $A_0 = 0.02$ mm; $A_{max} = 5$ mm and Resonant frequency $f = 100$ Hz

  (i) The quality factor (Q) of the system

$$= \omega_0\tau = \frac{\omega_0}{2k} = \frac{A_{max}}{A_0} = \frac{5}{0.02} = 250$$

  (ii) The damping constant (k)

$$\omega_0 = 2\pi\nu = 2\pi \times 100$$

$$k = \frac{\omega_0}{2Q} = \frac{2\pi \times 100}{2 \times 250} = 1.25 / s$$

  (iii) The relaxation time (τ)

$$\tau = \frac{1}{2k} = \frac{1}{2.5} = 0.4 \text{ s}$$

  (iv) Half width of resonance curve

$$\Delta p = \sqrt{3} \times 1.25 = 2.2 \text{ rad/s}$$

FORMULAE AT A GLANCE

6.1 Equation for simple harmonic oscillator: $\dfrac{d^2x}{dt^2} + \omega^2 x = 0$

where $\omega^2 = \dfrac{C}{m}$;

ω = angular frequency, m = mass of particle
The solution of above equation is
$$x = a\sin(\omega t - \theta)$$
$$T = \dfrac{2\pi}{\omega}$$

6.2 Energy of simple harmonic oscillator
$$E = \dfrac{1}{2} m\omega^2 a^2 = 2m\pi^2 a^2 \nu^2$$
$$\nu = \text{frequency} = \dfrac{\omega}{2\pi}$$
$$a = \text{amplitude}$$

6.3 Average values of K.E. of P.E.

For K.E., $(K_{av}) = \dfrac{1}{4} Ca^2$

For P.E., $(U_{av}) = \dfrac{1}{4} Ca^2$

$$\text{K.E.} = \text{P.E.}$$

6.4 For simple pendulum

(a) $\dfrac{d^2\theta}{dt^2} + \dfrac{g}{l}\theta = 0$

(b) $T = \dfrac{2\pi}{\omega} = 2\pi\sqrt{\dfrac{l}{g}}$

(c) $\theta = \theta_{\max} \sin(\omega t + \phi)$

6.5 Damping force equation

(a) $\dfrac{dv}{dt} + \dfrac{\gamma}{m} v = 0$

$\Rightarrow \dfrac{dv}{dt} + \dfrac{v}{\tau} = 0$

τ = relaxation time

(b) $v = v_0 e^{-t/\tau}$

(c) *Equation for damped harmonic oscillator*

$$\frac{d^2x}{dt^2} + \frac{\gamma}{m}\frac{dx}{dt} + \frac{C}{m}x = 0$$

or

$$\frac{d^2x}{dt^2} + 2k\frac{dx}{dt} + \omega_0^2 x = 0$$

where $2k = \frac{\gamma}{m} = \frac{1}{\tau}$ and $\sqrt{\frac{C}{m}}$

(d) *Solution of equation*

$$x = e^{-kt}\left[A_1 e^{\sqrt{(k^2 - \omega_0^2)}t} + A_2 e^{-\sqrt{(k^2 - \omega_0^2)}t}\right]$$

Case:

(i) If $k < \omega_0$, then $\sqrt{k^2 - \omega_0^2}$ = imaginary = underdamped case

(ii) If $k > \omega_0$, then $\sqrt{k^2 - \omega_0^2}$ = real = overdamped case

(iii) If $k = \omega_0$, then $\sqrt{k^2 - \omega_0^2}$ = 0 = critically damped

(e) *Underdamped case*

$$x = a_0 e^{-kt}\sin(\omega t + \phi)$$

$$T = \frac{2\pi}{\omega} = \frac{2\pi}{(\omega_0^2 - k^2)^{1/2}}$$

(i) Decrement

$$\frac{a_0}{a_1} = \frac{a_1}{a_2} = \frac{a_2}{a_3} = \cdots e^{kT} = e^{\lambda} \quad [kT = \lambda]$$

(ii) Logarithmic decrement

$$\lambda = \log_e \frac{a_0}{a_1} = \log_e \frac{a_1}{a_2} = \log_e \frac{a_2}{a_3} = \cdots$$

(f) *Overdamped*

$$x = [A_1 e^{-(k-\beta)t} + A_2 e^{-(k+\beta)t}]$$

As $k > \beta$ [Non-oscillatory, dead beat or aperiodic]

(g) *Critical damped or just aperiodic*

$$x = [x_0 + (v_0 + kx_0)t]e^{-kt}$$

6.6 Power dissipation

$$P = -\frac{dE}{dt} = ma_0^2 \omega^2 k e^{-2kt} = 2kE = \frac{2E}{\tau}$$

6.7 Average rate of doing work

$$= <Fv^2> = <r\left(\frac{dx}{dt}\right)^2> = \frac{2kma_0^2 e^{-2kt}\omega_0^2}{2} = 2kE = \frac{E}{\tau}$$

6.8 Quality factor (Q)

$$Q = 2\pi \times \frac{\text{Energy stored in system}}{\text{Enenrgy lost per second}} = \omega\tau$$

6.9 Equation for forced oscillator

$$\frac{d^2x}{dt^2} + 2k\frac{dx}{dt} + \omega_0^2 x = f_0 \sin \omega t$$

where $2k = \frac{\gamma}{m}, \omega_0 = \sqrt{\frac{C}{m}}$

ω_0 = natural frequency in absence of damping and driven force
k = damping constant
C = force constant

$$f_0 = \frac{F_0}{m}$$

$F_p = F_0 \sin pt$ = periodic force
Solution of above equation is

$$x = A \sin(pt - \theta)$$

then $\quad \dfrac{dx}{dt} = pA \cos(pt - \theta)$

and $\quad \dfrac{d^2x}{dt^2} = -p^2 A \cos(pt - \theta)$

$$A = \frac{f_0}{[(\omega_0^2 - p^2)^2 + 4k^2 p^2]^{1/2}}$$

$$\tan \theta = \frac{2kp}{\omega_0^2 - p^2}$$

$$\theta = \tan^{-1} \frac{2kp}{\omega_0^2 - p^2}$$

$$x = \frac{f_0}{[(\omega_0^2 - p^2)^2 + 4k^2 p^2]^{1/2}} \sin\left[pt - \tan^{-1} \frac{2kp}{\omega_0^2 - p^2}\right]$$

Case:
(i) For low driving frequency when

$$\omega_0 \gg p; \tan \theta \to 0 \text{ or } \theta \to 0$$

$$A = \frac{f_0}{[(\omega_0^2 - 2k^2 - p^2)^2 + 4k^2\omega_0^2 - 4k^4]^{1/2}}$$

For maximum value of A

$$\omega_0^2 - 2k^2 - p^2 = 0$$

or

$$p^2 = \sqrt{\omega_0^2 - 2k^2} = p_r \text{ (say)}$$

$$A_{max} = \frac{f_0}{2k\sqrt{\omega_0^2 - k^2}}$$

Hence, for certain value of driven frequency (p_r), the amplitude of the oscillatory is maximum.

$$\omega_0 = p_r$$

$$A_{max} = \frac{f_0}{2k\omega_0} \text{ or } \frac{f_0\tau}{\omega_0}$$

$$\frac{\text{Response at resonance}}{\text{Response at zero driven frequency}} = \frac{f/2k\omega_0}{f/\omega_0^2} = \frac{\omega_0}{2k} = \omega_0\tau = Q$$

(ii) When $p = \omega_0$

$$A = \frac{f_0}{2k\omega_0} \quad \frac{f_0\tau}{\omega_0}$$

and

$$\theta = \tan^{-1}\left(\frac{kp}{0}\right) = \tan^{-1}(\infty) = \frac{\pi}{2}$$

(iii) For high driving frequency when $\omega_0 \ll p$

$$A = \frac{f_0}{\sqrt{p^4 - 4k^2p^2}} = \frac{f_0}{p^2} = \frac{F}{mp^2}$$

and

$$\theta = \tan^{-1}\left[\frac{2kp}{\omega_0^2 - p^2}\right]$$

$$= \tan^{-1}\left(-\frac{2k}{p}\right) \approx \tan^{-1}(0) = \pi$$

6.10 Half width of resonance curve

Half width $\quad \Delta p = |p_h - p_r| = \sqrt{3}k$

6.11 Velocity resonance

$$v = \frac{dx}{dt} = \frac{f_0 p}{[(\omega_0^2 - p^2)^2 + 4k^2p^2]^{1/2}} \cos(pt - \theta)$$

$$= v_0 \sin\left(pt - \theta + \frac{\pi}{2}\right)$$

6.12 Power absorption

(a) $P_{av} = \dfrac{1}{2} mf_0^2 \dfrac{2kp^2}{(\omega_0^2 - p^2)^2 + 4k^2 p^2} = mkv_0^2 = \dfrac{\gamma v_0}{2}$

(b) $\dfrac{1}{2}(P_{av})_{res} = (P_{av})_{\text{half power frequency}}$

(c) Half width $\Delta p = |\omega_0 - p| = k = \dfrac{1}{2\tau}$

(d) (i) Quality factor $Q = \dfrac{\omega_0}{2\Delta p}$

$= \dfrac{\text{Frequency at resonance}}{\text{Full width at half maximum power}}$

$= 2\pi \times \dfrac{\text{Average energy stored}}{\text{Energy dissipated per second}}$

(ii) $Q = \dfrac{1}{2}\left[1 + \left(\dfrac{\omega_0}{p}\right)^2\right] p\tau$, at resonance $Q = \omega_0 \tau$

SOLVED NUMERICAL PROBLEMS

PROBLEM 6.1 A particle is making simple harmonic motion along the x-axis. If at distances x_1 and x_2, velocities of the particles are v_1 and v_2, calculate the time period.

Solution Simple harmonic motion can be represented by

$$x = a \sin(\omega t + \phi)$$

Velocity, $\dfrac{dx}{dt} = a\omega \cos(\omega t + \phi)$

$\Rightarrow \qquad v = a\omega\sqrt{(1 - \sin^2(\omega t + \phi))}$

$\Rightarrow \qquad v = \omega\sqrt{(a^2 - x^2)} \qquad$ (i)

Now according to question,

$v_1 = \omega\sqrt{(a^2 - x_1^2)} \qquad$ (ii)

and $\qquad v_2 = \omega\sqrt{(a^2 - x_2^2)} \qquad$ (iii)

Squaring Eqs. (ii) and (iii) and then subtracting them, we get

$$v_1^2 - v_2^2 = \omega^2(a^2 - x_1^2 - a^2 + x_2^2) = \omega^2(x_2^2 - x_1^2)$$

$\therefore \qquad \omega^2 = \dfrac{v_1^2 - v_2^2}{x_2^2 - x_1^2}$

The time period, $\qquad T = \dfrac{2\pi}{\omega} = 2\pi \sqrt{\dfrac{x_2^2 - x_1^2}{v_1^2 - v_2^2}}$

PROBLEM 6.2 A particle is initially displaced by 5 cm from its equilibrium position and then left. The particle executes damped linear oscillations with a logarithmic decrement $\theta = 0.02$. Find the total distance at the particle covers before it finally stops.

Solution In each oscillation, an oscillating particle moves the distance $4a$, where a is amplitude. A damped oscillator, theoretically speaking stops after infinite oscillations, its amplitude at time t is

$$a(t) = a_0 e^{-kt}$$

$$\Rightarrow \quad a(t) = 0 \text{ when } t \to \infty.$$

Therefore, the total distance moved by the particle is

$$l = 4a_0 + 4a_1 + 4a_2 + \cdots \infty$$

Where a_n is amplitude after $(n - 1)$ complete oscillations.

$$a_n = a_0 e^{-(n-1)kT} = a_0 e^{-(n-1)\lambda}$$

where $\lambda = kT$ is logarithmic decrement.

Hence, we get

$$l = 4a_0(1 + e^{-\lambda} + e^{-2\lambda} + \cdots \infty) = 4a_0 \frac{1}{1 - e^{-\lambda}}$$

For $\lambda = 0.02$, $\quad e^{-\lambda} = 1.02$ (approx); $a_0 = 5$ cm

Therefore, $\quad l = 4 \times 5 \times \dfrac{1.02}{0.02} = 1020$ cm $= 10.2$ m

PROBLEM 6.3 A simple pendulum of length 50 cm performs oscillations which slowly decay with time. If the total mechanical energy of the system falls to 10^3 times in time interval of 300 s, then, find the quality factor of the oscillator.

Solution The average total energy of a weakly damped oscillator at time t is given as

$$<E(t)> = \frac{1}{2} m a_0^2 \omega^2 e^{-2kt}$$

After time τ, the average energy becomes

$$<E(t + \tau)> = \frac{1}{2} m a_0^2 \omega^2 e^{-2k(t-\tau)} = <E(t)> e^{-2k\tau}$$

Hence

$$\omega^2 = \frac{<E(t)>}{<E(t+\tau)>} = e^{2k\tau}$$

It is given that in $\tau = 300$ s, energy falls to $\dfrac{1}{10^3}$ of its initial value i.e.,

$$10^3 = e^{2k \times 300}$$

or

$$k = \frac{\ln 10^3}{600} = \frac{2.303}{600} \text{s}^{-1}$$

The quality factor is $Q = \dfrac{\pi}{kT}$ where $T = 2\pi\sqrt{\dfrac{l}{g}}$ in the time period of the pendulum.

Thus, we get $Q = 43.5\sqrt{\dfrac{9.8}{0.5}} = \dfrac{700}{\sqrt{2.5}} = 193$ (approximately).

PROBLEM 6.4 If mass m is attached to a weightless spring and it has a time period T when oscillating in the horizontal position, show that the time period will not change if the system is turned into a vertical direction.

Solution Let C be the force constant of the spring. When the mass m is displaced through a small distance x horizontally, the restoring force is $-Cx$. The equation of motion is

$$m = \frac{d^2x}{dt^2} = -Cx \qquad \text{(i)}$$

or

$$\frac{d^2x}{dt^2} = -\frac{C}{m}x$$

which is simple harmonic motion with time period $T = 2\pi\sqrt{\dfrac{m}{C}}$.

When the system is made vertical, the force gravity mg, extends the spring in the downward direction, say, through a distance x_0.

At equilibrium position,

$$mg = Cx_0$$

If the spring is displaced further through a distance x, the restoring force due to spring upward is

$$= -C(x + x_0)$$

The force due to gravity (mg) is still acting downwards and the net restoring force

$$= -[C(x + x_0) - mg] = -C(x + x_0) + mg$$
$$= -Cx \qquad \text{(ii)}$$

which is identical with Eq. (i). It represents simple harmonic motion with the same period.

$$T = 2\pi\sqrt{\frac{m}{C}}$$

Thus, the time period of oscillation of the mass m, is unaltered, when the system is made vertical from the horizontal position.

PROBLEM 6.5 An electron behaving as a highly damped oscillator emits light of wavelength 6000 Å and has a Q-value of 5×10^7. Show that the width of the emitted spectral line is 1.2×10^{-14} m.

Solution Considering the power absorption curve (Figure 6.14).

The difference $(\omega_2 - \omega_1)$ is known as bandwidth of the spectral line. The curve gives the profile of the spectral line emitted by the atom, with the maximum emission occurring at the frequency $\omega = \omega_0$.

The quality factor,

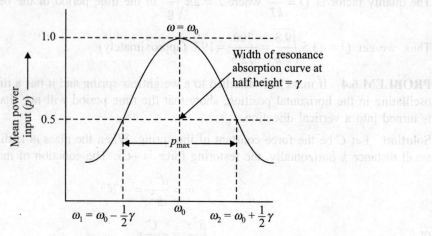

Figure 6.14

$$Q = \frac{\omega_0}{\omega_2 - \omega_1} = 5 \times 10^7$$

and
$$\omega_2 - \omega_1 = \omega/Q = (2\pi c/\lambda)/Q = \frac{2\pi c}{\lambda Q}$$

$$= \frac{2\pi \times 3 \times 10^8}{6 \times 10^{-7} \times 5 \times 10^7} = 2\pi \times 10^7$$

i.e.,
$$2\pi c \left(\frac{1}{\lambda_2} - \frac{1}{\lambda_1} \right) = 2\pi \times 10^7$$

or
$$\frac{\lambda_1 - \lambda_2}{\lambda_1 \lambda_2} = \frac{2\pi \times 10^7}{2\pi \times 3 \times 10^8} = \frac{1}{30}$$

or
$$(\lambda_1 - \lambda_2) = \frac{\lambda_1 \lambda_2}{30} \cong \frac{36 \times 10^{-14}}{30}$$

Taking $\lambda_1 = \lambda_2 \cong 6 \times 10^{-7}$

$$\Delta \lambda = 12 \times 10^{-14} \text{ m}$$

PROBLEM 6.6 An alternating voltage of amplitude V_0 is applied across an LCR series circuit. Show that the voltage at current resonance across either the inductance or the capacitor has a maximum value of qV_0.

Solution The charge variation in the LCR circuit is given by

$$L\frac{d^2q}{dt^2} + R\frac{dq}{dt} + \frac{q}{C} = V_0 \cos pt \qquad \text{(i)}$$

Comparing the above equation with the differential equation for forced oscillator.

$$m\frac{d^2x}{dt^2} + \gamma\frac{dx}{dt} + C'x = F_0 \cos pt \qquad \text{(ii)}$$

We have $L = m$, $R = \gamma$ and $\frac{1}{C} = C'$; q behaves like the displacement x, and current $i = \frac{dq}{dt}$ varies like velocity.

We know that the current at resonance is given by

$$\frac{V_0}{R} \cos(\omega t - \phi) \text{ with } \omega_0 = \frac{1}{\sqrt{LC}}$$

i.e.,
$$\frac{dq}{dt} = \frac{V_0}{R} \cos(\omega_0 t - \phi) \qquad \text{(iii)}$$

Voltage across the inductance at resonance is

$$L\frac{d^2q}{dt^2} = \left(-\frac{L\omega_0}{R}\right) V_0 \sin(\omega_0 t - \phi) = -qV_0 \sin(\omega_0 t - \phi) \qquad \text{(iv)}$$

with $q = \frac{\omega_0 L}{R}$

Voltage across the capacitor at resonance is

$$\frac{q}{C} = \frac{V_0}{\omega_0 CR} \sin(\omega_0 t - \phi)$$

As
$$L\omega_0 = \frac{1}{C\omega_0}$$

We have $\frac{q}{C} = \frac{L\omega_0}{R} V_0 \sin(\omega_0 t - \phi) = qV_0 \sin(\omega_0 t - \phi)$

Thus for the current resonance, the voltage across L and C oscillate with maximum value of qV_0.

PROBLEM 6.7 A forced harmonic oscillator shows equal amplitude of oscillations frequencies $p_1 = 300$ rad/s and $p_2 = 400$ rad/s. Find the resonance frequency p_r at which amplitude becomes maximum.

Solution The expression for amplitude of a forced harmonic oscillator

$$A = \frac{f_0}{[(\omega_0^2 - p^2)^2 + 4k^2 p^2]^{1/2}}$$

Let $A = A_1$ at $p_1 = 300$ rad/s and $A = A_2$ at $p_2 = 400$ rad/s

$$A_1 = A_2$$

$$(\omega_0^2 - p_1^2)^2 + 4k^2 p_1^2 = (\omega_0^2 - p_2^2)^2 + 4k^2 p_2^2$$

$$\Rightarrow \qquad (\omega_0^2 - p_1^2)^2 - (\omega_0^2 - p_2^2)^2 = 4k^2(p_2^2 - p_1^2)$$

Hence, we find $p_r^2 = \omega_0^2 - 2k^2 = \dfrac{p_1^2 + p_2^2}{2}$

$$p_r = \sqrt{\dfrac{p_1^2 + p_2^2}{2}} = \sqrt{\dfrac{(300)^2 + (400)^2}{2}} = 353.5 \text{ rad/s}$$

CONCEPTUAL QUESTIONS

6.1 What do you mean by resonance?

Ans It is a phenomenon of vibratory motion in which a body is set into vibration with its natural frequency by another body vibrating with the same frequency.

6.2 Define restoring force.

Ans In the equilibrium position of the oscillating particle, no net force acts on it. When the particle is displaced from its equilibrium position, a periodic force acts on it in such a direction as to bring the particle to its equilibrium position. This force is called restoring force F.

6.3 What is epoch?

Ans In equation $x = a \sin(\omega t + \phi)$; the value of phase when $t = 0$ is called the phase or epoch. In present case ϕ is the epoch.

6.4 If a hollow pipe passes across the centre of gravity the earth then what changes would take place in the velocity and acceleration of a ball dropped in the pipe?

Ans The ball will execute SHM to and fro about the centre of the earth. At the centre, the velocity of the ball, will be maximum (acceleration zero) and at the earth's surface, the velocity will be zero (acceleration maximum).

6.5 How is the period of a pendulum affected when its period of suspension is (i) moved horizontally when acceleration a, (ii) moved vertically upward with acceleration a, (iii) moved vertically downward with acceleration $a < g$? Which case, if any applies to a pendulum mounted on a cart rolling down an inclined plane?

Ans (i) The time period of pendulum will decrease as the effective acceleration would be $\sqrt{a^2 + g^2}$ and not g.

(ii) The time period will decrease because the value of g will be $(g + a)$.

(iii) The time period will decrease because g will be replaced by $(g - a)$.

Case (iii) applies to a pendulum.

6.6 What fraction of total energy is kinetic and what fraction is potential, when the displacement is one half of amplitude?

Ans Total energy

$$E = \dfrac{1}{2} m\omega^2 a^2, \text{ K.E.} = \dfrac{1}{2} m\omega^2(a^2 - x^2) \text{ and P.E.} = \dfrac{1}{2} m\omega^2 x^2$$

when $x = \dfrac{a}{2}, \dfrac{\text{K.E.}}{E} = \dfrac{3}{4}$ and $\dfrac{\text{P.E.}}{E} = \dfrac{1}{4}$

6.7 At what displacement the kinetic and potential energies are equal?

Ans $\frac{1}{2}m\omega^2 x^2 = \frac{1}{2}m\omega^2(a^2 - x^2)$ or $x = \frac{a}{\sqrt{2}}$

6.8 How can you differentiate free and forced vibrations?

Ans
(i) Free vibrations of a body take place under influence of its own elastic force without being acted upon by any external force. On the other hand, the force vibrations of a body take place due to action of a periodic force applied externally.
(ii) Due to damping effect of frictional forces, the free vibrations die in course of time while the forced vibrations persist as long as the applied periodic force acts on the body.
(iii) The frequency of free vibrations depends upon the mass, shape and elasticity of the body while the frequency of forced vibrations is independent of these factors and is equal to the frequency of applied periodic force.
(iv) Free vibrations may occur with any amplitude (small or large). In the presence of damping force, the amplitude goes on decreasing. In case of forced vibrations, the amplitude is small except in the special stages when resonance takes place.

6.9 Differentiate mechanical and electrical oscillators.

Ans In case of freely oscillating pendulum, the amplitude decreases continuously with time. The mechanical energy of the oscillator is spent in overcoming air resistance, resistance and capacitance are connected in series with a source of e.m.f. In this case, a part of the electrical energy is slowly dissipated in the form of heat in the resistance. So, the resistance causes the damping. Thus, the electrical oscillations are damped. This is an example of electrical oscillator.

6.10 Define sharpness of resonance.

Ans The sharpness of resonance means the rate of fall in amplitude, with the change of forcing frequency on each side of resonance frequency.

6.11 What do you mean by quality factor and bandwidth of resonance?

Ans *Quality factor:* The quality factor is defined as 2π times the ratio of the energy stored in the system to the energy lost per period.

i.e., $$Q = 2\pi \times \frac{\text{Energy stored in system}}{\text{Energy lost per second}}$$

Bandwidth of resonance: The difference in values of the driving frequency at which the average power absorbed drops to half its maximum (resonant) value, is called bandwidth of resonance.

6.12 Classify the vibrations.

Ans Vibrations are classified into the following types.
(i) *Free vibrations:* These refer to vibrations when the system is allowed to vibrate on its own. The system is not subjected to a periodic force.
(ii) *Forced vibrations:* If a system is subjected to a periodic force, the resulting vibrations are known as forced vibrations.
(iii) *Undamped vibrations:* If no energy is lost or dissipated in friction during vibration, the vibration is termed as undamped vibration.

(iv) *Damped vibrations:* This refers to vibration in which there is loss of energy during vibration. Consequently a system which is set to vibrate will come to rest after due course of time.

(v) *Linear/non-linear vibration:* If the restoring force in proportional to displacement or when the frictional force is linearly proportional to velocity the vibrations are termed linear, otherwise they are non-linear.

(vi) *Transient vibrations:* When a system is subjected to a periodic force, it takes a finite though a short time, for the onset of steady state vibrations. The vibrations during this interim period are known as transient vibrations.

6.13 What is the physical significance of Q-factor of a forced oscillator?

Ans Quality factor or Q-factor gives the measure of sharpness of resonance. Higher the quality factor, smaller is the bandwidth and hence the tuning will be sharper. This is used to increase of selectivity of the 'radio sets'. The sharpness of response of a circuit allows the radio signals to be reproduced without any interference from signals of frequencies close to it.

It also accounts for the amplification factor. At displacement resonance, the displacement at low frequency is amplified by Q times.

6.14 Is the energy stored in a forced oscillator? Explain.

Ans No, energy is not stored in the forced oscillator. In each cycle, the energy of the oscillator is lost due to damping offered by the medium. The energy supplied by the external periodic force is exactly same to this loss of energy. The external energy maintains the oscillations of the forced oscillator. In steady state, the amplitude and period of the force oscillator are so adjusted that the power supplied by the external periodic force is equal to the energy dissipated against the frictional forces.

EXERCISES

Theoretical Questions

6.1 Derive the expression for displacement of a particle in a damped harmonic oscillator.

6.2 Friction damps out the free motion of an oscillator. Show that the velocity of a particle decreases exponentially with time when it (the particle) is being acted upon by only the frictional force.

6.3 Draw a graph showing the variation of amplitude with time in case of heavily, lightly and critically damped oscillators.

6.4 Justify the statement "The lower the damping higher the Q-value".

6.5 Set up the equation of motion for a damped SHO. Discuss the heavily, lightly and critically damped conditions.

6.6 Set up the equation of motion in differential form for a harmonic oscillator driven by a sinusoidal force. Obtain expression for its maximum amplitude.

6.7 Write down and solve the differential equation of a driven harmonic oscillator subjected to a sinusoidal force and obtain expression for its maximum amplitude.

6.8 Draw a labelled plot of amplitude versus angular frequency of a force oscillator. What is the significance of the peak?

6.9 Solve the differential equation of a driven harmonic oscillator subjected to sinusoidal force and obtain the expression for maximum amplitude.

6.10 Show that in case of forced oscillation lower limit of displacement amplitude is zero and the resonance occurs at a frequency lower than that of corresponding simple harmonic oscillator.

6.11 Establish the equation for a forced harmonic oscillator. Solve the equation of motion for the steady state displacement response function. Also show that the displacement response function is
  (i) in phase with the driving for low driving limit.
  (ii) 180° (or π radian) out of phase with the force for high driving frequency limit.
  (iii) 90° (or π radian) out of phase with the driving force for the resonant frequency.

6.12 Define resonance in the context of SHM.

Numerical Problems

6.1 A solid cylinder of radius 8 cm is suspended by a vertical wire as a torsion pendulum. The axis of the cylinder is along the wire. Find the moment of torsion of the wire, if the mass of the cylinder is 7.0 kg and period of vibration is 5.0 s.

[**Ans.** 3.54×10^{-2} Nm/rad]

6.2 A condenser of capacity 3 µF is discharged through an inductance of 3H. Calculate the quality factor and the time in which the amplitude of oscillations will be reduced to 10% of its initial value. The resistance is 2Ω.

[**Ans.** $Q = 500$; $t = 7$s]

6.3 A damped vibrating system from rest reaches a first amplitude of 500 mm, which reduces to 50 mm after 100 oscillations, each of period 2.3 s. Find the damping constant, relaxation time correction for the first displacement for damping.

[**Ans.** $k = 0.01$/s; $\tau = 50$ s and $a_0 = 502.9$ mm]

6.4 Find the natural frequency of a circuit containing inductance of 144 µH and a capacity of 0.0025 µF. To which wavelength, its response will be maximum?

[**Ans.** $v = 2.653 \times 10^5$ Hz; $= 1131$ m]

6.5 If the quality factor of an undamped tuning fork of frequency 256 Hz is 10^3. What is the time in which the energy is reduced to $1/e$ of its energy in the absence of damping?

[**Ans.** $t = 0.62$ s]

6.6 If the quality factor of an undamped tuning fork of frequency 256 Hz is 103, calculate the time in which its energy is reduced to $1/e$ of its energy is absence of damping. How many oscillations the tuning fork will take in this time?

[**Ans.** Number of oscillations made by tuning fork = 159]

6.7 A uniform bar of 2 m oscillates about a knife edge 0.5 m from one end. What is the period of oscillations?

[**Ans.** $T = 2.46$ s]

6.8 In case of a forced harmonic oscillator, the amplitude of vibrations increases from 0.05 mm at very low frequencies to a value 7.5 mm at the frequency 210 Hz. Find
  (i) Q-factor of the system
  (ii) damping constant
  (iii) the relaxation time
  (iv) half width of resonance curve

[**Ans.** (i) $Q = 150$; (ii) $k = 4.40$/s; (iii) $\tau = \dfrac{1}{2 \times 4.4}$ s; (iv) $\Delta p = 7.62$ rad/s]

6.9 The total energy of a particle executing SHM of period π s is 2560 ergs and the displacement of the particle at $\pi/4$ s is 2 cm. Calculate the amplitude of vibrations and mass of the particle. [**Ans.** 2 cm; 320 gm]

6.10 A particle of mass m kg lies in a potential field given by $V = 200x^2 + 300$ J/kg. Find the frequency and time period. [**Ans.** 3.2 Hz; 0.31 s]

6.11 A particle is dropped down in a deep hole which extends to the centre of the earth. Calculate its velocity at a depth of one kilometer from the surface of the earth. Assume that $g = 10$ m/s^2 and radius of the earth = 6400 km. [**Ans.** 141.4 m/s]

6.12 At a certain place the value of g is 9.8 m/s^2 and the length of the pendulum in so adjusted that the period is 1s. When the period of the pendulum is measured on an elevator undergoing uniform acceleration, it is found to be 1.025 s.

  (i) For an acceleration $a(\ll g)$ of the elevator, show that the period measured on the elevator is given by $T = T_0\left(1 - \dfrac{a}{2g}\right)$, where T_0 is the period of the unaccelerated pendulum [an upward acceleration is assumed to be positive].

  (ii) What is the acceleration of the elevator?

[**Ans.** (i) An upward acceleration is assumed to be positive, (ii) 0.49 m/s^2 downward]

6.13 A particle is moving in a straight line with SHM. Its velocity has the value 5 ft/s and 4 ft/s when its distance from the centre point of its motion are 2 ft and 3 ft respectively. Find the length of the path, the frequency of oscillation and the phase of motion when the point is 2 ft from the centre. [**Ans.** 8.46 ft, 0.2135 cycle/s; sin^{-1} (0.473)]

6.14 Show that the maximum tension in the string of a simple pendulum, when the angular amplitude θ is small, is $mg(1 + \theta^2)$. At what position of the pendulum has the tension a maximum. [**Ans.** At equilibrium position ($\theta = 0°$)]

6.15 A condenser of capacity 20 µF and an inductance of 10 mH are joined in series. Calculate the frequency of electrical oscillations if the tension is maximum.

[**Ans.** 3.5×10^{-2} Hz]

6.16 A disc of radius r oscillates as a pendulum about a horizontal axis perpendicular to the plane of the disc through a point $r/4$ from its centre. Calculate the period of oscillation.

$\left[\textbf{Ans. } T = 2\pi\sqrt{\dfrac{3r}{2g}}\right]$

6.17 Using Kater's pendulum, find the value of g from the following:
Distance of one knife edge from C.G. = 23.5 cm
Distance of second knife edge from C.G. = 76.3 cm
Time period about first edge = 2.002 s
Time period about second edge = 2.004 s [**Ans.** 9.831 m/s^2]

6.18 What is the frequency of a simple pendulum 2.0 m long? Assuming small amplitude, what would its frequency be in an elevator accelerating upward at a rate of 2.0 m/s^2? What will be its frequency in case of free fall?
[**Ans.** $v_1 = 0.3524$ Hz, $v_2 = 0.3867$ Hz and $v_3 = 0$]

6.19 The differential equation for a certain system is $\dfrac{d^2x}{dt^2} + 2k\dfrac{dx}{dt} + \omega_0^2 x = 0$. If $\omega_0 \gg$ 'k, find the time, in which
  (i) amplitude falls to $1/e$ times of initial value.
  (ii) energy of the system falls to $1/e$ times the initial value.
  (iii) energy falls to $1/e^4$ times the initial value.
$$\left[\text{Ans. (i) } t = \frac{1}{k}\text{s, (ii) } t = \frac{1}{2k}\text{s, (iii) } t = \frac{2}{k}\text{s}\right]$$

6.20 A simple pendulum has a period of 2 s and an amplitude of 2°. After 10 complete oscillations its amplitude is reduced to 1.5°. Find the relaxation time and damping constant. [**Ans.** 34.76 s; 0.01438 s^{-1}]

6.21 Find whether the discharge of a condenser for following inductive circuit is oscillator.
$C = 0.1$ μF, $L = 10$ mH and $R = 200$ Ω
If the circuit is oscillatory, calculate its frequency. [**Ans.** 4772 Hz]

6.22 Give the theory of oscillations in a series LCR circuit with small damping. Deduce the frequency and quality factor for a circuit with $L = 2$ mH, $C = 20$ μF and $R = 0\,2$ Ω.
[**Ans.** 1.59×10^3 Hz, 100]

6.23 In an LCR circuit, $L = 10$ mH, $C = 1$ μF and $R = 0.1$ Ω. How long a time does the charge oscillation take to decay half amplitude? What is the quality factor for the circuit?
[**Ans.** 0.14 s, 1000]

6.24 A mass of 0.1 kg suspended from a spring oscillates freely with a time period of 1 s. When it is immersed in oil and made to oscillate, its time period becomes 1.01 s. Calculate the damping constant due to oil medium. [**Ans.** 0.88 kg/s]

6.25 A ballistic galvanometer has a period of 4.0 s and a Q-value of 5.0. Calculate the time taken after impulse for the amplitude to reach its first maximum.
[**Ans.** $T = 1.005$ s]

6.26 The logarithmic damping decrement of a mathematical pendulum is $kT = 0.2$. How will the amplitude of oscillations decrease during one full oscillation of pendulum?
[**Ans.** 1.22]

6.27 The frequency of an underdamped harmonic oscillator is adjusted to be equal to half the frequency experienced by the oscillator without damping. What is the logarithmic decrement of this system? [**Ans.** $4\pi r/\omega_0$]

6.28 Find the natural frequency of a circuit containing inductance of 50×10^{-3} H and a capacitance of 500×10^{-12} F. Find the wavelength of the ratio waves to which it will respond. [**Ans.** (i) $\nu = 3.18 \times 10^4$ Hz, $\lambda = 9.43 \times 10^2$ m]

6.29 A simple harmonic oscillator is subjected to a sinusoidal driving force where frequency is altered but amplitude kept constant. It is found that the amplitude of the oscillator increases from 0.02 mm at very low driving frequency to 8.0 mm at a frequency of 200 Hz. Obtain the value of

(i) the quality factor
(ii) the relaxation time
(iii) the full width of the resonance

[**Ans.** (i) $Q = 400$, (ii) $\tau = 0.62$ s, (iii) 5.59 rad/s]

6.30 Show that when a coil of inductance L and resistance R is attached to an external source of voltage, $E_0 \sin pt$, the average rate of consumption of energy is $\dfrac{1}{2} \times \dfrac{E_0^2}{R^2 + L^2 p^2}$.

6.31 A torsional pendulum with moment of inertia 900 gm-cm² and the torsional constant 4×10^{-3} N-m/rad, executes resonant vibrations under the influence of an external torque. Calculate the frequency of the applied torque. What should be the *quality factor* of the oscillator so that the amplitude of oscillation may reduce to 20%, when the frequency of the applied torque increased by 0.2%? [**Ans.** 1.06 Hz; 187]

6.32 In an experiment on forced oscillations, the frequency of a sinusoidal driving force is changed while its amplitude is kept constant. It is found that the amplitude of vibration is 0.1 mm at very low frequency of the driver and goes up to a maximum of 0.5 mm at driving frequency 200 s⁻¹. Calculate:

(i) Q-value of the system
(ii) relaxation time
(iii) half width of resonance curve. [**Ans.** (i) 500, (ii) 8.4 s, (iii) 2.16 rad/s]

6.33 An electric lamp which runs at 40 V D.C. and consumes 10 A current is connected to A.C. mains at 100 V, 50 cycles. Calculate the impedance of the choke. [**Ans.** 0.02918 H]

6.34 A coil of 3 mH is given, calculate the capacity which will make the resonance 1000 kHz. [**Ans.** 8.85 pF]

6.35 A coil having inductance of 100 μH and a condenser having capacitance of 0.0001 μF form a circuit having a measured Q-value of 150 at resonance. Find the effective resistance of the circuit. [**Ans.** 6.67 Ω]

6.36 An rms volts of 100 V is applied to a series LCR circuit having $R = 10$ Ω, $L = 10 \times 10^{-3}$ H and $C = 1.0 \times 10^{-6}$ F. Calculate the following:

(i) the natural frequency of the circuit
(ii) current at resonance
(iii) Q-value of the circuit at resonance
(iv) bandwidth of the circuit

[**Ans.** (i) 1592 s⁻¹, (ii) 10 A, (iii) 10, (iv) 10^3 rad/s]

6.37 An emf of 100 V and 500 Hz is applied to a circuit having $R = 10\ \Omega$, $L = \dfrac{2}{\pi}$ H and $C = \dfrac{1}{\pi} \times 10^{-6}$ F, in series, calculate:

(i) the reactance (ii) the impedance
(iii) the current (iv) the phase angle

[**Ans.** (i) 1000, (ii) 1000 Ω, (iii) 0.1 A, (iv) $\tan^{-1}$ 100 phase lag of current from the applied emf.]

Multiple Choice Questions

6.1 A particle in simple harmonic motion is described by the displacement function $x(t) = A\cos(\omega t + \theta)$. If the initial ($t = 0$) position of the particle is 1 cm, its initial velocity is π cm/s and its angular speed is π rad/s, then its amplitude is

(a) 1 cm (b) $\sqrt{2}$ cm
(c) 2 cm (d) 2.5 cm

6.2 An ideal spring with spring constant k is hung from the ceiling and a block of mass M is attached to its lower end. The mass is released with the spring lower end. The mass of released with the spring initially unstretched. Then maximum extension in the spring

(a) $4\dfrac{Mg}{k}$ (b) $2\dfrac{Mg}{k}$
(c) $\dfrac{Mg}{k}$ (d) $\dfrac{Mg}{2k}$

6.3 Two springs of force constant k and $2k$ are connected to a mass as shown in Figure 6.15. The frequency of oscillation of the mass is

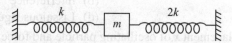

Figure 6.15

(a) $\dfrac{1}{2\pi}\sqrt{\dfrac{k}{m}}$ (b) $\dfrac{1}{2\pi}\sqrt{\dfrac{2k}{m}}$
(c) $\dfrac{1}{2\pi}\sqrt{\dfrac{3k}{m}}$ (d) $\dfrac{1}{2\pi}\sqrt{\dfrac{m}{k}}$

6.4 Total energy of a particle having mass m executing SHM of amplitude a, velocity v and angular velocity ω is

(a) 0 (b) $\dfrac{1}{2}mv^2$
(c) $\dfrac{1}{2}ma^2\omega^2 \cos\omega t$ (d) $\dfrac{1}{2}ma^2\omega^2$

6.5 Two linear simple harmonic oscillators of the same frequency and amplitude are combined at right angles on a particle. The resulting motion of the particle is circular when the phase difference between them is:
(a) 0
(b) $\frac{\pi}{4}$
(c) $\frac{\pi}{2}$
(d) π

6.6 The equation of motion of body of mass m attached to one end of a spring of force constant k is $m\frac{d^2x}{dt^2} + kx = 0$. The time period of oscillation of the body is
(a) $T = 2\pi\sqrt{\frac{m}{k}}$
(b) $T = 2\pi\sqrt{\frac{k}{m}}$
(c) $T = 2\pi\sqrt{km}$
(d) $T = 2\pi\frac{k}{m}$

6.7 Displacement at time t of a particle is $y = 5 \sin(100\pi t + \phi)$. Frequency of oscillation of the particle is
(a) 100 Hz
(b) 200 Hz
(c) 50 Hz
(d) 25 Hz

6.8 The displacement of a particle is given by $x = 6 \cos \omega t + 8 \sin \omega t$. This equation represents a simple harmonic oscillation having amplitude
(a) 14 m
(b) 2 m
(c) 10 m
(d) 5 m

6.9 When two sinusoidal waves moving at right angles to each other superimpose, they produce
(a) beat
(b) interference
(c) stationary
(d) Lissajous figure

6.10 For SHM, the displacement x of oscillating particle and force F acting on it are related by
(a) $F = kx$
(b) $F = \frac{1}{2}kx^2$
(c) $F = \frac{k^2}{x}$
(d) $F = -kx$

6.11 The total energy of a particle executing simple harmonic motion depends upon its
(a) initial state
(b) amplitude only
(c) frequency only
(d) frequency and amplitude

6.12 The time in which the amplitude of a damped system falls to $(1/e)$ of its initial value is called
(a) damping time
(b) reverberation time
(c) relaxation time
(d) none of these

6.13 A graph plotted between T^2 and M. T being the periodic time and M the mass suspended, for a mass spring system with small damping as shown in Figure 6.16. It can be stated that

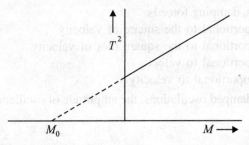

Figure 6.16

(a) mass of spring is negligible
(b) A_0 represents $\frac{1}{3}$ (mass of the spring)
(c) A_0 represents $\frac{2}{3}$ (mass of the spring)
(d) A_0 represents $\frac{1}{2}$ (mass of the spring)

6.14 The differential equation of a damped mechanical oscillator is

(a) $m\dfrac{d^2x}{dt^2} + kx = 0$
(b) $m\dfrac{d^2x}{dt^2} + \gamma\dfrac{dx}{dt} + kx = 0$
(c) $m\dfrac{d^2x}{dt^2} + \gamma\dfrac{dx}{dt} + kx = F_0 \sin \omega t$
(d) $m\dfrac{d^2x}{dt^2} + \gamma\dfrac{dx}{dt} + kx = F_0 \cos \omega t$

6.15 The differential equation of motion of a damped mechanical oscillator is $m\dfrac{d^2x}{dt^2} + \gamma\dfrac{dx}{dt} + kx = 0$. If the motion is oscillatory, the angular frequency is given by

(a) $\omega = \sqrt{\dfrac{k}{m}}$
(b) $\omega = \sqrt{\dfrac{k}{m} + \dfrac{\gamma^2}{4m^2}}$
(c) $\omega = \sqrt{\dfrac{k}{m} - \dfrac{\gamma^2}{4m^2}}$
(d) $\omega = \sqrt{\dfrac{k}{m} - \dfrac{\gamma^2}{2m^2}}$

6.16 The equation of motion for a torsional pendulum is $I\ddot{\theta} + \Gamma\dot{\theta} + k\theta = 0$ where I, Γ and k are constants for pendulum. If $\tau^2 = 4kI$, the solution of given equation will be
(a) $\theta = Ae^{-\alpha t} \cos(\omega t + B)$
(b) $\theta = Ae^{-\alpha t} + Be^{-\beta t}$
(c) $\theta = (A + Bt)e^{-\alpha t}$
(d) $\theta = Ae^{-\alpha t}$

6.17 Quality factor (Q) of an oscillator is formulated as
(a) $Q = 2\pi \times \dfrac{\text{Energy loss per second}}{\text{Energy stored}}$
(b) $Q = 2\pi \times \dfrac{\text{Energy stored}}{\text{Energy loss per second}}$
(c) $Q = \sqrt{2\pi} \times \dfrac{\text{Energy loss per second}}{\text{Energy stored}}$
(d) $Q = \sqrt{2\pi} \times \dfrac{\text{Energy stored}}{\text{Energy loss per second}}$

6.18 At low velocities, damping force is
(a) directly proportional to the square of velocity
(b) directly proportional to the square root of velocity
(c) directly proportional to velocity
(d) inversely proportional to velocity

6.19 In case of underdamped oscillations, the amplitude of oscillations of an object with time.
(a) decreases
(b) increases
(c) remains constant
(d) may decrease or increase

6.20 The amplitude of damped harmonic oscillator
(a) remains constant
(b) decreases exponentially
(c) suddenly becomes zero without any oscillation
(d) increases exponentially

6.21 The meaning of high quality factor of an oscillator
(a) damping is more
(b) damping is infinite
(c) damping is zero
(d) damping is small

6.22 The relation of the energy of damped harmonic oscillator is
(a) $E = E_0 e^{-t/\tau}$
(b) $E = E_0 e^{-2t/\tau}$
(c) $E = E_0 e^{-t/2\tau}$
(d) none of these

6.23 $\dfrac{d^2 y}{dt^2} + 2b \dfrac{dy}{dt} + \omega_0^2 y = 0$ is the equation of a damped harmonic oscillator. The oscillation, will be underdamped, if
(a) $b > \omega_0$
(b) $b = \omega_0$
(c) $b < \omega_0$
(d) $b \gg \omega_0^2$

6.24 The quality factor of LCR damped oscillator of zero resistance is given by
(a) $\dfrac{R}{L\omega}$
(b) $\dfrac{\omega}{RL}$
(c) $\dfrac{L\omega}{R}$
(d) $L\omega R$

6.25 A capacitor was discharged through a ballistic galvanometer and a throw θ, was observed in the lamp and scale arrangement. The correct value of throw is
(a) $\theta_1 \left(1 + \dfrac{\lambda}{4} \right)$
(b) $\theta_1 \left(1 + \dfrac{\lambda}{2} \right)$
(c) $\theta_1 \left(1 - \dfrac{\lambda}{2} \right)$
(d) $\theta_1 \left(1 - \dfrac{\lambda}{4} \right)$

6.26 Vibrations of the screen of a microphone are
(a) free oscillations
(b) damped oscillations
(c) resonant oscillations
(d) forced oscillations

6.27 In an LCR series circuit to which an alternating e.m.f. is applied, the sharpness of resonance or quality factor is defined by

(a) $Q = \dfrac{R}{\omega_0^2}$ (b) $Q = \omega_0 CR$

(c) $Q = \dfrac{V_0}{V_L}$ (d) $Q = \dfrac{V_c}{V_0}$

6.28 The phase of the displacement of a forced mechanical oscillator:

(a) lags the driving force by $\left(\dfrac{\pi}{2}+\theta\right)$ (b) lags the driving force by $\left(\dfrac{\pi}{2}-\theta\right)$

(c) leads the driving force by $\left(\dfrac{\pi}{2}-\theta\right)$ (d) leads the driving force by $\left(\dfrac{\pi}{2}+\theta\right)$

6.29 In an LCR circuit the sharpness of the resonance curve
(a) decreases with the increase of the inductance L
(b) decreases with the decrease of the capacitance C
(c) does not depend upon the value of L, C and R
(d) decreases with the increase of the resistance R

6.30 The differential equation of a forced mechanical oscillator is

(a) $m\dfrac{d^2 x}{dt^2} + \gamma \dfrac{dx}{dt} + Cx = F_0 \sin pt$ (b) $m\dfrac{d^2 x}{dt^2} + \gamma \dfrac{dx}{dt} + Cx = 0$

(c) $m\dfrac{d^2 x}{dt^2} + Cx = 0$ (d) $\dfrac{d^2 x}{dt^2} + \omega^2 x = 0$

6.31 The phenomenon in which the amplitude of driven oscillator becomes maximum at a particular driven frequency is called amplitude resonance and this frequency is known as
(a) driven frequency (b) resonant frequency
(c) threshold frequency (d) Larmor frequency

6.32 The rate of fall in amplitude with the change of forcing frequency on each side of resonance frequency is referred as
(a) resonance (b) sharpness of resonance
(c) half width of resonance curve (d) none of the above

6.33 The steady state solution of equation $\dfrac{d^2 x}{dt^2} + 2k\dfrac{dx}{dt} + \omega_0^2 x = F_0 \sin pt$ is

(a) $x = A \sin(pt - \theta)$ (b) $x = A \cos(pt - \theta)$
(c) $x = A \tan(pt - \theta)$ (d) $x = A \cot(pt - \theta)$

6.34 If $Q = 50$, then the value of $\dfrac{A}{A_{max}}$; when $\dfrac{p}{\omega_0} = 0.98$

(a) 0.32 (b) 0.50
(c) 0.45 (d) 0.60

6.35 A coil of 3 mH is given. Then the capacity, which will make resonance frequency 1000 kHz is
(a) 4.46 pF (b) 2.23 pF
(c) 17.61 pF (d) 8.85 pF

Answers

6.1 (b) 6.2 (b) 6.3 (c) 6.4 (d) 6.5 (c) 6.6 (a) 6.7 (c) 6.8 (c)
6.9 (d) 6.10 (d) 6.11 (d) 6.12 (c) 6.13 (b) 6.14 (b) 6.15 (c) 6.16 (a)
6.17 (b) 6.18 (c) 6.19 (a) 6.20 (b) 6.21 (d) 6.22 (a) 6.23 (c) 6.24 (c)
6.25 (b) 6.26 (d) 6.27 (d) 6.28 (a) 6.29 (d) 6.30 (a) 6.31 (b) 6.32 (b)
6.33 (a) 6.34 (c) 6.35 (a)

CHAPTER 7

Wave Motion

"In a wave, information and energy in the form of signals, propagate from one point to another but no materials object makes the journey."

—Sher Singh

IN THIS CHAPTER

- What Propagates in Wave Motion?
- Characteristics of Wave Motion
- Types of Wave Motion
- Important Definitions
- Differential Equation of Wave Motion
- Plane Progressive Wave in Fluid Media
- Pressure Variation in Longitudinal Wave (Acoustic Pressure)
- Energy of Plane Progressive Wave
- Reflection and Transmission of Transverse Waves in a String
- Reflection and Transmission of Longitudinal Waves at Discontinuity
- Superposition of Progressive Waves— Stationary Wave
- Comparison between Progressive Waves and Stationary Waves

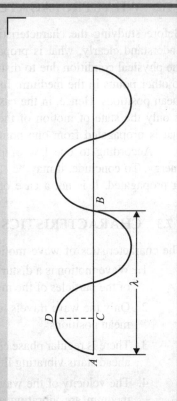

7.1 INTRODUCTION

Wave motion is a form of disturbance which travels through the medium due to the repeated periodic motion of the particles of the medium about their mean positions, the disturbance being handed over from one particle to the next. When a stone is dropped into a pond containing water, waves are produced at the point where the stone strikes the water in the pond. The waves travel outward, the particles of water vibrate only up and down about their mean positions. Water particles do not travel along the wave. Similarly, when a tuning fork is set into vibration, it produces waves in air. The wave travels from one particle to the next but the particles of air vibrate about their mean position.

It is essential to understand the concept of wave motion in the study of various branches of physics and engineering. Wave motion, in general, refers to the transfer of energy from one point to another point of the medium. Transference of various forms of energy like sound, heat, light, X-rays, γ-rays, radio-wave, etc., takes place in the form of wave motion. For the transference of energy through a medium must possess the properties of elasticity, inertia and negligible frictional resistance.

7.2 WHAT PROPAGATES IN WAVE MOTION?

Before studying the characteristics of the different forms of wave motion, it is essential to understand clearly, what is propagated in a wave motion? The answer to this question is that the physical condition due to disturbance generated at some points in the medium is propagated to other points in the medium. In all the waves, the particles of the medium vibrate about their mean positions. Hence, in the case of wave motion, it is not the matter that is propagated but it is only the state of motion of the matter that is propagated. It is a form of dynamic condition that is propagated from one point to the other point in the medium.

According to one law of physics, any dynamic condition is related to momentum and energy. To conclude, it may be said that in wave motion momentum and energy are transferred or propagated. It is not a case of propagation of matter as a whole.

7.3 CHARACTERISTICS OF WAVE MOTION

The characteristics of wave motion are given as follows:

1. Wave motion is a disturbance produced in the medium by the repeated periodic motion of the particles of the medium.
2. Only the wave travels forward whereas the particles of the medium vibrate about their mean positions.
3. There is regular phase change between the various particles of the medium. The particle ahead starts vibrating little later than the particles just preceding it.
4. The velocity of the wave is different from the velocity with which the particles of the medium are vibrating about their mean positions. The wave travels with a uniform velocity whereas the velocity of the particles is different at different positions. It is maximum at mean position and zero at extreme position of the particles.

Wave Motion

7.4 TYPES OF WAVE MOTION

There are two types of wave motions:
   (i) Transverse wave motion
   (ii) Longitudinal wave motion

Sound waves are longitudinal waves and light waves are transverse waves.

7.4.1 Transverse Wave Motion

In this type of motion, the particles of the medium vibrate at right angles to the direction of propagation of the wave.

To understand the propagation of transverse waves in a medium consider nine particles of the medium and the circle of reference as shown in Figure 7.1. The particles are vibrating about their mean positions up and down and the wave is travelling from left to right. The disturbance takes $T/8$ seconds to travel from one point to the next.

1. At $t = 0$, all particles are at their mean positions.
2. After $T/8$ seconds, the particle 1 travels a certain distance upward and the disturbance reaches particle 2.
3. After $2T/8$ seconds, particle 1 has reached its extreme position and the disturbance has reached particle 3.

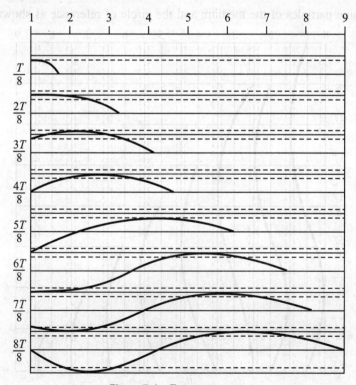

Figure 7.1 Transverse wave.

4. After 3T/8 seconds, particle 1 has completed 3/8 of its vibration and the disturbance has reached particle 4. The positions of particles 2 and 3 are also shown in Figure 7.1.
5. In this way after T/2 seconds, particle 1 has come back to its means position and the particles 2, 3 and 4 are at the positions shown in Figure 7.1. The disturbance has reached particle 5.

In this way the process continues and the positions of the particles after 5T/8, 6T/8, 7T/8 and T seconds are shown in Figure 7.1.

After T seconds, the particles 1, 5 and 9 are at their mean positions. The wave has reached particle 9. The particles 1 and 9 are in the same phase. The wave has travelled a distance between particles 1 and 9 in the time in which the particle 1 has completed on vibration.

The top point on the wave at the maximum distance from the mean position is called *crest*, while the point at the maximum distance below the mean position is called *trough*. Thus, in a transverse wave, crests and troughs are alternately formed. The contour of the displaced particles of the medium represents the wave. In case of transverse (or longitudinal) progressive waves, this contour continuously changes position in space and the wave seems to advance in the direction of propagation.

7.4.2 Longitudinal Wave Motion

In this type of wave motion, particles of the medium vibrate along the direction of propagation of the wave.

Consider nine particles of the medium and the circle of reference as shown in Figure 7.2.

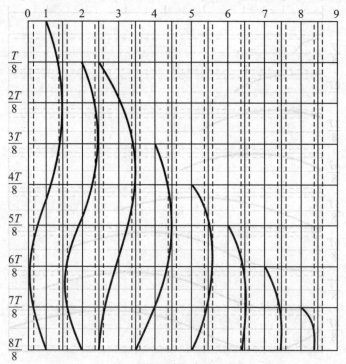

Figure 7.2 Longitudinal wave.

The wave travels from left to right and the particles vibrate about their mean positions. After $T/8$ seconds, the particle 1 goes to the right and completes 1/8 of its vibration. The disturbance reaches the particle 2. After $T/4$ seconds the particle 1 has reached its extreme right position and completes 1/4 of its vibration and particle 2 completes 1/8 of its vibration. The disturbance reaches the particle 3. The process continues.

After one complete time period, the positions of the various particles are as shown in Figure 7.2. The wave has reached particle 9. Here, particles 1 and 9 are again in the same phase. Here, particles 1, 5 and 9 are at their mean positions. The particles 1 and 3 are close to the particle 2. This is position of condensation. Similarly, particles 9 and 8 are close to the particle 7. This is also position of condensation or compression. On the other hand, the particles 4 and 6 are far away from the particle 5. This is position of rarefaction. Hence, in a longitudinal wave motion, condensation and rarefaction are alternately formed.

7.5 IMPORTANT DEFINITIONS

There are following important terms, which are defined as given as follows:

1. *Wavelength:* It is the distance travelled by the wave in the time in which the particle of the medium completes one vibration. It is also defined as the distance between two nearest particles in the same phase.

   The distance AB is equal to the wavelength (λ) as shown in Figure 7.3.

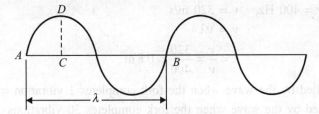

Figure 7.3 Wave motion.

2. *Frequency:* It is the number of vibrations made by the particle in one second.
3. *Amplitude:* It is maximum displacement of the particle from its mean of the rest. CD is amplitude in Figure 7.3.
4. *Time period:* It is the time taken by a particle to complete one vibration.
   Suppose frequency $= \nu$
   Time taken to complete ν vibration $= 1$ s
   Time taken to complete 1 vibration $= 1/\nu$ s
   From the definition of time period, time taken to complete one vibration is the time period (T),

   $\therefore$ $\qquad\qquad T = 1/\nu \quad$ or $\quad \nu \times T = 1$

   Frequency $\times$ Time period $= 1$

5. *Vibration:* It is the to and fro motion of a particle from one extreme position to the other and back again. It is also equal to the motion of a particle from the mean position to one extreme position, then to the other extreme position and finally back to the mean position.

6. *Phase:* It is defined as the ratio of the displacement of the vibrating particles at any instant to the amplitude of the vibrating particle or it is defined as the fraction of the time interval that has elapsed since the particle crossed the mean position or it is equal to the angle swept by the radius vector since the last vibrating particle crossed its mean position of rest.

7. *Relation between frequency and wavelength:* Velocity of the wave is the distance travelled by the wave in one second.

$$\text{Velocity} = \text{Distance/Time}$$

Wavelength (λ) is the distance travelled by the wave in one time period (T),

$$\text{Velocity} = \text{Wavelength/Time period}$$

$$v = \lambda/T$$

But Frequency × Time period = 1

$$\nu \times T = 1$$

$$T = 1/\nu$$

$$v = \lambda/T$$

$$\therefore \quad \text{Velocity } (v) = \nu\lambda \tag{7.1}$$

EXAMPLE 7.1 If the frequency of a tuning fork is 400 Hz and the velocity of sound in air is 320 m/s, find how far does the sound travel while the fork completes 30 vibrations?

Solution Given $\nu = 400$ Hz, $v = 320$ m/s

$$v = \nu\lambda$$

or

$$\lambda = \frac{v}{\nu} = \frac{320}{400} = 0.8 \, \text{m}$$

$\therefore$ Distance travelled by the wave when the fork completes 1 vibration = 0.8 m.
So, distance travelled by the wave when the fork completes 30 vibrations
$$= 0.8 \times 30$$
$$= 24 \, \text{m}$$

7.6 DIFFERENTIAL EQUATION OF WAVE MOTION

We will obtain the wave equation for one-dimensional. Consider a harmonic wave travelling in a medium with velocity v. Consider two planes A and B separated by a distance x as shown in Figure 7.4.

Let us assume that the wave is created at $x = 0$ (plane P) by giving harmonic oscillations of amplitude A and period T to particles at the plane A be given by

$$\psi(0, t) = A \sin \frac{2\pi t}{T} \tag{7.2}$$

where $\psi(0, t)$ stands for the displacement of particles located at plane P at time t. Let us now consider the motion of particles located at plane Q at a distance x from plane P (Figure 7.4). The wave created at A will reach Q in x/v seconds.

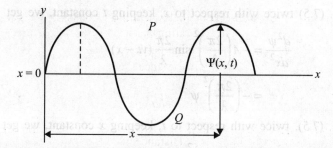

Figure 7.4 Motion of particles at planes A and B separated by a distance x.

Displacement of particles at plane Q located at x must be same as the displacement particles at a plane located at $x = 0$ had x/v seconds earlier. In other words, displacement of particle at x at time t = displacement of particles at $x = 0$ at time $t = t - x/v$.

$$\therefore \quad \psi(x,t) = A \sin \frac{2\pi t}{T} = A \sin \left\{ \frac{2\pi}{T} \left(t - \frac{x}{v} \right) \right\}$$

or
$$\psi(x,t) = A \sin \left\{ \frac{2\pi}{T} \left(t - \frac{x}{v} \right) \right\} \qquad (7.3)$$

Equation (7.3) gives the displacements of the particles of the continuous medium as a function of x and t as a harmonic wave travels in the +x-direction with a velocity v.

For a wave travelling in the negative x-direction, the corresponding equation is

$$\psi(x,t) = A \sin \frac{2\pi}{T} \left(t + \frac{x}{v} \right) \qquad (7.4)$$

Equation (7.3) can be written in a more compact form, by defining $k = \frac{2\pi}{\lambda}$, $v = \frac{\lambda}{T}$ and $\omega = \frac{2\pi}{T}$, thus

$$\psi(x,t) = A \sin \frac{2\pi}{T} (vt - x) \qquad (7.5)$$

$$\psi(x, t) = A \sin(\omega t - kx) \qquad (7.6)$$

This is the displacement of wave travelling in the positive x-direction, k is called the wave number of the wave and ω is the angular frequency of particle oscillations in a wave. The corresponding displacement for the wave travelling in the negative x-direction [Eq. (7.3)] is

$$\psi(x,t) = A \sin \frac{2\pi}{\lambda} \left(t + \frac{x}{v} \right)$$

or
$$\psi(x,t) = A \sin \frac{2\pi}{\lambda} (vt + x)$$

or
$$\psi(x, t) = A \sin(\omega t + kx) \qquad (7.7)$$

It may be mentioned that the above equations can also be represented by cosine function instead of sine function.

Differentiating Eq. (7.5) twice with respect to x, keeping t constant, we get

$$\frac{d^2\psi}{dx^2} = -A\left(\frac{2\pi}{\lambda}\right)^2 \sin\frac{2\pi}{\lambda}(vt-x)$$

$$= -\left(\frac{2\pi}{\lambda}\right)^2 \psi \qquad (7.8)$$

Differentiating Eq. (7.5), twice with respect to t, keeping x constant, we get

$$\frac{d^2\psi}{dt^2} = -A\left(\frac{2\pi v}{\lambda}\right)^2 \sin\frac{2\pi}{\lambda}(vt-x)$$

$$= -\left(\frac{2\pi v}{\lambda}\right)^2 \psi \qquad (7.9)$$

From Eqs. (7.8) and (7.9), we have

$$\frac{d^2\psi}{dt^2} = v^2 \frac{d^2\psi}{dx^2} \qquad (7.10)$$

Equation (7.10) is said to be differential equation for wave motion.

We may draw the following points on the basis of Eq. (7.10):

(i) Whenever the second order time derivative $d^2\psi/dt^2$ of any physical quantity is related to the second order space derivative $d^2\psi/dx^2$, a wave of some sort must travel in the medium.
(ii) The velocity v of that wave is given by the square root of the coefficient of the second-order space derivative.

Further, the individual particles which make up medium do not progress through the medium with the wave, they merely oscillate (transversely or longitudinally) about their equilibrium positions. It is their phase relationships, which we observe as waves. Therefore, the wave velocity is also called *phase velocity*. It is the velocity with which planes of equal phase travel through the medium. The phase velocity of a wave is given by

$$v = \nu\lambda = 2\pi\nu \times \frac{\lambda}{2\pi}$$

or

$$v = \frac{\omega}{k} \qquad (7.11)$$

$$\left[\text{Here } \omega = 2\pi\nu \text{ and } k = \frac{2\pi}{\lambda}\right]$$

The complete of one-dimensional wave Eq. (7.10) can be written as

$$\frac{d^2\psi(x,t)}{dt^2} = v^2 \frac{d^2\psi(x,t)}{dx^2} \qquad (7.12)$$

The solution of Eq. (7.12) will be a function of variable x and t. We will show that any function of the form $\psi(x, t) = f(vt - x)$ or of the form $\psi(x, t) = f(vt + x)$ is the possible solution. Let us consider the function

$$\psi(x, t) = f(vt - x) = f(z)$$

where $z = vt - x$.

Differentiating with respect to t, keeping x fixed, we have

$$\frac{d\psi}{dt} = v\frac{df}{dz}$$

or

$$\frac{d^2\psi}{dt^2} = v^2\frac{d^2f}{dz^2}$$

Now, differentiating with respect to x, keeping t fixed, we have

$$\frac{d\psi}{dt} = -\frac{df}{dz}$$

and

$$\frac{d^2\psi}{dx^2} = +\frac{d^2f}{dz^2}$$

so that

$$\frac{d^2\psi}{dt^2} = v^2\frac{d^2\psi}{dx^2}$$

This is the differential wave equation. Hence, $\psi(x, t) = f(vt - x)$ is a solution of the wave equation. We can similarly show that $\psi(x, t) = f(vt + x)$ is also the solution of the same equation.

If $\psi(x, t)$ is a simple harmonic displacement of an oscillator at position x and time t, we would expect this function to be a sine or cosine function, $(vt - x)$ has the dimensions of a length. Since the argument of a sine or cosine function must have the dimension of radians, the $(vt - x)$ must be multiplied by a factor $2\pi/\lambda$, where λ is a length. Thus, we may write

$$\psi(x,t) = A \sin \frac{2\pi}{\lambda}(vt - x)$$

where A has dimensions of ψ. We recognize that the length λ is the wavelength and this solution represents a wave travelling in the positive x-direction. Similar arguments hold for the function $\psi(x, t) = f(vt + x)$ which represents the wave travelling in the negative x-direction.

EXAMPLE 7.2 When a plane wave traverses a medium, the displacement of particles is given by $\psi(x, t) = 0.01 \sin(4\pi t - 0.02\pi x)$ when ψ and x are expressed in metres and t in seconds. Calculate: (i) the amplitude, wavelength, velocity and frequency of the wave, and (ii) the phase difference between two positions of the same particle at a time interval of 0.25 s.

Solution The given equation is

$$\psi(x, t) = 0.01 \sin(4\pi t - 0.02\pi x)$$

This may be written as

$$\psi(x,t) = 0.01 \sin \frac{2\pi}{100}(200t - x)$$

Comparing it with the wave equation

$$\psi(x,t) = A \sin \frac{2\pi}{\lambda}(vt - x)$$

which gives

(a) Amplitude $A = 0.01$ m, wavelength $\lambda = 100$ m

velocity $v = 200$ m/s, frequency $\nu = \dfrac{v}{x} 2$ Hz

(b) Phase change in time interval Δt is

$$\frac{2\pi}{T} = 2\pi \nu \Delta t$$

or phase difference $= 2\pi \times 2 \times 0.25 = \pi = 180°$.

The particle phase is reversed in a time 0.25 s. It is evident as its period

$$T = \frac{1}{\nu} = \frac{1}{2} = 0.2 \text{ s}$$

EXAMPLE 7.3 A transverse harmonic wave of amplitude 0.02 m is generated at one end ($x = 0$) along horizontal string by a tuning fork of frequency 500 Hz. At a given instant of time, the displacement of the particle at $x = 0.1$ m is 0.005 m and that of the particle at $x = 0.2$ m is 0.005 m. Calculate the wavelength and the velocity of the wave. Obtain the equation of the wave assuming that the wave is travelling along the positive x-direction and the end $x = 0$ is at equilibrium position at $t = 0$.

Solution The general equation of this wave is given by

$$\psi(x,t) = A \sin \frac{2\pi}{\lambda}(vt - x)$$

Here $A = 0.02$ m

Again, when $x = 0.1$ m, $\psi = -0.005$ m

$\therefore$
$$-0.005 = 0.02 \sin \frac{2\pi}{\lambda}(vt - 0.1)$$

or
$$\sin \frac{2\pi}{\lambda}(vt - 0.1) = -\frac{0.005}{0.02} \text{ m}$$

$$= -0.25 \text{ m}$$

$\therefore$ Phase $\phi_1 = \dfrac{2\pi}{\lambda}(vt - 0.1) = \sin^{-1}(-0.25)$

Again $x = 0.2$ m, $\psi = 0.005$ m

$$0.005 = 0.02 \sin \frac{2\pi}{\lambda}(vt - 0.2)$$

$\therefore$ Phase $\phi_2 = \dfrac{2\pi}{\lambda}(vt - 0.2) = \sin^{-1}\dfrac{0.005}{0.02}$

$$= \sin^{-1}(0.25)$$

Phase difference $\Delta\phi = \phi_1 - \phi_2$

$$= \sin^{-1}(-0.25) - \sin^{-1}(0.25) = \pi$$

But
$$\Delta\phi = \frac{2\pi}{\lambda}\Delta x$$

$$\pi = -\frac{2\pi}{\lambda}(0.1 - 0.2)$$

or $\quad\quad\quad\quad\quad\quad\quad\quad \lambda = 0.2$ m

Frequency $v = 500$ Hz (given)

$\therefore\quad\quad\quad\quad$ Velocity $v = v\lambda = 500 \times 0.2$ m $= 100$ m/s

The equation of the wave is given by

$$\psi = 0.02 \sin\left[\frac{2\pi}{0.2}(100t - x)\right]$$
$$= 0.02 \sin[10\pi(100t - x)]$$

7.7 PLANE PROGRESSIVE WAVE IN FLUID MEDIA

In a fluid media, only longitudinal wave can propagate as a gas or a liquid cannot sustain the transverse shear necessary for the propagation of transverse waves, we know the sound waves propagate in the form of compressions (particles come closer) and rarefaction (particle move apart). Naturally, as the sound waves propagate, variation in pressure and density of the medium takes place.

Let us consider a given (constant) mass of a fluid at equilibrium pressure P_0, volume V_0 and density ρ_0. Suppose, when the wave propagates, the pressure changes to $P = P_0 + p$, the volume becomes equal to $V = V_0 + v$, while the new density becomes $\rho = \rho_0 + \rho_d$, where $p = dP$ and $v = dV$.

Let the maximum pressure amplitude be denoted by P_m and dP is the fluctuating component superimposed on the equilibrium pressure P_0.

When the sound waves propagate, the changes in the medium are of an extremely small order and sets the limit within which the wave equation is approximate. The fraction change in volume $\frac{V}{V_0} = \delta$, is called the *dilation* while the fractional change in density $\frac{\rho_d}{\rho} = s$, is called *condensation*. For normal sound waves, the value of δ and s are of the order of 10^{-3} and a value of $P_m = 2 \times 10^{-5}$ Nm^{-2} ($\cong 10^{-10}$ atmosphere) gives a sound wave, which is audible at 1 kHz.

As the mass of the fluid is constant

$$\rho_0 V_0 = \rho V = \rho_0 V_0 (1 + \delta)(1 + s)$$

or $\quad\quad\quad\quad\quad\quad (1 + \delta)(1 + s) = 1$

giving to a close approximation

$$\delta = -s \quad\quad\quad\quad\quad\quad (7.13)$$

That is, dilation is equal and opposite to condensation.

The elastic property of a gas is a measure of its compressibility and is defined in terms of its bulk modulus B, given by

$$B = -\frac{dP}{dV/V} = -V\frac{dP}{dV} \tag{7.14}$$

Thus, the bulk modulus is the ratio of change in pressure for a fractional change in volume and the negative sign indicates that an increase in volume is accompanied by a fall of pressure. Laplace had correctly argued that when sound waves propagate through fluid, the compression and rarefaction, succeed each other rapidly. As gas (air) is a poor conductor of heat, the equalization of temperature is not possible across the region of compression and rarefaction, (i.e., the condition is not isothermal). However, the total heat content of the system remains constant, i.e., the condition is adiabatic, in the medium during the propagation of sound waves.

$$PV^\gamma = \text{constant} \tag{7.15}$$

where γ is the ratio of specific heats at constant pressure and constant volume.

Differentiate Eq. (7.15) we have

$$V^\gamma dP + \gamma P V^{\gamma-1} dV = 0$$

or

$$V^\gamma dP = -\gamma P V^{\gamma-1} dP$$

or

$$-V\frac{dP}{dV} = \gamma P = B_a \tag{7.16}$$

The subscript a denotes adiabatic. As for a given gas, its bulk modulus is constant, the elastic property of the gas γP is also a constant. We know

$$P = P_0 + p \text{ or } dP = p, \text{ the excess pressure, giving}$$

$$B_a = -\frac{p}{V/V_0}$$

or

$$p = -B_a \delta$$

$$= B_a s \tag{7.17}$$

7.7.1 Wave Equation of Propagation of Sound Waves in Fluid Media

In a sound wave, the particle displacements and velocities are along the direction of wave propagation, say along x-axis.

Let us consider the displacement is $\psi(x, t)$ and the motion of an element of the fluid medium of thickness Δx and unit cross-section under the influence of the sound waves. Figure 7.5 depicts the displacements of the element. The particles in the layer x are displaced by a distance ψ, while those at $x + \Delta x$ are displaced by distance $\psi + \Delta\psi$. Thus, the increase in thickness Δx of the element of unit cross-section, which equals the increase in volume, i.e.,

$$\Delta\Psi = \frac{d\Psi}{dx}\Delta x$$

and

$$\delta = \frac{V}{V_0} = \left(\frac{d\Psi}{dx}\right)\frac{\Delta x}{\Delta x} = \frac{d\Psi}{dx} = -s$$

where $\frac{d\Psi}{dx}$ is strain.

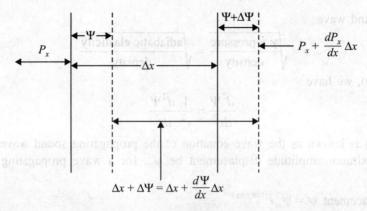

Figure 7.5 Wave propagation in fluid medium.

Because the pressure along the x-axis on either sides of the thin element are not balance, the medium gets deformed. The net force acting on the element is given by

$$P_x - P_{x+\Delta x} = \left[P_x - \left(P_x + \frac{dP_x}{dx}\Delta x\right)\right] = -\frac{dP_x}{dx}\Delta x$$

$$= -\frac{d}{dx}(P_0 + p)\Delta x = -\frac{dp}{dx}\Delta x$$

The mass of the considered element is equal to $\rho_0 \Delta x$ and its acceleration is $\frac{d^2\psi}{dx^2}$. Using Newton's second law of motion, we get

$$-\frac{dP}{dx}\Delta x = \rho_0 \Delta x \frac{d^2\psi}{dt^2} \qquad (7.18)$$

where

$$P = -B_a \delta = -B_a \frac{d\psi}{dx}$$

so that,

$$-\frac{dP}{dx} = B_a \frac{d^2\psi}{dt^2} \qquad (7.19)$$

From Eqs. (7.18) and (7.19), we have

$$B_a \frac{d^2\psi}{dt^2} = \rho_0 \frac{d^2\psi}{dt} \qquad (7.20)$$

As, $\frac{B_a}{\rho_0} = \frac{\gamma P}{\rho_0}$ is the ratio of the elasticity to the inertia or density of the gas and this ratio has dimension

$$\frac{\text{force}}{\text{area}} \times \frac{\text{volume}}{\text{mass}} = (\text{velocity})^2$$

∴

$$\frac{\gamma p}{\rho_0} = v^2$$

or speed of sound wave

$$v = \sqrt{\frac{\gamma \times \text{pressure}}{\text{density}}} = \sqrt{\frac{\text{adiabatic elasticity}}{\text{density}}} \tag{7.21}$$

From Eq. (7.20), we have

$$\frac{d^2\Psi}{dx^2} = \frac{1}{v^2}\frac{d^2\Psi}{dt^2} \tag{7.22}$$

Equation (7.22) is known as the wave equation of the propagating sound wave through fluid media. Let maximum amplitude displacement be ψ_m, for a wave propagating along x-axis, then we have

(i) Displacement $\psi = \psi_m e^{i(\omega t - kx)}$

(ii) Particle velocity $v = \dot{\psi} = \dfrac{d\psi}{dt} = i\omega \psi_m e^{i(\omega t - kx)} = i\omega\psi$

(iii) Condensation $\delta = \dfrac{\delta\psi}{dx} = -ik\psi = -s$

(iv) Dilation $s = ik\psi$

(v) Excess pressure $P = B_a s = iB_a k\psi$

The phase relationships between these parameters are visualized in Figure 7.6, when the wave is propagating in positive x-direction. The excess pressure P, the fractional density increase and the particle velocity v are all $\pi/2$ radian ahead in phase compared to displacement ψ. On the other hand, the volume change (π radian out of phase with density change) is $\pi/2$ radian behind the displacement.

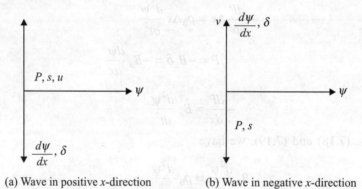

(a) Wave in positive x-direction   (b) Wave in negative x-direction

Figure 7.6 Phase relationships between the particle ψ, particle velocity v, excess pressure P and condensation $s = -\delta$ (dilation) with maximum condensation s_n.

In the case of wave propagating along negative x-direction, the expressions for various parameters will be given as

(i) Displacement $\psi = \psi_m e^{i(\omega t - kx)}$

(ii) Particle velocity $v = \dot{\psi} = \dfrac{d\psi}{dt} = i\omega \psi_m e^{i(\omega t - kx)} = i\omega\psi$

(iii) Condensation $\delta = \dfrac{\delta \psi}{dx} = +ik\psi$

(iv) Dilation $s = -\delta = -k\psi$

(v) Excess pressure $P = B_a s = -iB_a k\psi$

The particle displacement in both the cases is measured in the positive x-direction and the thin element of the gaseous (fluid) medium Δx, oscillates about the value $\psi = 0$, which defines its equilibrium or central position. For a wave in the positive r-direction, the value of $\psi = 0$ with $\dot{\psi}$ a maximum in the positive x-direction gives a maximum positive excess pressure (or compression) with maximum condensation s_n (maximum density) and maximum volume. For a wave propagating in negative x-direction, the value of $\psi = 0$, with a $\dot{\psi}$ maximum in positive x-direction gives a maximum negative excess pressure (or rarefaction), a maximum volume and a minimum density.

7.8 PRESSURE VARIATION IN LONGITUDINAL WAVE (ACOUSTIC PRESSURE)

Let us consider a simple harmonic longitudinal wave travelling along the positive x-direction with velocity v in a medium. The displacement at x is given by

$$\psi = A \sin \dfrac{2\pi}{\lambda}(vt - x)$$

The excess pressure over that of atmosphere is

$$\delta P = -k \dfrac{d\psi}{dx} = \dfrac{2\pi A k}{\lambda} \cos \dfrac{2\pi}{\lambda}(vt - x) \qquad (7.23)$$

where $k = \dfrac{\delta P}{-\dfrac{dy}{dx}}$ = bulk modulus of elasticity in medium

Now, the maximum acoustic pressure is

$$(\delta P)_m = \dfrac{2\pi A k}{\lambda} \qquad (7.24)$$

$(\delta P)_m$ is known as the acoustic pressure.

The root mean square acoustic pressure is found out as follows:

$$\text{Mean squared pressure} = \dfrac{1}{T} \int_0^T \left(\dfrac{2\pi A k}{\lambda}\right)^2 \cos^2 \dfrac{2\pi}{\lambda}(vt - x)\, dt$$

$$= \dfrac{1}{T}\left(\dfrac{2\pi A k}{\lambda}\right)^2 \dfrac{T}{2} = \dfrac{1}{2}\left(\dfrac{2\pi A k}{\lambda}\right)^2$$

$\therefore$ Root mean square acoustic pressure

$$[(\delta P)_m]_{\text{RMS}} = \sqrt{2}\, \dfrac{\pi A k}{\lambda} \qquad (7.25)$$

7.8.1 Relation between ψ and (δP)

The displacement ψ of a particle in a progressive wave at any time t is given by

$$\psi = A \sin \frac{2\pi}{\lambda}(vt - x) \tag{7.26}$$

and excess pressure may be expressed as

$$(\delta P) = -k \frac{d\psi}{dx} = \frac{2\pi A k}{\lambda} \cos\left[\frac{2\pi}{\lambda}(vt - x)\right] \quad \left[\because \frac{d\psi}{dx} = -\frac{2\pi A}{\lambda} \cos\frac{2\pi}{\lambda}(vt - x)\right]$$

$$\frac{2\pi A k}{\lambda} \sin \frac{2\pi}{\lambda}\left(vt - x + \frac{\pi}{2}\right) \tag{7.27}$$

Comparing Eqs. (7.26) and (7.27), we note that the pressure wave is $\pi/2$ ahead of the sound wave as shown in Figure 7.7. It depicts the variation of Ψ and δP with x.

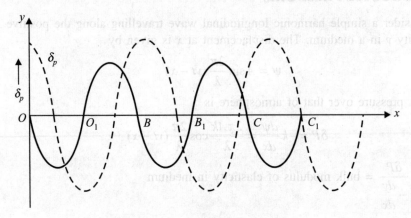

Figure 7.7 Variation of Ψ and δp with x.

It is evident that SP is maximum at O, B, C, at which ψ is minimum and δp is minimum at O, B, C, etc., at which $\psi = 0$.

7.9 ENERGY OF PLANE PROGRESSIVE WAVE

During the propagation of a progressive wave carries energy. By the energy of the wave, we mean the sum of energy of all the particles of the medium through which it travels. If ψ is the displacement of particle at any time t, then

$$\psi = A \sin \frac{2\pi}{\lambda}(vt - x)$$

If ρ is the density of the medium, then kinetic energy per unit volume is

$$E_k = \frac{1}{2}\rho\left(\frac{d\Psi}{dt}\right)^2$$

$$= \frac{1}{2}\rho\frac{4\pi^2}{\lambda^2}v^2 A^2 \cos^2\frac{2\pi}{\lambda}(vt-x)$$

$$= \frac{2\pi^2 v^2 A^2 \rho}{\lambda^2}\cos^2\frac{2\pi}{\lambda}(vt-x) \tag{7.28}$$

Now, potential energy

= Work done against the restoring force due to small displacement
= $d\psi \times$ opposing force

$$= d\psi \times \rho\frac{d^2\psi}{dt^2}$$

$$= \rho\frac{4\pi^2}{\lambda^2}v^2 A \sin\frac{2\pi}{\lambda}(vt-x)d\psi$$

$$= \rho\frac{4\pi^2 v^2}{\lambda^2}\psi\, d\psi$$

So, potential energy per unit volume due to displacement ψ is

$$E_P = \int_0^\psi \rho\frac{4\pi^2 v^2}{\lambda^2}\psi\, d\psi = \frac{4\pi^2 v^2 \rho}{\lambda^2}\frac{\psi^2}{2}$$

$$= \frac{1}{2}\rho\frac{4\pi^2}{\lambda^2}v^2 A^2 \sin^2\frac{2\pi}{\lambda}(vt-x)$$

$$= \frac{2\pi^2 v^2 A^2 \rho}{\lambda^2}\sin^2\frac{2\pi}{\lambda}(vt-x) \tag{7.29}$$

∴ Total energy per unit volume is

$$E = (KE) + (PE)$$

$$= E_K + E_P$$

$$= \frac{2\pi^2}{\lambda^2}v^2 A^2 \rho \cos^2\frac{2\pi}{\lambda}(vt-x) + \frac{2\pi^2}{\lambda^2}v^2 A^2 \rho \sin^2\frac{2\pi}{\lambda}(vt-x)$$

$$= \frac{2\pi^2}{\lambda^2}v^2 A^2 \rho \qquad [\because v = v\lambda]$$

$$\Rightarrow \qquad E = 2\rho\pi^2 v^2 A^2$$

∴ Energy density = $2\pi\rho v^2 A^2$ \qquad (7.30)

214 Fundamentals of Optics

Though the kinetic energy and potential energy of the wave depend on x and t, and the total energy is constant. Further, we see that the energy density E depends directly on: (i) density of the medium (ρ), (ii) square of the frequency (ν^2) and (iii) the square of the amplitude (A^2).

7.9.1 Intensity

Intensity of a wave is defined as the energy passing through the unit area normally in one second. It is also called the *energy current*.

If v is the velocity of the wave, then the energy contained in a cylinder of length l and unit cross-section will cross the unit area normally in one second.

$\therefore$ $\qquad$ Intensity I = energy density × velocity

$$= 2\rho\pi^2\nu^2A^2 v \qquad (7.31)$$

Thus, the wave intensity I depends directly on:

(i) the density of the medium (ρ)
(ii) the square of the frequency (ν^2)
(iii) the square of the amplitude (A^2)
(iv) the velocity of the wave (v)

EXAMPLE 7.4 Plane harmonic waves of frequency 500 Hz are produced in air with displacement amplitude 1×10^{-3} cm. Deduce: (i) pressure amplitude, (ii) energy flux in the wave (density of air = 1.29 gm/lit, speed of sound in air = 340 m/s).

Solution Given:

$\qquad A = 1 \times 10^{-3}$ cm $= 10^{-5}$ m, $\nu = 500$ Hz, $v = 340$ m/s, $\rho = 1.29$ g/lit $= 1.29$ kg/m^3

(i) Pressure amplitude

$$= \frac{2\pi AE}{\lambda} = \frac{2\pi^2 A v^2 \rho}{\lambda} \qquad [\because E = v^2\rho]$$

$$= 2\pi A v \nu \rho \qquad [\because v = \nu\lambda]$$

$$= 2 \times 3.14 \times 10^{-5} \times 500 \times 340 \times 1.29$$

$$= 13.8 \text{ N/m}^2$$

(ii) Energy flux

$$= 2\rho^2 A^2 \nu^2 v$$

$$= 2 \times 1.29 \times (3.14)^2 \times (10^{-5})^2 \times (500)^2 \times 340$$

$$= 0.22 \text{ J/m}^2$$

EXAMPLE 7.5 A source of sound has a frequency of 512 Hz and amplitude of 0.25 cm. What is the flow of energy across a square cm per second, if the velocity of sound in air is 340 m/s and density of air is 0.00129 gm/cm^3?

Solution $\qquad \nu = 512$ Hz, $A = 0.25$ cm, $\rho = 0.00129$ gm/cm^3
and $\qquad v = 340$ m/s = 34000 cm/s

Now, total energy per unit volume = $2\pi^2 \rho v^2 A^2$

Energy flow per square cm per second

$$= 2\pi^2 \rho v^2 A^2 v$$
$$= 2 \times (3.14)^2 \times 0.00129 \times (512)^2 \times (0.25)^2 \times 34000$$
$$= 1.417 \times 10^7 \text{ erg/cm}^2/\text{s}$$

7.10 REFLECTION AND TRANSMISSION OF TRANSVERSE WAVES IN A STRING

When a travelling wave meets discontinuity, i.e., a sudden change of impedance, a part of its amplitude is reflected and the rest is transmitted. The relative magnitude of the reflected and transmitted wave depends on the impedance of the media on the two sides of discontinuity. Now, consider two strings with different impedances joined together at point $x = 0$. The string lies left to $x = 0$ has linear density ρ_1 and that at right to $x = 0$ has linear density ρ_2. Both the strings have the same tension T_0. The speed of the wave in the first string is v_1 and in the second speed is v_2. The impedance of the first string is $\rho_1 v_1$ and that of the second is $\rho_2 v_2$ as shown in Figure 7.8.

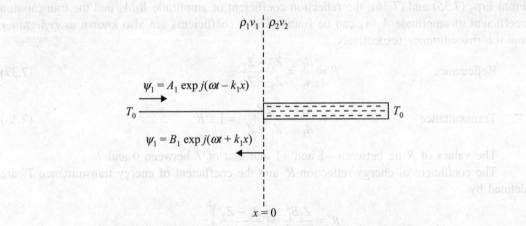

Figure 7.8 Reflection and transmission of transverse wave in a string.

Let the incident, reflected and the transmitted wave be

$$\psi_i = A_1 \exp j(\omega t - k_1 x) \tag{7.32}$$

$$\psi_r = B_1 \exp j(\omega t + k_1 x) \tag{7.33}$$

$$\psi_t = A_2 \exp j(\omega t - k_2 x) \tag{7.34}$$

Now, our target is to get the reflected and the transmitted amplitudes in terms of the incident amplitude. This can be accomplished by satisfying two boundary conditions at the point of discontinuity.

(i) The displacement is continuous at $x = 0$. That is, the displacement immediately to the left and to the right of $x = 0$ are the same.

(ii) The transverse force $T_0 \dfrac{d\psi}{dx}$ is continuous across the boundary.

Applying the first condition, we have

$$\psi_i + \psi_r = \psi_t$$

$$A_1 \exp j(\omega t - k_1 x) + B_1 \exp j(\omega t + k_1 x) = A_2 \exp j(\omega t - k_2 x)$$

This condition is hold at $x = 0$ for all the time

Hence, $\qquad A_1 + B_1 = A_2 \qquad$ (7.35)

Now, on the application of second boundary condition,

$$T_0 \frac{d\psi_i}{dx} + T_0 \frac{d\psi_r}{dx} = T_0 \frac{d\psi_t}{dx}$$

$$-T_0 A_1 k_1 + T_0 B_1 k_1 = -T_0 A_2 k_2$$

or $\qquad -\dfrac{T_0 \omega}{v} A_1 + \dfrac{T_0 \omega}{v} B_1 = -\dfrac{T_0 \omega}{v} A_2$

or $\qquad -Z_1 A_1 + Z_2 B_2 = -Z_2 A_2 \qquad$ (7.36)

From Eqs. (7.35) and (7.36), the reflection coefficient of amplitude B_1/A_1 and the transmission coefficient of amplitude A_2/A_1 can be found. These coefficients are also known as *reflectance* and the *transmittance* respectively.

Reflectance $\qquad R = \dfrac{B_1}{A_1} = \dfrac{Z_1 - Z_2}{Z_1 + Z_2} \qquad$ (7.37)

Transmittance $\qquad T = \dfrac{A_2}{A_1} = \dfrac{2Z_1}{Z_1 + Z_2} = 1 + R \qquad$ (7.38)

The values of R lie between -1 and $+1$ and that of T between 0 and 2.

The coefficient of energy reflection R' and the coefficient of energy transmittance T' are defined by

$$R' = \frac{Z_1 B_1^2}{Z_1 A_1^2} = \left(\frac{Z_1 - Z_2}{Z_1 + Z_2} \right)^2$$

$$T' = \frac{Z_2 A_2^2}{Z_1 A_1^2} = \frac{Z_2}{Z_1} \times \frac{4Z_1^2}{(Z_1 + Z_2)^2}$$

$$= \frac{4 Z_1 Z_2}{(Z_1 + Z_2)^2}$$

R' and T' satisfy the equation

$$R' + T' = 1 \qquad (7.39)$$

Which is in accordance with the law of conservation of energy.

Special cases

Following are the special cases:

(i) If the impedance of the second string is less than that of the first, i.e., $Z_2 < Z_1$, then the reflectance is positive. This means that the reflected wave suffers no change of phase.

(ii) If impedance of the second strip is greater than that of the first, i.e., $Z_2 > Z_1$, the reflectance is negative. The reflected wave, in this case, suffers a phase change of π. The displacement in the incident wave is opposite to that in the reflected wave at the point of discontinuity.

(iii) If $Z_1 = Z_2$, i.e., there is no discontinuity, then the reflectance is zero and the transmittance is unity. There is no reflection in this case. The incident wave is as such transmitted into the second medium (string). When $Z_1 = Z_2$, the impedances of the strings are said to be matched at that point.

(iv) If $Z_2 = \infty$, the reflectance is -1 and the transmittance is zero. The incident wave is totally reflected within the phase of π.

7.11 REFLECTION AND TRANSMISSION OF LONGITUDINAL WAVES AT DISCONTINUITY

When acoustic waves meet a sudden change in impedance, they suffer reflection. Let the plane separating the two media lie at $x = 0$. The medium left to the plane $x = 0$ has impedance $Z_1 = \rho_1 v_1$ and that to right has impedance $Z_2 = \rho_2 v_2$ is shown in Figure 7.9. We assume that the wave is incident normally to the boundary. Let the equations of the incident, reflected and transmitted waves be

$$\psi_i = A_1 \exp j(\omega t - k_1 x) \tag{7.40a}$$

$$\psi_r = B_1 \exp j(\omega t + k_1 x) \tag{7.40b}$$

$$\psi_t = A_2 \exp j(\omega t - k_2 x) \tag{7.40c}$$

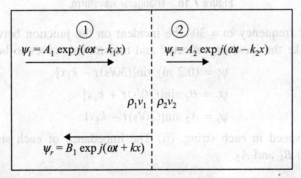

Figure 7.9 Reflection and transmission of longitudinal waves at discontinuity.

At the boundary separating the two media, two conditions must be satisfied

(i) The displacement must be continuous, i.e.,

$$\psi_i - \psi_r = \psi_t \tag{7.41}$$

(ii) The excess pressure must be continuous at the boundary $p_i + p_r = p_t$

or
$$\rho_1 v_1^2 \frac{d\psi_i}{dx} + \rho_1 v_1^2 \frac{d\psi_r}{dx} = \rho_2 v_2^2 \frac{d\psi t}{dt} \tag{7.42}$$

The first condition gives
$$A_1 + B_1 = A_2 \tag{7.43}$$

and the second condition gives
$$-\rho_1 v_1 A_1 + \rho_1 v_1 B_1 = -\rho_2 v_2 A_2$$

or
$$-Z_1 A_1 + Z_1 B_1 = -Z_2 A_2 \tag{7.44}$$

Equation (7.44) shows that if the impedance of the second medium Z_2 is greater than that of the first, the coefficient R is negative. This means that the reflected wave suffers phase change of π radian. This happens when the sound waves are incident from air to water surface. In the absence of any discontinuity (i.e., $Z_1 = Z_2$); $R = 0$ and $T = 1$, which means that the incident wave is transmitted as such without any reflection. When $Z_1 < Z_2$, the reflected wave has no change of phase.

EXAMPLE 7.6 Consider the waveform (triangular) as shown in Figure 7.10 heading towards a boundary between two strings. Let the string have a mass per unit length of $m_1 = 0.05$ kg/m and let string 2 have a mass per unit length $m_2 = 0.02$ kg/m. Let the tension in the string be $T = 100$ T.

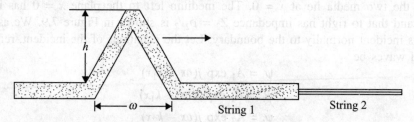

Figure 7.10 Triangular waveform.

Let a harmonic frequency $\omega = 30/\text{s}$ be incident on the junction between the two strings (as given above). Take the incident, reflected and transmitted waves to be of the form

$$\psi_i = (0.2 \text{ m}) \sin[(30/\text{s})t - k_1 x]$$
$$\psi_r = B_1 \sin[(30/\text{s})t + k_1 x]$$
and
$$\psi_t = A_2 \sin[(30/\text{s})t - k_1 x]$$

Find: (i) The wave speed in each string, (ii) The impedance of each string, (iii) k_1 and k_2, (iv) λ_i, λ_r and λ_t, (v) B_1 and A_2.

Solution

(i) $v_1 = \sqrt{\dfrac{T}{m_1}} = \sqrt{\left(\dfrac{100}{0.05}\right)}$ m/s $= 44.7$ m/s

$v_2 = \sqrt{\dfrac{T}{m_2}} = \sqrt{\left(\dfrac{100}{0.02}\right)}$ m/s $= 70.7$ m/s

(ii) $\quad Z_1 = \dfrac{T}{v_1} = \dfrac{100}{44.7}$ kg/s = 2.24 kg/s

$\quad\quad Z_2 = \dfrac{T}{v_2} = \dfrac{100}{70.7}$ kg/s = 1.41 kg/s

$\quad\quad \rho = \dfrac{Z_1 - Z_2}{Z_1 + Z_2} = 0.227$

(iii) $\quad k_1 = \dfrac{\omega}{v_1} = \dfrac{30}{44.7}$ m = 0.67 m

and $\quad k_2 = \dfrac{\omega}{v_2} = \dfrac{30}{70.7}$ m/s = 0.424 m

(iv) $\quad \lambda_i = \lambda_r = \dfrac{2\pi}{k_1} = \dfrac{2\pi}{0.67}$ m = 9.38 m

and $\quad \lambda_t = \dfrac{2\pi}{k_2} = \dfrac{2\pi}{0.424}$ m = 14.8 m

(v) Using reflection and transmission coefficient and applying condition $x = 0$

$$\psi_r = \dfrac{Z_1 - Z_2}{Z_1 + Z_2}\psi_i$$

$B_1 \sin [(30)t] = (0.227)(0.2) \sin [(30)t]$

$B_1 = 0.227 \times 0.2$ m $= 0.0454$ m

$$\psi_t = \left(1 + \dfrac{Z_1 - Z_2}{Z_1 + Z_2}\right)\psi_i$$

$A_2 \sin (30)t = 1.227 \times 0.2 \sin (30) t$

$\Rightarrow \quad\quad A_2 = 0.245$ m

7.12 SUPERPOSITION OF PROGRESSIVE WAVES: STATIONARY WAVE

When two identical progressive waves travelling in a medium with same velocity but in opposite directions along the same straight line are superposed, they give rise to a system of alternative rarefaction and compression that cannot be moved in any direction of medium. This resultant wave is called *stationary* or *standing wave*.

They are called stationary because there is no flow of energy along the waves. There are certain points, half a wavelength apart, which are permanently at rest are known as *nodes* and there are some other points midway between the nodes where the displacement is maximum, are known as *antinodes*.

7.12.1 Analytical Treatment of Stationary Waves and their Properties

Consider two plane progressive waves—one travelling along positive x-axis and another along negative x-axis with amplitude A, wave velocity v and wavelength λ.

So, the superposing progressive waves are:

$$\psi_1 = A \sin \frac{2\pi}{\pi}(vt - x) \text{ and } \psi_2 = \sin \frac{2\pi}{\pi}(vt + x)$$

But by the principle of superposition, the resultant displacement of the particle at x at time t will be

$$\psi = \psi_1 + \psi_2$$

$$= A \sin \frac{2\pi}{\lambda}(vt - x) + A \sin \frac{2\pi}{\lambda}(vt + x)$$

$$= 2A \sin \frac{2\pi}{\lambda} vt \cos \frac{2\pi}{\lambda} x$$

$$= 2A \cos \frac{2\pi}{\lambda} x \sin \frac{2\pi}{\lambda} vt \tag{7.45}$$

where

$$R = 2A \cos \frac{2\pi}{\lambda} x$$

Equation (7.45) represents a simple harmonic motion of same wavelength of the superposed wave but not of same amplitude. Moreover, the amplitude $R = 2A \cos \frac{2\pi}{\lambda} x$, is not a constant. For different values of x, R will have different values. Equation of motion given in Eq. (7.45) is not a progressive wave motion, since its phase does not contain any term like $(vt - x)$ or $(vt + x)$. So, Eq. (7.45) represents a stationary wave.

(a) Position of nodes: The nodes will be obtained, when

$$\cos \frac{2\pi}{\lambda} x = 0$$

or

$$\frac{2\pi}{\lambda} x = (2n \pm 1) \frac{\pi}{4}$$

or

$$x = (2n \pm 1) \frac{\lambda}{4} \tag{7.46}$$

where $n = 0, 2, 3, 4, \ldots$, etc.

So, nodes will be at $x = \pm \frac{\lambda}{4}, \pm \frac{3\lambda}{4}, \pm \frac{5\lambda}{4} \ldots$

The particle at these points will be at rest. The distance between any two successive nodes is $\frac{\lambda}{2}$.

(b) Position of antinodes: Antinodes are the points of maximum vibration, so the antinodes will be obtained when

$$\cos \frac{2\pi}{\lambda} x = 1$$

or

$$\frac{2\pi}{\lambda} x = \pm n\pi$$

or
$$x = \pm \frac{n\lambda}{2} \tag{7.47}$$

where $n = 0, 1, 2, 3, 4, \ldots$, etc.

So, antinodes will be obtained at $x = 0, \pm \frac{\lambda}{2}, \pm \lambda, \pm \frac{3\lambda}{2}$ at these points the amplitude of vibration of the particle will be $\pm 2A$. The distance between any two successive antinodes is $\frac{\lambda}{2}$. The distance between one node and next antinodes is $\frac{\lambda}{4}$.

7.12.2 Energy of Stationary Waves

According to Eq. (7.45), the stationary wave is represented by

$$\psi = 2A \sin \frac{2\pi}{\lambda} vt \cos \frac{2\pi}{\lambda} x$$

The particle velocity at x is given by

$$\frac{d\psi}{dt} = 2A \frac{2\pi v}{\lambda} \cos \frac{2\pi}{\lambda} vt \cos \frac{2\pi}{\lambda} x$$

$$= \frac{4\pi A v}{\lambda} \cos \frac{2\pi}{\lambda} vt \cos \frac{2\pi}{\lambda} x \tag{7.48}$$

Consider a unit area of a thin layer of medium dx at a distance x from the origin and perpendicular to the direction of propagation of the wave. Let ρ_n be the normal density of the medium, then the mass of the layer will be $\rho_n dx$. Then the kinetic energy of this layer is given by

$$dK = \frac{1}{2} \rho_n dx \left(\frac{d\psi}{dt}\right)^2$$

$$= \frac{1}{2} \rho_n dx \left(\frac{4\pi A v}{\lambda}\right)^2 \cos^2 \frac{2\pi vt}{\lambda} \cos^2 \frac{2\pi x}{\lambda}$$

$$= \frac{8\pi^2 A^2 v^2 \rho_n}{\lambda^2} \cos^2 \frac{2\pi x}{\lambda} \cos^2 \omega t \, dx$$

$$= 2\omega^2 A^2 \rho_n \cos^2 \frac{2\pi x}{\lambda} \cos^2 \omega t \, dx \tag{7.49}$$

where, velocity of the wave $= v = v\lambda$ and angular frequency $\omega = 2\pi v$. We know that total energy (dE) is nothing but the maximum kinetic energy. Then by using Eq. (7.49), the total energy will be

$$dE = 2\omega^2 A^2 \rho_n \cos^2 \frac{2\pi x}{\lambda} dx \tag{7.50}$$

Therefore, the potential energy (d_P) = Total energy – Kinetic energy

Hence,
$$d_P = dE - dK$$

$$= 2\omega^2 A^2 \rho_n \cos^2 \frac{2\pi x}{\lambda}(1 - \cos^2 \omega t)dx$$

$$= 2\omega^2 A^2 \rho_n \cos^2 \frac{2\pi x}{\lambda}\sin^2 \omega t dx \tag{7.51}$$

It is clear from Eq. (7.51), the total energy of a layer of medium is dependent on the position of layer. At nodes, $dE = 0$ and at antinodes dE is maximum.

The kinetic energy per wavelength of a stationary wave is

$$K_\lambda = 2\omega^2 A^2 \rho_n \cos^2 \omega t \int_0^\lambda \cos^2 \frac{2\pi x}{\lambda} dx$$

$$= \omega^2 A^2 \rho_n \lambda \cos^2 \omega t \tag{7.52}$$

The potential energy per wavelength of a stationary wave is

$$(E_P)_\lambda = 2\omega^2 A^2 \rho_n^2 \sin^2 \omega t \int_0^\lambda \cos^2 \frac{2\pi x}{\lambda} dx$$

$$= \omega^2 A^2 \rho_n \lambda \sin^2 \omega t \tag{7.53}$$

The total energy per wavelength of a stationary wave is

$$E_\lambda = 2\omega^2 A^2 \rho_n \int_0^\lambda \cos^2 \frac{2\pi x}{\lambda} dx = \omega^2 A^2 \rho_n \lambda \tag{7.54}$$

Hence, we found that the total energy per unit wavelength in stationary wave is double that for progressive wave. This again proves that the stationary wave is formed due to the superposition of two similar progressive waves travelling in opposite directions.

There is no transmission of energy in any plane in either direction in a stationary wave. Since the energy due to the two progressive waves travelling in the opposite directions are equal, the resultant flow of energy is zero.

According to Eq. (7.45), the stationary wave is represented by

$$\psi = 2A \cos \frac{2\pi x}{\lambda} \sin \frac{2\pi v t}{\lambda}$$

$$= 2A \cos \frac{2\pi x}{\lambda} \sin \omega t$$

The particle velocity is given by

$$\frac{d\psi}{dt} = 2A\omega \cos \frac{2\pi x}{\lambda} \cos \omega t \tag{7.55}$$

The excess pressure (dP) in the medium due to stationary wave is given by $dP = -K\frac{d\psi}{dx}$, where k is the bulk modulus and $\frac{d\psi}{dx}$ = volume strain

$$= \frac{4\pi KA}{\lambda} \sin \frac{2\pi x}{\lambda} \sin \omega t \tag{7.56}$$

Now, the work done against the excess pressure or energy transferred per unit area in small time dt is equal to $dP\left(\dfrac{d\psi}{dx}\right)dt$.

Thus, the energy transferred in one period (T) across a unit area is equal to $\int_0^T dP\left(\dfrac{d\psi}{dx}\right)dt$.

Therefore, using Eqs. (7.55) and (7.56), the rate of transfer of energy across the unit area is

$$= \dfrac{1}{T}\int_0^T dP\left(\dfrac{d\psi}{dx}\right)dt$$

$$= \dfrac{1}{T}\int_0^T \dfrac{4\pi KA}{\lambda}\sin\dfrac{2\pi x}{\lambda}\sin\omega t \times 2a\omega\cos\dfrac{2\pi x}{\lambda}\cos\omega t\,dt$$

$$= \dfrac{8\pi A^2 \omega K}{\lambda T}\sin\dfrac{2\pi x}{\lambda}\cos\dfrac{2\pi x}{\lambda}\int_0^T \dfrac{1}{2}\times 2\sin\omega t\cos\omega t\,dt$$

$$= \dfrac{4\pi A^2 \omega K}{\lambda T}\sin\dfrac{2\pi x}{\lambda}\cos\dfrac{2\pi x}{\lambda}\int_0^T \sin(2\omega t)\,dt$$

$$= \dfrac{4\pi A^2 \omega K}{\lambda T}\sin\dfrac{2\pi x}{\lambda}\cos\dfrac{2\pi x}{\lambda}\times 0$$

$$= 0 \qquad \left[\because \int_0^T \sin 2\omega t\,dt = 0\right]$$

Hence, for a stationary wave, there is no flow of energy across any section.

7.12.3 Characteristics of Stationary Wave

The stationary waves consists following properties:

(a) Stationary waves are produced when two identical waves travelling along same straight line but in opposite directions are superposed.

(b) Crests and troughs do not progress through the medium but simply appear or disappear at the same place alternatively.

(c) All the particles, except those at the nodes, follow simple harmonic motion. The amplitude of the oscillation is zero at nodes and maximum at antinodes. The distance between two successive antinodes and the node is equal to half of the wavelength.

(d) The particles between two successive nodes are in the same phase of vibration, while the particle on opposite sides of a node are in opposite phase of vibration.

(e) Stationary waves can be produced both by longitudinal and transverse waves.

(f) All the particles pass through their mean positions or reach their outermost positions simultaneously, twice in a periodic time.

(g) There is no advancement of the wave and no flow of energy in any direction.

EXAMPLE 7.7 Show that in a stationary wave, all the particles between any two consecutive nodes are in phase but they are in opposite phase with the particles between the next pair of nodes.

Solution The instantaneous displacement of stationary/standing wave is given by

$$\psi(x, t) = 2A \cos kx \sin \omega t$$

For any two points $x = x_1$ and $x = x_2$ between consecutive nodes, the phase is same, equal to ωt. But if $x_2 = x_1 + \dfrac{\lambda}{2}$, i.e., the points belong to adjacent loops (which mean that they lie between consecutive pairs of nodes). The corresponding displacements are

$$\psi_1 = 2A \cos kx_1 \sin \omega t \text{ and } \psi_2 = 2A \cos kx_2 \sin \omega t$$

$$\psi_1 = 2A \cos\left(kx_1 + \frac{k\lambda}{2}\right) \sin \omega t$$

$$= 2A \cos(kx_1 + \pi) \sin \omega t$$

$$= -2A \cos kx_1 \sin \omega t = 2A \cos kx_1 (\omega t + \pi)$$

This result shows that the phases of Ψ_1 and Ψ_2 differ by π.

7.13 COMPARISON BETWEEN PROGRESSIVE WAVES AND STATIONARY WAVES

S.No.	Progressive waves	Stationary waves
1.	The disturbance produced in the medium travels onward, it being handed over from one particle to the next. Each particle executes the same type of vibration as the preceding one, though not at the same time.	There is no onward motion of disturbance as no particle transfers its motion to the next. Each particle has its own characteristic.
2.	The amplitude of each particle is the same but the phase changes continuously.	The amplitudes of the different particles are different, ranging from zero at the nodes to maximum at the antinodes. All the particles in a given segment vibrate in phase but in opposite phase relative to the particles in the adjacent segment.
3.	No particle is permanently at rest. Different particles attain the state of momentary rest at different instants.	The particles at the nodes are permanently at rest but other particles attain theirs position of momentary rest simultaneously.
4.	All the particles attain the same maximum velocity when they pass through their mean positions.	All the particles attain their own maximum velocity at the same time, when they pass through their mean positions.
5.	In case of a longitudinal progressive wave, all the parts of the medium undergo similar variation of density one after the other. At every point there will be a density variation.	In case of a longitudinal stationary wave, the variation of density is different at different points being maximum at the node and zero at antinodes.
6.	There is a flow of energy across every plane in the direction of propagation.	Energy is not transported across any plane.

FORMULAE AT A GLANCE

7.1 Time period $T = \dfrac{1}{\nu}$, where $\nu =$ frequency.

7.2 Velocity $= \dfrac{\text{Wavelength}}{\text{Time period}}$

$\nu T =$ Wavelength (λ)

or $\quad \nu \times \dfrac{1}{\nu} = \lambda \quad$ or $\quad v = \nu\lambda$

7.3 *Differential equation of motion*

$$\frac{d^2\psi}{dt^2} = v^2 \frac{d^2\psi}{dx^2}$$

where $\Psi(x, t) = A \sin \dfrac{2\pi}{T}\left(t - \dfrac{x}{v}\right)$

$A =$ amplitude
$T =$ Time period, $t =$ time
$x =$ distance, $v =$ velocity

7.4 *Plane progressive wave in media*

(a) $B = -\dfrac{dP}{dV/V} = -V\dfrac{dP}{dV}$

where $B =$ bulk modulus
$P =$ pressure
$V =$ volume

(b) $PV^\gamma =$ constant

7.5 Wave equation of propagation of sound waves in fluid media

$$B\frac{d^2\psi}{dt^2} = \rho_0 \frac{d^2\psi}{dx^2}$$

$$\frac{B}{\rho_0} = \frac{\gamma P}{\rho_0} = v^2$$

Then $\quad \dfrac{d^2\psi}{dx^2} = \dfrac{1}{v^2}\dfrac{d^2\psi}{dt^2}$

(a) Displacement $\psi = \psi_m\, e^{i(\omega t - kx)}$

$\psi_m =$ maximum amplitude displacement

(b) Particle velocity

$v = \dot\psi = \dfrac{d\psi}{dt} = i\omega\, \psi_m e^{i(\omega t - kx)} = i\omega\,\psi$

(c) Condensation $\delta = \dfrac{d\psi}{dx} = -ik\psi = -s$

(d) Dilation $s = ik\psi$

(e) Excess pressure $P = B_a s = iB_a k\psi$

7.6 Acoustic pressure

(a) $(\delta P)_m = \dfrac{2\pi Ak}{\lambda}$

(b) RMS acoustic pressure

$$[(\delta P)_m]_{\text{RMS}} = \sqrt{2}\,\dfrac{\pi Ak}{\lambda}$$

(c) $\delta P = -k\dfrac{d\psi}{dx} = \dfrac{2\pi Ak}{\lambda}\sin\dfrac{2\pi}{\lambda}\left(vt - x + \dfrac{\pi}{2}\right)$

7.7 Energy of plane progressive wave

(a) K.E. $= \dfrac{2\pi v^2 A^2 \rho}{\lambda^2}\cos^2\dfrac{2\pi}{\lambda}(vt - x)$

(b) P.E. $= \dfrac{2\pi v^2 A^2 \rho}{\lambda^2}\sin^2\dfrac{2\pi}{\lambda}(vt - x)$

(c) Total energy $E = 2\rho\pi^2 v^2 A^2$

(d) Energy density $= 2\pi\rho v^2 A^2$

(e) Intensity $=$ energy density $\times$ velocity $= 2\rho\pi^2 v^2 A^2 v^2$

7.8 Reflection and transmission of transverse waves in string

(a) Reflection coefficient of amplitude

Reflectance $R = \dfrac{B_1}{A_1} = \dfrac{Z_1 - Z_2}{Z_1 + Z_2}$

(b) Transmission coefficient of amplitude or transmittance

$$T = \dfrac{A_2}{A_1} = \dfrac{2Z_1}{Z_1 + Z_2}$$

(c) $T = 1 + R$

(d) The coefficient of energy reflection R'

$$R' = \dfrac{Z_1 B_1^2}{Z_1 A_1^2} = \dfrac{(Z_1 - Z_2)^2}{(Z_1 + Z_2)^2}$$

(e) The coefficient of energy transmittance T'

$$T' = \dfrac{Z_2 A_2^2}{Z_1 A_1^2} = \dfrac{4Z_1 Z_2}{(Z_1 + Z_2)^2}$$

(f) $R' + T' = 1$

7.9 *Superposition of progressive waves: stationary wave*

Let
$$\psi_1 = A \sin \frac{2\pi}{\lambda}(vt - x)$$
$$\psi_2 = \sin \frac{2\pi}{\lambda}(vt + x)$$

Then
$$\psi = \psi_1 + \psi_2$$
$$= 2A \sin \frac{2\pi}{\lambda} vt \cos \frac{2\pi}{\lambda} x = R \cos \frac{2\pi vt}{\pi}$$
$$R = 2A \cos \frac{2\pi}{\lambda} x$$

(a) *Position of nodes*
$$\cos \frac{2\pi}{\lambda} x = 0$$
$$\frac{2\pi}{\lambda} x = (2n \pm 1)\frac{\pi}{4}$$
$$x = (2n \pm 1)\frac{\lambda}{4}$$

(b) *Position of antinodes*
$$\cos \frac{2\pi}{\lambda} x = 1$$
$$\frac{2\pi}{\lambda} x = \pm n\pi$$
$$x = \pm \frac{n\lambda}{2}$$

7.10 *Energy of stationary wave*
(a) Kinetic energy of layer is
$$dK = 2\omega^2 A^2 \rho_n \cos^2 \frac{2\pi x}{\lambda} \cos^2 \omega t\, dx$$

(b) Total energy of layer
$$dE = 2\omega^2 A^2 \rho_n \cos^2 \frac{2\pi x}{\lambda} dx$$

(c) Potential energy of layer
$$dE_P = 2\omega^2 A^2 \lambda_n \cos^2 \frac{2\pi x}{\lambda} \sin^2 \omega t\, dx$$

(d) K.E. per wavelength
$$K_\lambda = \omega^2 A^2 \rho_n \lambda \cos^2 \omega t$$

(e) P.E. per wavelength
$$(E_P)_\lambda = \omega^2 A^2 \rho_n \lambda \sin^2 \omega t$$

(f) Total energy per wavelength
$$E_\lambda = \omega^2 A^2 \rho_n \lambda$$

SOLVED NUMERICAL PROBLEMS

PROBLEM 7.1 The vibrations of a string of length 60 cm fixed at both ends are represented by the equation

$$\psi = 4\sin\left(\frac{\pi}{15}x\right)\cos(96\pi t)$$

where x and ψ are in cm and t in seconds

(i) What is the maximum displacement of point $x = 5$ cm?
(ii) Where are the nodes located along the string?
(iii) What is the velocity at the particle at $x = 7.5$ cm at $t = 0.25$s?
(iv) Write down the equations of the component waves whose superposition gives the above wave.

Solution The given equation is

$$\psi = 4\sin\frac{\pi}{15}x \cos 96\pi t$$

This can be written as

$$\psi = 2 \times 2 \sin\frac{2\pi x}{30}\cos 2\pi \frac{1440}{30}t$$

This shows that $A = 2$ cm, $\lambda = 30$ cm and $v = 1440$ cm/s

(i) The maximum displacement is given by

$$\psi_{max} = 4\sin\frac{2\pi x}{30}$$

For $\quad x = 5$ cm

$$\psi_{max} = 4\sin\frac{2\pi 5}{30} = \frac{4\sqrt{3}}{2} = 2\sqrt{3}$$

(ii) As $\lambda = 30$ cm, the nodes are located along the string at places 0, 15 cm, 30 cm, 45 cm, 60 cm.

(iii) The velocity of particle is given by $\dfrac{d\psi}{dx} = -4\sin\dfrac{\pi \times 7.5}{15}\sin 96\pi \times 0.25 \times 96\pi$

(iv) The equations of component wave are

$$\psi_1 = a\sin\frac{2\pi}{\lambda}(x - vt) = 2\sin\frac{2\pi}{30}(x - 1440t)$$

$$= 2\sin 2\pi\left(\frac{x}{30} - 48t\right)$$

and

$$\psi_2 = a\sin\frac{2\pi}{\lambda}(x + vt)$$

$$= 2 \sin \frac{2\pi}{30}(x + 1440t)$$

$$= 2 \sin 2\pi \left[\frac{x}{30} + 48t\right]$$

PROBLEM 7.2 A simple harmonic wave travelling x-axis is given by

$$\psi = 5 \sin 2\pi (0.2t - 0.5x) \quad [x \text{ is in metre and } t \text{ in second}]$$

Calculate the amplitude, frequency, wavelength, wave velocity, particle velocity, velocity amplitude, particle acceleration and acceleration amplitude.

Solution Here $\quad\quad\quad \psi = 5 \sin 2\pi (0.2t - 0.5x)$ \hfill (i)

But the standard progressive wave equation is

$$\psi = A \sin \frac{2\pi}{\lambda}(vt - x) = A \sin 2\pi \left(\frac{v}{\lambda}t - \frac{x}{\lambda}\right)$$

$$A \sin 2\pi \left(vt - \frac{x}{\lambda}\right) \quad\quad\quad \text{(ii)}$$

Comparing Eqs. (i) and (ii), we have

$$A = 5 \text{ cm}, \nu = 0.2/\text{s}, \frac{1}{\lambda} = 0.5, \lambda = \frac{1}{0.5} = 2\text{m}$$

Velocity of wave $v = \nu\lambda = 0.2 \times 2 = 0.4$ m/s

Particle velocity $u = \dfrac{d\psi}{dx} = 5 \cos 2\pi(0.2t - 0.5x) \times 0.2$

$$= \cos 2\pi(0.2t - 0.5x) \text{ m/s}$$

Particle velocity amplitude is 1 m/s
Particle acceleration

$$\frac{d^2\psi}{dx^2} = -\sin 2\pi(0.2t - 0.5x) \times 0.2 = -0.2 \sin 2\pi(0.2t - 0.5x)$$

∴ Particle acceleration amplitude = -0.2 m/s^2

PROBLEM 7.3 Equation of a plane progressive wave is given below:

$$\psi = 10 \sin \pi (0.01x - 2.00t)$$

where ψ and x are expressed in cm and t in second. Determine:
  (i) Amplitude of the wave
  (ii) Frequency of the wave
  (iii) Phase difference at an instant between two points 40 cm apart

Solution The given equation can be written as

$$\psi = -10 \sin 2\pi [1.00t - 0.005x] \quad\quad\quad \text{(i)}$$

Then compare with following equation

$$\psi = A \sin 2\pi \left(vt - \frac{x}{\pi}\right) \quad\quad\quad \text{(ii)}$$

We get $A = 10$ cm, $v = 1s^{-1}$, $\lambda = \dfrac{1}{0.005} = 200$ cm

Hence, phase difference $\phi = \dfrac{2\pi}{\lambda} x = \dfrac{2\pi}{200} \times 40 = \dfrac{2}{5}\pi$ rad

PROBLEM 7.4 A wave of frequency 400 Hz is travelling with a velocity 800 m/s. How far two points are situated whose displacement differs in phase by $\pi/4$?

Solution Given $v = 400$ Hz, $v = 800$ m/s

$\therefore$ Wavelength $\lambda = \dfrac{v}{v} = \dfrac{800}{400} = 2$ m

Phase difference between two points at distance x apart

$$\phi = \dfrac{2\pi}{\lambda} x$$

or

$$x = \dfrac{\lambda}{2\pi} \phi$$

$$= \dfrac{\lambda}{2\pi} \times \dfrac{\pi}{4}$$

$$= \dfrac{2}{2 \times 4} \text{ m} = 0.25 \text{ m}$$

$$= 25 \text{ m}$$

PROBLEM 7.5 The displacement wave is represented by

$$\psi = 0.25 \times 10^{-3} \sin(500t - 0.025x)$$

where ψ, t and x are in cm, s and m respectively. Deduce: (i) amplitude, (ii) Time period, (iii) angular frequency, and (iv) wavelength. Also deduce the amplitude of particle velocity and particle acceleration.

Solution Given $\psi = 0.25 \times 10^{-3} \sin(500t - 0.025x)$

Comparing it with standard equation

$$\psi = A \sin 2\pi \left(\dfrac{t}{T} - \dfrac{x}{\lambda} \right), \text{ we get}$$

(i) Amplitude $A = 0.25 \times 10^{-3}$ cm

(ii) Time period $\dfrac{2\pi}{T} = 500$ or $T = \dfrac{2\pi}{500} = \dfrac{\pi}{250} = 0.01257$ s

(iii) Angular frequency

$$\omega = \dfrac{2\pi}{T} = \dfrac{2\pi}{\pi} \times 250 = 500 \text{ rad/s}$$

(iv) Wavelength

$$\dfrac{2\pi}{\lambda} = 0.025 \Rightarrow \lambda = \dfrac{2\pi}{0.025} = 251.2 \text{ m}$$

(v) Velocity amplitude

$$\omega A = 500 \times 0.25 \times 10^{-3} = 0.125 \text{ cm/s}$$

(vi) Acceleration amplitude

$$\omega^2 A = (500)^2 \times 0.25 \times 10^{-3}$$
$$= 62.5 \text{ cm/s}^2$$

CONCEPTUAL QUESTIONS

7.1 What are two most essential requisites of a medium for the propagation of a wave through it?

Ans (i) Medium must be elastic and (ii) medium must possess the property of inertia of motion.

7.2 What is a phase?

Ans Phase is a physical quantity associated with a wave which described the state of vibrations.

7.3 What are pressure wave?

Ans These are the waves in which medium particles vibrate to and fro about their mean positions along a straight line parallel to the direction of propagation. In these waves alternate region of high pressure (compression) and low pressure (rarefaction) are formed.

7.4 Why are stationary waves named so?

Ans Stationary waves are named so because in these waves there is no transfer of energy. The crest and troughs and zone of compression and rarefaction merely appear and disappear in fixed positions.

7.5 Sound can be heard over longer distance on a summer-rainy day. Why?

Ans On the summer-rainy day, the speed of sound in air is high due to higher temperature and higher moisture content. Due to higher speed, the sound wave can cover a longer distance in shorter time and as a result of it sound can be heard over longer distance on a summer-rainy day.

7.6 What are the three main effects of superposition of waves?

Ans The superposition of waves gives rise to the following three important effects:

(a) When two waves of same nature and frequency superimpose in the same direction and interference takes place.

(b) When two waves of same nature and slightly different frequency superimpose in same direction and beats are formed.

(c) When waves of same nature and frequency superimpose in opposite direction and stationary waves are formed.

7.7 State the conditions that must be prevail for the formation of stationary waves.

Ans (a) There must be superposition of two progressive waves of similar nature.

(b) The frequency or wavelengths of the two superposing waves must be equal.

(c) The amplitude of two superposing waves must be equal.

(d) The two superposing waves must travel along a straight line in mutually opposite direction.

7.8 When we start filling an empty bucket with water, the pitch of the sound produced goes on increasing. Explain why?

Ans When we start filling an empty bucket with water by replacing it below running water tap, the pitch of the sound produced due to vibration in air column present in the bucket above the water surface are heard. As water gradually fills the bucket, the length of air column in the bucket gradually decreases. As frequency in air column is inversely proportional to the length of the air column, hence, frequency of sound note produced gradually increased.

EXERCISES

Theoretical Problems

7.1 What do you mean by wave motion? Differentiate between mechanical and non-mechanical waves.

7.2 What propagates in wave motion?

7.3 What are the characteristics of wave motion?

7.4 What are transverse waves? Briefly discuss their characteristics.

7.5 What are longitudinal waves? Briefly discuss their characteristics.

7.6 Define the terms:
  (i) Wavelength (ii) Frequency
  (iii) Wave number (iv) Phase velocity
  (v) Particle velocity

7.7 Explain the terms:
  (i) Compression (ii) Rarefaction
  (iii) Dilation (iv) Condensation

7.8 What are differences between plane progressive waves and stationary waves?

7.9 Define reflection of waves.

7.10 In reference to a wave motion, define
  (i) Amplitude (ii) Time period
  (iii) Frequency (iv) Angular frequency
  (v) Wavelength (vi) Wave number

7.11 What are standing waves? Explain formation of stationary waves graphically.

7.12 What are standing waves? Derive an expression for standing waves. Also define the term node and antinode.

7.13 Derive the differential equation of motion of a plane progressive wave.

7.14 Particle displacement ψ in a plane longitudinal wave in a medium of density ρ is given by

$$\psi = A \cos 2\pi v \left(t - \frac{x}{v}\right)$$

Deduce expressions for: (i) pressure amplitude and (ii) energy flux.

7.15 Justify the statement, "sound waves are elastic waves".

7.16 Prove that the energy of a progressive wave in unit cross-sectional area is $\rho \pi^2 A^2 \nu^2 \lambda$ where ρ is the density of the medium, A is the amplitude, ν as the frequency and λ is the wavelength of the wave.

7.17 Write down expression for the particle displacement for a plane progressive harmonic wave? What is the essential difference between the two expressions?

7.18 Deduce the equation for a stationary wave and from it derive its characteristics properties. Compare those with those of progressive wave?

7.19 Derive expressions for the reflection and transmission coefficient for a wave travelling across a boundary between two media of different impedances Z_1 and Z_2. Discuss the cases: (i) $Z_2 < Z_1$ (ii) $Z_2 > Z_1$ (iii) $Z_2 = Z_1$.

7.20 For a stationary wave on a string show that the energy transmission is: (i) zero on average at all points, (ii) zero at all instants at nodes and antinodes.

Numerical Problems

7.1 Satisfy yourself that the following equations can be used to describe the same progressive wave

$$\psi = A \sin 2\pi \frac{(x - vt)}{\lambda}$$

$$\psi = A \sin 2\pi(kx - vt)$$

$$\psi = A \sin 2\pi \left(\frac{x}{\lambda} - \frac{t}{T} \right)$$

$$\psi = -A \sin \omega \left(\frac{t - x}{v} \right)$$

and
$$\psi = A \exp [j^2 \pi (kx - vt)] \quad \text{(imaginary part)}$$

7.2 The following two waves in a medium are superposed

$$\psi_1 = A \sin(5x - 10t)$$
$$\psi_2 = A \sin(4x - 9t)$$

where x is in m and t in s.

(i) Write an equation for the combined disturbance.

(ii) What is the distance between points of zero amplitude in the combined disturbance?

$$\left[\text{Ans. (i) } \psi(x, t) = 2A \cos\left(\frac{x}{2} - \frac{t}{2}\right) \sin\left(\frac{9}{2}x - \frac{19}{2}t\right), \text{ (ii) } 2\pi \text{ m} \right]$$

7.3 The equation of the transverse wave travelling along a stretched string is given by

$$\psi(x, t) = 10 \sin \pi(2t - 0.01x)$$

where ψ and x are expressed in cm and t in seconds. Find the amplitude, frequency, velocity, wavelength and maximum transverse velocity in the string.

(Kanpur University, 1999)

[Ans. $a = 10$ cm; $\nu = 1$ Hz; $v = 200$ cm/s, λ 200 cm and $v_{max} = 62.8$ cm/s]

7.4 A train of simple harmonic wave travelling in a gas along the positive x-direction with an amplitude 2 cm, velocity 45 m/s and frequency 75 Hz. Calculate the displacement, particle velocity and acceleration at a distance of 135 cm from the origin after a time interval of 3 s. [**Ans.** $y = -2$ cm, $v = 0$, $a = 4.437 \times 10^5$ cm/s²]

7.5 The string of densities ρ_1 and ρ_2 are joined together and stretched with tension T. A transverse wave is incident on the boundary. Find the fraction of incident amplitude reflected and transmitted at boundary if $\dfrac{\rho_2}{\rho_1} = 4$ and $\dfrac{1}{4}$. $\left[\textbf{Ans.} \ \dfrac{A_t}{A_i} = \dfrac{2}{3}, \dfrac{A_r}{A_i} = \dfrac{4}{3}\right]$

7.6 Standing waves are produced by the superposition of two waves, $y_1 = 10 \sin(3\pi t - 4x)$ and $y_2 = 10 \sin(3\pi t + 4x)$. Find the amplitude of motion at $x = 18$.
[**Ans.** $A = 19.35$ unit of length]

7.7 A travelling wave propagates according to the expression

$$y = 0.003 \sin(3x - 2t)$$

where y is the displacement at position x at time t. Taking the unit to be in SI, determine: (i) the amplitude, (ii) the wavelength, (iii) the frequency and (iv) the period of the wave. [**Ans.** (i) $A = 0.03$ m, (ii) $\lambda = 2.09$ m, (iii) $\nu = 0.31$ Hz, (iv) $T = 3.14$ s]

7.8 A simple harmonic wave is represented by

$$\psi = 8 \sin 2\pi \left(\dfrac{t}{0.05} - 0.05x\right)$$

Find the wavelength, amplitude, frequency, velocity of wave. Also find the displacement of the particle 40 cm from the origin and 2 s after the start of motion.
[**Ans.** $\lambda = 20$ cm, $A = 8$ cm, $v = 400$ cm/s, $\nu = 20$ Hz and $\psi = 8 \sin 76 \pi t$]

Multiple Choice Questions

7.1 The differential equation of wave motion is

(a) $\dfrac{d^2\psi}{dx^2} = \dfrac{1}{v} \dfrac{d^2\psi}{dt^2}$ (b) $\dfrac{d^2\psi}{dx^2} = v^2 \dfrac{d^2\psi}{dt^2}$

(c) $\dfrac{d^2\psi}{dx^2} = \dfrac{1}{v^2} \dfrac{d^2\psi}{dt^2}$ (d) none of above

7.2 The displacement represented by the equation $\Psi(x, t) = A \cos(\omega t + kx)$ represents
 (a) transverse wave propagating along $+x$ direction
 (b) transverse wave propagating along $-x$ direction
 (c) longitudinal wave propagating along $-x$ direction
 (d) longitudinal wave propagating along $+x$ direction

7.3 When a plane wave travels in a medium in the positive x direction, the displacement of the particles are given by

$$\psi(x, t) = 0.01 \sin 2\pi(t - 0.1x)$$

where ψ and x are measured in metres and t in seconds. What is the wavelength of the wave?
(a) 0.1 m (b) 0.01 m
(c) 10.0 m (d) none of above

7.4 The amplitude of stationary wave is zero at
(a) an antinode (b) node
(c) at a point midway between the node and antinode
(d) none of these

7.5 Which of the following properties of wave is independent of the other?
(a) velocity (b) wavelength
(c) amplitude (d) frequency

7.6 Which of the phenomenon can take place with sound wave?
(a) reflection (b) diffraction
(c) polarization (d) interference

7.7 A transverse wave is described by the equation $\psi = \psi_0 \sin 2\pi(vt - x/\lambda)$. The maximum particle velocity is equal to four times the wave velocity if
(a) $\lambda = \dfrac{\pi \psi_0}{4}$ (b) $\lambda = \dfrac{\pi \psi_0}{2}$
(c) $\lambda = \pi \psi_0$ (d) $\lambda = 2\pi \psi_0$

7.8 In a progressive wave, the average kinetic energy is
(a) $\pi^2 v^2 A^2$ (b) $\dfrac{\pi^2 v^2 A^2}{\rho}$
(c) $\pi v^2 A^2 \rho$ (d) $\dfrac{\rho}{\pi^2 v^2 A^2}$

7.9 The intensity I of a sound wave is
(a) $I = \dfrac{1}{2}\omega^2 A^2 \rho v$ (b) $I = \omega^2 A^2 \rho v$
(c) $I = \dfrac{\omega^2 A^2}{\rho v}$ (d) $\dfrac{\rho v}{\omega^2 A^2}$

7.10 Standing waves can be produced
(a) on a string clamped at both the ends
(b) on a string clamped at one end and free at the other
(c) when the incident wave gets reflected from the wall
(d) all of the above

Answers
7.1 (c)   **7.2** (b)   **7.3** (c)   **7.4** (b)   **7.5** (a)   **7.6** (b)   **7.7** (b)   **7.8** (c)
7.9 (a)   **7.10** (d)

Physical Optics

Chapter 8   Interference of Light Waves
Chapter 9   Diffraction of Light Waves
Chapter 10   Polarization of Light Waves

CHAPTER 8

Interference of Light Waves

"Light + Light does not always give more light, but may in certain circumstances give darkness."
—Max Born

IN THIS CHAPTER

- Wavefront and Rays
- Huygen's Principle of Secondary Wavelets
- Principle of Superposition of Light Waves
- Groups of Interference
- Young's Double Slit Experiment (YDSE)
- Coherence
- Phase Difference and Path Difference
- Conditions for Constructive and Destructive Interference
- Theory of Interference Fringes
- Conditions for Interference of Light Waves
- Interference Fringes with Fresnel's Biprism
- Lloyd's Single Mirror
- Fresnel's Double Mirror
- Stoke's Law
- Interference from Parallel Thin Films or Colour of Thin Films
- Interference in Non-uniform Thick Film: Wedge-Shaped Film
- Newton's Rings
- Michelson's Interferometer
- Multiple Beam Interferometry
- Interference Refractometers
- Interference in Optical Technology

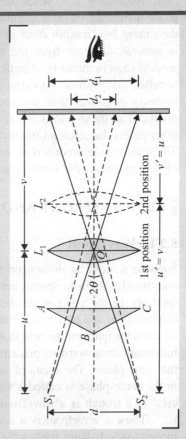

8.1 INTRODUCTION

The wave theory of light was first put forward by Huygen in 1678. On the basis of his theory, Huygen explained satisfactorily the phenomena of reflection, refraction and total internal reflection. According to his theory, a luminous body is a source of disturbance in a hypothetical medium called ether. This medium pervades all space. The disturbance from the source is propagated in the form of waves through space and the energy is distributed equally in all directions. When these waves carrying energy are incident on the eye, the optic nerves are excited and the sensation of vision is produced. Huygen's theory predicted that the velocity of light in medium shall be less than the velocity of light in free space, which is just converse of the prediction made from Newton's corpuscular theory. The experimental evidence for the wave theory in Huygen's time was very small. In 1801, however, Thomas Young obtained evidence that light could produce wave effects. Very shortly, diffraction was explained by Fresnel and Fraunhofer, while the transverse nature of light was explained by polarization experiments. The subject of interference, diffraction and polarization is called *physical optics* or *wave optics* and should be explained by using wave theory of light.

The phenomenon of interference of light has proved the validity of the wave theory of light. According to it, when the two light waves of the same frequency and having a constant phase difference traverse simultaneously in the same region of a medium and cross each other, then there is a modification in the intensity of light in the region of superposition, which is in general, different from the sum of intensities due to individual waves at that point. This modification in intensity of light resulting from the superposition of two (or more) waves of light is called *interference*. At certain points, the wave superimpose in such a way that the resultant intensity is greater than the sum of the intensities due to individual waves. The interference produced at these points is called *constructive interference* or *reinforcement*, while at certain other points the resultant intensity is less than the sum of the intensities due to individual waves. The interference produced at these points is called *destructive interference*. Beyond the region of superposition, the waves come through completely uninfluenced by each other.

8.2 WAVEFRONT AND RAYS

8.2.1 Wavefront

Suppose a stone is thrown on the surface of still water. Circular patterns of alternate crests and troughs begin to spread out from the point of impact. Clearly, all the particles lying on a crest are in the position of their maximum upward displacement and hence in the same phase. Similarly, all particles lying on the trough are in position of their maximum downward displacement and therefore, in the same phase. The locus of such points oscillating in the same phase is called a *wavefront*. Thus, every crest or a trough is a wavefront.

> **DEFINITION**
> A *wavefront* is defined as the continuous locus of all such particles of the medium which are vibrating in the same phase at any instant.

Thus, a wavefront is a surface of constant phase. The speed with which the wavefront moves towards from the source is called the *phase speed*.

Types of Wavefronts

The geometrical shape of a wavefront depends on the source of disturbance. Some of the common shapes are given as:

(i) **Spherical wavefront:** In case of waves travelling in all directions from a point source, the wavefronts are spherical in shape. This is because all such points which are equidistant from the point source will lie on a sphere as shown in Figure 8.1 and the disturbance starting from the source S will reach all these points simultaneously.

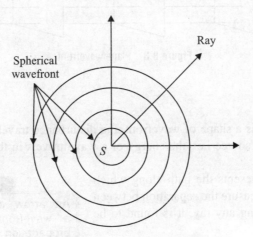

Figure 8.1 Spherical wavefront.

(ii) **Cylindrical wavefront:** When a source of light is linear in shape, such as a fine rectangular slit, the wavefront is cylindrical in shape. This is because the locus of all such points which are equidistant from the linear source will be cylinder as shown in Figure 8.2.

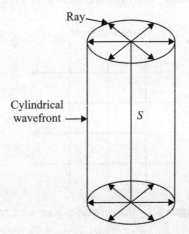

Figure 8.2 Cylindrical wavefront.

(iii) **Plane wavefront:** As a spherical or cylindrical wavefront advances, its curvature decreases progressively. So a small portion of such a wavefront at a large distance from the source will be a plane wavefront as shown in Figure 8.3.

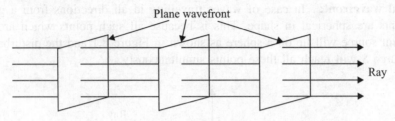

Figure 8.3 Plane wavefront.

8.2.2 Ray of Light

It is seen that whatever is a shape of wavefront, the disturbance travels outwards along straight lines emerging from the source i.e., the energy of a wave travels in the direction perpendicular to the wavefront.

A ray of light represents the path along which the light travels. If we measure the separation between a pair of wavefronts along any ray, it is found to be constant.

> **DEFINITION**
> An arrow drawn perpendicular to a wavefront in the direction of propagation of wave is called a *ray*.

This illustrates two general principles:

1. Rays are perpendicular to wavefronts.
2. The time taken for light to travel from one wavefront to another is the same along any ray.

In case of a plane wavefront, the rays are shown in Figure 8.4.

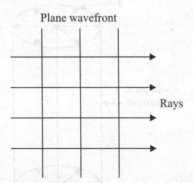

Figure 8.4 Rays in case of plane wavefront.

A group of parallel rays is called a *beam of light*. In case of a spherical wavefront, the rays either converge to a point [Figure 8.5] or diverge from a point [Figure 8.6].

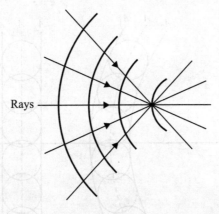

Figure 8.5 Rays in converging spherical wavefront.

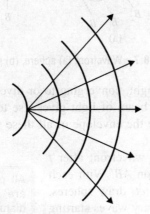

Figure 8.6 Rays in diverging spherical wavefront.

8.3 HUYGEN'S PRINCIPLE OF SECONDARY WAVELETS

According to Huygen, "A source of light in a homogeneous hypothetical medium called ether sends out waves in all directions. These waves carry energy which is transmitted in all directions."

If S is the source of light, it sends energy in the form of waves in all directions. After an interval of time t, all the particles of the medium lying on the surface AB are vibrating in the same phase. AB is thus the portion which has been drawn with S as centre and radius SA equals to ct, where c is the velocity of propagation of waves. The surface AB is called the primary wavefront.

In a homogeneous medium, for a point source of light, if distance is small, the wavefront is sphere as shown in Figure 8.7(a). If source is at a large distance, then the small portion of the wavefront can be considered to be plane as shown in Figure 8.7(b).

244 Fundamentals of Optics

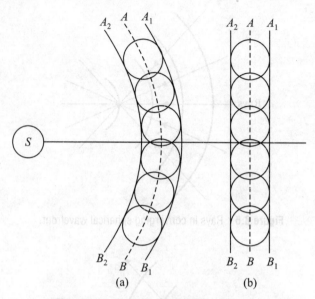

Figure 8.7 Wavefront (a) sphere, (b) plane.

This shows that the rays of light, converging to or diverging from a point, give rise to a spherical wavefront and a parallel beam of light gives rise to a plane wavefront.

After a given interval of time the envelope of all these secondary waves gives rise to the secondary wavefront.

To find the position of new wavefront after t seconds, take a number of points on AB. With each point in turn as centre and radius ct, draw spheres. These spheres represent the secondary waves starting from these points respectively. A surface A_1B_1 touching all these spheres in all forward direction in the new wavefront.

> **STATEMENT**
> All points on the primary wavefront are considered to be centres of disturbance and sends out secondary waves in all directions which travel through space with the same velocity in an isotropic medium.

8.4 PRINCIPLE OF SUPERPOSITION OF LIGHT WAVES

When a number of waves travel through a medium simultaneously, each wave travels independently of the others, i.e. as if all other waves were absent. An important consequence of this independent behaviour of the waves is that the effects of all these waves get added together. The resultant wave is obtained by the *principle of superposition of waves*, which can be stated as follows:

When a number of waves travelling through a medium superpose on each other, the resultant displacement at any point at a given instant is equal to the vector sum of the displacements due to individual waves at that point.

If y_1, y_2, y_3, ..., y_n be the displacements due to different waves acting separately, then according to the principle of superposition, the resultant displacement, when all the waves act together, is given by the vector sum:

Interference of Light Waves   245

$$Y = y_1 + y_2 + y_3 + \cdots + y_n \qquad (8.1)$$

When the two superposing waves are in the same phase, i.e. the crest of one falls over the crest of the another [Figure 8.1(a)] or the trough of one falls over the trough of another [Figure 8.1(b)], their displacements get added. When the two waves meet in the opposite phase, i.e. the crest of one falls over the trough of another [Figure 8.8(c)], their displacements get sbtracted.

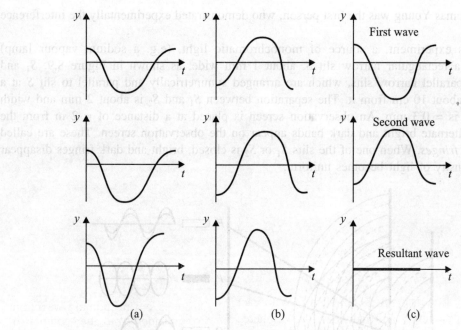

Figure 8.8   Illustration of principle of superposition of waves.

8.5   GROUPS OF INTERFERENCE

The phenomenon of interference may be grouped into two categories depending upon the formation of two coherent sources in practice:

Division of wavefront

Under this category, the coherent sources are obtained by dividing the wavefront, originating from a common source, by employing mirrors, biprisms or lenses. This class of interference requires essentially a point source or a narrow slit source. The instruments used to obtain coherent sources with interference by division of wavefront are *Young's double slit experiment*; *Fresnel's biprism, Fresnel mirrors, Lloyd's mirror, Laser*, etc.

Division of amplitude

In this category, the amplitude of the incident beam is divided into two or more parts either by partial reflection, or refraction. Thus we have coherent beams produced by division of amplitude. Thus beams travel through different paths and are finally brought together to as *two*

246 Fundamentals of Optics

beam interference and those resulting from superposition of more than two beams are referred to as *multiple beam interference*. The interference in *thin films*, *Newton's rings* and *Michelson's interferometer* are the examples of two beam interference and *Fabry-Perot interferometer* and *LG plate* are the example of multiple beam interference.

8.6 YOUNG'S DOUBLE SLIT EXPERIMENT (YDSE)

In 1801, Thomas Young was the first person, who demonstrated experimentally the interference of light.

In this experiment, a source of monochromatic light, (e.g. a sodium vapour lamp) illuminated a rectangular narrow slit S, about 1 mm wide, as shown in Figure 8.9. S_1 and S_2 are two parallel narrow slits, which are arranged symmetrically and parallel to slit S at a distance of about 10 cm from it. The separation between S_1 and S_2 is about 2 mm and width of each slit is ≈ 0.3 mm. An observation screen is placed at a distance of ≈ 2 m from the two slits. Alternate bright and dark bands appear on the observation screen. These are called *interference fringes*. When one of the slits S_1 or S_2 is closed, bright and dark fringes disappear and the intensity of light becomes uniform.

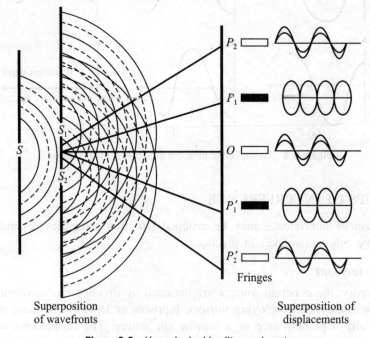

Figure 8.9 Young's double slit experiment.

Explanation

Figure 8.10 shows a section of Young's experiment in the plane of paper. According to Huygen's principle, cylindrical wavefronts emerge out from slit S, whose sections have been shown by circular arcs. The solid curves represent crests and the dashed curves represent troughs. As $SS_1 = SS_2$, these waves fall on the slits S_1 and S_2 simultaneously so that the wave spreading out from S_1 and S_2 are in the same phase. Thus S_1 and S_2 acts as two coherent sources of monochromatic light. Interference takes place between the waves diverging from these sources.

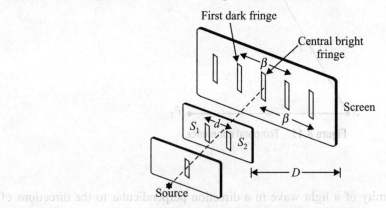

Figure 8.10 Interference of two light beams.

At the lines leading to O, P_2 and P_2', the crest of one wave falls over the crest of the other wave or the trough of one wave falls over the trough of other wave, the amplitudes of the two waves get added up and hence the intensity ($I \propto a^2$) becomes maximum. This is called *constructive interference*. At the lines leading to P_1 and P_1', the crest of one wave falls over the trough of other or the trough of wave falls over the crest of the other wave, the amplitudes of two waves get subtracted and, hence, the intensity becomes minimum. This is called *destructive interference*. So on the observation screen, we obtain a number of alternate bright and dark fringes, parallel to the two slits.

8.7 COHERENCE

If a fixed and predictable phase difference between several light waves travelling in a particular direction be maintained, then we may say the motion is *coordinated* or *coherence*. The corresponding waves are called *coherence waves* and sources emitting them are the *coherent sources*.

Coherence effects are mainly two types:
(i) temporal coherence and (ii) spatial coherence

(i) Temporal coherence

If a phase difference at a single point in the bundles of light waves propagating in space, at the beginning and end of a fixed time interval does not change with time then the waves are said to have *temporal coherence*. The phase difference between any two fixed points P_1 and P_2 as shown in Figure 8.11 along any ray will be independent of time but depends on $P_1 P_2$ and the coherence length (l_c) of light beam i.e., the distance $P_1 P_2 < l_c$. The waves will correlated in their rise and full maintaining *constant phase difference*.

If $P_1 P_2 > l_c$, then the points P_1 and P_2 would not maintain any phase relationship. In that case many wave will span the distance $P_1 P_2$. At any instant of time the plane at P_1 and P_2 will be *time independent*. The degree of correlation of phases is the amount of *longitudinal coherence*.

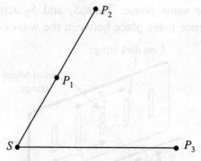

Figure 8.11 Temporal coherence.

(ii) Spatial coherence

The continuity and uniformity of a light wave in a direction perpendicular to the directions of propagation refers to *spatial coherence*. The move is said to have spatial coherence if the phase difference for any two fixed points in a plane normal to the wave propagation does not vary with time. In Figure 8.11, if $SP_1 = SP_3$, then the field points P_1 and P_3 would have phase as shown in Figure 8.12. Since the wave produced by an ideal source exhibits spatial coherence as the phases of the waves at any two points, which are equidistants from the source are equal. An extended source, however, exhibits less lateral coherence.

The *degree of contrast* of interference fringes is a measure of the degree of spatial coherence of the source resulting the waves. Spatial coherence is better if the contrast is higher.

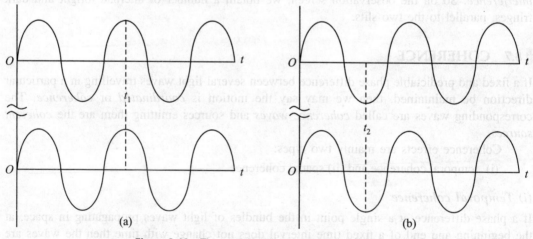

Figure 8.12 The spatial coherence between two waves.

(iii) Coherent and incoherent sources

Two or more sources, which are continuously emit light waves of same frequency (or wavelength) with a zero or constant phase difference between them are called *coherent sources*.

Two or more sources of light which do not emit light waves with a constant phase difference are called *incoherent sources*.

8.8 PHASE DIFFERENCE AND PATH DIFFERENCE

The difference between optical path of two rays, which are in constant phase difference with each other remitting at a particular point is known as *path difference*. For example, let the two coherent sources traversed different paths and meet at a particular point P as shown in Figure 8.13.

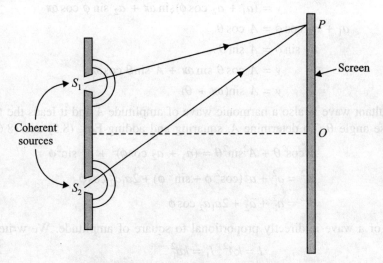

Figure 8.13 Illustration for path difference.

The path difference between two rays S_2P and S_1P is given by

$$\Delta = S_2P - S_1P$$

Suppose for a path difference λ, the phase difference is ϕ.

∵ For a path difference λ, the phase difference = 2π

∴ For a path difference x, the phase difference = $\dfrac{2\pi}{\lambda} x$

∴ Phase difference $(\phi) = \dfrac{2\pi}{\lambda} x = \dfrac{2\pi}{\lambda} \times$ path difference

or Phase difference = $\dfrac{2\pi}{\lambda} \times$ path difference (8.2)

8.9 CONDITIONS FOR CONSTRUCTIVE AND DESTRUCTIVE INTERFERENCE

8.9.1 Expression for Constructive and Destructive Interference Pattern

Suppose the displacements of two light waves from two coherent sources S_1 and S_2 at point P on the observation screen at any time t are given by

$$y_1 = a_1 \sin \omega t \tag{8.3}$$

and
$$y_2 = a_2 \sin(\omega t + \phi) \qquad (8.4)$$

where a_1 and a_2 are the amplitudes of the two waves, ϕ is the constant phase difference between the two waves. By the superposition principle, the resultant displacement at point P is

$$y = y_1 + y_2 = a_1 \sin \omega t + a_2 \sin(\omega t + \phi)$$
$$= a_1 \sin \omega t + a_2 \sin \omega t \cos \phi + a_2 \cos \omega t \sin \phi$$

or
$$y = (a_1 + a_2 \cos \phi)\sin \omega t + a_2 \sin \phi \cos \omega t$$

Put
$$a_1 + a_2 \cos \phi = A \cos \theta \qquad (8.5)$$

and
$$a_2 \sin \phi = A \sin \theta \qquad (8.6)$$

Then,
$$y = A \cos \theta \sin \omega t + A \sin \theta \cos \omega t$$

or
$$y = A \sin(\omega t + \theta)$$

Thus, the resultant wave is also a harmonic wave of amplitude A and it leads the first harmonic wave by phase angle θ. To determine A, squaring and adding Eqs. (8.5) and (8.6), we have

$$A^2 \cos^2 \theta + A^2 \sin^2 \theta = (a_1 + a_2 \cos \phi)^2 + a_2^2 \sin^2 \phi$$

or
$$A^2 = a_1^2 + a_2^2(\cos^2 \phi + \sin^2 \phi) + 2a_1 a_2 \cos \phi$$

or
$$A^2 = a_1^2 + a_2^2 + 2a_1 a_2 \cos \phi \qquad (8.7)$$

But intensity of a wave is directly proportional to square of amplitude. We write

$$I = kA^2, I_1 = ka_1^2$$

and
$$I_2 = ka_2^2$$

where k is proportionality constant. Equation (8.7) can be written as

$$kA^2 = ka_1^2 + ka_2^2 + 2\sqrt{ka_1}\sqrt{ka_2} \cos \phi$$

or
$$I = I_1 + I_2 + 2\sqrt{I_1 I_2} \cos \phi \qquad (8.8)$$

Equation (8.8) gives the total intensity at a point, where the phase difference is ϕ. Here I_1 and I_2 be the intensities which the two individual sources produce on their own. The total intensity is also contains a third term $2\sqrt{I_1 I_2} \cos \phi$. It is called *interference term*.

8.9.2 Constructive Interference

The resultant intensity at the point P will be maximum when $\cos \phi = 1$ or $\phi = 0, 2\pi, 4\pi, ...$

Since a phase different of 2π corresponds to a path difference of λ, therefore, if Δ is the path difference between the two superposing waves, then

$$\frac{2\pi \Delta}{\lambda} = 0, 2\pi, 4\pi, ...$$

or
$$\Delta = 0, \lambda, 2\lambda, 3\lambda, 4\lambda, ... = m\lambda$$

Hence the resultant intensity at a point is maximum when the different between the two superposing waves is an even multiple of π or path difference is an integral multiple of wavelength λ. This is the condition of constructive interference. In this case

$$I_{max} = I_1 + I_2 + 2\sqrt{I_1 I_2}$$

$$a_1^2 + a_2^2 + 2a_1 a_2 = (a_1 + a_2)^2$$

or $\qquad\qquad I_{max} > I_1 + I_2 \qquad\qquad (8.9)$

8.9.3 Destructive Interference

The resultant intensity at the point P will be minimum when $\cos\phi = -1$ or $\phi = \pi, 3\pi, 5\pi, ...$

or $\qquad\qquad \dfrac{2\pi \Delta}{\lambda} = \pi, 3\pi, 5\pi$

or $\qquad\qquad \Delta = \dfrac{\lambda}{2}, \dfrac{3\lambda}{2}, \dfrac{5\lambda}{2}, \cdots = (2m+1)\dfrac{\lambda}{2}$

Hence the resultant intensity at a point is minimum when the phase difference between the two superposing waves is an odd multiple of π or the path difference is an odd multiple of $\lambda/2$. This is condition of destructive interference.

In this case

$$I_{min} = I_1 + I_2 - 2\sqrt{I_1 I_2}$$

$$= a_1^2 + a_2^2 - 2a_1 a_2 = (a_1 - a_2)^2$$

$$I_{min} < I_1 + I_2 \qquad\qquad (8.10)$$

8.9.4 Conservation of Energy in Interference: Average Intensity

The average intensity is given by

$$I_{av} = \dfrac{\int_0^{2\pi} I d\phi}{\int_0^{2\pi} d\phi} = \dfrac{\int_0^{2\pi} (a_1^2 + a_2^2 + 2a_1 a_2 \cos\phi) d\phi}{\int_0^{2\pi} d\phi}$$

$$= \dfrac{[a_1^2 \phi + a_2^2 \phi + 2a_1 a_2 \sin\phi]_0^{2\pi}}{2\pi} = \dfrac{2\pi(a_1^2 + a_2^2)}{2\pi}$$

$$= a_1^2 + a_2^2$$

i.e., $\qquad\qquad I_{av} = I_1 + I_2 \qquad\qquad (8.11a)$

If $\qquad\qquad a_1 = a_2 = a$

$$I_{av} = 2a^2 \qquad\qquad (8.11b)$$

Thus, the average intensity is equal to the sum of the separate intensities. That is, whether energy is apparently disappears at the minima is actually present at maxima. Thus, *there is no violation of the law of conservation of energy in the phenomenon of interference.*

8.9.5 Comparison of Intensities at Maxima and Minima

Let a_1 and a_2 be the amplitudes and I_1 and I_2 be the intensities of light waves from two different sources.

As intensity is directly proportional to square of the amplitude.

$$\frac{I_1}{I_2} = \frac{a_1^2}{a_2^2}$$

Amplitude at a maximum in interference pattern = $(a_1 + a_2)$
Amplitude at a minimum in interference pattern = $(a_1 - a_2)$

Therefore, the ratio of intensities at maxima and minima is

$$\frac{I_{max}}{I_{min}} = \frac{(a_1 + a_2)^2}{(a_1 - a_2)^2} = \frac{\left(\frac{a_1}{a_2} + 1\right)^2}{\left(\frac{a_1}{a_2} - 1\right)^2} = \left(\frac{r+1}{r-1}\right)^2 \tag{8.12}$$

where $r = \frac{a_1}{a_2} = \sqrt{\frac{I_1}{I_2}}$ = amplitude ratio of two waves.

8.9.6 Visibility of Fringes

The quality of fringes produced by interferometric system can be described quantitatively using visibility (V), which is given by

$$V = \frac{I_{max} - I_{min}}{I_{max} + I_{min}} \tag{8.13a}$$

It was first formulated by Michelson. Here I_{max} and I_{min} are the intensities corresponding to the maximum and adjacent minimum in the fringe system. There will be best contrast in fringes or visibility when the difference I_{max} and I_{min} is maximum. As we know

$$I_{max} = I_1 + I_2 + 2\sqrt{I_1 I_2}$$

and

$$I_{min} = I_1 + I_2 - 2\sqrt{I_1 I_2}$$

Then visibility will be

$$V = \frac{2\sqrt{I_1 I_2}}{I_1 + I_2} \tag{8.13b}$$

8.9.7 Intensity Variation in Interference

Due to interference of two coherent waves, the resultant intensity at a point depends on the phase difference between the waves at that point. As the phase difference is a function of the position of that point, there shall be variation in the resultant intensity from point to point. Further the intensity is a measure of net energy of disturbance at the given point so that there shall be a *non-uniform distribution* in space. Corresponding to a path difference Δ, the resultant intensity at that point in given by

$$I = a_1^2 + a_2^2 + 2a_1 a_2 \cos\phi = I_1 + I_2 + 2\sqrt{I_1 I_2} \cos\phi$$

If we plot a graph between resultant intensity I and phase difference ϕ. The variation of I and ϕ is shown in Figure 8.14. When amplitudes a_1 and a_2 are different, the intensity maxima will have a value $(a_1 + a_2)^2$ or $(\sqrt{I_1} + \sqrt{I_2})^2$, while the intensity minima will have value $(a_1 - a_2)^2$ or $(\sqrt{I_1} - \sqrt{I_2})^2$ as shown in Figure 8.14(a).

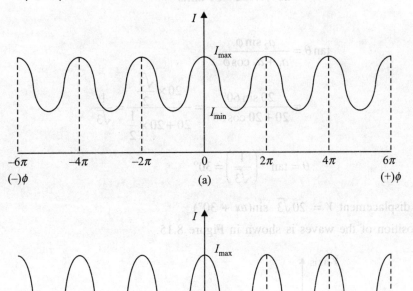

Figure 8.14 (a) Energy distribution curve for different valued amplitudes.
(b) Energy distribution curve for equal amplitudes.

When the amplitudes are equal i.e., $a_1 = a_2 = a$ or $I_1 = I_2 = I_0$, then $I_{max} = 4a^2$ or $I_{max} = 4I_0$ and $I_{min} = 0$ as shown in Figure 8.14(b).

EXAMPLE 8.1 Superimpose the following waves

$$y_1 = 20 \sin \omega t \text{ and } y_2 = 20 \sin (\omega t + 60°)$$

and also show the superimposition diagrammatically.

Solution Given $a_1 = 20$, $a_2 = 20$ and $\phi = 60°$

The resultant amplitude

$$R = \sqrt{a_1^2 + a_2^2 + 2a_1a_2 \cos\phi}$$
$$= \sqrt{(20)^2 + (20)^2 + 2 \times 20 \times 20 \times \cos 60°}$$
$$= \sqrt{400 + 400 + 400}$$
$$= 20\sqrt{3}$$
$$= 20 \times 1.732 = 35 \text{ units}$$

Direction

$$\tan\theta = \frac{a_2 \sin\phi}{a_1 + a_2 \cos\phi}$$

$$= \frac{20 \sin 60°}{20 + 20 \cos 60°} = \frac{20 \times \frac{\sqrt{3}}{2}}{20 + 20 \times \frac{1}{2}} = \frac{1}{\sqrt{3}}$$

$$\Rightarrow \qquad \theta = \tan^{-1}\left(\frac{1}{\sqrt{3}}\right) = 30°$$

Resultant displacement $Y = 20\sqrt{3} \sin(\omega t + 30°)$

Superimposition of the waves is shown in Figure 8.15.

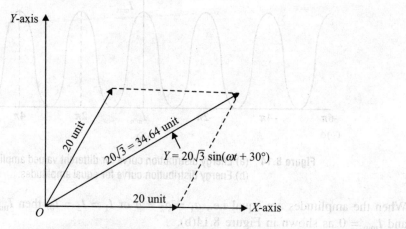

Figure 8.15 Superimposition of given two light waves.

EXAMPLE 8.2 Two waves of same frequency have amplitudes 1.0 and 2.0. They interfere at a point, where the phase difference is 60°. What is the resultant amplitude?

Solution Given $a_1 = 1.0$, $a_2 = 2.0$ and $\phi = 60°$

Then, the resultant amplitude

$$R = \sqrt{a_1^2 + a_2^2 + 2a_1a_2 \cos\phi}$$
$$= \sqrt{1^2 + 2^2 + 2 \times 1 \times 2 \cos 60°}$$
$$= \sqrt{1+4+2}$$
$$= \sqrt{7} = 2.65 \text{ units}$$

EXAMPLE 8.3 The coherent sources of intensity ratio α interfere. Prove that in the interfere pattern,

$$\frac{I_{max} - I_{min}}{I_{max} + I_{min}} = \frac{2\sqrt{\alpha}}{1+\alpha}$$

where symbols have their usual meanings.

Solution We know that the resultant intensity at a point due to two waves of amplitudes a_1 and a_2 is given by

$$I = a_1^2 + a_2^2 + 2a_1a_2 \cos\phi$$

and
$$I_{max} = (a_1 + a_2)^2$$
$$I_{min} = (a_1 - a_2)^2$$

Given
$$\alpha = \frac{I_1}{I_2} = \frac{a_1^2}{a_2^2}$$

where I_1 and I_2 are the intensities of two sources of respective amplitudes a_1 and a_2, Then

$$\frac{I_{max} - I_{min}}{I_{max} + I_{min}} = \frac{(a_1+a_2)^2 - (a_1-a_2)^2}{(a_1+a_2)^2 + (a_1-a_2)^2} = \frac{4a_1a_2}{2(a_1^2 + a_2^2)}$$

$$= \frac{2a_1a_2}{a_1^2 + a_2^2} = \frac{2\sqrt{I_1 I_2}}{I_1 + I_2}$$

$$= \frac{2\sqrt{\frac{I_1}{I_2}}}{1 + \frac{I_1}{I_2}} = \frac{2\sqrt{\alpha}}{1+\alpha}$$

8.10 THEORY OF INTERFERENCE FRINGES

8.10.1 Expression for Fringe Width

As shown in Figure 8.16, suppose a narrow slit S is illuminated by monochromatic light of wavelength λ, S_1 and S_2 be two narrow slits at equal distance from S. Being derived from the same parent source S, the slits S_1 and S_2 act as two *coherent sources*, separated by a small distance d. Interference fringes are obtained on a screen placed at distance D from the sources S_1 and S_2.

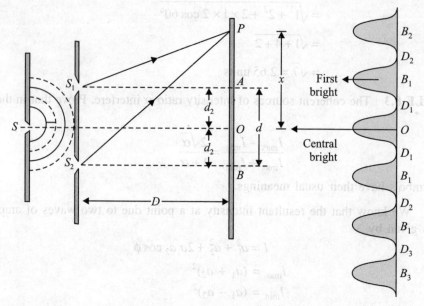

Figure 8.16 Position of bright and dark fringes in, Young's double slit experiment.

Consider a point P on the screen at distance x from the centre O. The nature of the interference at the point P depends on path difference

$$\Delta = S_2P - S_1P$$

From right angled ΔS_2BP and ΔS_1AP

$$(S_2P)^2 - (S_1P)^2 = [(S_2B)^2 + (BP)^2] - [(S_1A)^2 + (AP)^2]$$

$$= \left[D^2 + \left(x + \frac{d}{2}\right)^2\right] - \left[D^2 + \left(x - \frac{d}{2}\right)^2\right]$$

or $\qquad (S_2P - S_1P)(S_2P + S_1P) = 2xd$

or $\qquad S_2P - S_1P = \dfrac{2xd}{S_2P + S_1P}$

In practice, the point P lies very close to O, therefore, $S_1P \cong S_2P \cong D$. Hence,

$$\Delta = S_2P - S_1P = \frac{2xd}{2D}$$

or $\qquad \Delta = \dfrac{xd}{D} \qquad\qquad (8.14)$

8.10.2 Positions of Bright Fringes

For constructive interference

$$\Delta = \frac{xd}{D} = m\lambda$$

or $$x = \frac{mD\lambda}{d}$$

where $m = 0, 1, 2, 3, ...$

Clearly, the positions of various bright fringes are as follows:

For $m = 0$, $\quad x_0 = 0 \quad$ Central bright fringe

For $m = 1$, $\quad x_1 = \dfrac{D\lambda}{d} \quad$ First bright fringe

For $m = 2$, $\quad x_2 = \dfrac{2D\lambda}{d} \quad$ Second bright fringe

$\vdots \qquad \vdots \qquad \vdots$

For $m = m$ $\quad x_m = \dfrac{mD\lambda}{d} \quad$ mth bright fringe

8.10.3 Positions of Dark Fringes

For destructive interference,

$$\Delta = \frac{xd}{D} = (2m-1)\frac{\lambda}{2}$$

or $$x = (2m-1)\frac{D\lambda}{2d}$$

where $m = 1, 2, 3, ...$

Clearly, the positions of various dark fringes are as follows:

For $m = 1$, $\quad x'_1 = \dfrac{1}{2}\dfrac{D\lambda}{d} \quad$ First dark fringe

For $m = 2$, $\quad x'_2 = \dfrac{3}{2}\dfrac{D\lambda}{d} \quad$ Second dark fringe

$\vdots \qquad \vdots \qquad \vdots$

For $m = m$ $\quad x'_m = (2m-1)\dfrac{D\lambda}{d} \quad$ mth bright fringe

Since the central point O is equidistant from S_1 and S_2, the path difference Δ for it is zero. There will be a bright fringe at the centre O. But as we move from O upwards or downwards, alternate dark and bright fringes are obtained.

8.10.4 Fringe Width

It is a separation between two successive bright or dark fringe.

Width of a dark fringe = Separation between two consecutive bright fringes

$$x_m - x_{m-1} = \frac{mD\lambda}{d} - \frac{(m-1)D\lambda}{d} = \frac{D\lambda}{d}$$

Width of a bright fringe = Separation between two consecutive dark fringes

$$= x'_m - x'_{m-1}$$

$$= (2m-1)\frac{D\lambda}{2d} - [2(m-1)-1]\frac{D\lambda}{2d}$$

$$= \frac{D\lambda}{d}$$

Clearly, both the bright and dark fringes are of equal width. Hence the expression for the fringe width in Young's double slit experiment can be written as

$$\beta = \frac{D\lambda}{d} \tag{8.15}$$

As β is independent of m (the order of fringe), therefore all the fringes are of equal width. In the case of light, λ is extremely small, D should be much larger than d, so that the fringe width β may be appreciable, and hence, observable.

8.10.5 Measurement of Wavelength

Young's double slit experiment can be used to determine the wavelength of a monochromatic light. The interference pattern is obtained in the focal plane of a micrometer eyepiece and with it fringe width β is measured. By measuring the distance d between the two coherent sources and their distance D from the eyepiece, the value of wavelength λ can be calculated as

$$\lambda = \frac{\beta d}{D} \tag{8.16}$$

8.10.6 Interference Pattern with White Light

White light consists of colours from violet to red with wavelength range from 400 nm to 700 nm. Different component colours of white light produce their own interference pattern. At the centre of the screen, the path difference is zero for all such components. So bright fringes of different colours overlap at the centre. Consequently, the central fringe is white.

Now fringe width $\beta = \frac{D\lambda}{d}$, i.e. $\beta \propto \lambda$. Since the violet colour has the lowest λ, the closest fringe on either side of the central fringe is violet, while the farthest ring is red. After few fringes, the interference pattern is lost due to large overlapping of fringes and uniform white illumination is seen on the screen.

8.10.7 Shape of the Interference Fringes

Let S_1 and S_2 be the two coherent sources. At point P, there is maximum and minimum intensity, according to following conditions

$$S_2P - S_1P = m\lambda \quad \text{(maximum)}$$
$$S_2P - S_1P = (2m + 1)\frac{\lambda}{2} \quad \text{(minimum)}$$

Thus, for a given value of m, locus of points of maximum and minimum intensity is given by

$$S_2P - S_1P = \text{constant} \tag{8.17}$$

which is the equation of hyperbola with S_1 and S_2 as foci of hyperbola. This establishes that the interferences fringes are hyperbolas as shown in Figure 8.17.

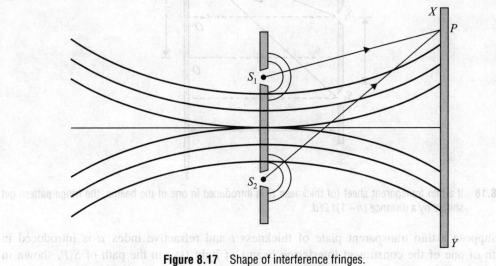

Figure 8.17 Shape of interference fringes.

Since the wavelength of light is extremely small ($\cong 10^{-7}$ m), the value of $S_2P - S_1P$ is also of that order. Therefore, the eccentricity of fringes is quite large and hence these hyperbolas appear more or less as straight line.

8.10.8 Angular Fringe Width

The angular fringe width is defined as the angular separation between consecutive bright and dark fringes and is denoted by θ.

$$\text{As Angle} = \frac{\text{Arc}}{\text{Radius}}$$

$$\theta = \theta_{m+1} - \theta_m = \frac{x_{m+1}}{D} - \frac{x_m}{D} = \frac{x_{m+1} - x_m}{D} = \frac{\beta}{D} = \frac{\frac{D\lambda}{d}}{D} = \frac{\lambda}{d}$$

$$\therefore \qquad \theta = \frac{\lambda}{d} \tag{8.18}$$

8.10.9 Displacement of Fringes

We will now discuss the change in the interference pattern produced when a thin transparent plate, say of glass or mica, is introduced in the path of one of the two interfering beams, as shown in Figure 8.18. It is observed that the entire fringe pattern is displaced to a point towards the beam in path of which the plate is introduced. If the displacement is measured, the thickness of the plate can be obtained provided the refractive index of the plate and the wavelength of the light are known.

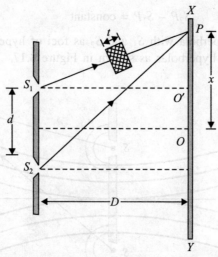

Figure 8.18 If a thin transparent sheet (of thickness t) is introduced in one of the beams, the fringe pattern get shifted by a distance $(n-1)t\, D/d$.

Suppose a thin transparent plate of thickness t and refractive index μ is introduced in the path of one of the constituent interfering beams of light (say in the path of S_1P, shown in Figure 8.18). Now, light from S_1 travel partly in air and partly in the plate. For the light path from S_1 to P, the distance travelled in air is $(S_1P - t)$, and that in the plate is t. Suppose, c and v are the velocities of light in air and in the plate, respectively. If the time taken by light beam to reach from S_1 to P is T, then

$$T = \frac{S_1P - t}{c} + \frac{t}{v}$$

or
$$T = \frac{S_1P - t}{c} + \frac{nt}{c} \qquad \left[\because v = \frac{c}{n}\right]$$

$$= \frac{S_1P + (n-1)t}{c}$$

Thus the effective path in air from S_1 to P is $[S_1P + (n-1)t]$, i.e. the air path S_1P is increased by an amount $(n-1)t$, due to the introduction of the plate of material of refractive index, n.

Let O be the position of the central bright fringe in absence of the plate, the optical paths S_1O and S_2O being equal. On introducing the plate, the two optical paths become unequal.

Therefore, the central fringe is shifted to O', such that at O' the two optical paths become equal. A similar argument applies to all the fringes. Now, at any point P, the effective path difference is given by

$$S_2P - [S_1P + (n-1)t] = S_2P - S_1P - (n-1)t$$

From Eq. (8.14),

$$S_2P - S_1P = \frac{d}{D}x$$

∴ Effective path difference at P is

$$P = \frac{d}{D}x - (n-1)t$$

If the point P is to be the centre of mth bright fringe, the effective path difference should be equal to $m\lambda$, i.e.

$$\frac{d}{D}x_m - (n-1)t = m\lambda$$

or

$$x_m = \frac{D}{d}[m\lambda + (n-1)t] \tag{8.19}$$

In absence of the plate ($t = 0$), the distance of the mth bright fringe from O is $\frac{D}{d}m\lambda$.

∴ Displacement x_o of the mth bright fringe is given by

$$x_o = \frac{D}{d}[m\lambda + (n-1)t] - \frac{D}{d}m\lambda$$

or

$$x_o = \frac{D}{d}(n-1)t \tag{8.20}$$

The shift is independent of the order of the fringe, showing that shift is the same for all the bright fringes. Similarly, it can be shown that the displacement of any dark fringe is also given by Eq. (8.20). Thus the entire fringe system is displaced through a distance $\frac{D}{d}(n-1)t$ towards the side on which the plate is placed. The fringe width is given by

$$\beta = x_{m+1} - x_m$$

$$= \frac{D}{d}(m+1)\lambda + (n-1)t] - \frac{D}{d}[m\lambda + (n-1)t]$$

$$= \frac{D\lambda}{d} \tag{8.21}$$

which is the same as before the introduction of the plate.

EXAMPLE 8.4 In Young's two slit experiment, the distance between the slits is 0.2 mm and screen is at a distance 1.0 m. The third bright fringe is at a distance 7.5 mm from the central fringe. Find the wavelength of light used.

Solution Given $d = 0.2$ mm, $D = 1.0$ m, $m = 3$, $x = 7.5$ mm

We know that $x = \dfrac{m\lambda D}{d}$

Then
$$\lambda = \dfrac{x(d)}{mD} = \dfrac{(7.5\times 10^{-3}\,\text{m})\times(0.2\times 10^{-3}\,\text{m})}{3\times 1.0\,\text{m}}$$

$$= 5.0 \times 10^{-7}\,\text{m}$$

$$= 500\,\text{nm}$$

EXAMPLE 8.5 Show that in a two-slit interferences pattern, the intensity at a point is given by

$$I = A + B\cos^2 \dfrac{kx}{2}$$

where A, B and k are constants of the set up and x is the linear distance of this point from the central fringe.

Solution The resultant intensity at a point due to two waves of amplitudes a_1 and a_2 are given by

$$I = a_1^2 + a_2^2 + 2a_1 a_2 \cos\phi$$

where ϕ is the phase difference at the point

$$\phi = \dfrac{2\pi}{\lambda}\times \text{path difference}$$

$$= \dfrac{2\pi}{\lambda}\times(S_2 P - S_1 P)$$

$$= \dfrac{2\pi}{\lambda}\times \dfrac{xd}{D}$$

where x is the linear distance of the point P from the central ring, d is the separation between the sources and D is the distance between slit and screen

$$\phi = \dfrac{2\pi}{\lambda}\times \dfrac{xd}{D} = kx$$

where k is a constant and its value is $\dfrac{2xd}{\lambda D}$.

$\therefore$
$$I = a_1^2 + a_2^2 + 2a_1 a_2 \cos kx$$

$$= a_1^2 + a_2^2 + 2a_1 a_2 \left(2\cos^2 \dfrac{kx}{2} - 1\right)$$

$$= (a_1 + a_2)^2 + 4a_1 a_2 \cos^2 \dfrac{kx}{2}$$

$$= A + B\cos^2 \dfrac{kx}{2}$$

where $A = (a_1 - a_2)^2$ and $B = 4a_1 a_2$ are constants.

EXAMPLE 8.6 In Young's double slit experiment, the angular width of a fringe formed on a distant screen is 0.1°. The wavelength of light used is 600 nm. What is the spacing between the slits?

Solution Given $\theta = 0.1° = \dfrac{0.1 \times 3.14}{180} = 1.74 \times 10^{-3}$ radian

$$\lambda = 600 \text{ nm} = 6.0 \times 10^{-7} \text{ m}$$

We know the angular width

Then
$$\theta = \dfrac{\lambda}{d}$$

$$d = \dfrac{\lambda}{\theta} = \dfrac{6.0 \times 10^{-7}}{1.74 \times 10^{-3}}$$

$$= 3.4 \times 10^{-4} \text{ m}$$
$$= 0.34 \text{ mm}$$

EXAMPLE 8.7 Interference fringes are produced by monochromatic light of wavelength 5460Å, when a thin sheet of transparent material of thickness 6.3×10^{-4} cm is introduced in the path of one of the interfering beams, the central fringe shifts a position occupied by 6th bright fringe. Compute refractive index of the sheet.

Solution Given $\lambda = 5460\text{Å} = 5.46 \times 10^{-7}$ m, $t = 6.3 \times 10^{-4}$ cm $= 6.3 \times 10^{-6}$ m,

$$m = 6, n = ?$$

we know that
$$(n - 1)t = m\lambda$$

or
$$n = \dfrac{m\lambda}{t} + 1$$

$$= \dfrac{6 \times 5.46 \times 10^{-7}}{6.3 \times 10^{-6}} + 1$$

$$= 0.52 + 1$$
$$= 1.52$$

8.11 CONDITIONS FOR INTERFERENCE OF LIGHT WAVES

To obtain a well defined observable interference pattern, the following conditions must be fulfilled.

8.11.1 Conditions for Sustained Interference

By sustained interference, we mean that the nature and order of interference at a point of the medium should remain unchanged with time.

For this to happen, there are two conditions:

(i) The two sources must be *monochromatic*, i.e., they must emit light of same wavelength or frequency.

(ii) The two sources must have either no phase difference or if there is a phase difference, it must remain unchanged with time.

If the above conditions are not satisfied, the phase difference between interfering waves at a point will go on changing and hence the resultant amplitude (or resultant intensity) at a point will go on changing with time. This will result in either uniform intensity or fluctuating intensity at the point.

8.11.2 Conditions for Good Visibility

(i) The separation between two coherent sources i.e., d should be small so that the width of bright and dark fringes formed will increase giving rise to increase resolving power and hence good visibility of fringes.
(ii) The separation of screen from the two coherent sources i.e., D should be large so that the width of fringes increases, and hence they are clearly seen.
(iii) The background in which the fringes are seen should be dark.

8.11.3 Conditions for Good Contrast

By good contrast, we mean the difference between maximum and minimum intensity or the difference between the intensities of bright and dark fringes should be as large as possible.

For this following conditions are necessary:

(i) The amplitude of the two interfering waves must be nearly the same or equal. In this case

$$a_1 = a_2 = a$$
$$I_{max} = (a + a)^2 = 4a^2$$
$$I_{min} = (a - a)^2 = 0$$

so that the difference between I_{max} and I_{min} is maximum and equal to $4a^2$.

(ii) The two light sources should be very narrow. If the sources are wide, they contain a large number of narrow sources giving rise to many interference patterns which overlap on the screen resulting in the decrease contrast.
(iii) Light sources should be monochromatic or should have wavelengths with smaller difference, otherwise due to overlapping of interference fringes of different colours, the interference pattern is seen white.

8.12 INTERFERENCE FRINGES WITH FRESNEL'S BIPRISM

Critics of the Thomas Young's experiment argued that the dark and bright fringes were probably due to some complicated modifications of the light by the edges of the slits and not due to the true interference. In order to answer the critics of Young's experiment, Fresnel thought of a new experiment for demonstrating the interference of light. He made use of a biprism for this purpose.

A schematic diagram of the biprism experiment is shown in Figure 8.19. Biprism is actually a simple prism, the base angles of which are extremely small (1/2°). The base of the prism is shown in Figure 8.19 and the prism is assumed to stand perpendicular to the plane of paper. S represents the slit which is also perpendicular to the plane of paper. Light from slit S gets refracted by the prism and produces two virtual images S_1 and S_2. These images act as coherent sources and produce interference fringes on the right of the prism. The fringes can be viewed through an eyepiece.

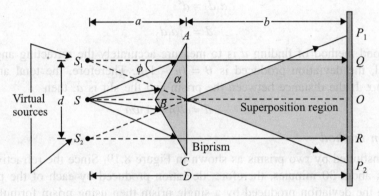

Figure 8.19 Fresnel's biprism.

The theory of the biprism experiment is the same as that of Thomas Young's double slit experiment. Of course, the distance between two sources d has to be determined and this can be evaluated by two methods:

(i) *Displacement method*

To determine the distance between the virtual sources of light in the biprism experiment, one makes use of what is called the *displacement method*. A lens with a focal length less than one-fourth of the distance between the biprism and eyepiece is mounted between the biprism and eyepiece as shown in Figure 8.20. The lens is adjusted in two positions L_1 and L_2 till sharp images of S_1 and S_2 are obtained in the field of view of the eyepiece. The distances d_1 and d_2 between the real images in two cases are measured.

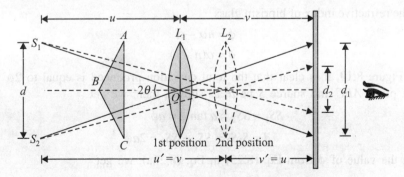

Figure 8.20 Fresnel's biprism arrangement showing the position of lens.

Thus, we have

$$\frac{d_1}{d} = \frac{v}{u}$$

and

$$\frac{d_2}{d} = \frac{u}{v}$$

where u and v are the object and image distances, i.e.

$$d_1 d_2 = d^2$$

or
$$d = \sqrt{d_1 d_2} \qquad (8.22)$$

The second method of finding d is to measure accurately the refracting angle α. As the angle is small, the deviation produced is $\theta = (n - 1)\alpha$. Therefore, the total angle $S_1 O S_2$ is $2\theta = 2(n - 1)\alpha$. If the distance between the prism and the slit is a, then

$$d = 2(n - 1)\alpha a \qquad (8.23)$$

(ii) *Deviation method*

Biprism is constituted by two prisms as shown in Figure 8.19. Since the refractive angle α of the biprism is about 30 minutes, therefore, deviation produced by each of the prism is quite small. If ϕ be the deviation produced by a single prism then using prism formula, we have

$$n = \frac{\sin\frac{\alpha + \phi}{2}}{\sin\frac{\alpha}{2}}$$

since $\frac{\alpha + \phi}{2}$ and $\frac{\alpha}{2}$ are very small then $\sin\frac{\alpha + \phi}{2}$ and $\sin\frac{\alpha}{2}$ are equal to $\frac{\alpha + \phi}{2}$ and $\frac{\alpha}{2}$ respectively.

Then
$$n = \frac{\frac{\alpha + \phi}{2}}{\frac{\alpha}{2}} = \frac{\alpha + \phi}{\alpha} \qquad (8.24)$$

where n is the refractive index of biprism glass

$$\phi = n\alpha - \alpha$$

or
$$\phi = \alpha(n - 1) \qquad (8.25)$$

From Figure 8.19, it is clear that the total deviation produced is equal to 2ϕ. If distance between the prism ABC and source S be a, as shown, then, we get

$$SS_1 = SS_2 = a \tan \phi = a\phi$$

or
$$d = S_1 S_2 = SS_1 + SS_2 = 2a\phi \qquad (8.26a)$$

Substituting the value of ϕ from Eq. (8.25) in Eq. (8.26a), we get

$$d = 2a(n - 1)\alpha \qquad (8.26b)$$

8.12.1 Interference Fringes with White Light

When white light is employed, the centre of the fringe at O is white, while the fringes on both sides of O are coloured because the fringe width depends on wavelength. In the biprism, the two coherent virtual sources are produces by refraction and the distance between two sources depends on the refractive index which, in turn, depends on wavelength of the light used. Therefore, for blue colour light, the distance between two virtual sources is different from that for the red. The distance x_m of the mth fringe from the centre

$$x_m = \frac{m\lambda D}{d}$$

where $d = 2(n - 1)\alpha\, a$

or
$$x_m = \frac{m\lambda D}{2(n-1)\alpha a} \tag{8.27}$$

Therefore, for blue and red rays the mth fringe will be

$$x_{mB} = \frac{m\lambda_B D}{2(n_B - 1)\alpha a} \tag{8.28}$$

and
$$x_{mR} = \frac{n\lambda_R D}{2(n_R - 1)\alpha a} \tag{8.29}$$

8.12.2 Effect of Increasing the Slit Width on Fresnel's Fringes

When in biprism experiment, the width of the slit is gradually increased, the contrast between the bright and dark fringes become poorer and poorer. Ultimately, the fringes disappear, leaving a uniform illumination everywhere.

Explanation

On increasing the slit width, the two virtual source slits are correspondingly widened. They are then equivalent to a large number of pairs of narrow slits. All pairs produce their fringe patterns, which are relatively shifted. This causes partial overlapping of maxima and minima due to different pairs, resulting in indistinctness. Greater the width of source of slits, greater the overlapping of maxima and minima and fringes disappear.

8.12.3 Effect of Increasing the Angle of Biprism on Fringes

If the angle α of the biprism be increased, the distance d between the virtual sources would increase because $d = 2a(n - 1)\alpha$. This, in turn, would reduce the fringe width ($\beta = D\lambda/d$). The fringes will not be separately visible and may disappear ultimately.

8.12.4 Location of Zero Order Fringe in Biprism Experiment

When the monochromatic light is used in biprism, alternate bright and dark fringes are obtained in which all the bright fringes are exactly similar in appearance. Hence, it is not possible to locate the zero order fringe.

To locate zero order fringe, monochromatic light is replaced by a source of white light. Now, the zero order (control) fringe is white and all other fringes are coloured. Now, the vertical crosswise is adjusted on zero order (white) fringes and white light is again replaced by monochromatic light. The vertical crosswise will still be on zero order fringes and thus zero order fringe is located.

EXAMPLE 8.8 A biprism is placed at a distance of 5 cm from slit illuminated by sodium light of wavelength 589 nm. Find the width of fringes observed in eyepiece at a distance of 75 cm from biprism, given the distance between virtual sources is 0.005 cm.

Solution Given $a = 5$ cm, $\lambda = 589$ nm $= 5.89 \times 10^{-5}$ cm, $\beta = ?$, $b = 75$ cm, $d = 0.005$ cm. We know that the fringe width (β) is

$$\beta = \frac{\lambda D}{d} = \frac{\lambda(a+b)}{d}$$

$$= \frac{5.89 \times 10^{-5} \text{cm} \times (5+75)\text{cm}}{0.005 \text{ cm}}$$

$$= \frac{5.89 \times 10^{-5} \times 80}{0.005} \text{cm}$$

$$= 4.712 \text{ cm}$$

EXAMPLE 8.9 The inclined faces of a biprism ($n = 1.5$) make angles of $1°$ with the base of the prism. The slit is 10 cm from the biprism and is illuminated by the wavelength 590 nm. Find the fringe width observed at a distance of one metre.

Solution Given $n = 1.5$, $\alpha = 1° = \dfrac{\pi}{180}$ radian, $D = 1$ m $+ 0.10$ m $= 1.1$ m, $\lambda = 590$ nm $= 5.9 \times 10^{-7}$ m, $a = 0.1$ m.

Then we may find the d by following formula

$$d = 2a(n-1)\alpha = 2 \times 0.1 \times (1.5 - 1)\frac{\pi}{180}$$

Then fringe width will be

$$\beta = \frac{\lambda D}{d}$$

$$= \frac{(5.9 \times 10^{-7}) \times (1.1) \times 180}{2 \times 0.1 \times 0.5 \times \pi}$$

$$= 0.000372 \text{ m}$$

$$= 3.72 \times 10^{-4} \text{ m}$$

EXAMPLE 8.10 Fringes are formed by a Fresnel's biprism in the focal plane of a reading microscope, which is 100 cm from the slit. A lens is inserted between the biprism and the microscope gives two images of the slit in two positions. In one case, the two images of the slit are 4.05 mm and in other, they are 2.90 mm apart. If sodium light (λ = 5893 Å) is used, find the distance between the interference fringes.

Solution *Given:*

$$d_1 = 4.05 \text{ mm}, d_2 = 2.09 \text{ mm}, D = 100 \text{ cm} = 1 \text{ m}, \lambda = 5893 \text{ Å} = 5.893 \times 10^{-7} \text{ m}$$

Then distance between two virtual sources i.e.,

$$d = \sqrt{d_1 d_2}$$
$$= \sqrt{4.05 \times 2.90 \times 10^{-6}}$$
$$= 3.427 \times 10^{-3} \text{ m}$$

So fringe width (β)

$$\beta = \frac{\lambda D}{d}$$
$$= \frac{5893 \times 10^{-7} \times 1}{3.427 \times 10^{-3}}$$
$$= 1.72 \times 10^{-4} \text{ m}$$

8.13 LLOYD'S SINGLE MIRROR

Lloyd's mirror is another device developed during the early part of the 19th century to demonstrate interference. Here, light from a monochromatic slit source(S_1) reflects from a glass surface and a small angle, appearing to come from a virtual source S_2. The reflected light interferes with a direct light from the surface forming interference fringes, Humphry Lloyd published an article about the demonstration in *Proceedings of the Royal Irish Society of Science* in 1837.

A schematic arrangement of Fresnel's single mirror is shown in Figure 8.21.

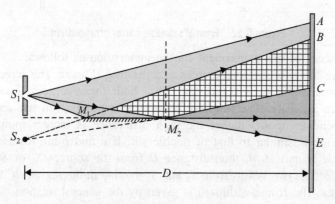

Figure 8.21 Lloyd's mirror arrangement.

It has a plane mirror M_1M_2, to avoid multiple reflection, M_1M_2 is polished on the front surface and blackened at the back. Light from narrow slit S_1 is allowed to incident at mirror almost at grazing angle. The reflected beam appears to diverge from S_2, which is virtual image of S_1. Thus S_2 and S_1 act as two coherent sources. The direct cone of light AS_1E and reflected cone of light BS_2C superimposed over each other and produces interference fringes in overlapping region BC of the screen.

8.14 FRESNEL'S DOUBLE MIRROR

Fresnel used two mirrors placed at angle of nearly 180° to produce coherent sources and thereby obtained interference fringes. A schematic arrangement of Fresnel's double mirror is shown in Figure 8.22.

A beam of light from a monochromatic source S is incident on two mirrors M_1M and M_2M and after reflection, appears to come out from two virtual sources S_1 and S_2 as shown in the Figure 8.22. S_1 and S_2 are at some distance (say d) apart, therefore, S_1 and S_2 act as two virtual coherent sources and produce interference fringes on the screen.

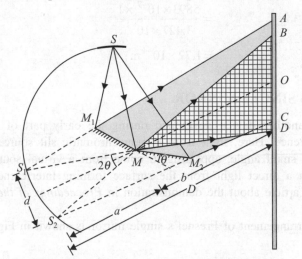

Figure 8.22 Fresnel's double mirror arrangement.

The basic theory of this experiment can be understood as follows:

S_1 and S_2 are the two coherent sources at a distance d apart. The screen is at a distance D apart. The central point O is equidistant from both these sources, hence, becoming the point of maximum intensity. The intensity at any other point P on the screen is maximum or minimum depending upon whether the path difference is integer multiple of λ or not. The basic discussion is similar to that of double slit. If a and b are distances of the joining point of mirrors M_1M and M_2M, then distance D from the sources S_1 or S_2 from the screen becomes $D = a + b$. The two beams from S_1 and S_2 overlap in the region BC and, thus, produce interference fringes. The fringe width (β) is given by the general relation

$$\beta = \frac{\lambda D}{d}$$

The distance of point on the centre of bright fringe from O, x_m is given by the expression

$$x_m = \frac{m\lambda D}{d} \quad \text{(with } m = 0, 1, 2, 3, 4, \ldots \text{ so on)} \tag{8.30}$$

and the distance of the point on the centre of dark fringe from O, x'_m is given by the expression

$$x'_m = \frac{(2m+1)\lambda D}{2d} \quad \text{(with } m = 0, 1, 2, 3, 4, \ldots \text{ so on)} \tag{8.31}$$

The distance between the virtual sources is estimated from the known values of a and θ using the relation

$$d = 2a\theta \tag{8.32}$$

when white light is used, the central fringe at O is white whereas the other fringes on both side of O are coloured because the fringe width β depends upon λ. It is possible to see only few coloured fringes as other fringes overlap.

8.15 STOKE'S LAW

To investigate the phase change in the reflection of light at an interface between two media, G.C. Stoke used the principle of optical reversibility.

> **OPTICAL REVERSIBILITY PRINCIPLE**
> A light ray, that is reflected or refracted, will retrace its original path, if its direction is reversed, provided there is no absorption of light.

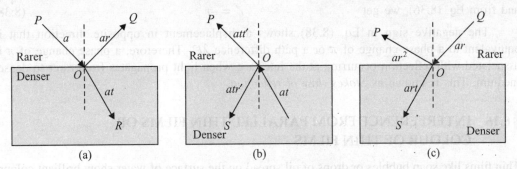

Figure 8.23 (a) Reflection and refraction of light wave from rarer to denser medium; (b) Reflection and refraction of light wave with amplitude of (a) and with incidence angle equal to angle of refraction of (a) from denser to rarer medium; (c) Reflection and refraction of light wave with amplitude of reflected wave of (a) and with incidence angle equal to angle of reflection of (a) from rarer to denser medium.

Consider a light wave PO with amplitude a falls on the interface of a denser medium from a rarer medium as shown in Figure 8.23. Now, we can define coefficients of reflection and refraction as:

Coefficient of reflection,

$$r = \frac{\text{Amplitude of reflected wave}}{\text{Amplitude of incident wave}} \tag{8.33}$$

and coefficient of refraction,

$$t = \frac{\text{Amplitude of refracted wave}}{\text{Amplitude of incident wave}} \tag{8.34}$$

> **STATEMENT**
> Stokes' law states that if waves are reflected at a rarer to denser medium interface (for example air-glass interface), the reflected waves have a phase difference of π (or path difference $\lambda/2$) compared to the incident wave. This also occurs in elastic waves such as sound waves.

Therefore, the amplitude of the reflected wave OQ is ar and that of refracted wave OR is at as shown in Figure 8.23(a).

Now, consider the situation when the directions of reflected and refracted waves are reversed. To do so, first, it is considered that a light wave of amplitude at is allowed to fall on interface from denser to rarer medium along RO. Then one has a reflected ray along OS with amplitude atr' and a refracted wave with amplitude att' along OP as shown in Figure 8.23(b), where r' and t' are the coefficients of reflection and refraction from denser to rarer medium respectively. Thereafter, it is allowed to fall the light wave of amplitude of ar on the interface from rarer to denser medium along QO. Now, there is a reflected wave along OP with amplitude ar^2 and a refracted wave with amplitude art along OS as shown in Figure 8.23(c).

Now, superposition of these two cases of propagation of light waves gives a light wave with amplitude $(ar^2 + att')$ along OP and another one with amplitude $(art + atr')$ along OS as shown in Figures 8.23(b) and 8.23(c). The reversal of reflected (with amplitude ar) and refracted (with amplitude at) light wave must produce a light wave with amplitude OP, and no wave along OS because when we have considered propagation of light wave from rarer to denser medium along PO, there is no wave along OS.

Therefore $$ar^2 + att' = a \qquad (8.35)$$
and $$art + ar't = 0 \qquad (8.36)$$
From Eq. (8.35), we have $$tt' = 1 - r^2 \qquad (8.37)$$
and from Eq. (8.36), we get $$r' = -r \qquad (8.38)$$

The negative sign in Eq. (8.38) shows a displacement in opposite direction that is equivalent to a phase change of π or a path difference $\lambda/2$. Therefore, a phase change of π is associated with reflection occurring at the interface when light propagates from rarer to denser medium. This is known as *Stokes' law of reflection*.

8.16 INTERFERENCE FROM PARALLEL THIN FILMS OR COLOUR OF THIN FILMS

Thin films like soap bubbles or drops of oil spread on the surface of water show brilliant colours when exposed to sunlight or to some other extended source of light. This is an example of interference due to multiple reflections from thin films. In this case, the interference pattern is produced by the division of amplitude and not by the division of wavefront.

Robert Hooke observed colours in thin film of mica and certain other thin transparent plates. Newton produced interference rings by enclosing a thin wedge-shaped film between a convex lens and a plane glass plate. Thomas Young explained that the interference is between light reflected from top and bottom surfaces of a thin film.

8.16.1 Interference Due to Reflected Light

Let GH and $G'H'$ be the parallel faces of a thin transparent film of thickness t and refractive index n as shown in Figure 8.24. Suppose a ray (single wave train) AB from a monochromatic source A of light is incident on the surface of GH at an angle i. A part of the incident ray will be reflected along BR. Another part will be refracted, into the film, along BC at an angle r which will be known as angle of refraction.

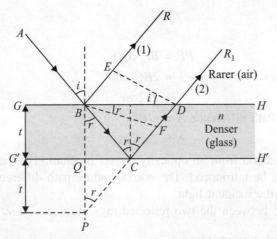

Figure 8.24 Path difference created by a thin film (reflected case).

The ray BC is incident on the bottom surface $G'H'$ of the film at an angle r. At C, a part will be reflected along CD. At D, the ray CD will again divided and this division will continue. A part of CD will be refracted along DR_1. Because of the reduction in amplitudes from one ray to next, there will be a rapid decrease in intensity from one ray to the next.

Let us now find path difference between two successive rays (1) and (2) in reflected system.

Draw DE perpendicular on BR and BF perpendicular on CD. $\angle i$ and $\angle r$ the angles of incidence and refraction respectively. DC is produced in backward direction to meet BQ at P.

Now $\angle BDE = i$, $\angle DBF = \angle CBQ = \angle CPQ = \angle r$

Beyond D and E, the rays (1) and (2) travel equal optical paths

∴ Path difference between rays (1) and (2)

$$\Delta = \text{Path } (BC + CD) \text{ in film} - \text{Path } BE \text{ in air}$$

$$\Delta = n(BC + CD) - BE \quad [\because \text{Optical path} = n \times \text{path in medium}] \quad (8.39)$$

Now

$$n = \frac{\sin i}{\sin r}$$

In ΔBED; $\sin i = \dfrac{BE}{BD}$ and in ΔBFD, $\sin r = \dfrac{FD}{BD}$

∴ $$n = \frac{\sin i}{\sin r} = \frac{BE/BD}{FD/BD} = \frac{BE}{FD}$$

∴ $BE = nFD$ \hfill (8.40)

∴ Path difference $\Delta = n(BC + CD) - nFD = n(BC + CD - FD)$

$$= n(PC + CD - FD) \quad (\because BC = PC)$$

$$= n(PC + CF + FD - FD)$$

$$= n(PC + CF)$$

$$= nPF \quad (8.41)$$

In $\triangle BPF$, $\dfrac{PF}{BP} = \cos r$

$\therefore$
$$PF = BP \cos r$$
$$= 2BQ \cos r \quad (\because BP = 2BQ)$$
$$= 2t \cos r \quad (\because BQ = 2t) \quad (8.42)$$

$\therefore$ Path difference $\Delta = n(PF)$

or $$\Delta = 2nt \cos r \quad (8.43)$$

The ray (1) suffers reflection from an optically denser medium (1). Therefore, an extra phase change of π radian will be introduced. The corresponding path difference will be $\lambda/2$, where λ is the wavelength of the incident light.

$\therefore$ Net path difference between the two reflected rays = $2nt \cos r - \lambda/2$

Condition for brightness or maxima

Constructive interference will be take place if $\Delta = m\lambda$
where $m = 0, 1, 2, 3, 4, \ldots$
then,

$$2nt \cos r - \dfrac{\lambda}{2} = m\lambda$$

or $$2nt \cos r = (2m + 1)\dfrac{\lambda}{2} \quad (8.44)$$

The film will appear bright in reflected light to satisfy condition (8.44).

Condition for darkness or minima

Destructive interference will take place if $\Delta = (2m - 1)\dfrac{\lambda}{2}$
where $m = 0, 1, 2, 3, 4, \ldots$
then,

$$2nt \cos r - \dfrac{\lambda}{2} = (2m - 1)\dfrac{\lambda}{2}$$

or $$2nt \cos r = m\lambda \quad (8.45)$$

The film will appear dark in reflected light to satisfy condition (8.45).

8.16.2 Interference due to Transmitted Light

Incident ray AB is refracted along BC as shown in Figure 8.25. At C, it is partly reflected along CD and partly transmitted as say (1) (CT). At D, a part of CD is partly reflected along DE. At E, the ray DE is partly transmitted as ray (2). Path difference between transmitted rays (1) and (2) can be found in the same manner as in the case of reflected system.

$$\text{Path difference } \Delta = n(CD + DE) - CD \quad (8.46)$$

In $\triangle CEP$, $\sin i = \dfrac{CP}{CE}$ and in $\triangle CEQ$, $\sin r = \dfrac{QE}{CE}$

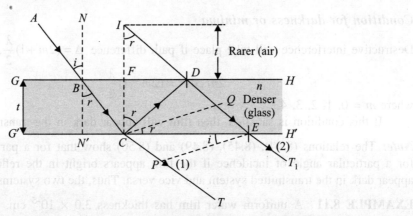

Figure 8.25 Path difference created by a thin film (transmitted case).

Then from Snell's law

$$n = \frac{\sin i}{\sin r} = \frac{CP/CE}{QE/CE}$$

∴ $CP = nQE$ (8.47)

∴ Path difference $\Delta = n(CD + DE) - nQE$
$= n(CD + DE - QE)$
$= n(ID + DE - QE)$ $(\because CD = ID)$
$= n(ID + DQ + QE - QE)$
$= n(IQ)$ [As $ID + DQ = IQ$]

In ΔIQC,

$$\frac{IQ}{IC} = \cos r$$

∴ $IQ = IC \cos r$
$= 2 CF \cos r$ [As $IC = 2CF$]
$= 2t \cos r$

Path difference $\Delta = n(IQ)$
$= 2nt \cos r$ (8.48)

Here reflection always takes place from the surface of a rarer medium. Thus no extra phase difference and thereby no extra path difference is produced.

∴   Net path difference = $2nt \cos r$

Condition for brightness or maxima

Constructive interference will take place if path difference ($\Delta = m\lambda$)

$$2nt \cos r = m\lambda$$

where $m = 0, 1, 2, 3, 4, \ldots$ (8.49)

If this condition is satisfied, then film will appear bright in the transmitted system.

Condition for darkness or minima

Destructive interference will take place if path difference $\Delta = (2m+1)\dfrac{\lambda}{2}$, i.e.

$$2nt\cos r = (2m+1)\dfrac{\lambda}{2} \qquad (8.50)$$

where $m = 0, 1, 2, 3, 4, ...$

If this condition is satisfied, then film will appear dark in the transmitted system.

Note: The relations (8.44), (8.45), (8.49) and (8.50) show that for a particular thickness and for a particular angle of incidence if the film appears bright in the reflected system, it will appear dark in the transmitted system and vice versa. Thus, the two systems are complementary.

EXAMPLE 8.11 A uniform water film has thickness 3.0×10^{-5} cm. What colour does it show when seen in reflected white light along the normal? ($n_{\text{water}} = 4/3$)

Solution The wavelength for destructive interference in reflected light (condition for minima) are given by

$$2nt = m\lambda$$

$\Rightarrow \qquad 2 \times \dfrac{4}{3} \times 3.0 \times 10^{-5} = m\lambda \quad (m = 1, 2, 3, ...)$

$\therefore \qquad \lambda = 8 \times 10^{-5}$ cm, 4×10^{-5} m, 2.7×10^{-5} cm, ...

The possible λ values in the visible range are 8×10^{-5} cm (red), 4×10^{-5} cm (violet). Hence these parts of the spectral colours are absent in reflected light. Intermediate wavelengths correspond to yellow and green colour will be seen.

EXAMPLE 8.12 Light of wavelength 5893 Å is reflected at nearly normal incidence from a soap film of refractive index $(n) = 1.42$. What is the least thickness of the film that will disappear (i) dark and (ii) bright?

Solution Given: $\lambda = 5893$ Å $= 5.893 \times 10^{-7}$ m, $n = 1.42$

(i) For the film to appear bright in reflected light at normal incidence

$$2nt = (2m+1)\dfrac{\lambda}{2}$$

$\Rightarrow \qquad t = \dfrac{(2m+1)\lambda}{4n}$

For least thickness $m = 0$

$$t = \dfrac{\lambda}{4n} = \dfrac{5.893 \times 10^{-7}}{4 \times 1.42}$$

$$= 1.0375 \times 10^{-7} \text{ m}$$

$$= 1037.5 \text{ Å}$$

(ii) For the film to appear bright in reflected light at normal incidence:

$$2nt = m\lambda$$

For least thickness, $m = 1$, then

$$t = \frac{\lambda}{2n}$$

$$= \frac{5.893 \times 10^{-7}}{2 \times 1.42} \text{ m}$$

$$= 2075 \text{ Å}$$

EXAMPLE 8.13 A beam of parallel rays is incident at an angle of 30° with the normal to a plane parallel film of thickness 4×10^{-5} cm and refractive index 1.50. Show that the refracted light whose wavelength is 7.539×10^{-5} cm will be strengthened by reinforcement.

Solution Given $t = 4 \times 10^{-5}$ cm, $i = 30°$, $n = 1.5$
To show that: $\lambda = 7.539 \times 10^{-5}$ cm will be strengthened by reinforcement in reflected light.

$$n = \frac{\sin i}{\sin r}$$

$$\Rightarrow \qquad \sin r = \frac{\sin i}{n} = \frac{\sin 30°}{1.5} = \frac{0.5}{1.5} = 0.33$$

$$\therefore \qquad \cos r = \sqrt{1 - \sin^2 r} = \sqrt{1 - (0.33)^2} = 0.9437$$

∴   This film in reflected region

$$2nt \cos r = (2m - 1)\frac{\lambda}{2}; \ m = 1, 2, 3, 4, \ldots$$

$$\lambda = \frac{4nt \cos r}{(2m-1)} = \frac{4 \times 1.5 \times 4 \times 10^{-5} \times 0.9437}{(2m-1)}$$

$$= \frac{22.6488 \times 10^{-5}}{(2m-1)} \text{ cm}$$

For $n = 1$, $\lambda_1 = \frac{22.6488 \times 10^{-5}}{1} = 22.6488 \times 10^{-5}$ cm

For $n = 2$, $\lambda_2 = \frac{22.6488 \times 10^{-5}}{3} = 7.5496 \times 10^{-5}$ cm

For $n = 3$, $\lambda_3 = \frac{22.6488 \times 10^{-5}}{5} = 4.5297 \times 10^{-5}$ cm

For $n = 4$, $\lambda_4 = \frac{22.6488 \times 10^{-5}}{5} = 3.2355 \times 10^{-5}$ cm

So, the wavelengths of 7.5496×10^{-5} cm and 4.5297×10^{-5} cm will be strengthened by reinforcement.

8.16.3 Colours in Reflected and Transmitted Light be Complementary

The colours observed in thin film in case of reflected light will be complementary of those observed in transmitted light. This is because the conditions for maxima and minima in the reflected light is just the reverse of those in the transmitted light.

8.16.4 Origins of Colours in Thin Film

We know that when oil is spread over water, then
  (i) different colours will be seen, i.e. the thin film will appear as coloured, and
  (ii) those colours also will change according to the angle of vision. Now the question is why?

1. When sunlight which consists of seven colours (seven wavelengths) is incident on thin film, then some of them will satisfy destructive condition is interference after repeated reflection from the thin film. Since all the seven colours (seven wavelengths) cannot satisfy the constructive condition at a time for a particular thin film of particular thickness t and refractive index n, so as a result, the thin film will appear as coloured, due to missing of some wavelengths, out of seven.

2. Now at a particular point of the thin film and for a particular position of the eye, the angle of refraction r will also be fixed, so for that particular r (the angle of vision), thickness t and refractive index n, a particular wavelength (λ) will satisfy constructive condition [Eq. (8.44)] and only that colour will be seen in that angle of vision. *If the angle of vision, i.e. r is changed, then some other (λ) will be seen. Hence colour will change with angle of vision, and due to the different position of eye one can see different colours.*

3. The condition for minima and maxima is reverse in the case of transmitted light. So those colours which are absent or suppressed in reflected light appear as intense colours in transmitted light, so they are complementary to each other.

8.16.5 Colour in Thick Films

When the thickness of the film is large compared to the wavelength of light, the path difference at any point of the film will be large. Then the same point will have maximum intensity for a large number of wavelengths, and minimum intensity for another large number of wavelengths sending minimum intensity. The wavelengths sending maximum and minimum intensity will be distributed equally over all the colours in white light. Hence if a certain number of wavelengths say, in red colour, is sending maximum intensity at a point, the same number of wavelengths in red is sending minimum intensity at the same point. Consequently that point will receive average intensity due to red. The same holds for all colours. Hence the resultant effect at any point will be sum of all colour i.e., white.

8.16.6 Necessity of an Extended Source

Consider a thin uniform film which is illuminate a point source S as shown in Figure 8.26.

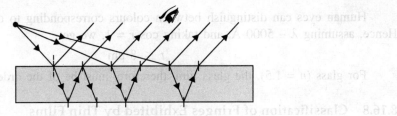

Figure 8.26 Point source.

It can be observed from the Figure 8.26 that for different angles of incidence, different pairs of interfering rays are obtained, of course, in a pair the interfering rays are parallel.

If the film is viewed by keeping eye in a particular position, the whole film cannot be viewed. In order to observe the whole film the eye will have to be shifted from one position to another position.

Let an extended source of light (which may be thought to be made up of large number of point sources S_1, S_2, S_3, ...) be used to illuminate the thin uniform film. The light reflected from every point of the film reaches the eye as shown in Figure 8.27. Hence, by placing the eye in a suitable position, one can see the entire film simultaneously. Due to this reason, an extended source of light is used to view a film.

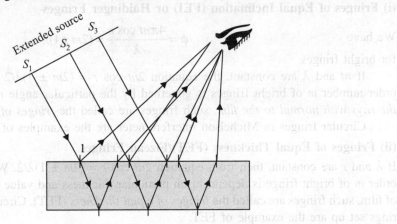

Figure 8.27 Extended source.

8.16.7 How Thin Must be a Thin Film?

We do not see interference colour, when thick layer of oil is illuminated in day light. In fact, even when a thin film is illuminated by so-called monochromatic light, the interference pattern disappears as the thickness of the film is increased beyond a certain limit.

The necessary condition for observing interference in thin films is that the path difference between two interfering beams must be less than longitudinal coherence l_c, otherwise they would be incoherent. Hence, for interference to be visible, we find that

$$2nt \cos r \leq \frac{\lambda^2}{\Delta \lambda} \quad \text{or} \quad t = \frac{(\lambda^2/\Delta \lambda)}{2n \cos r} \tag{8.51}$$

Human eyes can distinguish between colours corresponding to difference $\Delta\lambda \cong 100$ Å. Hence, assuming $\lambda \sim 5000$ Å, and taking $\cos r = 1$, we get

$$t \leq 8 \text{ μm}$$

For glass ($n = 1.5$), the glass film, therefore, must be of the order of few μm.

8.16.8 Classification of Fringes Exhibited by Thin Films

We have the path difference between two interfering rays of light from a transparent film of thickness t and refractive index n is

$$\Delta = 2nt \cos r \pm \frac{\lambda}{2}$$

So, the phase difference

$$\phi = \frac{2\pi}{\lambda} \times \left[2nt \cos r \pm \frac{\lambda}{2}\right] \quad (8.52)$$

Clearly ϕ depends on (i) λ, (ii) nt and (iii) r of a particular film.

Types of Fringes

(i) Fringes of Equal Inclination (FEI) or Haidinger Fringes

We have
$$\phi = \frac{4\pi nt \cos r}{\lambda} = (2m \pm 1)\pi \quad (8.53)$$
for bright fringes.

If nt and λ are constant, the equation $2nt \cos r = (2m \pm 1)\lambda/2$ shows that a particular order number m of bright fringes is governed by the particular angle r, i.e., the *inclination of the rays with normal to the film*, such fringes are called the *fringes of* FEI.

Circular fringes in Michelson interferometer are the examples of FEI.

(ii) Fringes of Equal Thickness (FET)/Fizeau Fringes

If λ and r are constant, then from equation $2nt \cos r = (2m \pm 1)\lambda/2$. We note that a particular order m of bright fringe is dependent on particular thickness and value of nt, i.e., the thickness of film, such fringes are called the *fringes of equal thickness* (FET). Circular fringes of Newton's rings set up are the example of FET.

(iii) Fringes of Equal Chromatic Order (FECO)

If nt and r are constant, then a particular order number m of a bright fringe is governed by λ, i.e., particular colour. A particular colour fringe satisfies the condition for particular wavelength. These fringes are called *fringes of equal chromatic order* (FECO).

8.17 INTERFERENCE IN NON-UNIFORM THICK FILM: WEDGE-SHAPED FILM

Let us consider two planes OX and OY which are inclined at angle α. Between these two planes, a medium of refractive index n is enclosed. When a light ray incidents on the inclined plane, then it is reflected from top and bottom surface of the thin film in the form of AR_1 and

CR_2. Let the angle of incidence and refraction be $\angle i$ and $\angle r$ respectively. If refractive index of medium is greater than refractive index of medium of incident ray, then AR_1 suffers an extra path difference of $\lambda/2$ (Figure 8.28).

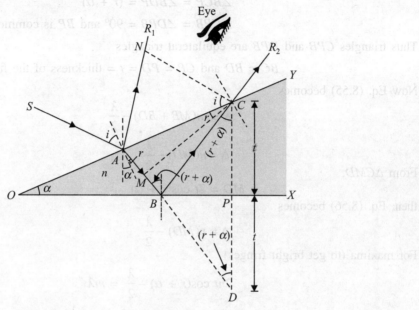

Figure 8.28 Interference produced by wedge-shaped films.

Since the time taken by the first light ray to go from AN is same as for the second ray to go from A to B, B to C, thus a path difference (Δ) between reflected rays AR_1 and CR_2 can be written as

$$\Delta = (AB + BC)_{med.} - \left(AN + \frac{\lambda}{2}\right)_{air}$$

$$= (AM + MB + BC)_{med.} - \left(AN + \frac{\lambda}{2}\right)_{air}$$

$$= n(AM + MB + BC) - AN - \frac{\lambda}{2} \qquad (8.54)$$

From Snell's law

$$n = \frac{\sin i}{\sin r}$$

From triangles ANC and AMC

$$n = \frac{AN/AC}{AM/AC} \;\Rightarrow\; AN = n(AM)$$

Putting the value of AN in Eq. (8.54), we get

$$\Delta = n(AM + MB + BC - AM) - \frac{\lambda}{2}$$

$$\Delta = n(MB + BC) - \frac{\lambda}{2} \qquad (8.55)$$

In triangles CPB and DPB

$$\angle BCP = \angle BDP = (r + \alpha)$$

$$\angle CPB = \angle DPB = 90° \text{ and } BP \text{ is common}$$

Thus triangles CPB and DPB are equilateral triangles

$$BC = BD \text{ and } CP = PD = t = \text{thickness of the film}$$

Now Eq. (8.55) becomes

$$\Delta = n(MB + BD) - \frac{\lambda}{2}$$

$$\Delta = n(MD) - \frac{\lambda}{2} \qquad (8.56)$$

From ΔCMD,

$$MD = 2t \cos(r + \alpha)$$

then, Eq. (8.56) becomes

$$\Delta = n(MD) - \frac{\lambda}{2} \qquad (8.57)$$

For maxima (to get bright fringes)

$$2nt \cos(r + \alpha) - \frac{\lambda}{2} = m\lambda$$

or

$$2nt \cos(r + \alpha) = (2m + 1)\frac{\lambda}{2} \qquad (8.58)$$

For minima (to get dark fringes)

$$2nt \cos(r + \alpha) - \frac{\lambda}{2} = (2m - 1)\frac{\lambda}{2}$$

or

$$2nt \cos(r + \alpha) = m\lambda \qquad (8.59)$$

At the edge of wedge, the thickness of film is zero, have it satisfies the minima condition for $m = 0$. Therefore, at point of contact the fringe will be dark.

Since locus of constant thickness from point of contact is a line, thus line fringes are formed in wedge-shaped film. The fringes are localized fringes, because they are formed within the film due to diverging nature of reflected light.

8.17.1 Spacing between Two Consecutive Dark Bands

Let x_1 be the distance of mth dark ring from the edge for thickness t_1 and x_2 be the distance of $(m + p)$th dark ring from the edge for thickness t_2 as shown in Figure 8.29. From Eq. (8.59), we know

$$2nt \cos(r + \alpha) = m\lambda$$

$$t_1 = \frac{m\lambda}{2n \cos(r + \alpha)} \qquad \text{[for mth dark band]} \qquad (8.60)$$

and

$$t_2 = \frac{(m + p)\lambda}{2n \cos(r + \alpha)} \qquad \text{[for } (m + p)\text{th dark band]} \qquad (8.61)$$

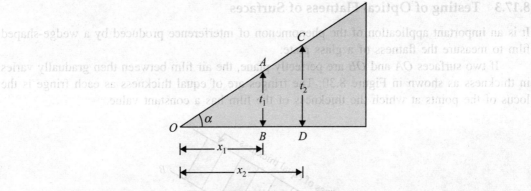

Figure 8.29 Wedge-shaped film.

From Figure (8.29), we have
$$t_1 = x_1 \tan \alpha$$
and
$$t_2 = x_2 \tan \alpha$$
Putting these values in Eqs. (8.60) and (8.61) and rearrange them, we get

$$x_1 = \frac{m\lambda}{2n \tan \alpha \cos(r + \alpha)} \quad (8.62)$$

and
$$x_2 = \frac{(m+p)\lambda}{2n \tan \alpha \cos(r + \alpha)} \quad (8.63)$$

Subtracting Eq. (8.62) from Eq. (8.63), we have

$$x_2 - x_1 = \frac{p\lambda}{2n \tan \alpha \cos(r + \alpha)}$$

The width of single band (fringe width)

$$\beta = \frac{x_2 - x_1}{p} = \frac{\lambda}{2 \tan \alpha \cos(r + \alpha)} \quad (8.64)$$

If α is very small, then $\tan \alpha = \alpha$, $r \gg \alpha \Rightarrow (r + \alpha) = r$

Then the fringe width is given as:

$$\beta = \frac{\lambda}{2n\alpha \cos r} \quad (8.65)$$

8.17.2 If White Light is Substituted for a Sodium Light

When the film is seen in white light, each colour (wavelength) produces its own interference fringes. The separation between two consecutive fringes will be least for violet, and greatest for red. At the edge of film $t = 0$ and $m = \lambda/2$. Hence, each wavelength gives minimum intensity at the edge. The edge will therefore be dark. As we move away from the edge in the direction of thickness increasing, we obtain a few coloured bands of mixed colour. For still greater thickness, the overlapping increases so much that uniform illumination is produced.

8.17.3 Testing of Optical Flatness of Surfaces

It is an important application of the phenomenon of interference produced by a wedge-shaped film to measure the flatness of a glass plate.

If two surfaces *OA* and *OB* are perfectly plane, the air film between then gradually varies in thickness as shown in Figure 8.30. The fringes are of equal thickness as each fringe is the locus of the points at which the thickness of the film has a constant value.

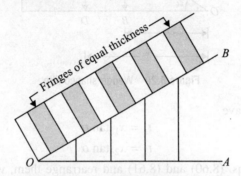

Figure 8.30 Testing of flatness of a glass plate.

If the fringes are not equal thickness, it means that the surface is not flat (plane).

To test the optical flatness of a surface, the specimen surface to be tested *OB* is placed over an optically plane surface *OA*. The fringes are observed in the field of view. If they are of equal thickness, the surface *OB* is plane. If not then surface *OB* is not plane. The surface *OB* is polished and the process is repeated. When the fringes observed are of equal width, it means the surface *OB* is plane.

EXAMPLE 8.14 A beam of monochromatic light of wavelength 5.82×10^{-7} m falls normally on a glass wedge with the wedge angle 20 seconds of an arc. If the refractive index of glass is 1.5, find the number of dark interference fringes per cm of the wedge length.

Solution Given $\lambda = 5.82 \times 10^{-7}$ m, $\alpha = 20''$, $n = 1.5$

The fringe width
$$\beta = \frac{\lambda}{2n\alpha}$$

where,
$$\alpha = \frac{20 \times \pi}{60 \times 60 \times 180} \text{ radian}$$

$$\therefore \quad \beta = \frac{5.82 \times 10^{-7} \times 60 \times 60 \times 180}{2 \times 1.5 \times 20 \times \pi}$$

$$= \frac{5.82 \times 6 \times 6 \times 18 \times 10^{-5}}{2 \times 1.5 \times 2 \times \pi}$$

$$= 2.0 \times 10^{-3} \text{ m} = 0.2 \text{ cm}$$

Number of dark interference fringes (p) per cm of the wedge length i.e.,

$$\frac{p}{x_2 - x_1} = \frac{1}{\beta} = \frac{1}{0.2 \text{ cm}} = \frac{10}{2} \text{ per cm}$$

$$= 5 \text{ fringes per cm}$$

EXAMPLE 8.15 Interference fringes are produced by monochromatic light falling normally on wedge-shaped film of cellophane whose refractive index is 1.4. The angle of wedge is 40″ and distance between successive fringes is 1.25 mm. Calculate the wavelength of light used.

Solution Given $n = 1.4$, $\alpha = 40'' = \dfrac{40}{60 \times 60} \times \dfrac{\pi}{180}$ radian, $\beta = 1.25$ mm $= 1.25 \times 10^{-3}$ m

For normal incidence, the fringe width (β) is

$$\beta = \frac{\lambda}{2n\alpha} \quad \text{[here } \lambda = \text{wavelength of monochromatic light]}$$

Then
$$\lambda = 2n\alpha\beta$$

$$= 2 \times 1.4 \times \frac{40 \times 3.14}{60 \times 60 \times 180} \times 1.25 \times 10^{-3}$$

$$= 6.784 \times 10^{-7} \text{ m}$$

$$= 6784 \text{ Å}$$

EXAMPLE 8.16 Light of wavelength 6000 Å falls normally on a wedge-shaped film of refractive index 1.4 forming fringes that are 2.0 mm apart. Find the angle of wedge in seconds.

Solution $\lambda = 6000$ Å $= 6.000 \times 10^{-7}$ m, $n = 1.4$, $\beta = 2$ mm $= 2 \times 10^{-3}$ m
We know that angle of wedge

$$\alpha = \frac{\lambda}{2n\beta} = \frac{6.000 \times 10^{-7}}{2 \times 1.4 \times 2 \times 10^{-3}}$$

$$= 1.071 \times 10^{-4} \text{ rad} = 22''$$

8.18 NEWTON'S RINGS

When a planoconvex lens of long focal length is placed on a glass plate, a thin film of air is enclosed between the lower surface of the lens and the upper surface of the plate. Thickness of the air film is negligible at the point of contact and it gradually increases from the centre towards periphery. When the air film is illuminated by a monochromatic light, the concentric dark and bright circular fringes with a dark centre are observed in the interference pattern, Since Newton was the first scientist who obtained experimentally these rings, therefore these rings are called *Newton's rings*.

8.18.1 Experimental Arrangement for Newton's Rings

The experimental arrangement for Newton's rings is shown in Figure 8.31. Light from a monochromatic source (sodium discharge lamp) is allowed to fall on a convex lens through a broad slit, which renders it into a nearly parallel beam. Not it falls on a glass plate inclined at an angle of 45° to the vertical, thus the parallel beam is reflected from the lower surface. Due to the air film formed by a glass plate and a planoconvex lens of large radius of curvature, interference fringes are formed which are observed directly through a travelling microscope. These rings are called *concentric circles*.

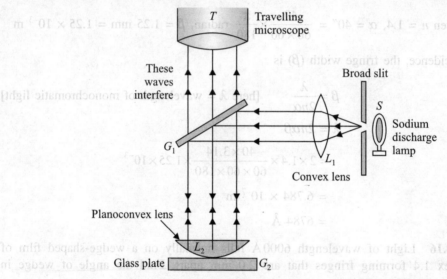

Figure 8.31 Experimental arrangement for Newton's rings.

8.18.2 Formation of Newton's Rings

Consider an optical arrangement as shown in Figure 8.32, which consists of a plane glass plate and a planoconvex lens with large radius of curvature, the latter placed over the former

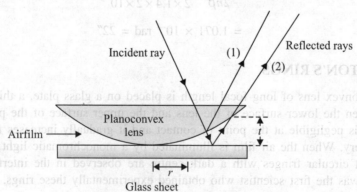

Figure 8.32 Formation of Newton's rings.

with convex surface being in contact with glass plate. The arrangement forms an airfilm between the glass plate and the planoconvex lens, the thickness of the film is zero at the centre (point of contact) and increases as one moves radially outwards. By taking the radius of curvature of the lens very large, the thickness of the airfilm becomes very small.

Now suppose that light is normally incident on the airfilm. A part of the amplitude of the incident light is reflected from the upper surface of the airfilm and rest goes through the airfilm and again a part of this is reflected from the upper surface of the glass plate. Thus we have two reflected light rays—one reflected from upper surface of the film and the other reflected from the glass plate. These reflected beams when superimpose produce interference.

Let us calculate the path difference between two interfering waves. The waves reflected from the glass plate travels an extra path $2nt$ (for air $n = 1$), where t is the thickness of the airfilm at the point of incidence. The extra path difference is called *geometrical path difference*. This wave is reflected from a denser medium, hence it suffers a phase change of π (which is equivalent to path difference $\lambda/2$). This path difference is called *optical path difference*. Thus the total path difference between the interfering waves is the sum of geometrical path difference and optical path difference. i.e.,

$$\Delta = 2nt + \frac{\lambda}{2}$$

The condition for constructive interference is

$$\Delta = 2nt + \frac{\lambda}{2} = m\lambda$$

or
$$2nt = (2m-1)\frac{\lambda}{2} \tag{8.66}$$

where $m = 1, 2, 3, 4, \ldots$

The point where thickness of airfilm is constant lie on a circle with centre at the point of contact, and the condition of constructive interference is satisfied at all these points, hence the fringes are circular (fringes are of the locus of points of equal path differences). Similarly, the condition of destructive interference is satisfied when

$$\Delta = 2nt + \frac{\lambda}{2} = (2m+1)\frac{\lambda}{2}$$

or
$$2nt = m\lambda \tag{8.67}$$

Again the dark fringes are also circular. The interference pattern consists of concentric bright and dark circles. At centre $t = 0$, the geometrical path difference is zero, but there is optical path difference $\lambda/2$. Hence, at the centre, condition of destructive interference is satisfied. So the central fringe is dark.

8.18.3 Production of Coherent Sources in Newton's Rings Experiment

In this experiment, a ray is partially reflected back from the lower surface of the planoconvex lens and partially refracted. The refracted ray is then partially reflected back from upper surface of the plane glass plate placed below the planoconvex lens. These two rays are derived from the same ray incident on the plane surface of the planoconvex lens and have a constant phase difference depending on the thickness of the airfilm at the point of reflection. In this way, one gets these two rays by means of *division of amplitude*. Therefore, these two rays are coherent in nature.

8.18.4 Theory of Newton's Rings

Newton's rings by reflected light

Diameter of Newton's rings: Let r be the radius of curvature of the mth bright rings as shown in Figure 8.33 and R the radius of curvature of the lens, then by geometry of a circle.

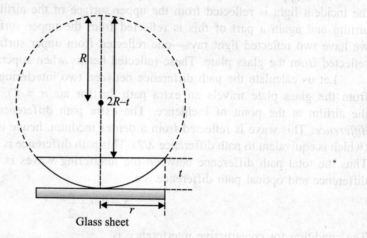

Figure 8.33 Illustration to find diameter or rings.

$$r \times r = t(2R - t)$$

or
$$r^2 = 2Rt - t^2$$
$$= 2Rt \text{ (approximately)}$$

$$\therefore \quad r^2 = 2Rt$$

or
$$t = \frac{r^2}{2R}$$

or
$$2t = \frac{r^2}{R} \tag{8.68}$$

Substituting the value of $2t$ in Eq. (8.66) and suppose the system in air (i.e. $n = 1$), then

$$2t = (2m-1)\frac{\lambda}{2}$$

or
$$\frac{r^2}{R} = (2m-1)\frac{\lambda}{2}$$

or
$$r^2 = \frac{(2m-1)\lambda R}{2} \tag{8.69}$$

Replace r as putting $r = \frac{D}{2}$, we get the diameter of mth bright ring

$$\frac{D^2}{4} = \frac{(2m-1)\lambda R}{2}$$

or
$$D = \sqrt{(2\lambda R)(2m-1)} \quad (8.70)$$
and
$$D \propto \sqrt{(2m-1)}$$

Hence the diameter of bright rings are proportional to the square root $(2m-1)$, i.e. odd number.

In the similar way, we can calculate the diameter of dark ring

$$\frac{r^2}{R} = m\lambda$$

or
$$r^2 = m\lambda R$$
or
$$D^2 = 4m\lambda R$$
or
$$D = 2\sqrt{m\lambda R} \quad (8.71)$$
and
$$D \propto \sqrt{m}$$

Hence the diameters of dark Newton's rings are proportional to the square roots of natural number, i.e. $\sqrt{m}$.

With the help of above expressions (8.70) and (8.71), it can be seen that fringe width decreases with the order of fringe and fringes get closer with increase in their order. The central ring will be dark in case of reflected light as shown in Figure 8.34.

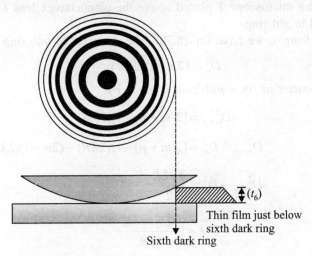

Figure 8.34 Newton's rings in reflected case.

Newton's rings by transmitted light

As we have discussed in section 8.16, that both reflected light and transmitted cases of thin film are just opposite to each other. Hence for bright rings in case of transmitted light

$$2t = m\lambda$$

or
$$\frac{r^2}{R} = m\lambda$$

or
$$r^2 = m\lambda R$$
or
$$D = 2\sqrt{m\lambda R} \qquad (8.72)$$
or
$$D \propto \sqrt{m}$$

and for dark rings in cases of transmitted light

$$2t = (2m-1)\frac{\lambda}{2} \Rightarrow \frac{r^2}{R} = (2m-1)\frac{\lambda}{2}$$

or
$$D = \sqrt{2\lambda R}\sqrt{2m-1}$$

$$D \propto \sqrt{2m-1}$$

In the case of transmitted light, the central ring is bright.

8.18.5 Determination of Wavelength of Monochromatic Light using Newton's Rings Experiment

As shown in Figure 8.31, light from a monochromatic extended source (sodium light) S is converted into parallel beam by a lens L_1, which falls on a glass plate G_1 oriented at an angle of 45° with the vertical. The light reflected by the plate G_1 turns through 90° and incident normally on the wedge-shaped film formed between the planoconvex lens L_1 and the glass plate G_2. A travelling microscope T placed above the planoconvex lens L_2 is used to see the concentric dark and bright rings.

To get bright fringes, we have Eq. (8.70). The diameter of mth ring is

$$D_m^2 = (2m-1)(2\lambda R)$$

In similarly, the diameter of $(m + p)$th ring is given by

$$D_{m+p}^2 = [2(m+p)-1](2\lambda R)$$

Then,
$$D_{m+p}^2 - D_m^2 = [2(m+p)-1](2\lambda R) - (2m-1)(2\lambda R)$$

or
$$D_{m+p}^2 - D_m^2 = \frac{4p\lambda R}{2} \qquad (8.73)$$

Hence
$$\lambda = \frac{D_{m+p}^2 - D_m^2}{4pR} \qquad (8.74)$$

If we plot a graph between square of the diameter of the ring and, the number of rings as shown in Figure 8.35, then this figure is very useful to find the wavelength of light.

From Figure 8.35,

$$\frac{D_{m+p}^2 - D_m^2}{p} = \frac{AB}{CD}$$

knowing the value of R, we can calculate the wavelength of monochromatic light by Eq. (8.74).

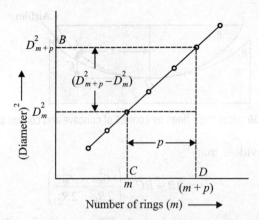

Figure 8.35 Number of rings versus square of the diameter of the rings.

8.18.6 Determination of Refractive Index of Transparent Liquid using Newton's Rings Experiment

As shown in Figure 8.31, light from a monochromatic extended source (sodium light) S is converted into parallel beam by a lens L_1, which falls on a glass plate G_1 oriented at an angle of 45° with the vertical. The light reflected by the glass plate G_1 turns through 90° and is incident normally on the wedge-shaped liquid film of refractive index n formed between the planoconvex lens L_2 and the glass plate G_2. A travelling microscope T placed above the planoconvex lens L_2 is used to see the concentric dark and bright rings.

For bright rings (reflected case), we have diameter of mth ring in liquid

$$D_m'^2 = (2m-1)\left(\frac{2\lambda R}{n}\right)$$

Similarly, for $(m + p)$th ring

$$D_{m+p}'^2 = [2(m-p)-1]\left(\frac{2\lambda R}{n}\right)$$

and

$$D_{m+p}'^2 - D_m'^2 = \frac{4p\lambda R}{n} \tag{8.75}$$

If the system is in air, then $n = 1$. Then we get same equation as Eq. (8.73). Hence from Eqs. (8.73) and (8.75)

$$n = \frac{D_{m+p}^2 - D_m^2}{D_{m+p}'^2 - D_m'^2} \tag{8.76}$$

8.18.7 Newton's Rings by Contact of Concave and Convex Surfaces

Let R_1 and R_2 be the radii of curvature of the convex and concave surfaces. It is essential that $R_2 > R_1$. In Figure 8.36, a convex surface of radius of curvature R_1 is in contact with a concave surface R_2 at point O. Let us consider a Newton's rings of radius r_n, where the thickness of the airfilm is t.

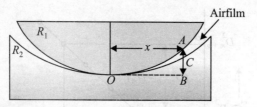

Figure 8.36 Newton's rings by contact of concave and convex surfaces.

From Figure 8.36, it is evident that

$$t = AB - BC = \left(\frac{r_m^2}{2R_1} - \frac{r_m^2}{2R_2}\right) \tag{8.77}$$

For the thin film and normal incidence, the path difference is given by

$$\Delta = 2nt = 2n\left(\frac{r_m^2}{2R_1} - \frac{r_m^2}{2R_2}\right) \tag{8.78}$$

For maxima or bright rings

$$\Delta = (2m+1)\frac{\lambda}{2}$$

Then

$$nr_m^2\left(\frac{1}{R_1} - \frac{1}{R_2}\right) = (2m+1)\frac{\lambda}{2}$$

or

$$nD_m^2\left(\frac{R_2 - R_1}{2R_1R_2}\right) = (2m+1)\lambda$$

where $D_m = 2r_m$ = diameter of mth ring

or

$$D_m^2 = \frac{2(2m+1)\lambda R_1 R_2}{n(R_2 - R_1)} \tag{8.79}$$

For air $n = 1$, Eq. (8.79) becomes

$$D_m^2 = \frac{2(2m+1)\lambda R_1 R_2}{(R_2 - R_1)} \tag{8.80}$$

For minima or dark rings

$$\Delta = m\lambda$$

or

$$nr_m^2\left[\frac{1}{R_1} - \frac{1}{R_2}\right] = m\lambda$$

or

$$nD_m^2\left[\frac{R_2 - R_1}{4R_1R_2}\right] = m\lambda$$

or

$$D_m^2 = \frac{4m\lambda R_1 R_2}{n(R_2 - R_1)} \tag{8.81}$$

For air $n = 1$, Eq. (8.81) becomes

$$D_m^2 = \frac{4m\lambda R_1 R_2}{R_2 - R_1} \tag{8.82}$$

8.18.8 Newton's Rings by Contact of Two Convex Surfaces

Let the radii of curvature of two convex surfaces in contact be R_1 and R_2, then thickness of airfilm for convex-convex surfaces as shown in Figure 8.37 is given by

$$t = t_1 + t_2 = \frac{r_m^2}{2R_1} + \frac{r_m^2}{2R_2} \tag{8.83}$$

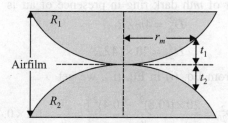

Figure 8.37 Newton's rings by contact of two convex surfaces.

Proceeding, exactly as previous case, one obtains the expression for diameter of mth bright ring

$$D_m^2 = \frac{2(2m+1)\lambda R_1 R_2}{n(R_1 + R_2)} \tag{8.84}$$

For air, put $n = 1$ in Eq. (8.84)

$$D_m^2 = \frac{2(2m+1)\lambda R_1 R_2}{R_1 + R_2} \tag{8.85}$$

The diameter of mth dark ring is given by

$$D_m^2 = \frac{4m\lambda R_1 R_2}{n(R_1 + R_2)} \tag{8.86}$$

For air, $n = 1$, Eq. (8.86) becomes

$$D_m^2 = \frac{4m\lambda R_1 R_2}{R_1 + R_2} \tag{8.87}$$

Wavelength relation can be expressed as

$$D_{m+p}^2 - D_m^2 = \frac{4p\lambda R_1 R_2}{n(R_1 + R_2)} \tag{8.88}$$

Here $(R_2 - R_1)$ is for concave–convex surfaces and $(R_2 + R_1)$ for convex–convex surfaces. Similarly, one can obtain expression for transmitted patterns.

EXAMPLE 8.17 In a Newton's ring experiment the diameter of 4th and 12th rings are 0.4 cm and 0.8 cm respectively. Deduce the diameter of 20th dark ring.

Solution Given $m = 4$, $(m + p) = 12$, $p = 8$, $D_m = 0.4$ cm and $D_{m+p} = 0.8$ cm
The wavelength of sodium light using Newton's rings set up is

$$\lambda = \frac{D_{m+p}^2 - D_m^2}{4pR}$$

or
$$4\lambda R = \frac{D_{m+p}^2 - D_m^2}{p}$$

$\Rightarrow$
$$4\lambda R = \frac{(0.8)^2 - (0.4)^2}{8} \quad \text{(i)}$$

We know that the diameter of mth dark ring in presence of air is
$$D_m^2 = 4m\lambda R$$

$\Rightarrow$
$$D_{20}^2 = 20 \times (4\lambda R) \quad \text{(ii)}$$

Putting the value of $4\lambda R$ from Eq. (i) in Eq. (ii), we get

$$D_{20}^2 = \frac{20 \times [(0.8)^2 - (0.4)^2]}{8} = \frac{20}{8} \times 1.2 \times 0.4$$

or $\quad D_{20} = 1.2$ cm

EXAMPLE 8.18 In a Newton's ring set up, diameter of 20th dark ring is found to be 7.25 mm. The space between spherical surface and the flat slab is then filled with water ($n = 1.33$). Calculate the diameter of 16th dark ring in new set up.

Solution In Newton's ring set up.

1st set up: Given $D_{20} = 7.25$ mm
We know that the diameter of mth dark ring in presence of air is
$$D_m^2 = 4m\lambda R$$

i.e.,
$$D_{20}^2 = 4 \times 20 \times \lambda R$$

or
$$4\lambda R = \frac{(7.25)^2}{20} \quad \text{(i)}$$

New set up: Now, liquid is introduced, then

$$D_m'^2 = \frac{4m\lambda R}{n}$$

or
$$D_m'^2 = \frac{4 \times 16 \times \lambda R}{1.33} = \frac{16 \times (4\lambda R)}{1.33} \quad \text{(ii)}$$

Putting the value of $4\lambda R$ from Eq. (i) in Eq. (ii), we get

$$D_{16}'^2 = \frac{16 \times (7.25)^2}{20 \times 1.33}$$

or
$$D_{16}' = \frac{4 \times 7.25}{\sqrt{20 \times 1.33}} = 5.62 \text{ mm}$$

EXAMPLE 8.19 Two plano-convex lenses each of radius of curvature 1.0 m are used to observe Newton's rings with their curved surfaces in contact with each other in the light of wavelength 600 nm. Find distance between 10th and 20th rings.

Solution In Figure 8.38

$$t = t_1 + t_2 = \frac{D}{8R_1} + \frac{D^2}{8R_2}$$

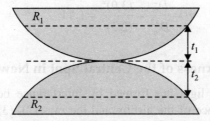

Figure 8.38

For mth dark ring

$$2t = m\lambda$$

$$2\left[\frac{D^2}{8R_1} - \frac{D^2}{8R_2}\right] = m\lambda$$

writing $D = 2r_m$
where r_m = radius of mth ring

$$r_m = \left[\frac{m\lambda R_1 R_2}{R_1 + R_2}\right]^{1/2}$$

The desired separation

$$r_{20} - r_{10} = \left[\frac{\lambda R_1 R_2}{R_1 + R_2}\right]^{1/2} [(20)^{1/2} - (10)^{1/2}]$$

$$= \left[\frac{6 \times 10^{-7} \times 1 \times 1}{1+1}\right]^{1/2} \times [4.472 - 3.162]$$

$$= 0.717 \times 10^{-3} \text{ m}$$

$$= 0.717 \text{ mm}$$

EXAMPLE 8.20 If the diameter of mth dark ring in an arrangement giving Newton's rings changes from 3 mm to 2.5 mm as a liquid is introduced between the lenses and plate, what is value of refractive index of the liquid?

Solution In Newton's rings arrangement,
Given D_m = 3 mm = 3×10^{-3} m, D'_m = 2.5 mm = 2.5×10^{-3} m, n = ?

We know that the diameter of mth ring in presence of liquid is

$$(D'_m)^2 = \frac{4m\lambda R}{n} \quad \text{(i)}$$

and the diameter of mth ring in air is

$$D_m^2 = 4m\lambda R \quad \text{(ii)}$$

Dividing Eq. (ii) by Eq. (i), we get

$$n = \frac{D_m^2}{D_m'^2} = \frac{(3.0)^2}{(2.5)^2} = 1.44$$

8.18.9 The Perfect Blackness of the Central Spot in Newton's Rings System

Newton's rings in reflected light are formed by interference between the ray (1) reflected directly from the upper surface of the airfilm and the rays (2), (3), ..., etc. which are obtained after one, three, five, ..., etc. internal reflections. These rays are shown in Figure 8.39. Near about the point of contact the thickness of airfilm almost zero and hence no path difference is introduced. But ray (2) reflected from the lower surface of film suffers a phase change of π, while the ray (1) reflected from the upper surface does not suppose such change. Thus, the two interfering waves at the centre are opposite in phase and destroy each other. The destruction is, however, not complete, since the amplitude of (2) is less than that of ray (1). But sum of the amplitudes of (2), (3), (4), ..., etc., which are all in phase* is exactly equal to amplitude of ray (1) as shown by Stoke's treatment. Hence, complete destructive interference is produced and centre of ring system is dark.

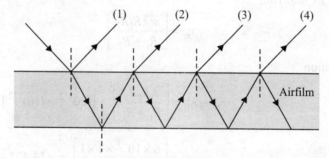

Figure 8.39 Illustration for perfect blackness of the central spot in Newton's rings system.

8.18.10 Newton's Rings are Circular but Air-wedge Fringes are Straight

In both the Newton's rings arrangement and the air-wedge fringes arrangement, each fringe is the locus of points of equal thickness of film lie on circles with the point of contact of the lens and plate as centre. Hence, the fringes are concentric circles. In case of wedge-shaped airfilm, the loci of focal thickness are straight lines parallel to the edge of wedge. Hence, the fringes are straight and parallel.

---

* The rays (2), (3), (4), etc. suffer one, three, five, etc. internal reflections and hence a change of π, 3π, 5π, ..., etc. in phase. Thus, any two consecutive rays have a phase difference of 2π, they are all in the same phase.

8.18.11 Newton's Rings with White Light

The diameter of the ring is a function of wavelength of light used. In case of white light, the diameter of the rings of different colours over each other. Thus only first few coloured rings will be clear while other rings cannot be viewed.

8.19 MICHELSON'S INTERFEROMETER

The American physicist Albert Abraham Michelson (1852–1931), then a novel instructor, took up the idea about interferometer at the age of 26, had already established a favourable reputation by performing an extremely precise determination of velocity of light. Michelson announced that "Physical discoveries in the future are the matter of the sixth decimal place". Michelson received a Nobel prize in 1907. He was the first American to do so.

There are good number of amplitude-splitting interferometers that utilize arrangements of mirrors and beam splitters. By far the best known and historically the most important is Michelson's interferometer.

Principle

The amplitude of light beam is divided into two parts of nearly equal intensity by partial reflection and refraction. They are sent in two directions at right angles and meet together by after reflection by plane mirror to produce interference.

Construction

This instrument is also based on thin film interference and this film is virtual. Figure 8.40 shows the construction schematics of Michelson's interferometer and Figure 8.41 shows its optical diagram.

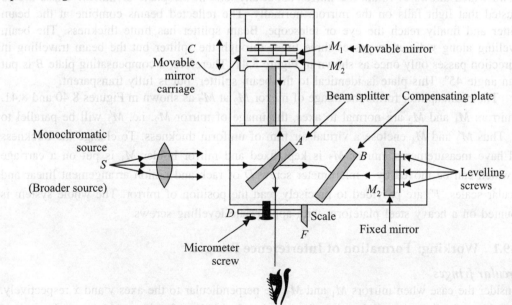

Figure 8.40 Schematic diagram of Michelson's interferometer.

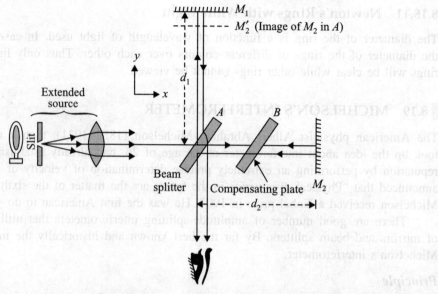

Figure 8.41 Optical diagram of Michelson's interferometer.

Light from a broader source S falls on a beam splitter A put at an angle of 45° to the light beam. This splits the incoming beam into two mutually perpendicular amplitude divided coherent beams travelling say in x- and y-directions. These beams fall on front silvered mirrors M_2 and M_1 respectively put perpendicular to the axes. Light beams fall on these mirrors normally, and hence, reflected back. Mirrors are put in holders provided with three travelling screws which allows to rotate the mirror about any axis in the plane of mirror. These are so adjusted that light falls on the mirrors normally. The reflected beams combine at the beam splitter and finally reach the eye or telescope. Beam splitter has finite thickness. The beam travelling along y-direction passes three times through the splitter but the beam travelling in x-direction passes only once as shown in Figure 8.41. Therefore, a compensating plate B is put at an angle 45°. This plate is identical to the beam splitter, but is fully transparent.

The beam splitter forms the image of mirror M_2 at M_2' as shown in Figures 8.40 and 8.41. If mirrors M_1 and M_2 are normal to axes, the image of mirror M_2, i.e. M_2' will be parallel to M_1. Thus M_2' and M_1 enclose a virtual air film of uniform thickness. To change the thickness and have measurements, mirror M_2 is kept fixed and mirror holder M_1 is put on a carriage C, which can be moved by a micrometer screw D of rack and pinion arrangement linear and circular scales 'F' are provided to precisely read the position of mirror. The whole system is mounted on a heavy steel plateform with appropriate levelling screws.

8.19.1 Working: Formation of Interference Fringes

Circular fringes

Consider the case when mirrors M_1 and M_2 are perpendicular to the axes y and x respectively. Then M_1 and M_2' are parallel and enclose a virtual air film of uniform thickness determined by the distances of the mirrors from the beam splitter. Let mirrors M_1 and M_2 be at respective distances d_1 and d_2 from the beam splitter. When we view, normally we see the centre of the mirror. In this case, the path difference between the two rays reaching the telescope is

$$\Delta = 2(d_1 \sim d_2) + \frac{\lambda}{2} \tag{8.89}$$

If we view the light rays say at an angle θ, then since thickness is same everywhere, hence interference fringes are locus of constant angle, i.e. circle. Therefore, circular fringes are observed. Viewing normally $\theta = 0$, let us now gradually increase the spacing between the mirrors.

1. If $d_1 = d_2$, then path difference $\Delta = \frac{\lambda}{2}$. The wave interfere destructively and field of view becomes dark as shown in Figure 8.42(i)

   $d =$ mirror separation $(d_1 \sim d_2)$

2. If $d_1 - d_2 = \frac{\lambda}{4}$, then $\Delta = 2 \times \frac{\lambda}{4} + \frac{\lambda}{2} = \lambda$. The wave interfere constructively. The central spot becomes bright and dark ring is pushed out to encircle it as shown in Figure 8.42(ii).

3. When d is increased to $\frac{\lambda}{2}$, the path difference becomes

$$\Delta = 2 \times \frac{\lambda}{2} + \frac{\lambda}{2} = \frac{3\lambda}{2} \tag{8.90}$$

   and so central spot is dark again and bright and dark rings encircle it. Note that mirror movement of $\frac{\lambda}{2}$ or creation of path difference of λ shifts the fringe by one, so the dark spot in Figure 8.42(iii) corresponds to $m = 1$. Here m is the order of fringes.

4. Thus when d becomes λ or $\Delta = 2 \times \frac{1}{2}\lambda$, the dark spot at the centre shall correspond to $m = 2$ as shown in Figure 8.42(iv).

5. Each increase of $\frac{\lambda}{2}$ in d creates a new fringe at the centre, hence for $d = 5\lambda$, $\Delta = 10 \cdot \frac{1}{2}\lambda$ and the fringe at centre corresponds to $m = 10$.

6. It is noted that ring at the centre has highest order and order decreases as we move away from the centre.

7. When we see off the centre, light falls on the film obliquely that reach our eye.
   As shown in Figure 8.43 through two-dimensional diagrams, the difference in path lengths between two beams coming after reflection from the two mirrors is $2d \cos \theta$, when $d = (d_1 \sim d_2)$ and, hence, the path difference becomes

$$\Delta = 2d \cos \theta + \frac{\lambda}{2} \tag{8.91}$$

300 Fundamentals of Optics

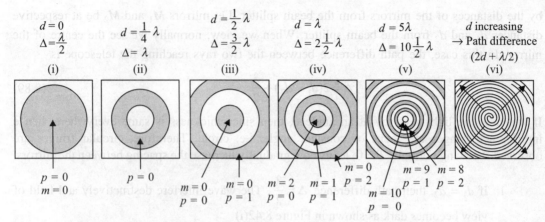

Figure 8.42 Origin of fringes of equal inclination (Hadinger fringes) as separation of mirrors is gradually increased from zero, d is mirror separation and Δ is the path difference [$\Delta = 2d + (\lambda/2)$].

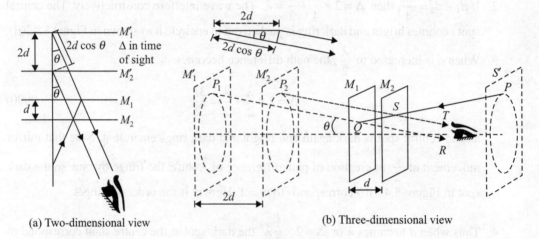

Figure 8.43 Oblique incidence.

Therefore, for bright and dark fringes, we have

Bright : $2d \cos\theta + \dfrac{\lambda}{2} = m\lambda$ or $2d \cos\theta = \left(m - \dfrac{1}{2}\right)\lambda$ \hfill (8.92)

Dark : $2d \cos\theta + \dfrac{\lambda}{2} = \left(m + \dfrac{1}{2}\right)\lambda$ or $2d \cos\theta = m\lambda$ \hfill (8.93)

where m is the fringe order.
For a given fringe order m, since film thickness d is constant, and so

$$2d \cos\theta = \text{constant} \hfill (8.94)$$

The fringes are of circular symmetry.

8. From Eq. (8.94), it is clear that as d increases, $\cos\theta$ decreases and so θ increases. Combining this observation with observations (i) to (vi), it becomes clear that as d increases the ring of the given order increases in size. Thus as d is gradually increased, new higher order ring appears at the centre and lower ordering increases in size.
9. Since $2d$ varies linearly with d, but $\cos\theta$ varies non-linearly, $\cos\theta$ decreases slowly near zero, but decreases rapidly for higher θ. Hence, the rings appear quickly at the centre, but disappear at the periphery slowly. Consequently, ring system is more crowded at periphery as shown in Figure 8.42(iv). In other words, ring thickness decreases as we move from centre to periphery.
10. Since the ring at the centre is highest order and its order is not generally known so we renumber the fringes. Say that the $p = 1, p = 2, p = 3, \ldots$ will respectively correspond to $(m-1), (m-2), (m-3), \ldots$. Therefore, the mth ring has order $(m-p)$. Equation (8.93) can be written as

$$2d\cos\theta = (m-p)\lambda \qquad (8.95)$$

when the ring at the centre has $p = 0$

$$2d = m\lambda \qquad (8.96)$$

Subtracting Eq. (8.95) from Eq. (8.96), we have

$$2d(1-\cos\theta_p) = p\lambda \qquad (8.97)$$

where $p = 0, 1, 2, 3, 4, \ldots$. If we count the ring at the centre as first ring, then

$$2d(1-\cos\theta_p) = (p-1)\lambda \qquad (8.98)$$

where $p = 1, 2, 3, \ldots$.

Radius of mth circular ring (with m = 0 at the centre)

If point of observation be at a distance D from the mirrors and mth ring corresponds to θ_m, then for small values of θ, we find the geometry of Figure 8.44.

$$r_p = D\theta_p \qquad (8.99)$$

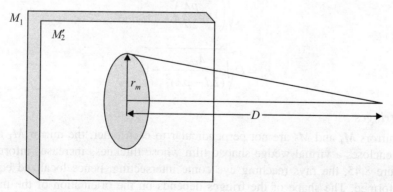

Figure 8.44 Calculation of radius of mth ring with $m = 0$ at the centre.

Now, θ_p is given by Eq. (8.97)

$$(1-\cos\theta_p) = \frac{p\lambda}{2d}$$

or

$$1 - 1 + 2\sin^2\frac{\theta_p}{2} = \frac{p\lambda}{2d}$$

for small θ, $\sin\theta \to \theta$

or

$$\frac{\theta_p^2}{2} = \frac{p\lambda}{2d}$$

or

$$\theta_p = \left(\frac{p\lambda}{d}\right)^{1/2} \tag{8.100}$$

Substituting the value of θ_p from Eq. (8.100) into Eq. (8.99), we get

$$r_p = D\left(\frac{p\lambda}{d}\right)^{1/2} \tag{8.101}$$

If θ_p is not small, then one has to find r_p using the following relation:

$$r_p = D\tan\theta_p$$

$$= D\sqrt{\sec^2\theta_p - 1}$$

$$= D\sqrt{\frac{1}{\cos^2\theta_p} - 1}$$

Substituting $\cos\theta_p = 1 - \frac{p\lambda}{2d}$ from Eq. (8.97), we get

$$r_p = D\sqrt{\frac{1}{\left(1-\frac{p\lambda}{2d}\right)^2} - 1}$$

$$r_p = D\sqrt{\frac{4d^2}{(2d-p\lambda)^2} - 1} \tag{8.102}$$

Localized fringes

When the mirrors M_1 and M_2 are not perpendicular to each other, the mirror M_1 and image of M_2, i.e. M_2' encloses a virtual wedge-shaped film whose thickness increases informly. Also, as seen in Figure 8.45, the rays reaching eye come intersecting, hence localized equally spaced fringes are formed. The shape of the fringes depends on the orientation of the mirrors. When M_1 and M_2' intersect, straight line fringes are formed.

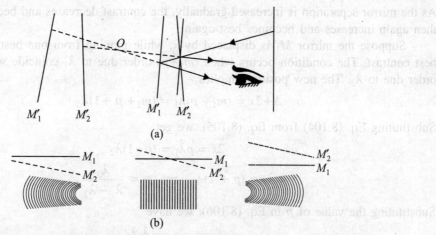

Figure 8.45 The localized fringes originate when M_1 and M_2' are inclined. Shape depends on orientation of mirrors M_1 and M_2.

8.19.2 Measurements with Michelson's Interferometer

Wavelength of light

1. The interferometer is adjusted, particularly orientation of mirrors M_1 and M_2 so as to obtain circular fringes of good contrast.
2. Crosswire of the telescope adjusted on the ring at the centre may be bright or dark. The reading of micrometer screw x_1 is noted.
3. Mirror M_1 is moved, the rings disappear or appear at the centre. These are counted and reading of micrometer screw x_2 is again noted for fixed number of fringes m (say 100 to 200).

The movement of mirror $(x_2 - x_1) = d$ introduces a path difference $2d$ between light beams travelling along x and y. Since each path differences of λ creates a new ring, hence

$$2d = 2(x_2 - x_1) = m\lambda$$

or
$$\lambda = \frac{2d}{m} = \frac{2(x_2 - x_1)}{m} \qquad (8.103)$$

λ is determined from Eq. (8.103).

Wavelength separation between closely spectral lines

Michelson's interferometer is extremely sensitive and versatile instrument. Its least count is as low as few micron and, hence, it can be used to measure the wavelength separation between close spectral lines, e.g. D_1 and D_2 lines of sodium.

Theory: Let two closely spaced spectral lines have wavelengths λ_1 and λ_2 and that $\lambda_1 > \lambda_2$. When interferometer is adjusted for circular fringes, each wavelengths λ_1 and λ_2 produce their own rings. The mirror is moved so that best contrast circular fringes are obtained. It will happen when path difference is such that maximum due to λ_1 coincides with maximum due to λ_2. Under this condition, say m_1 order of λ_1 coincides with m_2 order of λ_2,

$$\Delta = m_1\lambda_1 = m_2\lambda_2 \qquad (8.104)$$

304 Fundamentals of Optics

As the mirror separation is increased gradually, the contrast decreases and becomes worst, and then again increases and becomes best again.

Suppose the mirror M_1 is displaced by x, while moving from one best contrast to next best contrast. The condition occurs when (m_1+p) order due to λ_1 coincide with $(m_1 + p + 1)$ order due to λ_2. The new position implies

$$\Delta + 2x = (m_1 + p)\lambda_1 = (m_2 + p + 1)\lambda_2 \tag{8.105}$$

Substituting Eq. (8.104) from Eq. (8.105), we get

$$2x = p\lambda_1 = (p+1)\lambda_2 \tag{8.106}$$

$$p_1\lambda_1 = (p+1)\lambda_2 \implies p = \frac{\lambda_2}{\lambda_1 - \lambda_2} \tag{8.107}$$

Substituting the value of p in Eq. (8.106), we have

$$2x = p\lambda_1 = \frac{\lambda_1\lambda_2}{\lambda_1 - \lambda_2}$$

or

$$\lambda_1 - \lambda_2 = \Delta\lambda = \frac{\lambda_1\lambda_2}{2x}$$

Therefore, λ_1 and λ_2 are quite close, so $\sqrt{\lambda_1\lambda_2} = \lambda$

or

$$\Delta\lambda = \frac{\lambda^2}{2x} \tag{8.108}$$

Thus, measuring mirror movement between two best contrast positions of interference fringes, $\Delta\lambda$, can be determined.

Thickness of thin film

1. The interferometer is set for localized straight line fringes and monochromatic source is replaced by mercury lamp (white light source). The central fringe is dark line.
2. The crosswire is set in this dark line and reading the micrometer screw is noted say x_1.
3. The thin film having thickness t and refractive index n is introduced in the path of one of the rays. This shifts the dark line because introduction of film introduces a path difference

$$\Delta = 2(n-1)t \tag{8.109}$$

4. The mirror is moved so that crosswire coincides with the dark line again. The movement or mirror compensates the path difference. The mirror position is now say x_2. Then

$$\Delta = 2x \tag{8.110}$$

From Eqs. (8.109) and (8.110)

$$2x = 2(n-1)t$$

Therefore,

$$t = \frac{x}{n-1} \tag{8.111}$$

The thickness or reflective index, as the case may be, can be calculated by Eq. (8.111).

EXAMPLE 8.21 Michelson interferometer is set for straight fringes using light of $\lambda = 500$ nm. Calculate the number of fringes that moves across the field of view, when one of the mirrors is moved back by a distance of 0.1 mm.

Solution Given $\lambda = 5000$ Å, $(x_2 - x_1) = 0.1$ mm, $m = ?$

We know that $2(x_2 - x_1) = m\lambda$

$$m = \frac{2(x_2 - x_1)}{\lambda}$$

$$= \frac{2 \times 0.1 \times 10^{-3}}{5000 \times 10^{-10}} = \frac{2000}{5}$$

$$m = 400$$

EXAMPLE 8.22 A thin transparent sheet of refractive index $n = 1.6$ is introduced in one of the beams of Michelson interferometer and shift of 24 fringes for $\lambda = 6000$ Å is obtained. Calculate the thickness of the sheet.

Solution Given $\lambda = 6000$ Å $= 6.0 \times 10^{-7}$ m, $n = 1.6$, $m = 24$, $t = ?$.

We know that $2t(n - 1) = m\lambda$

or
$$t = \frac{m\lambda}{(n-1)}$$

$$= \frac{24 \times 6.0 \times 10^{-7}}{2 \times (1.6 - 1)}$$

$$= \frac{24 \times 6.0 \times 10^{-7}}{2 \times 0.6}$$

$$= 120 \times 10^{-7} \text{ m}$$

$$= 1.2 \times 10^{-5} \text{ m}$$

$$= 12 \text{ μm}$$

EXAMPLE 8.23 Calculate the distance between successive positions of the movable mirror of Michelson interferometer giving best fringes in case of sodium source having wavelengths 5896 Å and 5890 Å.

Solution Given $\lambda_1 = 5896$ Å $= 5.896 \times 10^{-7}$ m, $\lambda_2 = 5890$ Å $= 5.89 \times 10^{-7}$ m
We know that small difference in two wavelengths.

$$\Delta\lambda = \lambda_1 - \lambda_2 = \frac{\lambda_1 \lambda_2}{2(x_2 - x_1)}$$

or
$$(x_2 - x_1) = \frac{\lambda_1 \lambda_2}{\Delta\lambda}$$

$$= \frac{(5.896 \times 10^{-7}) \times (5.89 \times 10^{-7})}{2 \times (5.896 - 5.89) \times 10^{-7}} = 28.94 \text{ Å}$$

EXAMPLE 8.24 A shift of 100 circular fringes is observed, when movable mirror of Michelson interferometer is shifted by 0.295 mm. Calculate the wavelength of light.

Solution Given $(x_2 - x_1) = 0.295$ mm, $m = 200$, $\lambda = ?$

We know that $2(x_2 - x_1) = m\lambda$

or
$$\lambda = \frac{2(x_2 - x_1)}{m} = \frac{2 \times 0.295 \times 10^{-3}}{100}$$
$$= 5900 \times 10^{-9} \text{ m}$$
$$= 5900 \text{ nm}$$

EXAMPLE 8.25 A Michelson interferometer is adjusted for good fringes with monochromatic light of wavelength 605.78 nm. If the observer views the interference pattern through a telescope with a crosswire eyepiece, how many fringes pass the crosswire when l moves exactly one centimeter?

Solution Given $\lambda = 605.78$ nm $= 6.0578 \times 10^{-7}$ m, $l = 1$ cm $= 1 \times 10^{-2}$ m, $p = ?$

Let on moving l distance forward p fringes are displaced, then
$$2t + 2l = (m + p)\lambda$$
or
$$2l = p\lambda$$
or
$$p = \frac{2l}{\lambda} = \frac{2 \times 10^{-2}}{605.78 \times 10^{-9}}$$
$$= 33015$$

8.20 MULTIPLE BEAM INTERFEROMETRY

Important and useful interferometers have been designed using multiple reflection and transmission of incident beam on two parallel surfaces. If a beam of light is made to be incident on a film having parallel surfaces, then multiple reflections and transmissions occur. Due to this process, a set of interfering beams in reflected side and another set of interfering beams are generated as shown in Figure 8.46.

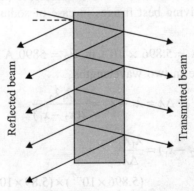

Figure 8.46 Multiple beam generation.

In this process of multiple reflections and transmissions, the amplitude of beams increases in each step. This process of multiple beam generation has been used in two very important interferometers. These are:

1. Fabry–Perot interferometer, and
2. Lummer–Gehrcke plate

In the first case, two silvered optically plane glass plates are used to get multiple beams to produce interference. In the second case, a glass slab of optically plane surface is used for the purpose.

8.20.1 Fabry–Perot Interferometer

Fabry–Perot interferometer consists of two glass plates A and B separated by a distance t. The inner surfaces of the plates are optically plane, exactly parallel and thin silvered so that about 70% of incident light gets reflected. The outer faces of the plates are also parallel to each other, but inclined to their respective inner faces. A schematic diagram of this interferometer is shown in Figure 8.47.

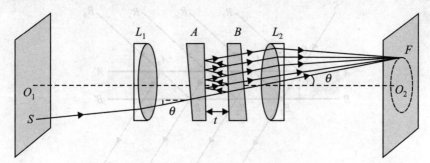

Figure 8.47 Fabry–Perot interferometer showing formation of circular fringes.

Monochromatic light of wavelength λ from a broad source S is incident on a collimating lens L_1, which renders the incident rays parallel. A ray of light entering the air film between the plates undergoes multiple reflections between silvered surfaces and emerges from the plate B as a parallel beam and they are made to interfere by converging them at F, the focal plane of the lens L_2. The fringes are concentric circles with O_2 as centre, having equal inclination. The fringes of constant inclination are called *Haidinger's fringes*.

If θ is the angle of incidence on the silvered face of A, then, the path difference between successive rays is $2nt \cos\theta$ or $2t \cos\theta$ as $n = 1$ for air. Therefore, for a bright fringe

$$2t \cos\theta = m\lambda, \quad m = 0, 1, 2, 3, 4, \ldots \tag{8.112}$$

For all points on a circle passing through F with centre at O_2 on the axis $O_1 O_2$.

In practice, one of the plates is kept fixed and the other is movable with rack and pinion arrangement, so that the film thickness t can be changed. The fringes can be obtained up to 10 cm of plate separation. Due to large separation, fringes are observed almost near the normal direction. If the distance t between the two plates is decreased, the value of θ decrease for a given value of m and λ. This means when t is decreased, the ring shrinks and disappears at the centre. When t is decreased by $\lambda/2$, one ring disappear at the centre.

In this interferometer, the radius of a ring is a function of the wavelength of light used and very sharp bright fringes are produced (better than Michelson interferometer), so this interferometer is used for resolving two wavelengths and analysis of a single spectral line.

Theory of intensity distribution

Consider a plane wave of unit amplitude incident at angle θ on the glass plate AB, as shown in Figure 8.48. Although, from an extended source, light is incident on the plate at all possible angles. Due to multiple reflections, a set of parallel reflected rays A_1R_1, A_2R_2, A_3R_3, ..., and a set of parallel transmitted rays B_1T_1, B_2T_2, B_3T_3, ... are produced. Let $\sqrt{R}$ and $\sqrt{T}$ be the reflection and transmission coefficients of amplitudes from the surfaces respectively. Thus the amplitude of A_1B_1 is $\sqrt{T}$ as the amplitude of incident ray SA_1 is unity. The amplitude of B_1T_1 is $T(=\sqrt{T} \times \sqrt{T})$. The amplitude of B_1A_2 is $(=\sqrt{T} \times R\sqrt{T})$. In this way, we can see that the amplitudes of B_1T_1, B_2T_2, B_3T_3, ... are T, RT, R^2T, R^3T, ... respectively. As these rays are obtained from same incident ray, they are coherent and produce sustained interference pattern.

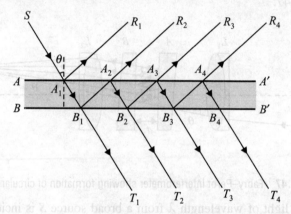

Figure 8.48 Illustration for intensity distribution for Fabry–Perot interferometer.

Neglecting the small phase change due to reflection from silvered surfaces, the path difference between two consecutive transmitted rays is $2t\cos\theta$ ($\mu = 1$ for air), the corresponding phase difference

$$\delta = \frac{2\pi}{\lambda} \times 2t\cos\theta = \frac{4\pi t \cos\theta}{\lambda}$$

Let the incident wave be represented by

$$y_1 = a\sin\omega t = \sin\omega t \ (a \text{ is assume to be unity})$$

As discussed above, the transmitted rays can be represented as

$$y_1 = T\sin\omega t$$
$$y_2 = RT\sin(\omega t - \delta)$$
$$y_3 = R^2T\sin(\omega t - 2\delta)$$
$$y_4 = R^3T\sin(\omega t - 3\delta), \text{ and so on}$$

Let A be the resultant amplitude of these rays (upon interfering) and ϕ the difference, so that it can be represented as
$$Y = A \sin(\omega t - \phi)$$
Also, from the superposition principle, we have
$$Y = y_1 + y_2 + y_3 + \cdots$$
$\therefore \qquad A \sin(\omega t - \phi) = T \sin \omega t + RT \sin(\omega t - \delta) + R^2 T \sin(\omega t - 2\delta) + \cdots$

or $\qquad A \sin \omega t \cos \phi - A \cos \omega t \sin \phi = T \sin \omega t + RT \sin \omega t \cos \delta - RT \cos \omega t \sin \delta$
$$+ R^2 T \sin \omega t \cos 2\delta + R^2 T \cos \omega t \sin 2\delta + \cdots$$

Equating the coefficients of $\sin \omega t$ and $\cos \omega t$ on both sides, we get

$$A \cos \phi = T + RT \cos \delta + R^2 T \cos 2\delta + \cdots \qquad (8.113)$$

and $\qquad A \sin \phi = RT \sin \delta + R^2 T \sin 2\delta + \cdots \qquad (8.114)$

The resultant intensity I is given by
$$I = A^2 = (A \cos \phi + iA \sin \phi) \times (A \cos \phi - iA \sin \phi) \qquad \text{(here } i = \sqrt{-1})$$
From Eqs. (8.113) and (8.114), we have
$$A \cos \phi + iA \sin \phi = T + RT(\cos \delta + i \sin \delta) + R^2 T(\cos 2\delta + i \sin 2\delta) + \cdots$$
which in terms of exponential can be put as
$$A \cos \phi + iA \sin \phi = T + RT \, e^{i\delta} + R^2 T \, e^{2i\delta} + \cdots$$
$$= T[1 + R e^{i\delta} + R^2 e^{2i\delta} + \cdots]$$
$$= \frac{T}{1 - R e^{i\delta}}$$
Similarly,
$$A \cos \phi - iA \sin \phi = \frac{T}{1 - R e^{-i\delta}}$$
The resultant intensity is, therefore, given by
$$I = A^2 = \frac{T}{1 - R e^{-i\delta}} \times \frac{T}{1 - R e^{-i\delta}}$$

$$= \frac{T^2}{1 + R^2 - 2R \cos \delta} \qquad (\because e^{i\delta} = \cos \delta + i \sin \delta, e^{-i\delta} = \cos \delta - i \sin \delta)$$

$$= \frac{T^2}{(1 - R)^2 + 2R - 2R \cos \delta}$$

$$= \frac{T^2}{(1 - R)^2 + 2R(1 - \cos \delta)} = \frac{T^2}{(1 - R)^2 + 4R \sin^2 \delta/2}$$

$$= \frac{T^2}{(1 - R)^2 \left[1 + \dfrac{4R}{(1 - R)^2} \sin^2 \dfrac{\delta}{2}\right]} \qquad (8.115)$$

This expression, known as *Airy's formula*, shows that the resultant intensity depends upon the properties of the silver coating and δ.

Condition for maxima
For maximum intensity of the fringe

$$\sin^2 \frac{\delta}{2} = 0 \Rightarrow \frac{\delta}{2} = m\pi$$

or
$$\delta = 2m\pi, \; m = 0, 1, 2, 3, \ldots$$

Maximum intensity

$$I_{max} = \frac{T^2}{(1-R)^2} \tag{8.116}$$

Condition for minima
The intensity is minimum when

$$\sin^2 \frac{\delta}{2} = 1$$

or
$$\frac{\delta}{2} = (2m+1)\frac{\pi}{2} \Rightarrow \delta = (2m+1)\pi, \; m = 0, 1, 2, 3, \ldots$$

Minimum intensity

$$I_{min} = \frac{T^2}{(1-R)^2 \left[1 + \dfrac{4R}{(1-R)^2}\right]}$$

$$= \frac{T^2}{(1+R)^2} \tag{8.117}$$

When there is no absorption at the reflecting surface, $T = (1-R)$.

$\therefore \qquad I_{max} = 1$

and $\qquad I_{min} = \dfrac{(1-R)^2}{(1+R)^2}$

$\therefore \qquad \dfrac{I_{max}}{I_{min}} = \left(\dfrac{1+R}{1-R}\right)^2 \tag{8.118}$

Also from Eqs. (8.115) and (8.116), we have

$$\frac{I}{I_{max}} = \frac{1}{1 + \dfrac{4R}{(1-R)^2}\sin^2\dfrac{\delta}{2}} = \frac{1}{\left(1 + F\sin^2\dfrac{\delta}{2}\right)} \tag{8.119}$$

where
$$F = \frac{4R}{(1-R)^2}$$

Circular shape of fringe

In terms of path difference,

For maxima: $2t \sin \theta = m\lambda$

For minima: $2t \cos \theta = (2m - 1)\dfrac{\lambda}{2}$

For constant t, as in this interferometer, for a particular m (order of fringe) and λ, θ is a constant. The locus of points having the same value of θ is a circle and, hence, circular fringes are obtained. The bright fringe in the interference pattern is very sharp as the intensity falls rapidly almost to zero on either side. The sharpness increases with reflecting power of silver coating.

Visibility of fringes

The visibility of the fringe is defined as

$$V = \dfrac{I_{\max} - I_{\min}}{I_{\max} + I_{\min}} \tag{8.120}$$

Thus, the visibility of the fringes depends only on the reflection coefficient of the silver coating and is independent of its transmission coefficient.

Sharpness of fringes—half fringe width

Figure 8.49 shows the variation of intensity I of a fringe with phase difference for a given value of R. The half image width of a fringe is defined as the width of fringe (in terms of phase difference) between the two points on either side of maxima, where the intensity is half of its maximum value as shown in Figure 8.49.

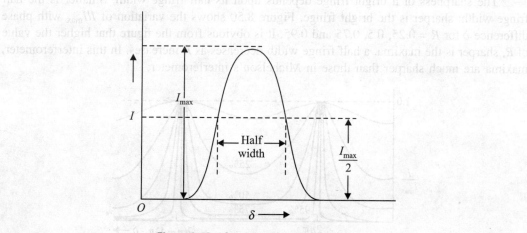

Figure 8.49 Graph between intensity and δ.

From Eq. (8.119)

$$\frac{I}{I_{max}} = \frac{1}{1 + F\sin^2\frac{\delta}{2}}$$

At half fringe width

$$\frac{I}{I_{max}} = \frac{1}{2}$$

$$\therefore \quad \frac{1}{1 + F\sin^2\frac{\delta}{2}} = \frac{1}{2}$$

or

$$1 + F\sin^2\frac{\delta}{2} = 2$$

or

$$F\sin^2\frac{\delta}{2} = 1$$

or

$$\sin\frac{\delta}{2} = \frac{1}{\sqrt{F}} = \frac{1}{\sqrt{\frac{4R}{(1-R)^2}}} = \frac{(1-R)}{2\sqrt{R}}$$

or

$$\delta = 2\sin^{-1}\frac{1-R}{2\sqrt{R}}$$

·For small value of δ

$$\sin^2\frac{\delta}{2} = \frac{\delta}{2}$$

$$\therefore \quad \delta = \frac{1-R}{\sqrt{R}}$$

The sharpness of a bright fringe depends upon its half fringe width. Smaller is the half fringe width; sharper is the bright fringe. Figure 8.50 shows the variation of I/I_{max} with phase difference δ for $R = 0.25, 0.5, 0.75$ and 0.95. It is obvious from the figure that higher the value of R, sharper is the maxima, a half fringe width decreases as R increases. In this interferometer, maxima are much sharper than those in Michelson's interferometer.

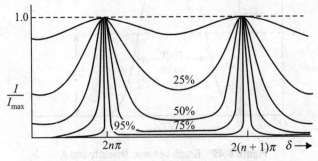

Figure 8.50 I/I_{max} versus δ for $R = 0.25, 0.50, 0.75$ and 0.95.

Determination of wavelength

The Fabry–Perot interferometer is adjusted so that circular fringes are obtained. Let m be the order of fringe at the centre (in this interferometer order, decreases from the centre), so that

$$2t = m\lambda \qquad (\because \quad \theta \text{ is zero at centre})$$

Now, the movable plate is moved from a known distance, say from x_1 to x_2 and let N be the number of fringes that disappear at the centre. Thus

$$N\frac{\lambda}{2} = x_2 - x_1$$

or
$$\lambda = \frac{2(x_2 - x_1)}{N} \qquad (8.121)$$

Thus, the wavelength λ, can be obtained.

EXAMPLE 8.26 A Fabry–Perot interferometer is used to determine the small difference in wavelength of two component of radiation of average wavelength 6000 Å. It is found that at a given point near the centre the two sets of fringes coincide for a given distance between the two plates, and again when the distance is increased by 0.9 mm. Find the difference in wavelengths.

Solution Given $\lambda = 6000$ Å $= 6 \times 10^{-7}$ m; $t_2 - t_1 = 0.9$ mm $= 0.9 \times 10^{-3}$ m $= 9.0 \times 10^{-4}$ m
The difference in wavelength $\Delta\lambda$ is given by the relation

$$\Delta\lambda = \lambda_1 - \lambda_2 = \frac{\lambda^2}{2(t_2 - t_1)}$$

$$= \frac{(6.0 \times 10^{-7})^2}{2 \times (9 \times 10^{-4})}$$

$$= 2 \times 10^{-10} \text{ m}$$

$$= 2 \text{ Å}$$

EXAMPLE 8.27 A shift of 100 fringes is observed when movable mirror of Fabry–Perot interferometer is shifted through 0.0295 mm. Calculate the wavelength of light used.

Solution $N = 100$, $t = 0.0295$ mm $= 2.95 \times 10^{-5}$ m
Now $2t = N\lambda$

$\therefore$
$$\lambda = \frac{2t}{N} = \frac{2 \times 2.95 \times 10^{-5}}{100}$$

$$= 5900 \times 10^{-10} \text{ m}$$

$$= 5900 \text{ Å}$$

8.20.2 Lummer–Gehrcke Plate

Lummer–Gehrcke (LG) plate is an optical device of very high resolving power. Its resolving power is approximately 20,000. It consists of a rectangular quartz plate of length ~15 cm, width ~1.5 cm and thickness ~0.5 cm. At one end of the plate is attached a small quartz prism with its two plane faces at the angle of 90°. The angle of prism is such that any light ray entering it through the plane perpendicular to its base, after reflection from opposite plane, enters the plate at critical angle and suffers multiple reflections as shown in Figure 8.51.

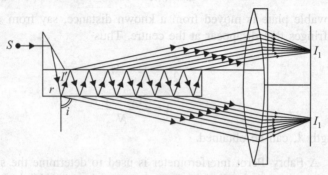

Figure 8.51 Lummer–Gehrcke plate and its action to incident light.

The rays come out of the plate at grazing angle. These multiple rays coming out of the surfaces interfere and are brought to focus by an achromatic convex lens on a screen put at suitable distance. Two sets of fringes are obtained; one from the rays coming out of the upper surface and the other coming out of the lower surface. Let us look at Figure 8.51. Let a monochromatic light from a source S be incident such that angle of emergence is i with the normal of the plate. The angle of refraction is r. Both are indicated in Figure 8.51. The ray entering the plate will suffer multiple reflections and, hence we have number of emergent beams. The two adjacent waves interfere and produce interference pattern with alternating intensities. The pattern consists of nearly straight lines in the focal plane of the objective of the telescope provided with micrometer to take measurement on them. Thus different orders come out of the plate above one another. As path difference increases, the angle of emergence decreases. Thus higher orders look at a place higher in the focal plane of lower one. It may be noted that various orders correspond to angle differing a little from that of grazing angle of emergence. If t be the thickness of the plate and n the refractive index of its material, then condition for maxima is realised if

$$2nt \cos r = m\lambda \quad (8.122)$$

since refractive index $n = \dfrac{\sin i}{\sin r}$. Therefore,

$$\cos r = \sqrt{1 - \sin^2 r} = \sqrt{1 - \dfrac{\sin^2 i}{n^2}}$$

Substituting the value of $\cos r$ in Eq. (8.122), we get

$$2t(n^2 - \sin^2 i)^{1/2} = m\lambda$$

or $\quad 4t^2(n^2 - \sin^2 i) = m^2\lambda^2 \quad (8.123)$

Differentiating Eq. (8.123) with respect to m, we get

$$-8t^2 \sin i \cos i \; \delta i = 2m\lambda^2 \delta m$$

If we make $\delta m = 1$, then we get the angle δi between successive order. Thus

$$\delta i = \frac{-m\lambda^2}{2t^2 \sin 2i} \qquad (8.124)$$

Here δi expresses the angle between two rays producing two successive order as shown in Figure 8.52. Putting the value of m from Eq. (8.123) into Eq. (8.124), we get

$$\delta i = -\frac{\lambda(2t)(n^2 - \sin^2 i)^{1/2}}{2t^2 \sin 2i}$$

$$= \frac{-\lambda(n^2 - \sin^2 i)^{1/2}}{t \sin 2i} \qquad (8.125)$$

Figure 8.52 Different modes.

Equation (8.125) infers that on approaching towards grazing emergence, the angle between successive orders increase. It is also inferred that this angle is inversely proportional to the thickness t of the plate. Now differentiating Eq. (8.123) with respect to λ, we have

$$8t^2 n \frac{dn}{d\lambda} - 8t^2 \sin i \cos i \frac{di}{d\lambda} = 2m^2 \lambda$$

or

$$\frac{di}{d\lambda} = \frac{4t^2 n \dfrac{dn}{d\lambda} - m^2 x}{2t^2 \sin 2i} \qquad (8.126)$$

After substituting the value of $m\lambda$ from Eq. (8.123) into Eq. (8.126), we get

$$\frac{di}{d\lambda} = \frac{4t^2 n \dfrac{dn}{d\lambda} - \dfrac{4t^2}{\lambda}(n^2 - \sin^2 i)}{2t^2 \sin 2i}$$

$$= \frac{2n\lambda \dfrac{dn}{d\lambda} - 2(n^2 - \sin^2 i)}{\lambda \sin 2i} \qquad (8.127)$$

Equation (8.127) gives the dispersion of the beam and suggests that the dispersion of the beam depends upon n of the material and its variation with λ, i.e. $dn/d\lambda$, the wavelength λ and the angle of convergence i.

If $\Delta \lambda$ be the change in the wavelengths of incident light when a given order of one line coincident with the next order of one line coincides with the next order of the other line, and Δi be the corresponding change between successive orders for a single wavelength, then from Eq. (8.126), we have

$$\Delta i = \frac{4t^2 n\, dn - m^2 \times \Delta\lambda}{2d^2 \sin 2i} = \frac{-m\lambda^2}{2d^2 \sin 2i}$$

The extreme right-hand term has been put in Eq. (8.125). This yields

$$\Delta\lambda = \frac{m\lambda^2}{m^2 \lambda - 4t^2 n \dfrac{dn}{d\lambda}} \qquad (8.128)$$

The intensity distribution in the fringes and also the resolving powers are not easy to determine is an exact manner. However, approximate value of resolving power at grazing emergence can be defined in the following manner. When the emergence disturbances are considered as single uniform plane wavefront, then the width of the wavefront is given by l', where l' is the length of the plate. This wavefront gives a central maxima in the focal plane of the lens placed suitably to focus the disturbance. Angular width of central maxima in this situation will be $\dfrac{2\lambda}{l' \cos i}$. Using Eq. (8.127) we have

$$\delta i = \dfrac{\left[2\lambda n \dfrac{dn}{d\lambda} - 2(n^2 - \sin^2 i) \right] d\lambda}{\lambda \sin 2i} = \dfrac{\lambda}{l' \cos i} \qquad (8.129)$$

or

$$\dfrac{\lambda}{d\lambda} = \dfrac{l'}{\lambda \sin i} \left[\lambda n \dfrac{dn}{d\lambda} - (n^2 - \sin^2 i) \right] \qquad (8.130)$$

$\dfrac{dn}{d\lambda}$ is very small quantity and can be neglected. Further, $i \cong \pi/2$, hence $\sin i = 1$, this yields

$$\left| \dfrac{\lambda}{d\lambda} \right| = \dfrac{l'(n^2 - 1)}{\lambda} \qquad (8.131)$$

Since λ is in denominator, resolving power $\lambda/d\lambda$ will be very-very high. Lummer–Gehrcke (LG) plate is primarily used in studying the fine structure of spectral lines and Zeeman effect.

EXAMPLE 8.28 In a Lummer–Gehrcke plate, the thickness of plate is 15 cm and refractive index of the material of the plate is 1.54. Evaluate the resolving power of the LG-plate.

Solution Given $l' = 15$ cm $= 0.15$ m, $n = 1.54$, $\lambda = 6000$ Å $= 6.0 \times 10^{-7}$ m
The resolving power of LG plate

$$R.P. = \dfrac{\lambda}{d\lambda} = \dfrac{l'(n^2 - 1)}{\lambda} = \dfrac{0.15 \times [(1.54)^2 - 1]}{6 \times 10^{-7}}$$

$$= \dfrac{0.15 \times (2.3716 - 1)}{6 \times 10^{-7}} = \dfrac{0.15 \times 1.3716}{6 \times 10^{-7}}$$

$$= 3.429 \times 10^5$$

$$= 342900$$

8.21 INTERFERENCE REFRACTOMETERS

In general, optical instrument employing interference phenomenon to measure the refractive index of transparent medium, particularly gases, are termed as interference refractometers. They are very accurate and sensitive. This is the reason that they are the prime optical instrument to study the variation of refractive index of gases with temperature and pressure. Various refractometers have been designed. Important amongst are:

1. Jamin's refractometer
2. Rayleigh's refractometer, and
3. Mach–Zehnder's refractometer

8.21.1 Jamin's Refractometer

The instrument consists of two exactly identical and optically plane glass blocks P_1 and P_2 (cut from the same block) silvered on their back surfaces and arranged with their faces parallel to each other as shown in Figure 8.53. Light coming from an extended monochromatic source S incident on P_1 at an angle of 45° is splitted into two parallel rays (1) and (2) by reflection at the upper and lower surfaces of P_1. These two rays recombine after suffering reflections at the two surfaces of P_2 as shown, and from interference fringes as observed in a telescope. If the plates P_1 and P_2 are parallel, the light paths will be identical.

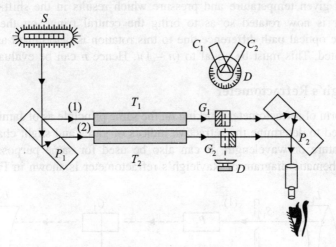

Figure 8.53 Jamin's refractometer.

Measurement of refractive index

To measure the refractive index of a gas at a given temperature and pressure, two similar evacuated tubes T_1 and T_2 are placed in the paths of two parallel beams. The gas, whose refractive index is to be measured is then shortly allowed to enter in one of the tubes. As the gas enters, the optical path of the beam passing through the tube increases. The fringes, therefore move fast the crosswire of the telescope. The fringes are counted until the gas entering the tube reaches the given pressure and temperature.

If l be the length of the tube and n the refractive index of the gas, then the change in the optical path due to presence of the gas is $(n - 1)l$. If N be the number of fringes passed across the field of view, then

$$(n - 1)l = N\lambda \tag{8.132}$$

Hence n can be calculated.

The refractivity of a gas is given by

$$(n - 1) = \frac{N\lambda}{l} \tag{8.133}$$

Compensator

To avoid counting of fringes, a device called the 'Jamin's compensator' is used. It consist of two equally thick glass plates G_1 and G_2 cut from the same piece of glass and inclined at a small angle. The plates can be rotated together about a horizontal axis and the rotation is read on a divided circle D. One plate of the compensator is placed in the path of each beam. When the plates are equally inclined to the incident beam, the optical paths through the plates are same. When the plates are rotated, the angle of incidence of two beams change and a relative phase difference is introduced which varies as compensator is rotated. By using monochromatic light and observing the passage of fringes across the field of view as the compensator is rotated, the scale can be calibrated to reach the optical path difference in terms of wavelength.

Now, in the actual experiment, white light is used and the central achromatic fringe is adjusted on the crosswire by adjusting the compensator. The gas is then introduced in one of the tubes at a given temperature and pressure which results in the shifting of the fringes. The compensator is now rotated so as to bring the central fringe on the crosswire again. The change in the optical path difference due to this rotation is determined as the compensator is already calibrated. This must be equal to $(n - 1)l$. Hence n can be evaluated.

8.21.2 Rayleigh's Refractometer

This is another form of refractometer working on the same principle as of Jamin's refractometer. It is primarily used to determine the refractive indices of gases and slight change in them with pressure, temperature or wavelength. It can also be used for same purposes for transparent solutions. The schematic diagram of Rayleigh's refractometer is shown in Figure 8.54.

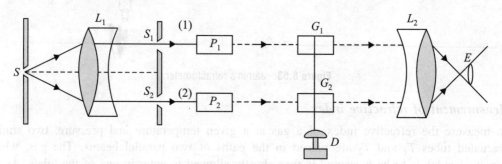

Figure 8.54 Rayleigh's refractometer.

Light from monochromatic source S is incident on lens L_1. The emerging parallel beam from lens is split up into two beams by the slits S_1 and S_2. The ray number (1) passes through tube P_1 and ray number (2) passes through P_2.

On emerging from tubes, these two rays passes through compensating plates G_1 and G_2 respectively. After passing through the lens L_2, interference fringes are observed with the help of eyepiece or telescope. The circular disc D attached to compensating plates G_1 and G_2 is calibrated in terms of wavelength and refractive index.

For the measurement of refractive index of a gas both tubes P_1 and P_2 are first evacuated and then central white fringe is observed in the field of eyepiece of telescope, when white light source is used at S. Now introduce the gas whose refractive index is to be determined into one of the tubes, say P_1, and its temperature and pressure is noted. Now plates G_1 and G_2 are rotated by rotating the circular disc D to bring back the central white fringe in the centre of field of view and on the basis of calibrated disc reading, refractive index is obtained.

8.21.3 Mach–Zehnder's Refractometer

It is better suited refractometer for the measurement of the change in refractive index of the gases as a result of variation of other parameters, like pressure and temperature. The schematic diagram of the Mach–Zehnder's (MZ) refractometer is shown in Figure 8.55.

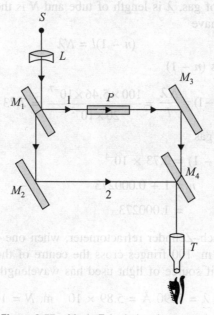

Figure 8.55 Mach–Zehnder's refractometer.

It consists of four mirrors M_1, M_2, M_3 and M_4. The light from monochromatic source S is made parallel by lens L. It falls on M_1 and divides into two parts. The reflected beam (1) passes through tube P and after reflection from M_3 falls on M_4 and is refracted to telescope. The ray (2) after reflection from mirror M_3 and M_4 goes to telescope. The beams (1) and (2) interfere after emerging from M_4 and form interference fringes. When gas is filled in P, set of fringes are obtained. Let the refractive index of gas in this case be n_1. If gas pressure is increased, then refractive index becomes n_2 and number of fringes cross the field of view of telescope. Let the length of tube P be l and N fringes cross the field of view when refractive index changes from n_1 to n_2. Then we must have

$$(n_2 l - n_1 l) = N\lambda$$

or

$$(n_2 - n_1) = \frac{N\lambda}{l}$$

or

$$\Delta n = \frac{N\lambda}{l} \qquad \text{(Here } \Delta n = n_2 - n_1 \text{)} \quad (8.134)$$

Using Eq. (8.134), the change is refractive index can be determined. This refractometer is particularly useful for studying the flow pattern in mine tunnels.

EXAMPLE 8.29 In Jamin's refractometer two evacuated tubes, each of length 20 cm are placed in the path of two beams. A gas at a known pressure and temperature is slowly admitted in one of the tubes and it is observed that 100 fringes cross the field of view. Evaluate (i) the refractivity and (ii) refractive index of the gas. The wavelength of light used was 5460 Å.

Solution Given $N = 100$, $\lambda = 5460 \times 10^{-10}$ m, $l = 20 \times 10^{-2}$ m

If n is the refractive index of gas, λ is length of tube and N is the number of fringes crossing the field of view, then we have

$$(n - 1)l = N\lambda$$

(i) The refractivity of gas is $(n - 1)$

$$\therefore \quad (n-1) = \frac{N\lambda}{l} = \frac{100 \times 5.46 \times 10^{-7}}{20 \times 10^{-2}} = 2.73 \times 10^{-4}$$

(ii) The refractive index of gas

$$(n - 1) = 2.73 \times 10^{-4}$$
$$n = 1 + 0.000273$$
$$= 1.000273$$

EXAMPLE 8.30 In a Mach–Zehnder refractometer, when one of the beams passes through a wind tunnel of length 10 m, 100 fringes cross the centre of the field of view. Evaluate the change in refractive index, if source of light used has wavelength 5890 Å.

Solution Given $l = 10$ m, $\lambda = 5890$ Å $= 5.89 \times 10^{-7}$ m, $N = 100$

In Mach–Zehnder refractometer, change in refractive index (Δn) is given by the expression

$$\Delta n = \frac{N\lambda}{l} = \frac{100 \times 5.89 \times 10^{-7}}{10}$$
$$= 5.89 \times 10^{-6}$$

8.22 INTERFERENCE IN OPTICAL TECHNOLOGY

With increasing use of lasers in various fields, the phenomenon of interference is widely used in various measuring and testing equipment. To have an overview let us consider two examples:

  (i) interference filters, and
  (ii) antireflection coatings

8.22.1 Interference Filters

Interference filter provides the selection of narrow band of light wavelengths say ≤100Å out of complete spectrum. In all spectroscopic applications this is needed. In its simplest form, an interference filter consists of a *multibeam interferometer*. As shown in Figure 8.56, a thin transparent dielectric like magnesium fluoride or cryolite is sandwiched between two optical plane parallel glass plates. The inner surfaces of glass plates are coated with thin layer of high quality reflecting material, like silver or dielectric of desired characteristics. As we know that, the path difference between two successive emergent parallel rays form normal incidence is

$$\Delta = 2nt \qquad (8.135)$$

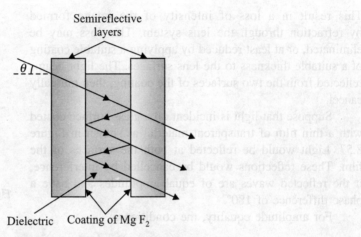

Figure 8.56 Typical interference filter.

For chromatic light, transparent beam shall interfere constructively, when Δ is integral multiple of wavelength λ, i.e.

$$2nt = m\lambda$$

or

$$\lambda = \frac{2nt}{m} \qquad (8.136)$$

where $m = 1, 2, 3, \ldots$

If $nt = \lambda_0$, then the transmitted or filtered wavelengths shall be $2\lambda_0$, λ_0, $2\lambda_0/3$, etc.

The desired wavelength is filtered by making undesired wavelengths as separation between $2\lambda_0$, λ_0, $2\lambda_0/3$ is quite large.

Interference filters can also be designed by reflecting certain wavelengths and transmitting others. The reflected ones undergo constructive interference. If we design a wedge-shaped film of varying thickness, then Eq. (8.136) is satisfied at different thicknesses for different wavelengths. Whole spectrum may be scanned from blue to red. A precisely moving slit can be used to select desired wavelengths.

8.22.2 Antireflection Coatings

When light passes through a lens system, a part is lost due to reflection at each lens surface. For example, of the light passing from air to glass ($n = 1.5$) about 4% is reflected and lost. The reflectivity R at a boundary between two media of refractive indices n_1 and n_2, at normal incidence, is

$$R = \left(\frac{n_2 - n_1}{n_2 + n_1}\right)^2 \qquad (8.137)$$

For air–glass boundary, $n_2 = 1.5$ and $n_1 = 1$.

$$\therefore \qquad R = \left(\frac{1.5 - 1}{1.5 + 1}\right)^2 = 0.04 = 4\%$$

This result in a loss of intensity of the image formed by refraction through the lens system. This loss may be eliminated, or at least reduced by applying a suitable coating of a suitable thickness to the lens surfaces. The light beams reflected from the two surfaces of the coating, then mutually cancel.

Suppose that light is incident on a glass surface coated with a thin film of transparent material as shown in Figure 8.57. Light would be reflected at both the surfaces of the film. These reflections would be cancelled by interference, if the reflected waves are of equal amplitudes and have a phase difference of 180°.

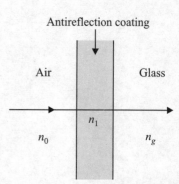

Figure 8.57 Antireflection coatings.

For amplitude equality, the condition is

$$\left(\frac{n_1 - n_0}{n_1 + n_0}\right)^2 = \left(\frac{n_g - n_1}{n_g + n_1}\right)^2$$

where n_0 is the outside refractive index, n_1 the index of the coating and n_g the index of the glass. For $n_0 = 1$ (air), we have

$$\frac{n_1^2 - 2n_1 + 1}{n_1^2 + 2n_1 + 1} = \frac{n_g^2 - 2n_1 n_g + n_1^2}{n_g^2 + 2n_1 n_g + n_1^2}$$

This reduce to

$$4n_1^3 n_g + 4n_1 n_g = 4n_1^3 + 4n_1 n_g^2$$

Dividing both sides by $4n_1$ and rearranging, we get

$$n_1^2 - n_1^2 n_g + n_g^2 - n_g = 0$$

or

$$n_1 \approx \sqrt{n_g} \qquad (8.138)$$

This means that the refractive index of the coating should be roughly equal to the square root of the refractive index of glass.

For normal incidence, the refracted waves are in opposite phase, if

$$2n_1 t = \frac{\lambda}{2} \qquad \text{or} \qquad t = \frac{\lambda}{4n_1} \qquad (8.139)$$

This is thickness required of the coating. A coating or such thickness required of the coating. A coating or such thickness and refractive index do not allow any light to be reflected.

Of course, the coating material should not only be transparent but also be insoluble and scratch-resistant. The best material known is magnesium fluoride (MgF_2), its refractive index is 1.38.

A single layer coating is effective for one particular wavelength, which is chosen near the centre of the visible spectrum. Thus the blue and red parts of the light will be partly reflected. A wider coverage of the spectrum can be achieved by multiple coatings.

EXAMPLE 8.31 Find the minimum thickness of a layer of magnesium fluoride ($n = 1.38$) required in an interference filter designed to isolate light of wavelength 5893 Å. How will peak transmittance change if the filter is titled by $10°$?

Solution Given $N = 1$ (for minimum thickness), $\lambda = 5893$ Å $= 5.893 \times 10^{-7}$ m, $i = 10°$.
Condition of maximum transmission from interference filter requires (for normal incidence)

$$t = \frac{N\lambda}{2n} = \frac{1 \times 5.893 \times 10^{-7}}{2 \times 1.38}$$

$$= 2.135 \times 10^{-7} \text{ m} = 0.2135 \text{ μm}$$

when filter is tilted by $10°$, then

$$\lambda' = \lambda_0 \sqrt{1 - \frac{\sin^2 i}{n^2}}$$

$$= 5.893 \times 10^{-7} \sqrt{1 - \frac{\sin^2 10°}{(1.38)^2}}$$

$$= 5.846 \times 10^{-7} \text{ m} = 5846 \text{ Å}$$

$$\Delta\lambda = \lambda' - \lambda = (5846 - 5893) \text{ Å} = -47 \text{ Å}$$

when filter is tilted, transmittance always change towards shorter wavelength.

EXAMPLE 8.32 Consider a antireflecting film of refractive index 1.36. Assume that its thickness is 2.7×10^{-6} cm. Compute wavelength in visible region for which the film will be antireflecting.

Solution Given $t = 2.7 \times 10^{-6}$ cm $= 2.7 \times 10^{-8}$ m, $n_1 = 1.36$
Since we know that the required thickness of the coating is

$$t = \frac{\lambda}{4n_1} \implies \lambda = 4n_1 t = 4 \times 1.36 \times 2.7 \times 10^{-8} = 1468 \text{ Å}$$

EXAMPLE 8.33 Find the reflectivity of a glass surface of refractive index 1.6, when light beam is incident normally on it.

Solution Given $n_1 = 1.0$ and $n_2 = 1.6$
Since, we know that the reflectivity R at a boundary between two media of refractive indices n_1 and n_2 at normal incidence is

$$R = \left(\frac{n_2 - n_1}{n_2 + n_1}\right)^2 = \left(\frac{1.6 - 1}{1.6 + 1}\right)^2 = 0.05 = 5\%$$

FORMULAE AT A GLANCE

8.1 Superposition of waves:
$$Y = y_1 + y_2 + y_3 + \ldots + y_n$$
where Y is the resultant displacement and $y_1, y_2, y_3, \ldots, y_n$ represents the displacements due to individual waves.

8.2 Relation between path difference and phase difference:
The phase difference ϕ
$$\phi = \frac{2\pi}{\lambda} \times x$$
where $x = (S_2P - S_1P)$ is path difference and ϕ is the phase difference between two interfering beams emanating from S_1 and S_2 reaching point P on the screen as shown in Figure 8.13.

8.3 *Young double slit experiment (YDSE):* In this case, the total intensity I is not just the sum of individual intensities $I_1(=a_1^2)$ and $I_2(=a_2^2)$ due to two sources but also includes an interference term where magnitude depends on the phase difference at given point.
$$I = a_1^2 + a_2^2 + 2a_1a_2 \cos\phi = I_1 + I_2 + 2\sqrt{I_1 I_2} \times \cos\phi$$

(a) For incoherent source (example, incandescent light bulb) of light, no interference pattern is observed because ϕ keep on changing with time and average of phase over a cycle is zero, i.e. $<\cos\phi> = 0$, then resultant intensity is given as:
$$I = I_1 + I_2$$

(b) Condition for maximum intensity (constructive interference) at maxima.
$$\left. \begin{array}{l} \phi = 2m\pi, x = m\lambda \\ I_{max} > (I_1 + I_2) \end{array} \right\}$$

(c) Condition for minimum intensity (destructive interference)
$$\begin{cases} \phi = (2m+1)\pi, x = (2m+1)\dfrac{\lambda}{2}, I_{min} < (I_1 + I_2) \\ \text{if } a_1 = a_2 \text{ then } I_{min} = 0 \end{cases}$$

(d) Average intensity = $I_{av} = I_1 + I_2$

8.4 In YDSE set up, path difference between two beams emanating from S_1 and S_2 reaching at any arbitrary point P is

(a) $S_2P - S_1P = \dfrac{xd}{D}$

where
$S_2P - S_1P$ = Path difference
d = Separation between two slits
D = Distance between slits and screen

(b) Position of mth bright fringes.
$$x_m = \frac{m\lambda D}{d}$$

(c) Position of mth dark fringes
$$x_m = \frac{(2m \pm 1)\lambda D}{2d}$$

(d) Fringe width
$$(x_m - x_{m-1}) = \beta = \frac{\lambda D}{d}$$

8.5 Fringe shift in YDSE:
$$x = \frac{(n-1)tD}{d}$$

where D = Distance between slits and screen.
d = Separation between the slits.

8.6 Fresnel's biprism experiment.
(a) Fringe width $\beta = \dfrac{D\lambda}{d}$ (as in YDSE)

(b) The wavelength of monochromatic source.
$$\lambda = \frac{d}{(a+b)} \cdot \beta$$

where d = Distance between 2 virtual sources
$a + b$ = Distance between biprism and screen.

(c) d can be obtained by
$$d = \sqrt{d_1 d_2}$$

where d_1 = size of image formed by lens L_1
d_2 = size of image formed by lens L_2 using displacement method.

8.7 Interference in thin films:
(a) When a light beam travelling from a medium of refractive index (n_1) toward another medium of refractive index (n_2), it undergoes a phase change of π (180°) on reflection if $n_2 > n_1$.

(b) Interference due to plane parallel thin film.
(i) *In reflected system:* The path difference between two reflected beams
$$\Delta = 2nt \cos r$$
where n = Refractive index of the parallel film
t = Thickness of the film
r = Angle of refraction

(ii) For maxima (constructive interference)
$$2nt \cos r = (2m \pm 1)\frac{\lambda}{2} \qquad \text{[Film will be bright]}$$

(iii) For minima (destructive) $2nt \cos r = m\lambda$ [Film will be dark]

(c) In transmitted system
   (i) For maxima path difference is $2nt \cos r = m\lambda$ [bright film]
   (ii) For minima path difference is $2nt \cos r = m\lambda$ [dark film]

8.8 Wedge-shaped film:

$$\text{Fringe width} = \frac{(x_2 - x_1)}{m} = \frac{\lambda}{2n\alpha}$$

where x_1 = Position of mth dark fringe from edge of the wedge
x_2 = Position of $(m + p)$th dark fringe from edge of the wedge
α = Angle of the wedge film
n = Refractive index of the wedge film
λ = Wavelength of the light used

8.9 Newton's rings

(a) Relation between thickness of the film (t) and radius of the ring (r).

$$t = \frac{r^2}{2R}$$

(i) For bright ring

$$2\frac{r_m^2}{2t} = (2m \pm 1)\frac{\lambda}{2}$$

or

$$r^2 = \frac{(2m \pm 1)\lambda}{2}$$

or

$$D = \sqrt{2\lambda R}\sqrt{(2m-1)}$$

(ii) For dark ring

$$2\frac{r^2}{2R} = m\lambda$$

or

$$r^2 = m\lambda R$$

or

$$D = 2\sqrt{m\lambda R}$$

(b) Newton's rings by transmitted light
   (i) For bright ring

$$2\frac{r^2}{2R} = m\lambda$$

or

$$r^2 = m\lambda R$$

or

$$D = 2\sqrt{m\lambda R}$$

(ii) For dark ring

$$2 \times \frac{r^2}{2h} = (2m \pm 1)\frac{\lambda}{2}$$

or

$$r^2 = \frac{(2m \pm 1)\lambda R}{2}$$

or

$$D = \sqrt{2\lambda R} \times \sqrt{(2m \pm 1)}$$

(c) Wavelength of sodium light

$$\lambda = \frac{D_{(m+p)}^2 - D_m^2}{4mR}$$

where D_{m+p} = Diameter of $(m + p)$th ring in air.
D_m = Diameter of mth ring in air.

(d) Refractive index of a liquid:

$$n = \frac{D_{m+p}^2 - D_m^2}{D'^2_{m+p} - D'^2_m}$$

D_{m+p} = Diameter of $(m + p)$th ring in medium of refractive index (n)
D_m = Diameter of mth ring in medium of refractive index (n)

8.10 Michelson Interferometer:

(a) Shape of fringes
   (i) Circular fringes

   For bright rings → $2d \cos \theta = \left(m - \dfrac{1}{2}\right)\lambda$

   d = film thickness
   θ = Angle subtended by circle
   For dark fringe → $2d \cos \theta = m\lambda$

(b) White light straight fringes
   t = Thickness of the film
   x = Displacement of mirror M to bring the image back their position $2x = 2(n - 1)t$
   n = refractive index of film
   $x = m\lambda = (n - 1)t$

(c) Wavelength of monochromatic light $\lambda = \dfrac{2d}{N}$

   where d = Displacement = $(x_2 - x_1)$
   N = Fringes are produced.

(d) Refractive index of thin plane sheet

$$n = 1 + \frac{N\lambda}{2d}$$

   n = Refractive index of sheet.

(e) Small difference in two wavelength from the same source

$$\Delta\lambda = \frac{\lambda^2}{2x}$$

$$\Delta\lambda = (\lambda_1 - \lambda_2) \qquad [\lambda^2 = \lambda_1 \lambda_2]$$

8.11 *Fabry–Perot interferometer:*
  (a) The path difference between successive rays
  $2t \cos \theta = N\lambda$, here t the distance, N, the number of fringes and λ is the wavelength of the light.
  (b) The resultant intensity (Airy Formula)
  $$I = \frac{T^2}{(1-R)^2 \left[1 + \frac{4R}{(1-R)^2} \sin^2 \frac{\delta}{2}\right]}$$
  here T is transmission coefficient; R the reflection coefficient and δ the phase difference
  (c) Condition of maxima
  $$I_{max} = \frac{T^2}{(1-R)^2}$$
  (d) Condition of minima
  $$I_{min} = \frac{T^2}{(1+R)^2}$$
  (e)
  $$\frac{I_{max}}{I_{min}} = \left(\frac{1+R}{1-R}\right)^2$$
  (f)
  $$\frac{I}{I_{max}} = \frac{1}{1 + F \sin^2 \frac{\delta}{2}}$$
  here
  $$F = \frac{4R}{(1-R)^2}$$
  (g) Visibility of fringes
  $$V = \frac{I_{max} - I_{min}}{I_{max} + I_{min}}$$

8.12 *Lummer–Gehrcke plate:*
$\left|\frac{\lambda}{d\lambda}\right| = \frac{l'(n^2 - 1)}{\lambda}$, where $\left|\frac{\lambda}{d\lambda}\right|$ is resolving power of the LG plate, l', the length of plate and n is the refractive index.

8.13 *Jamin's refractometer:*
$(n - 1)l = N\lambda$, here n = refractive index of the gas; λ be the length of the tube, N is the number of fringes and λ is the wavelength.

8.14 *Mach–Zehnder's refractometer:*
The change in refractive index $\Delta n = \frac{N\lambda}{l}$, here l = length of the tube, n = the refractive index, N = number of fringes.

SOLVED NUMERICAL PROBLEMS

PROBLEM 8.1 Two sources of intensity I and $4I$ are used in an interference experiment. Find the intensity at points where the waves from two sources superimpose with a phase difference
(a) zero
(b) $\pi/2$, and
(c) π.

Solution At a point with phase difference ϕ, the resultant intensity is given as:

$$I_R = I_1 + I_2 + 2\sqrt{I_1 I_2}\cos\phi$$

As $I_1 = I$ and $I_2 = 4I$
Therefore,

$$I_R = I + 4I + 2\sqrt{I.4I}\cos\phi$$
$$= 5I + 4I\cos\phi$$

(a) when $\phi = 0$, $I_R = 5I + 4I\cos 0 = 9I$
(b) when $\phi = \pi/2$, $I_R = 5I + 4I\cos 90° = 5I$
(c) when $\phi = \pi$, $I_R = 5I + 4I\cos\pi = I$

PROBLEM 8.2 Find the resultant of superposition of two waves

$$y_1 = 2.0\sin(\omega t), \text{ and}$$
$$y_2 = 5.0\sin(\omega t + 30°)$$

Symbols have their usual meanings.

Solution According to superposition principle

$$Y = y_1 + y_2 = 2.0\sin(\omega t) + 5.0\sin(\omega t + 30°)$$
$$= 2.0\sin\omega t + 5.0(\sin\omega t\cos 30°) + 5.0(\cos\omega t\sin 30°)$$
$$= 2.0\sin\omega t + \frac{5.0\sqrt{3}}{2}\sin\omega t + \frac{5.0}{2}\cos\omega t$$
$$= (2.0 + 2.5 \times 1.732)\sin\omega t + 2.5\cos\omega t$$

Here,
$$R\cos\theta = 6.33$$
and
$$R\sin\theta = 2.5$$
$$R^2(\sin^2\theta + \cos^2\theta) = 46.3189 \Rightarrow R = 6.8$$

and
$$\tan\theta = \frac{R\sin\theta}{R\cos\theta} = 0.394 \Rightarrow \theta = 21.55°$$

Then
$$Y = R\sin(\omega t + \theta) = 6.8\sin(\omega t + 21.55°)$$

PROBLEM 8.3 Laser light of wavelength 630 nm is incident on a pair of slits, produces an interference pattern in which the bright fringes are separated by 8.1 mm. A second light produces an interference pattern in which the fringes are separated by 7.2 mm. Calculated the wavelength of the second light.

Solution Here,

$$\lambda_1 = 630 \text{ nm},$$
$$\beta_1 = 8.1 \text{ mm},$$
$$\beta_2 = 7.2 \text{ mm, and}$$
$$\lambda_2 = ?$$

Fringe width $\beta = \dfrac{D\lambda}{d}$

For constant D and d, $\beta \propto \lambda$ or $\dfrac{\beta_2}{\beta_1} = \dfrac{\lambda_2}{\lambda_1}$

$$\therefore \quad \lambda_2 = \dfrac{\beta_2}{\beta_1} \times \lambda_1$$

$$= \dfrac{7.2 \text{ mm}}{8.1 \text{ mm}} \times 630 \text{ nm} = 560 \text{ nm}$$

PROBLEM 8.4 In Young's double slit experiment, a source of light of wavelength 4200 Å is used to obtain interference fringes of width 0.64×10^{-2} m. What should be the wavelength of the light source to obtain fringes 0.46×10^{-2} m wide, if the distance between screen and slits is reduced to half the initial value?

Solution In first case:
$$\lambda = 4200 \text{ Å} = 4200 \times 10^{-10} \text{ m}$$
$$\beta = 0.64 \times 10^{-2} \text{ m}$$

$$\therefore \quad 0.64 \times 10^{-2} = \dfrac{4200 \times 10^{-10} \times D}{2d} \qquad \left[\because \beta = \dfrac{\lambda D}{2d}\right] \quad \text{(i)}$$

In second case:
$$\beta = 0.46 \times 10^{-2} \text{ m}, \quad \lambda = ?, \quad D = \dfrac{D}{2}$$

$$0.46 \times 10^{-2} = \dfrac{\lambda \times \left(\dfrac{D}{2}\right)}{2d} = \dfrac{\lambda D}{2(2d)} \qquad \text{(ii)}$$

Dividing Eq. (i) by (ii)

$$\dfrac{0.64 \times 10^{-2}}{0.46 \times 10^{-2}} = \dfrac{4200 \times 10^{-10} \times D \times 2d \times 2}{(2d) \lambda D}$$

$$\therefore \quad \lambda = \dfrac{4200 \times 10^{-10} \times 2 \times 0.46}{0.64} = 6037.5 \text{ Å}$$

PROBLEM 8.5 In a double slit interference arrangement, one of the slits is covered by a thin mica sheet whose refractive index is 1.58. The distance between the slits is 0.1 cm and the distance between the slits and the screen is 50 cm. Due to introduction of the mica sheet, the central fringe gets shifted by 0.2 cm. Determine the thickness of the mica sheet.

Solution $x_0 = 0.2$ cm, $d = 0.1$ cm and $D = 50$ cm
Hence,
$$t = \frac{dx_0}{D(n-1)}$$
$$= \frac{0.1 \times 0.2}{50 \times 0.58} \text{ cm}$$
$$= 6.7 \times 10^{-4} \text{ cm} = 6.7 \text{ μm}$$

PROBLEM 8.6 In an experiment with Fresnel's biprism fringes for light of wavelength 5×10^{-7} m are observed 0.2×10^{-3} m apart at a distance of 1.75 m from the prism. The prism is made of glass of refractive index 1.50 and it is at a distance of 0.25 m from the illuminated slit. Calculate the angle of the vertex of the biprism.

Solution In Fresnel's biprism,
Given
$$\lambda = 5 \times 10^{-7} \text{ m}$$
$$\beta = 0.2 \times 10^{-3} \text{ m}$$
$$a = 0.25 \text{ m}$$
$$n = 1.50, \text{ and}$$
$$b = 1.75 \text{ m}$$
$$\alpha = ?$$

We know that distance between virtual sources in Fresnel's biprism
$$d = 2a(n-1)\alpha \qquad (i)$$
and fringe width
$$\beta = \frac{\lambda D}{d}$$
or
$$d = \frac{\lambda D}{\beta} \qquad (ii)$$
where $D = a + b = 1.75 + 0.25 = 2.00$ m
From Eqs. (i) and (ii)
$$\frac{\lambda D}{\beta} = 2a(n-1)\alpha$$
or
$$\alpha = \frac{\lambda D}{\beta[2a(n-1)]}$$
$$= \frac{5 \times 10^{-7} \times 2.00}{0.2 \times 10^{-3} \times 2 \times 0.25 \times (1.5-1)} = 0.02 \text{ radian}$$
Vertex angle $\phi = (\pi - 2\alpha) = (\pi - 0.04) = 177°42'$.

PROBLEM 8.7 The inclined faces of a biprism of refractive index 1.5 make an angle of 2° with the base. A slit illuminated by monochromatic light is placed at a distance of 10 cm from the biprism. If the distance between two dark fringes observed at a distance of 1 cm from the prism is 0.18 mm, find the wavelength of light used.

Solution In biprism experiment
Given

$$n = 1.5, \alpha = \text{angle of prism} = 2° = \frac{\pi}{90} \text{ radian}$$

$a = 10$ cm $= 0.10$ cm, $b = 1$ m, and $D = a + b = 1.1$ m
$d = 2(n - 1)\alpha a$

$$= 2 \times (1.5 - 1) \times \frac{\pi}{90} \times 0.10$$

$$= 3.49 \times 10^{-3} \text{ m}$$

$\lambda = ?$
$\beta = 0.18$ m $= 1.8 \times 10^{-4}$ m

$\therefore \quad \beta = \frac{\lambda D}{d} \Rightarrow \lambda = \frac{\beta d}{D}$

$$= \frac{1.8 \times 10^{-4} \times 3.49 \times 10^{-3}}{1.1}$$

$$= 5.711 \times 10^{-7} \text{ m}$$

$$= 5711 \text{ Å}$$

PROBLEM 8.8 White light is incident on a soap film at an angle $\sin^{-1}(4/5)$ and the reflected light shows dark bands. The consecutive dark bands correspond to wavelength 6.1×10^{-5} cm and 6.0×10^{-5} cm. If $n = 1.33$ for the film, calculate the thickness.

Solution In colour of thin film (soap film),
Given

$$i = \sin^{-1}(4/5) \Rightarrow \sin i = 4/5$$
$$\lambda_m = 6.1 \Rightarrow 10^{-5} \text{ m}$$
$$\lambda_{m+1} = 6.0 \times 10^{-5} \text{ cm and}$$
$$n = 1.33$$

The condition for dark bands in this film for reflected light

$$2nt \cos r = m\lambda$$

Now according to problem, consecutive dark bands for wavelength λ_m and λ_{m+1}; above condition will be

$$2nt \cos r = m\lambda_m \qquad \text{(i)}$$

and $$2nt \cos r = (m + 1)\lambda_{m+1} \qquad \text{(ii)}$$

From Eqs. (i) and (ii)

$$m\lambda_m = (m + 1)\lambda_{m+1}$$

or $$m(\lambda_m - \lambda_{m+1}) = \lambda_{m+1}$$

or $$m = \frac{\lambda_{m+1}}{\lambda_m - \lambda_{m+1}} \qquad \text{(iii)}$$

Putting this value of m in Eq. (i), we get

$$2nt \cos r = \frac{\lambda_{m+1} \lambda_m}{\lambda_m - \lambda_{m+1}}$$

or
$$t = \frac{\lambda_m \lambda_{m+1}}{(\lambda_m - \lambda_{m+1})(2n \cos r)} \qquad \text{(iv)}$$

But
$$\cos r = \sqrt{1 - \sin^2 r}; \quad \frac{\sin i}{\sin r} = n \quad \text{or} \quad \sin r = \frac{\sin i}{n}$$

$$\cos r = \sqrt{1 - \frac{\sin^2 i}{n^2}} = \frac{1}{n}\sqrt{(n^2 - \sin^2 i)} \qquad \text{(v)}$$

Putting the value of $\cos r$ from Eq. (v) in Eq. (iv)

$$t = \frac{\lambda_m \lambda_{m+1}}{2 \times (\lambda_m - \lambda_{m+1})\sqrt{(n^2 - \sin^2 i)}}$$

$$= \frac{6.1 \times 10^{-5} \times 6.0 \times 10^{-5}}{2 \times (6.1 - 6.0) \times 10^{-5} \sqrt{(1.33)^2 - (0.8)^2}}$$

$$= 0.00085 \text{ cm} = 8.5 \text{ μm}$$

PROBLEM 8.9 A glass wedge of angle 0.01 radian is illuminated by monochromatic light of wavelength 600 nm falling normally on it. At what distances from the edge of the wedge will be 10th fringe be observed by the reflected light?

Solution Given
$$\alpha = 0.01 \text{ radian}$$
$$m = 10 \text{ and}$$
$$\lambda = 600 \text{ nm} = 6.00 \times 10^{-7} \text{ m}$$

The condition for dark fringe $2t = m\lambda$

The angle of wedge $\alpha = \dfrac{t}{x}$ or $t = \alpha x$

$\therefore \qquad 2x\alpha = m\lambda \Rightarrow x = \dfrac{m\lambda}{2\alpha} = \dfrac{10 \times 6.00 \times 10^{-7}}{2 \times 0.01} = 3 \text{ mm}$

PROBLEM 8.10 A thin planoconvex lens of focal length 1.8 m and of refractive index 1.6 is used to obtain Newton's ring. The wavelength of the light used is 589 nm. Calculate the radius of 10th dark ring by

(i) reflection, and
(ii) transmission

Solution In Newton's ring,
Given
$$f = 1.8 \text{ m} = 180 \text{ cm}$$
$$n = 1.6 \text{ and } \lambda = 589 \text{ nm} = 5.89 \times 10^{-7} \text{ m}$$

Here we use the lens formula
$$\frac{1}{f} = (n-1)\left(\frac{1}{R_1} - \frac{1}{R_2}\right)$$

$R_1 = R$, and $R_2 = \infty$, then $\dfrac{1}{f} = (n-1)\dfrac{1}{R}$

or
$$R = (n-1)f$$
$$= (1.6 - 1) \times 180 \text{ cm} = 0.6 \times 180 \text{ cm}$$
$$= 108 \text{ cm} = 1.08 \text{ m}$$

(i) Radius of 10th dark ring (in case of reflection)
$$r_m^2 = m\lambda R$$
$$r_m = \sqrt{m\lambda R} = \sqrt{10 \times 5.89 \times 10^{-7} \times 10.8}$$
$$= 0.252 \text{ cm}$$

(ii) Radius of 10th dark ring (in case of transmission)
$$r_m^2 = \frac{(2m-1)\lambda R}{2}$$

or
$$r_m = \sqrt{\frac{(2m-1)\lambda R}{2}}$$
$$= \sqrt{\frac{(2 \times 10 - 1)}{2} \times 5.89 \times 10^{-7} \times 1.08}$$
$$= 0.245 \text{ cm}$$

PROBLEM 8.11 If the diameter of mth dark ring is an arrangement giving Newton's rings changes from 3 mm to 2.5 mm as a liquid is introduced between the lenses and plate, what is the value of refractive index of the liquid?

Solution In Newton's ring arrangement.
Given
$$D_m = 3 \text{ mm} = 3 \times 10^{-3} \text{ m}$$
$$D'_m = 2.5 \text{ mm} = 2.5 \times 10^{-3} \text{ and}$$
$$n = ?$$

We know that the diameter of mth ring in presence of liquid is
$$(D'_m)^2 = \frac{4m\lambda R}{n} \qquad \text{(i)}$$
and the diameter of mth ring in air is
$$(D_m)^2 = 4m\lambda R \qquad \text{(ii)}$$

Dividing Eq. (ii) by Eq. (i), we get

$$n = \left(\frac{D_m}{D'_m}\right)^2$$

$$= \left(\frac{3.0}{2.5}\right)^2 = (1.2)^2 = 1.44$$

PROBLEM 8.12 A Michelson's interferometer is set to form circular fringes with light of wavelength 500 nm. By changing path length of movable mirror slowly 50 fringes cross the centre of view. How much path length has been changed?

Solution If the change in path length is x, then

$$2x = N\lambda$$

$$x = \frac{N\lambda}{2} = \frac{50 \times 5.00 \times 10^{-7}}{2}$$

$$= 12.5 \times 10^{-5} \text{ m}$$

$$= 0.125 \times 10^{-3} \text{ m}$$

$$= 0.125 \text{ m}$$

PROBLEM 8.13 In Laser Michelson's interferometer, when a transparent thin glass plate of refractive index 1.5 is introduced in the path of one of the beams, 100 fringes cross the field of view at given point. If the wavelength of laser used is 6328 Å, find the thickness of the plate.

Solution

Here, $$2(n-1)t = N\lambda$$

$$n = 1.5$$

$$\lambda = 6328 \times 10^{-10} \text{ m, and}$$

$$N = 100$$

Therefore,

$$t = \frac{N\lambda}{2(n-1)}$$

$$= \frac{100 \times 6328 \times 10^{-10}}{2(1.5-1)}$$

$$= 6.328 \times 10^{-5} \text{ m}$$

PROBLEM 8.14 White light is incident normally on a Fabry–Perot's interferometer with a plate separation of 4×10^{-4} cm. Calculate the wavelengths for which there are interference maxima in the transmitted beam in the range 4000 Å to 5000 Å.

Solution For a Fabry–Perot's interferometer, the condition of maxima in the transmitted beam is

$$2t \cos\theta = m\lambda$$

where t = plate separation

For normal incidence $\theta = 0°$, so that
$$2t = m\lambda$$
For 4000 Å (= 4×10^{-5} cm) wavelength, the order of at the centre is
$$m = \frac{2t}{\lambda}$$
$$= \frac{2 \times 4 \times 10^{-4} \text{ cm}}{4 \times 10^{-5}} = 20$$
while for 5000 Å (= 5×10^{-5} cm), the order is
$$m = \frac{2 \times 4 \times 10^{-4}}{5 \times 10^{-5}}$$
$$= 16$$
For intermediate wavelengths, the order will be 19, 18 and 17. Therefore, the corresponding wavelengths are
$$\lambda = \frac{2t}{m}$$
$\Rightarrow$
$$\lambda_1 = \frac{2 \times 4 \times 10^{-6}}{19} \text{ cm}$$
$$= 4210 \text{ Å}$$
$$\lambda_2 = \frac{2 \times 4 \times 10^{-6}}{18} \text{ cm} = 4444 \text{ Å}$$
and
$$\lambda_3 = \frac{2 \times 4 \times 10^{-6}}{17} \text{ cm} = 4706 \text{ Å}$$

PROBLEM 8.15 Find minimum thickness required of a layer of cryolite ($n = 1.35$) in an interference filter designed to isolate light of wavelength 594 nm. How will the peak transmittance change if filter is tilted by 10°?

Solution The desired thickness for normal incidence is
$$t = \frac{m\lambda}{2n}$$
$$= \frac{(1)(594) \times 10^{-9}}{2 \times 1.35}$$
$$= 220 \times 10^{-9} = 220 \text{ nm}$$
When filter is tilted by 10°, then
$$t = \frac{m\lambda}{2n \cos r}$$
when filter is tilted by 10°, angle of incidence $i = 10°$.
From Snell's law:
$$n_0 \sin i = n \sin r$$

$n_0 = 1$ for air, $i = 10°$, $n = 1.35$, then

$$\sin r = \frac{\sin i}{n}$$

$$\cos r = \sqrt{1 - \sin^2 r} = \sqrt{1 - \frac{\sin^2 i}{n^2}}$$

The transmitted wavelength is

$$\lambda = \lambda_0 \sqrt{1 - \frac{\sin^2 i}{n^2}}$$

$$= 594 \sqrt{1 - \frac{\sin^2 10°}{(1.35)^2}}$$

$$= 589 \text{ nm}$$

So the change in wavelength is

$$\Delta\lambda = 594 - 589 = 5 \text{ nm}$$

PROBLEM 8.16 Two $\lambda/4$ thick layers are deposited on an ophthalmic glass ($n = 1.52$) to reduce reflection loss. The first layer is of magnesium fluoride, what is refractive index of second layer?

Solution

$$n_0 n_2^2 = n_1^2 n_g \quad \text{or} \quad n_2 = n_1 \sqrt{\frac{n_g}{n_0}}$$

Here, $n_1 = 1.38$, $n_g = 1.52$, and $n_0 = 1$ (for air)

$$\therefore \quad n_2 = 1.38 \sqrt{\frac{1.52}{1.00}} = 1.70$$

CONCEPTUAL QUESTIONS

8.1 "Any monochromatic light is necessarily coherent", true or false? Justify your answer.

Ans Monochromatic light is coherent, if the waves of the monochromatic light oscillates in the same direction and have the same frequency and the phase. In other words, the monochromatic light must be collimated. This requirement applies to a laser as nuclear ghost mentioned it. In contrast to that the waves of a light which come from the light bulb are incoherent, as the wave oscillates in different directions which have different frequencies and phases.

8.2 What is meant by coherent sources of light? Can two identical and independent sodium lamps act as coherent sources? Justify.

Ans Two light sources are said to be coherent if they continuously emit light waves of same frequency (or wavelength) with zero or constant phase difference between them. Two independent sources cannot act as coherent sources. The emission of light in them is due to millions of atoms in which electrons jump from higher to lower orbit. The process is occurs in 10^{-8} s. Thus, phase difference can remain constant for about 10^{-8} s only i.e., phase changes 10^8 times in one second. Such rapid changes in the positions of maxima and minima cannot be detected by our eyes. The interference pattern is lost and almost a uniform illumination is seen on the screen.

8.3 Interference can be observed with two independent tuning fork, but it cannot be observed with two independent bulbs. Why?

Ans When two tuning forks are struck simultaneously they produce sound wave almost in the same phase. Their phase difference, if any, varies slowly with time. Interference pattern also varies slowly with time. Such variations can be detected easily by the human ear. So, interference pattern is easily observable.

The phase difference between two independent light bulbs changes 108 times per second. The interference pattern also changes 10^8 times per second. Such rapid variations cannot be detected by our eyes. So, interference pattern is not observable.

8.4 What are bright and dark fringes in case of Young double slit experiment?

Ans The intensity maxima and minima in the interference are called bright and dark fringes. The fringes are neither image nor shadow of slit, but a locus of a point, which moves in such a way that point the path difference between the waves from the two sources remain constant, in case of bright fringes, it is integral multiple of the wavelength and in case of dark fringes, it is odd multiple of the wavelength. An array of fringes is called the interference pattern.

8.5 Can two independent point sources of light operating under similar conditions produce sustained interference?

Ans No, the two independent point sources of light operating under similar conditions may produce light of the same wavelength and amplitude, but they will not able to satisfy the most essential requirement of coherence for sustained interference i.e., constancy in phase relationship. The two independent sources emit light of the phase difference between the two interfering beams goes on varying randomly with time, it will not be possible to obtain sustained interference pattern.

8.6 What is difference between fringes obtained by Fresnel's biprism and those obtained by Newton's rings?

Ans (i) The biprism fringes are straight and equally spaced whereas the fringes in Newton's rings are circular and not equally spaces.

(ii) In biprism fringes are obtained by division of wavelength whereas in Newton's rings, they are obtained by division of amplitudes.

(iii) In biprism, fringes are non-localised while in Newton's rings they are localized.

8.7 Why two independent sources of light of the same wavelength cannot produce interference fringes?

Ans If two independent sources of light of same wavelength are placed side by side, no interference fringes or effect are observed because the light waves from one source are emitted independently of those from the other source. The emissions from the two independently do not maintain constant phase relationship with each other over time.

Light waves from an ordinary source such as light bulb undergo random phase changes in time intervals less than a nanosecond. Therefore, the conditions for constructive interference, destructive interference or some intermediate state are maintained only for such time intervals. Because the eye cannot follow such rapid changes, no interference fringes/effects are observed. Such light sources are said to be incoherent.

8.8 Why the colours of thin films in reflected and transmitted light are complementary?

Ans In the reflected system, there is an additional path difference of $\lambda/2$ between the two rays producing interference as one of the rays suffers reflection at a denser medium, while in the transmitted system it is not so. Thus for the same path difference in the reflected system, a bright band will correspond to a dark band in transmitted system and vice versa. Thus, the two systems are complementary.

8.9 Why does a soap bubble show beautiful colours, when illuminated by white light?

Ans Light waves reflected from the upper and lower surfaces of a thin film interfere. Since the conditions for bright and dark fringes are satisfied at different positions for different wavelengths, so coloured fringes are observed.

8.10 Why does the colour of the oil film on the surface of water continuously changes?

Ans The position of the bright and dark fringes produced by thin oil films depends upon the thickness of the film. The thickness of oil film on the surface of water continuously varies and as a result, the position of the coloured fringes also varies. This appears as a variation in the colour of the oil films.

8.11 Explain why excessively thin film seen in reflected light appears dark.

Ans Excessively thin film is dark in reflected system. The effective path difference between the interfering reflected rays is $2nt \cos r - \lambda/2$. When the film is excessively thin, so that t is practically zero, the effective path difference is $\lambda/2$. This is the condition for minimum intensity. Hence, the film appears dark.

8.12 In relation to Newton's rings pattern, what will happen if
  (a) we use plano-concave lens and place it with concave side on glass plate?
  (b) we use plano-convex lens of smaller radius of curvature?
  (c) the plano-convex lens is raised by height Δh from the surface of plane glass plate?

Ans (a) In this case, fringes are still circular. The lens touches the glass plate at the circumference hence dark rings shall be observed there. The thickness of air film increases as we move from circumference to centre and therefore order of rings is reversed i.e., highest order ring at the periphery of the lens. If t_H be the thickness of air film at the centre and t_m be thickness of film at position of mth ring, then radius of mth ring from centre is

$$r_m = [2(t_H - t_m)R]^{1/2}$$

For mth dark ring,

$$2t = m\lambda$$

Hence

$$r_m = [(2t_H - m\lambda)R]^{1/2}$$

Here R is radius of curvature of plano-convex lens. The spacing between two consecutive rings decreases as we move towards the centre of the lens.

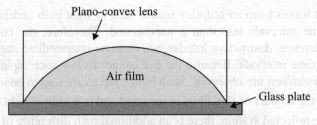

Figure 8.58

(b) The radius of given Newton's rings is proportional to square root of radius of curvature of plano-convex lens $[r_m = (m\lambda R)^{1/2}]$. Therefore, decrease in the radius of curvature of plano-convex lens shall reduce the radii of Newton's rings.

(c) Film thickness at the position of mth ring is

$$t_m = \Delta h + \frac{r_m^2}{2R}$$

where R is the radius of curvature of the plano-convex lens. For mth dark ring

$$2t_m = m\lambda$$

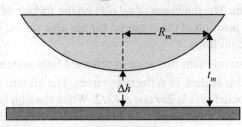

Figure 8.59

Which gives (for dark ring),

$$r_m = (m\lambda R - 2\Delta hR)^{1/2}$$

similarly, for the mth bright ring

$$2t_m = (2m-1)\frac{\lambda}{2}$$

and

$$r_m = \left[(2m-1)\frac{\lambda R}{2} - 2\Delta hR\right]^{1/2}$$

If $\Delta h = \frac{\lambda}{4}$, the centre becomes bright as $R_m = 0$ for $m = 0$. We also observe that as t_H increases; R decreases, the radius for fringes decreases while raising the lens.

8.13 The central part in Newton's rings seen in reflected light appears dark. Why?

Ans In Newton's rings experimental arrangement at the point of the contact between the interfering rays due to difference in the path lengths. But one of the rays suffers a phase change of π on reflection at the surface of the glass plate, i.e., denser medium. This is why the rays suffer destructive interference and centre appears dark.

8.14 What are Newton's rings? Why the central ring is dark when observed in reflected light?

Ans When a plano-convex lens radius of curvature is placed with its convex surface in contact with a plane glass plate, on air film is formed between the lower surface of the lens and the upper surface of the plate. If monochromatic light is allowed to fall normally on this film, a system of alternate bright and dark concentric rings with their centre dark concentric rings with their centre dark is formed. Since these rings were discovered by Newton, so these are called Newton's rings. The central ring is dark when observed in reflected light because effective path difference $\Delta = 2t + \frac{\lambda}{2}$. At the point of contact $t = 0$, then $\Delta = \frac{\lambda}{2}$. This is the condition for minimum intensity. Hence, the central ring is dark.

8.15 Why do the Newton's rings get closer as the order of the rings increases?

Ans Because radius of dark ring $D_m \propto \sqrt{m}$, while for bright ring $D_m \propto \sqrt{(2m+1)}$ i.e., square root of odd natural numbers.

8.16 In Newton's rings experiment, predict, what will happen?
 (a) if sodium lamp is replaced by white light source in Newton's rings?
 (b) if a few drops of a transparent liquid introduced between the lens and plate?

Ans (a) Few coloured fringes will be observed near the centre.
 (b) The diameter of fringes is contracted.

8.17 Explain, why interference fringes are circular in Newton's rings.

Or

Why Newton's rings are circular?

Ans In Newton's rings experiment, the path difference of two interfering waves is dependent on the thickness of the air firm. In the set up of this experiment, the locus of same thickness forms a circle with centre at the point of contact between the lens and plane glass plate. Therefore, the condition of constructive interference or destructive interference is satisfied over a circle and the fringe system becomes circular.

8.18 What happens to the ring system if a plane polished mirror is used instead of a glass plate in Newton's rings arrangement?

Ans If a plane polished mirror is used instead of a glass plate in Newton's ring arrangement, the interference pattern produced due to reflected and transmitted light will be superimposed and we get these two patterns, which are complimentary of their superposition. Hence, we get uniform illumination.

8.19 The interference fringes produced in the Newton's rings experiment are real or virtual? Justify.

Ans The interference fringes produced in the Newton's rings experiment are real because as we are observing the reflected geometry, the centre of the ring system is dark spot. These rings are formed in the plane of the film and these are observed by a microscope. So, these rings are real.

8.20 Why do we prefer a convex lens of large radius of curvature for producing Newton's rings?

Ans In Newton's rings arrangement, it is preferred to be large radius of curvature lens because (i) it results in increase the diameter of Newton's rings which increases the accuracy in the measurement of their diameters; (ii) large value of radius of curvature in the decrease of the thickness of the air film at any point and hence it is justified to neglect t^2 as compared to $2Rt$, (iii) the angle of wedge-shaped film enclosed between the glass plate and the lower surface of lens is very small and hence can be neglected.

8.21 What is an interferometer?

Ans An interferometer is an instrument which employs large path difference between the interfering beams to obtain interference. It is used to make precise measurement of wavelengths or distance between fine spectral lines.

8.22 What is Michelson interferometer?

Ans An important experimental device that uses interference is the Michelson interferometer. In the late of 19th century, it helped to provide one of the key experimental underpinnings of the theory of relativity. More recently, Michelson interferometers have been used to make precise measurements of wavelengths and of very small distances, such as minute changes in thickness of an axon when a nerve pulse propagate along its length.

8.23 What are the differences between the methods of division of wavefront and division of amplitude?

Ans

S. No.	Division of wavefront	Division of amplitude
1.	In this method, a narrow light source is used.	In this method, an extended source is used.
2.	In this method, the wavefront emitted by a narrow source is divided in two parts by reflection or refraction.	In this method, the amplitude of emitted by an extended source is divided in two parts by partial reflection or partial refraction.
3.	The coherent sources used in this method are imaginary.	The coherence sources used in this method are real.
4.	Examples: Young's double slit experiment, Fresnel's biprism, Fresnel's double mirror and Lloyd's mirror.	Example: Thin films, wedge-shaped films, Newton's rings, Michelson interferometer.

8.24 What is a Jamin's interferometer?

Ans One of the many variants of Michelson interferometer is Jamin's interferometer. It is particularly useful in determining the refractive index of a gas at different pressures.

8.25 What is Mach–Zehnder interferometer?

Ans Mach–Zehnder is other modification of Michelson interferometer. It is widely useful for plasma diagnostics and gas flow, say in wind tunnel.

8.26 What is Fabry–Perot interferometer?

Ans If we have interference involving by many beams. These beams are derived from a single beam by multiple reflections. Then, the interference fringes so formed are much sharper than those by two beam interference. The Fabry–Perot interferometer is based on multiple beam interferences exploit the high contrast interferences fringes in light transmitted by high reflectivity films and find wide applications in high resolution spectroscopy.

8.27 What are the main advantages of Fabry–Perot interferometer over Michelson interferometer?

Ans The main advantages of Fabry–Perot interferometer over Michelson interferometer are:
(i) The obtained fringes of Fabry–Perot interferometer are sharper than those of Michelson interferometer.
(ii) The resolving power of FP interferometer is very high so that it is used to observe the fine structure of spectral line while this is impossible in case of Michelson interferometer.

8.28 What is the Lummer–Gehrcke plate?

Ans The LG plate is based on the phenomenon of total internal reflection. It produces straight line fringes. It is particularly useful in the study of fine structure of spectral lines in the ultraviolet region.

8.29 What is blooming in antireflection coatings?

Ans For practically entire range of visible spectrum, the process of reducing reflectivity is called blooming.

8.30 What is interference filter?

Ans Interference filter provides the selection of narrow band of light wavelength say ≤ 100 Å out of complete spectrum. In all spectroscopic applications that is needed. It is simplest form of interference filter consists of a multiple beam interferometer. Interference filter can also be designed by reflecting certain wavelengths and transmitting others.

EXERCISES

Theoretical Questions

8.1 State and explain superposition principle of waves.

8.2 What is interference of light? Describe Young's double slit experiment for observing the interference of light.

8.3 What do you mean by coherent and incoherent sources of light? Why are coherent sources required to produce interference of light? Can two independent light sources be coherent?

8.4 Deduce an expression for fringe width in Young's double slit experiment. How can the wavelength of monochromatic light be found by this experiment?

8.5 What is a sustained interference pattern? State the necessary conditions for obtaining a sustained interference of light.

8.6 Show that the phenomenon of interference is inaccordance with the law of conservation of energy.

8.7 Deduce an expression for the intensity at a point in the region of superposition of two waves of same periods and wavelengths. Hence, establish the need of two coherent sources for the production of observable interference pattern.

[Kanpur University, 2007; Avadh University, 2008]

8.8 What do you understand by coherent sources? Give the conditions for sustained interference of light.
  [Bundelkhand University, 2007, 2011; Avadh University, 2010]

8.9 What are effects on the fringe width if the wavelength of light reduces in the Young's double slit experiment? **[Kanpur University, 2008]**

8.10 Give the conditions for well defined and sustained interference.
  [Kanpur University, 2010]

8.11 Define interference of light wave. Obtain expressions for constructive and destructive interference in Young's double slit experiment. **[Avadh University, 2013]**

8.12 Draw well labelled diagram and explain how you will measure thickness of thin silicon chip using Fresnel's biprism. Give the necessary theory.

8.13 Why are coherent sources necessary for formation of interference fringes?
  [Kanpur University, 2009]

8.14 How will you measure thickness of thin film using the phenomenon of interference?

8.15 Give the theory of Fabry–Perot's interferometer. Prove that in the fringe system formed, the ratio of the intensity of maxima to the intensity mid-way between minima is given by $[(1 + R)/(1 - R)]^2$, where R is the reflection coefficient. Discuss the effect of magnitude of the reflection coefficient on the sharpness of fringes.

8.16 Give the theory of interference fringes.

8.17 Prove that for thin films, the interference patterns in reflected and transmitted light are complementary.

8.18 Explain, why excessively thin films seen by reflected lightly appear dark.

8.19 Explain with the help of a diagram an experimental arrangement to produce Newton's rings. What is the difference between these rings and those produced by biprism? How will you use Newton's rings to measure wavelength of light?

8.20 Describe and explain the formation of Newton's rings is reflected light. How can these be used to determine the refractive index of a liquid? Derive the formula used.

8.21 Explain the formation of Newton's rings in reflected monochromatic light. Prove that in reflected light:
  (i) diameters of bright rings are proportional to the square roots of odd numbers; and
  (ii) diameters of dark rings are proportional to the square roots of natural numbers.
  [Kanpur University, 2006; Avadh University, 2010, 2012]

8.22 Explain how Newton's rings are formed and describe the method for the determination of wavelength of light with their use. **[Avadh University, 2011]**

8.23 Describe an interference method for determining the refractive index of transparent liquid given in a very small quantity. Derive the formula use. **[Avadh University, 2012]**

8.24 Why Newton's rings are circular and explain why do we get black spot in Newton's ring? **[Kanpur University, 2011]**

8.25 Distinguish between Newton's rings and Haidinger fringes.
  [Bundelkhand University, 2005, 2008]

8.26 Show that in Newton's ring, the square of the diameters of successive rings have equal differences. **[Bundelkhand University, 2006]**

8.27 Describe construction, working and applications of Michelson's interferometer. How would you use it to measure the wavelength of monochromatic light.

8.28 Describe Michelson interferometer with a diagram and explain the formation of fringes in it. How would you use it to measure the difference in wavelength of sodium D-line? How will you use it to determine the thickness of a thin transparent film?
[Bundelkhand University, 2008, 2011; Kanpur University, 2007; Avadh University, 2009]

8.29 Describe the construction of Michelson interferometer. How does it function?
[Avadh University, 2013]

8.30 What is the function of compensating plate in Michelson interferometer?
[Bundelkhand University, 2012; Kanpur University, 2009, 2011]

8.31 How will you use a Michelson interferometer for finding the difference between two close wavelengths? **[Kanpur University, 2010]**

8.32 Explain the difference between the localized and non-localized fringes.
[Kanpur University, 2012]

8.33 Describe the construction and working of Michelson's interferometer. Explain with expression. How can it use to find the wavelength difference of two nearby wavelengths of sodium light? **[Agra University, 2006]**

8.34 Describe the working of Michelson interferometer. State the condition for obtaining white light fringes. Show with necessary theory how this interferometer can be used to measure the wavelength of light. **[Agra University, 2007]**

8.35 How can we get coloured fringes of white light in Michelson's interferometer?
[Agra University, 2010]

8.36 Give the theory of formation of fringes in Michelson interferometer. How will you use to determine the difference of two close wavelengths? **[Agra University, 2013]**

8.37 How are the interference fringes used to measure the displacement?
[Agra University, 2013]

8.38 Explain the phenomenon of interference of the transmitted beam from an air film between two plane silvered plates. What are the factors which determine the sharpness of fringes?

8.39 Describe construction and function of Fabry–Perot interferometer with the help of ray diagram. **[Avadh University, 2013]**

8.40 Explain the principle of the Fabry–Perot interferometer. Obtain expression for the intensity distribution in the transmitted light. **[Bundelkhand University, 2006, 2010]**

8.41 Write short notes on following:
  (i) Lloyd's single mirror
  (ii) Fresnel's double mirror
  (iii) Newton's rings
  (iv) Michelson interferometer
  (v) Fabry–Perot interferometer **[Agra University, 2008]**

8.42 Explain the principle of Fabry–Perot interferometer. Obtain an expression for the intensity distribution in the transmitted light and explain the sharpness of the fringes.
[Agra University, 2009, 2012]

8.43 Write a short note on Lummer–Gehrcke plate.

8.44 Explain the principle of Rayleigh's refractometer and write down its uses.
[Kanpur University, 2008]

8.45 What is interference of light? Explain with the help of neat diagram. What is meant by the same state of polarization in interference? **[Kanpur University, 2012]**

8.46 Describe any interference refractometer and explain its use in determining the refractive index of a gas.

8.47 Write short notes on:
(a) Interference filters
(b) Antireflection coatings

8.48 Explain the principle of interference refractometer. How would you use it to determine the refractive index of a gas at different temperature?

Numerical Problems

8.1 Two coherent monochromatic light beams of intensities I and $4I$ are superposed. What will be the maximum and minimum possible intensities? **[Ans. $9I, I$]**

8.2 Find the ratio of intensities of two points P and Q on a screen in Young's double slit experiment when waves from sources S_1 and S_2 have phase difference of
(i) 0° and
(ii) $\pi/2$ respectively **[Ans. 2:1]**

8.3 In Young's double slit experiment the fringe width obtained is 0.6 cm, when light of wavelength 4800 Å is used. If the distance between the screen and the slit is reduced to half, what should be the wavelength of light used to obtain fringes 0.045 m wide?
[Ans. 7.2×10^{-6} m]

8.4 Two waves travelling along the same line by
$$y_1 = 5 \sin(\omega t + \pi/2) \text{ and } y_2 = 7 \sin(\omega t + \pi/3)$$
Find
(a) resultant amplitude
(b) the initial phase angle of resultant, and
(c) the resultant equation of motion
[Ans. (a) 11.60, (b) 72.4°, (c) $Y = 10.60 \sin(\omega t + 72.4°)$]

8.5 Two waves travelling together along the same line are represented by
$$y_1 = 25 \sin\left(\omega t - \frac{\pi}{4}\right), \text{ and } y_2 = 15 \sin\left(\omega t - \frac{\pi}{6}\right)$$
Find
(a) the resultant amplitude,
(b) the initial phase angle of the resultant, and
(c) the resultant equation for the sum of two motion.
[Ans. (a) 39.69, (b) 39.28°, (c) $y = 39.69 \sin(\omega t - 39.38°)$]

8.6 Young's experiment is performed with orange light. If the fringes are measured with a micrometer eyepiece at a distance 100 cm from the double slit, it is found that 25 of them occupy a distance of 12.87 mm between centres. Find the distance between the centres of the two slits. [**Ans.** 1.1297 mm]

8.7 In Young's double slit experiment, the separation of the slits is 1.9 mm and the fringe spacing is 0.31 mm at a distance of 1 m from the slits. Find the wavelength of the light.
[**Ans.** $\lambda = 589$ nm]

8.8 The inclined faces of a biprism of glass ($n = 1.5$) make angle of 2° with the base. The slit is at 10 cm from the biprism and is illuminated by light of wavelength 550 nm. Find the fringe width at a distance of 1 m from the biprism. [**Ans.** $\beta = 0.0173$ cm]

8.9 In a biprism experiment, the distance between the slit and the screen is 180 cm. The biprism is 60 cm away from the slit and its refractive index is 1.52. When the source of wavelength 5893 Å is used, the fringe width is found to be 0.010 cm. Find the angle between the two refracting surfaces of the biprism. [**Ans.** $\alpha \cong 1°$]

8.10 Interference fringes are produced by Fresnel's biprism in the focal plane of eyepiece 200 cm away from the slit. The two images of the slit that are formed for each of the two positions of a convex lens placed between the biprism and eyepiece are found to be separately by 4.5 mm and 2.9 mm respectively. If the width of the interference fringes be 0.326 mm, find the wavelength of the light used. [**Ans.** $\lambda = 587$ nm]

8.11 Fringes are formed by a Fresnel's biprism in the focal plane of a reading microscope which is 100 cm from the slit. A lens is inserted between the biprism and the microscope gives two images of the slit in two positions. In case the two images of the slits are 4.05 mm and in other, they are 2.9 mm apart. If sodium light ($\lambda = 5893$ Å) is used, find the distance between the interference fringes. [**Ans.** $B = 1.72 \times 10^{-3}$]

8.12 Green light of wavelength 5100Å from a narrow slit is incident on a double slit. If the overall separation of 10 fringes on a screen 2 m away is 0.02 m. Find the double slit separation. [**Ans.** $d = 51 \times 10^{-5}$ m]

8.13 In a biprism experiment, bands of width 0.0195 cm are observed at 100 cm from the slit. On introducing a convex lens 30 cm from the slit, the two images of the slit are seen 0.7 cm apart, at 100 cm distance from the slit. Calculate the wavelength of light used.
[**Ans.** $\lambda = 5850$ Å]

8.14 In a biprism experiment, the eyepiece is placed at a distance of 0.8 m from the source. The distance between the virtual sources is found to be 0.0005 m, calculate the wavelength of the source. [**Ans.** $\lambda = 5893$ Å]

8.15 On introducing a thin sheet of mica (thickness 12×10^{-5} cm) in the path of one of the interfering beams in a biprism arrangement, the central fringe is shifted through a distance equal to the spacing between successive bright fringes. Calculate the refractive index of mica. [**Ans.** $n = 1.5$]

8.16 White light falls normally upon a film of soapy water whose thickness is 5×10^{-5} cm and refractive index is 1.33. What wavelength in visible region will be reflected more strongly? [**Ans.** 5320 Å]

8.17 A parallel beam of sodium light (589 nm) strikes a film of oil ($n = 1.46$) floating on water ($n = 1.33$). When viewed at an angle of 30° from the normal, the eighth dark band is seen. Find the thickness of the film. [**Ans.** $t = 1.7 \times 10^{-4}$ cm]

8.18 White light is incident on a soap film at an angle $\sin^{-1}$ (4/5) and the reflected light shows dark bands. The consecutive dark bands correspond to wavelength 6.1×10^{-5} cm and 6.0×10^{-5} cm. If $n = 1.33$ for film, calculate the thickness. [**Ans.** $t = 0.0017$ cm]

8.19 A soap film surrounded by air has an index of refraction of 1.34. If a region of the film appears bright red ($\lambda = 633$ nm) in normally reflected light, what is its minimum thickness there? [**Ans.** $t = 118$ nm]

8.20 A thin parallel film of refractive index 1.4 is illuminated with white light. Two consecutive dark bands are observed corresponding to wavelength 459 nm and 450 nm when observed at an angle of 35°. Determine the thickness of the film. [**Ans.** $t = 8.98$ nm]

8.21 Soap film of refractive index 1.33 is illuminated with light of different wavelengths at an angle 45°. There is complete destructive interference for $\lambda = 5890$ Å. Find the thickness of the film. [**Ans.** $t = 313.2$ nm]

8.22 Two glass plates enclose a wedge-shaped air film touching at one edge are separated by a wire of 0.03 mm diameter at distance 15 cm from the edge. Monochromatic light ($\lambda = 600$ nm) from a broad source falls normally on the film. Calculate the fringe width. [**Ans.** $\beta = 0.15$ cm]

8.23 Light of wavelength 6000 Å falls normally on a thin wedge-shaped film of refractive index 1.4 forming fringes that are 2 mm apart. Find the angle of wedge. [**Ans.** $\alpha = 1.07 \times 10^{-4}$ radians]

8.24 Newton's rings are observed normally in reflected light of wavelength 5.9×10^{-5} cm. The diameter of the 10th dark ring is 0.50 cm. Find the radius of curvature of the lens and the thickness of the film. [**Ans.** $R = 106$ cm, $t = 3 \times 10^{-4}$ cm]

8.25 In a Newton's rings experiment the radius of curvature R of the lens is 5.0 m and its diameter is 2.0 cm.
(a) How many rings are produced?
(b) How many rings would be seen if the arrangement were immersed in water ($n = 1.33$)? Given $\lambda = 5890$ Å. [**Ans.** (a) $N = 33.45 = 33$ (b) $N' = 44$]

8.26 The diameter of the 10th bright ring in a Newton's ring apparatus changes from 1.40 to 1.27 cm as a liquid is introduced between the lens and the plate. Find index of refraction of the liquid? [**Ans.** $n = 1.21$]

8.27 In Newton's ring experiment the diameter of 15th ring was found to be 0.590 cm and that of the 5th ring was 0.336 cm. If the radius of plano-convex lens is 100 cm. Calculate the wavelength of light used. [**Ans.** $\lambda = 5880$ Å]

8.28 Light containing two wavelengths λ_1 and λ_2 falls normally on a plano-convex lens of radius of curvature R resting on a glass plate. If the mth dark ring due to λ_1 coincides with $(m-1)$th dark ring for λ_2. Prove that the radius of the mth dark ring of λ_1 is

$$\sqrt{\frac{\lambda_1 \lambda_2 R}{(\lambda_1 - \lambda_2)}}$$

8.29 When space between the flat disc and convex surface in a Newton's rings setup is filled with a liquid, the radius of fringes is reduced to 80% of the original value. Calculate refractive index of the liquid. **[Ans.** $n = 1.56$**]**

8.30 In Newton's ring experiment, the diameter of the 15th ring was found to be 0.617 cm and that of the 5th ring was 0.341 cm. If the radius of the curvature of the plano-convex lens is 100 cm. Compute the wavelength of light used. **[Ans.** $\lambda = 661$ nm**]**

8.31 How far must the movable mirror of a Michelson's interferometer be displaced for 2500 fringes of red cadmium light ($\lambda = 6438$ Å) to cross the field of view.
[Ans. 0.80475 mm**]**

8.32 Michelson's interferometer is adjusted for distinct straight line fringes, using sodium light having wavelengths 5896 Å and 5890 Å. The pattern changes if one of the mirrors is slowly shifted. Calculate the distance through which the mirror should be shifted so that distinct fringes are again obtained. **[Ans.** 0.02894 cm**]**

8.33 In an experiment for determination of the refractive index of air with Michelson's interferometer, a shift of 150 fringes is observed when all the air was removed from the tube. Fringes were obtained with light of wavelength 4000 Å, if length of the tube is 20 cm. Find refractive index of air. **[Ans.** $n = 1.00015$**]**

8.34 When a thin film of a transparent material of a refractive index 1.45 for $\lambda = 589$ nm is inserted in one of the arms of a Michelson's interferometer, a shift of 65 circular fringes is observed. Calculate the thickness of the film. **[Ans.** $t = 0.00425$ cm**]**

8.35 In Michelson's interferometer, the scale readings for two successive maxima indistinctness of fringes were found to be 0.6939 mm and 0.9884 mm. If the mean wavelength of the two components of light 5893 Å, deduce the difference between the wavelengths of the components. **[Ans.** $\Delta\lambda = 3$ Å**]**

8.36 When a movable mirror of Michelson's interferometer is moved through 0.05896 mm, a shift of 200 fringes is observed, what is the wavelength of light used?
[Ans. $\lambda = 5896$ Å**]**

8.37 The initial and final readings of a Michelson interferometer screw are 10.7347 mm and 10.7051 mm respectively when 100 fringes pass through the field of view. Calculate the wavelength of light used. **[Ans.** $\lambda = 5920$ Å**]**

8.38 Calculate the visibility of the fringes for a reflection of 80% is multiple beam interferometer.
[Ans. $V = 0.976$**]**

8.39 Two Fabry–Perot interferometers have equal plate separations but the coefficients of intensity reflection are 0.80 and 0.90. Deduce the relative width of the maxima in the two cases. **[Ans.** $\delta_1/\delta_2 = \sqrt{9/2}$**]**

Multiple Choice Questions

8.1 Two coherent sources of light emit waves of amplitudes a and $2a$. They meet at a point P equidistant from the two sources. If the intensity of the first wave is I, the resultant intensity at P will be
(a) zero (b) $5I$
(c) $9I$ (d) $3I$

8.2 To demonstrate the phenomenon of interference, we required two sources which emit
   (a) radiation of the same frequency
   (b) radiation of nearly the same frequency
   (c) radiation of the same frequency and have a definite phase relationship
   (d) radiation of the same wavelength

8.3 Which of the following must be same for two rays of light to be considered coherent?
   (a) wave frequency   (b) wavelength
   (c) wave speed   (d) all of the above

8.4 Two waves having intensities in the ratio of 9 : 1 produce interference. The ratio of the maximum to minimum intensity is equal to
   (a) 4:1   (b) 9:1
   (c) 16:1   (d) 3:1

8.5 If one of the two slits of a Young's double slit experiment is painted so that it transmits half the light intensity of the other, then
   (a) the fringe system will disappear
   (b) the bright fringes will be brighter and the dark fringes will be darker
   (c) the bright fringes will be darker and the dark fringes will be brighter
   (d) both the dark and bright fringes will be darker

8.6 When a thin sheet of mica is introduced in the path of one of the interfering beams, then the fringe width
   (a) increases   (b) decreases
   (c) remains unchanged   (d) none of these

8.7 If Young's experimental set up is displaced from air and immersed in water, the fringe width will
   (a) decrease   (b) increase
   (c) remain unchanged   (d) be zero

8.8 In Young's double slit experiment the distance between two slits is 2 mm and the screen is at a distance 120 cm from the slits. The smallest distance from the central maxima where the brightest fringes due to light of wavelength 6500 Å and 5200 Å would coincide is
   (a) 0.117 cm   (b) 0.156 cm
   (c) 0.234 cm   (d) 0.20 cm

8.9 Two waves of equal amplitude and wavelength but differing in phase are superposed. Amplitude of the resultant wave is maximum, when phase difference is
   (a) $\dfrac{\pi}{2}$ (b) $\dfrac{3\pi}{2}$
   (c) 2π (d) π

8.10 Interference may be seen using two independent
   (a) sodium   (b) fluorescent tubes
   (c) lasers   (d) mercury vapour lamps

8.11 Two coherent sources of intensity ratio 25:4 are used in an interference experiment. The ratio of intensities of maxima and minima in the interference pattern is

(a) 25:16 (b) 49:4
(c) 4:9 (d) 7:3

8.12 In a Fresnel's biprism experiment if the width of the slits is increased, then
(a) the contrast between the bright and dark fringes becomes poorer and poorer
(b) the contrast between the bright and dark fringes becomes richer and richer
(c) the contrast between the bright and dark fringes remains unchanged
(d) the screen appears completely dark

8.13 When monochromatic light is replaced by white light in Fresnel's biprism arrangement, the central fringe is
(a) dark (b) white
(c) coloured (d) none of these

8.14 Two light sources are coherent when
(a) their amplitudes are same
(b) their frequencies are same
(c) their wavelengths are same
(d) their frequencies are same and their phase difference is constant

8.15 A biprism of refracting angle 1° is made up of a material of refractive index 1.55. The biprism is placed at a distance of 13 cm from the slit (source). The separation between the coherent source formed by it is
(a) 0.35 cm (b) 0.25 cm
(c) 0.5 cm (d) 0.45 cm

8.16 A Fresnel's biprism arrangement is set with sodium light ($\lambda = 5893$ Å) and in the field of eyepiece, 62 fringes are seen. Now the source is replaced with a mercury source and a green filter of ($\lambda = 5461$ Å) is placed in front of it. The number of fringes now seen will be
(a) 54 (b) 71
(c) 67 (d) 81

8.17 In a biprism experiment when a glass plate of thickness t and refractive index n is placed in the path of one of the interfering ray (wave) the entire system shifts through a distance given by
(a) $\dfrac{2d}{D}(n-1)t$ (b) $\dfrac{2d}{D}(n+1)t$
(c) $\dfrac{D}{2d}(n-1)t$ (d) $\dfrac{D}{2d}(n+1)t$

8.18 When light of wavelength λ falls on a thin parallel film of constant thickness t and refractive index n, the essential condition for the production of constructive interference fringes by the rays reflected from the upper and lower edge of the film is
(a) $2nt \cos r = (2m - 1) \lambda/2$ (b) $2nt \cos r = 2m\ \lambda/2$
(c) $2nt \cos(r + \theta) = (2m - 1) \lambda/2$ (d) $2nt \cos(r + \theta) = 2m - \lambda/2$

8.19 Oil floating on water surface is seen coloured because of interference of light. The possible thickness of the oil film is
(a) 100 Å (b) 1000 Å
(c) 1 mm (d) 1 cm

8.20 The central fringe of the interference pattern produced by the light of wavelength 6000Å is found to shift to the position of 4th bright fringe after a glass plate of refractive index 1.5 is introduced. The thickness of the glass plate would be
(a) 4.80×10^{-6} m
(b) 8.23×10^{-6} m
(c) 14.98×10^{-6} m
(d) 3.78×10^{-6} m

8.21 The phenomenon which produces colours in a soap bubble is due to
(a) diffraction
(b) dispersion
(c) interference
(d) polarization

8.22 A thin air film between a plane glass plate and a convex lens is irradiated with a parallel beam of monochromatic light and is observed under a microscope, one finds
(a) uniform brightness
(b) complete darkness
(c) field crossed over by concentric bright and dark ring
(d) field crossed over by many coloured fringes

8.23 A wedge-shaped film is observed in reflected sunlight first through a red glass and then through a blue glass. The number of fringes in the later case is
(a) less
(b) more
(c) equal in both case
(d) none of these

8.24 Two ordinary glass plates are placed one over another and illuminated by monochromatic light, the interference fringes will
(a) be straight lines
(b) be circular
(c) be irregular shaped
(d) not be formed

8.25 In the Newton's rings arrangement, if the distance between the lens and the plate is increased, the order of the ring at a given point
(a) decreases
(b) increases
(c) remains unchanged
(d) none of the above

8.26 In the Newton's rings arrangements, the diameters of bright rings are
(a) directly proportional to the square root of natural number
(b) inversely proportional to the square root of odd natural numbers
(c) directly proportional to the square root of odd natural numbers
(d) directly proportional to the square root of even natural numbers

8.27 Fringe of width 1.474 mm is observed at distance of 50 cm inside the geometrical shadow of a wire, when illuminated by light of wavelength 5896 Å. The diameter of wire is
(a) 4×10^{-4} m
(b) 3×10^{-4} m
(c) 2×10^{-4} m
(d) 1×10^{-4} m

8.28 In Newton's rings experiment the diameter of the bright rings are proportional to the square root of
(a) natural number
(b) odd natural number
(c) even natural number
(d) half integral multiples of natural number

8.29 Newton's rings are fringes of
(a) equal inclination
(b) equal thickness
(c) both equal inclination and equal thickness
(d) equal radii

8.30 In Michelson's interferometer the shape of fringes depends upon
(a) distance of source from compensating plate
(b) distance of semisilvered plate from compensating plate
(c) the inclination between the two mirrors
(d) distance of semisilvered plate from the telescope

8.31 In a Michelson interferometer, when the screw is moved through d mm, 100 and 80 circular fringes are observed for lights of wavelength λ_1 and λ_2 respectively. The ratio is λ_1/λ_2 equals
(a) 0.8
(b) 1.25
(c) 1.8
(d) 4.5

8.32 Michelson's interferometer is based on the principle of
(a) division of amplitude
(b) division of wavefront
(c) addition of amplitude
(d) none of the above

8.33 One leg of a Michelson's interferometer is lengthened so that the mirror is shifted by 0.020 mm. If the light used has $\lambda = 5000$ Å, then number of dark fringes sweeping through the field of view is
(a) 80
(b) 100
(c) 150
(d) 200

8.34 When a thin glass plate ($n = 1.5$) is introduced in one of the arms of Michelson interferometer using light of wavelength 5890 Å, there is a shift of 10 fringes. The thickness of the plate will be
(a) 5.89×10^{-6} m
(b) 5.48×10^{-4} m
(c) 5.89×10^{-5} m
(d) None of these

8.35 In the Fabry–Perot's interferometer, the sharpness of maxima depends upon
(a) the distance between the glass plates
(b) the inclination between the two plates
(c) the reflectivity of the glass plate
(d) the resolving power of a telescope

8.36 The most suitable method for determination of the refractive index of gases and liquids is
(a) Michelson's interferometer
(b) Newton's rings
(c) Fabry–Perot's interferometer
(d) Jamin's refractometer

8.37 Two interfering light waves have their amplitudes in the ratio 3:2. The ratio of the intensity of maxima to that of minima will be
(a) 3:2
(b) 5:1
(c) 9:4
(d) 25:1

Answers

8.1 (c)	**8.2** (c)	**8.3** (c)	**8.4** (a)	**8.5** (c)	**8.6** (c)	**8.7** (a)	**8.8** (c)		
8.9 (c)	**8.10** (c)	**8.11** (b)	**8.12** (a)	**8.13** (b)	**8.14** (d)	**8.15** (b)	**8.16** (c)		
8.17 (c)	**8.18** (a)	**8.19** (b)	**8.20** (a)	**8.21** (c)	**8.22** (c)	**8.23** (b)	**8.24** (a)		
8.25 (b)	**8.26** (c)	**8.27** (c)	**8.28** (b)	**8.29** (b)	**8.30** (c)	**8.31** (a)	**8.32** (a)		
8.33 (a)	**8.34** (a)	**8.35** (d)	**8.36** (d)	**8.37** (d)					

CHAPTER 9

Diffraction of Light Waves

"What I give form to in daylight is only one percent of what I have seen in darkness."
—M.C. Escher

IN THIS CHAPTER

- Diffraction and Huygen's Principle
- Distinction Between Interference and Diffraction
- Fresnel's Explanation of Rectilinear Propagation of Light
- Zone Plate
- Comparison Between the Action of a Zone Plate and that of a Convex Lens
- Fresnel's Diffraction Due to a Straight Edge
- Fresnel's Diffraction at a Circular Disc
- Fresnel's Diffraction at a Circular Aperture
- Fraunhofer's Diffraction at a Single Slit
- Fraunhofer's Diffraction at Double Slit
- Plane Transmission Diffraction Grating: Fraunhofer's Diffraction at N Parallel Slits
- Resolving Power of an Optical Instrument
- Rayleigh Criterion for the Limit of Resolution
- Resolving Power of a Plane Diffraction Grating
- Resolving Power of a Prism
- Theory of Concave Grating
- Mountings of Concave Grating
- Echelon Grating
- Difference Between Prism Spectrum and Grating Spectrum
- Difference Between Dispersive Power and Resolving Power of a Grating

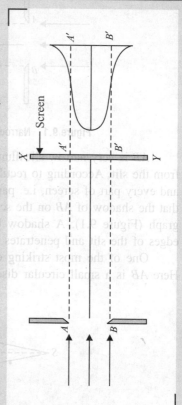

9.1 INTRODUCTION

The phenomenon of diffraction was discovered by Grimaldi in 1665. Diffraction is the characteristic of wave motion. All types of waves undergo diffraction. In simple terms, *diffraction is the process of bending of waves around an obstacle and their penetration into the region of the geometrical shadow*. Diffraction is very easy to observe with waves on the surface of water in a ripple tank or sound waves. It is due to diffraction that we are able to hear the sound of a man standing behind a wall. However, diffraction is difficult to observe in case of light. This is because wavelength of light is very small as compared to the wavelength of water waves or sound waves. In this chapter, various aspects of diffraction phenomenon would be discussed in systematic way.

Grimaldi observed that the shadow of a very *thin wire* placed in the path of a beam of light *much broader* than expected, according to laws of ray optics. The significance of this was realized by Huygen. According to Huygen, the broader shadow is due to bending of light around the edges of the wire.

Another experiment that demonstrates the diffraction of light is shown in Figure 9.1.

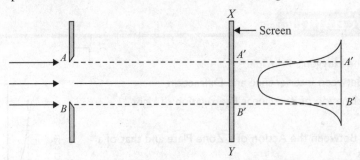

Figure 9.1 Narrow slit experiment for demonstration of diffraction of light.

AB is very narrow slit illuminated with a parallel beam. XY is a screen at a large distance from the slit. According to rectilinear propagation of light, $A'B'$ on the screen should be bright and every part of screen, i.e. part $A'X$ or $B'Y$ should be completely dark. However, it is found that the shadow of AB on the screen is much broader than $A'B'$. This is shown in the intensity graph (Figure 9.1). A shadow broader than $A'B'$ is only possible if light bends around the edges of the slit and penetrates into the region of geometrical shadow *i.e.*, shows diffraction.

One of the most striking demonstrations of diffraction of light is shown in Figure 9.2. Here AB is a small circular disc placed in the path of light from a source S. On the screen,

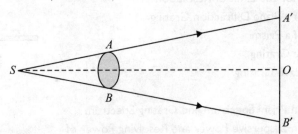

Figure 9.2 Small circular disc experiment for diffraction of light.

a dark shadow of diameter $A'B'$ is expected. The point O, symmetrically situated with respect to the disc, is definitely expected to dark on the basis of rectilinear propagation of light. It is, however, found that point O is maximum bright. In addition, concentric circular rings that are alternatively bright and dark, as shown in Figure 9.3, are observed on the screen. The fact that point O is maximum bright clearly shows that light bends around the edges of the disc, penetrating into the region of the geometrical shadow *i.e.*, light exhibits diffraction.

Figure 9.3 Bright and dark circular fringes.

9.2 DIFFRACTION AND HUYGEN'S PRINCIPLE

According to Huygen's wave theory of light, each point of wavefront acts as an independent source of secondary waves. When the secondary waves starting from different points of the same wavefront get superimposed, they suffer interference.

Diffraction is due to such an interference between the secondary waves starting from the same wavefront.

Diffraction phenomenon is divided into two groups:

1. Fresnel's diffraction, and
2. Fraunhofer's diffraction

In the Fresnel's class of diffraction, the source and screen are at finite distance from the obstacle as visualized in Figure 9.4. Here *wavefront incident on obstacle is either spherical or cylindrical* depending on whether S is a point of the obstacle at the same time. The secondary wavelets from different points of the obstacle are not in same phase.

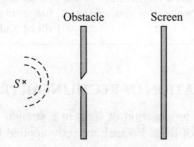

Figure 9.4 Fresnel's diffraction.

In the *Fraunhofer's class of diffraction*, the source and the screen are effectively at infinity (Figure 9.5). This is achieved using lens.

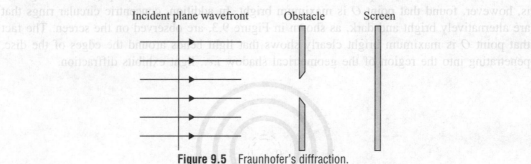

Figure 9.5 Fraunhofer's diffraction.

The incident wavefront on the obstacle is a *plane wavefront*. Since the incident wavefront is plane, all points in the obstacle become a source of secondary wavelets simultaneously. In other words, the secondary wavelets starting from the obstacle are in the phase.

9.3 DISTINCTION BETWEEN INTERFERENCE AND DIFFRACTION

S.No.	Interference	Diffraction
1.	Interference is the result of superposition of secondary waves starting from two different wavefronts originating from two coherent sources.	Diffraction is the result of superposition of secondary waves starting from different parts of the same wavefront.
2.	All bright and dark fringes are of equal width.	The width of central bright fringe is much larger than the width of any secondary maximum.
3.	All bright fringes are of same intensity.	Intensity of bright fringes decreases as we move away from central bright fringe on either side.
4.	Regions of dark fringes are perfectly dark. So there is a good contrast between bright and dark fringes.	Regions of dark fringes are not perfectly dark. So there is a poor contrast between bright and dark fringes.
5.	At an angle of l/d, we get a bright fringe in the interference pattern of two narrow slits separated by a distance d.	At an angle of l/d, we get the first dark fringes in the diffraction pattern of a single slit of width d.

9.4 FRESNEL'S EXPLANATION OF RECTILINEAR PROPAGATION OF LIGHT

The explanation of the fact of propagation of light in a straight line was very tedious for the supporters of the wave theory of light. Fresnel correctly applied Huygen's principle to explain diffraction of light waves. Now, it is necessary to know Fresnel's method of finding the effect of a plane wavefront at any point.

9.4.1 Fresnel's Half-period Zones

Let us assume that *UVWX* is a wavefront of the wavelength λ travelling from the left to the right (in the direction of arrow) as shown in Figure 9.6. The result intensity is to be computed at point *Z*.

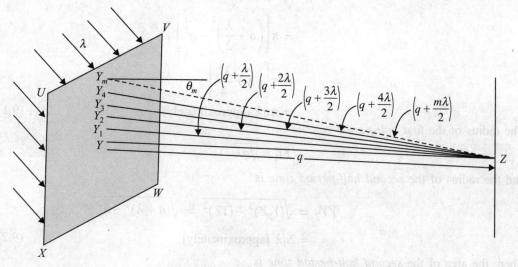

Figure 9.6 Fresnel's half-period zones.

To make simple computation of resultant intensity, Fresnel divided the incident wavefront into a number of zones, which are said to be *Fresnel's half-period zones*.

Suppose ZY $(=q)$ is normal to wavefront with Z as centre and λ is the wavelength of light.

Now, let us make concentric sphere of radii $\left(q + \dfrac{\lambda}{2}\right), \left(q + \dfrac{2\lambda}{2}\right), \left(q + \dfrac{3\lambda}{2}\right), \cdots \left(q + \dfrac{n\lambda}{2}\right)$.

The plane *UVWX* cuts these spheres in concentric circles of radii YY_1, YY_2, YY_3, YY_3, YY_4, ... The area covered by the first innermost circle is *first half-period zone*. Similarly the area covered between the first and second circles is *second half-period zone* and so on. The area surrounded by $(n-1)$th and nth circles is known as nth *half-period zone*.

Every zone differs from its nearest zone by a phase difference of π because there is a path difference of $\lambda/2$ for any consecutive ray as shown in Figure 9.6. Thus, the successive zones differ by a phase difference of π or by a half period $(T/2)$. Hence, these are known as *half-period zones*.

9.4.2 Governing Factors of Amplitude

The amplitude of the disturbances at *Z* due to zone depends upon the following factors:

1. Area of half-period zone (HPZ)
2. Average distance of half-period zone from *Z*, and
3. Obliquity factor $(1 + \cos\theta_m)$.

1. Area of the half-period zone (HPZ)

The amplitude is directly proportional to the area of the zone.

In Figure 9.6, the area of the *first half-period zone* is given by

$$\pi(YY_1)^2 = \pi[(Y_1Z)^2 - ZY)^2]$$

$$= \pi\left[\left(q + \frac{\lambda}{2}\right)^2 - q^2\right]$$

$$= \pi\left[q\lambda + \frac{\lambda^4}{4}\right]$$

$$= \pi q\lambda \text{ (approximately)} \qquad (9.1)$$

The radius of the *first half-period zone* is

$$YY_1 = \sqrt{q\lambda}$$

and the radius of the *second half-period zone* is

$$YY_2 = \sqrt{(Y_2Z)^2 - (YZ)^2} = \sqrt{(q+\lambda)^2 - q^2}$$

$$= 2q\lambda \text{ [approximately]} \qquad (9.2)$$

Then, the area of the *second half-period zone* is

$$= \pi[(YY_2)^2 - (YY_1)^2]$$

$$= \pi[2q\lambda - q\lambda] = \pi q\lambda \qquad (9.3)$$

Similarly, the area of the *n*th half-period zone is

$$= \pi[(YY_m)^2 - (YY_{m-1})^2]$$

$$= \pi q\lambda \qquad (9.4)$$

Hence, we can say that the area of each half-period zone is $\pi q\lambda$. We also conclude that the area of the half-period zone is directly proportional to:

(i) The wavelength of incident light.
(ii) The distance of point Z from the wavefront.

2. Average distance of half-period zone from Z

The amplitude of distance is inversely proportional to the average distance of the zone from Z.

The average distance of the *n*th zone from point Z is

$$= \frac{\left(q + \frac{m\lambda}{2}\right) + \left[q + (m-1)\frac{\lambda}{2}\right]}{2}$$

$$= q + (2m-1)\frac{\lambda}{4} \qquad (9.5)$$

3. Obliquity factor $(1 + \cos\theta_m)$

The amplitude is directly proportional to the obliquity factor $(1 + \cos\theta_m)$, where θ_m is the angle between the normal to the zone and the line joining to the zone of Z.

The amplitude of the successive zone depends upon obliquity factor $(1 + \cos\theta_m)$, as θ increases from zero and the amplitude of the disturbance decreases slowly first and more quickly for higher value of θ_m.

Hence, the successive amplitudes $A_1, A_2, A_3, ..., A_m$ decrease slowly first and rapidly for larger value of m.

At last, the expression for the amplitude due to the mth *half-period zone* can be written as

$$A_m = \frac{\pi\left[q\lambda + \frac{\lambda^2}{4}(m-1)\right]}{q + (2m-1)\frac{\lambda}{4}}(1 + \cos\theta_m) \qquad (9.6)$$

or
$$A_m = km\lambda(1 + \cos\theta_m) \qquad (9.7)$$

where k is the constant of proportionality.

9.4.3 Resultant Amplitude Due to Wavefront UVWX

Suppose that $A_1, A_2, A_3, ...$ are the amplitudes of the secondary wavelets originating from the first, second, third, ... half-period zones. Figure 9.6 depicts that on increasing of the obliquity, the amplitude decreases.

In simple form Figure 9.6 can be shown in Figure 9.7 for our implicity.

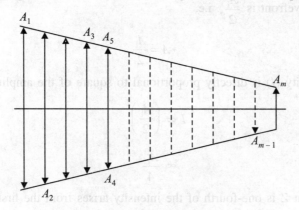

Figure 9.7 Resultant amplitude of the wave.

Let us assume that A_1 is slightly greater than A_2, A_2 is slightly greater than A_3, A_3 is slightly greater than A_4, and so on. We again assume that the amplitude of the disturbance at Z is almost correctly the mean of the amplitude of its preceding and succeeding *half-period zones*. Hence

$$A_2 = \frac{A_1 + A_3}{2}$$

$$A_3 = \frac{A_2 + A_4}{2} \text{ and so on.}$$

It is clear from the figure that the successive amplitudes have reverse directions as it is known that there is π phase difference between two consecutive zones.

The resultant amplitude due to superposition is given by

$$A = A_1 - A_2 + A_3 - A_4 + \cdots + A_m \quad [\text{if } m \text{ is odd}]$$

and

$$A = A_1 - A_2 + A_3 - A_4 + \cdots - A_m \quad [\text{if } m \text{ is even}]$$

$\therefore$

$$A = \frac{A_1}{2} + \left[\frac{A_1}{2} - A_2 + \frac{A_3}{2}\right] + \left[\frac{A_3}{2} - A_4 + \frac{A_5}{2}\right] + \cdots + \frac{A_m}{2} \quad [\text{if } m \text{ is odd}]$$

and

$$A = \frac{A_1}{2} + \left[\frac{A_1}{2} - A_2 + \frac{A_3}{2}\right] + \left[\frac{A_3}{2} - A_4 + \frac{A_5}{2}\right] + \cdots - A_m \quad [\text{if } m \text{ is even}]$$

But

$$A_2 = \frac{A_1}{2} + \frac{A_m}{2}, \quad A_4 = \frac{A_3}{2} + \frac{A_5}{2}, \cdots \text{ so on}$$

$\therefore$

$$A = \frac{A_1}{2} + \frac{A_m}{2} \quad [\text{if } m \text{ is odd}]$$

and

$$A = \frac{A_1}{2} - A_m \quad [\text{if } m \text{ is even}]$$

As $m \to \infty$, A_m tends to zero, as the amplitudes gradually decreasing. So the amplitude at Z due to the whole wavefront is $\frac{A_1}{2}$, i.e.

$$A = \frac{A_1}{2}$$

We know that intensity (I) is directly proportional to square of the amplitude then

$$I \propto \left(\frac{A_1}{2}\right)^2$$

$$\propto \frac{A_1^2}{4}$$

Hence the intensity at Z is one-fourth of the intensity arises from the firstzone, i.e.

$$I = \frac{I_1}{4} \tag{9.8}$$

Here, the constant of proportionality is one.

EXAMPLE 9.1 A screen is placed at a distance of 100 cm from a circular hole illuminated by a parallel beam of light of wavelength 6400Å. Compute the radius of the fourth half-period zone.

Solution If q be the distance of a point from the pole of the plane wavefront, then the radii of the spheres whose sections cut by wavefront from the half-period zones are

$\left(q+\dfrac{\lambda}{2}\right), \left(q+\dfrac{2\lambda}{2}\right), \cdots$. Hence the radius of the fourth half-period zone is given by

$$r_4 = \sqrt{\left(q+\dfrac{4\lambda}{2}\right)^2 - q^2} = \sqrt{4\lambda^2 + 4p\lambda} = \sqrt{4p\lambda}$$

Here $\quad q = 100$ cm, $\lambda = 6400$ Å $= 6400 \times 10^{-8}$ cm

∴ $\quad r_4 = \sqrt{4 \times 100 \times 6400 \times 10^{-8}} = 0.16$ cm

9.5 ZONE PLATE

The proof of Fresnel's method of dividing a plane wavefront into half-period zones can be explained with an optical device, known as a *zone plate*.

A zone plate is a kind of highly focused convex lens with multiple radii. Hence there is multiple focal lengths in a zone plate.

For the construction of a zone plate, concentric circles are drawn on white paper, such that the radii are proportional to the square root of any natural number, i.e.

$$r_m \propto \sqrt{m} \tag{9.9}$$

Then

$$r_1 \propto \sqrt{1} = 1, \quad r_2 \propto \sqrt{2} = 1.414, \quad r_3 \propto \sqrt{3} = 1.732\ldots$$

The alternate zones are pointed black and a reduced photograph is taken on glass plate, which is known as a zone plate. If the odd zones are transparent, then it is called positive zone plate and when even zones are transparent it is called negative zone plate as shown in Figure 9.8.

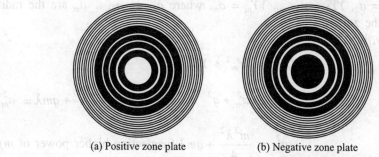

(a) Positive zone plate   (b) Negative zone plate

Figure 9.8   Zone plates.

9.5.1 Theory of the Zone Plate

Light is coming from source at infinity

We conceive that the different zones on the zone plate held vertically occupy the same position with half-period zone (Figure 9.9).

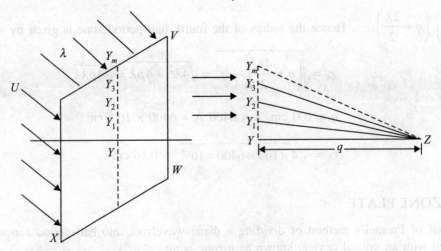

Figure 9.9 An illustration for the theory of a zone plate when source is at infinity.

So in this case, if $YZ = q$. Then

$$Y_1Z = \left(q + \frac{\lambda}{2}\right)$$

$$Y_2Z = \left(q + \frac{2\lambda}{2}\right)$$

$$\vdots$$

$$Y_mZ = \left(q + \frac{m\lambda}{2}\right)$$

Let $YY_1 = a_1$, $YY_2 = a_2$, $YY_3 = a_3$, ..., $YY_m = a_m$, where $a_1, a_2, a_3, ... a_m$ are the radii of the different zones on the zone plate.

Hypothetically, in ΔYY_mZ

$$Y_mZ^2 = YY_m^2 + YZ^2$$

$\Rightarrow$ $\left(q + \dfrac{m\lambda}{2}\right)^2 = a_m^2 + q^2$ $\Rightarrow$ $q^2 + \dfrac{m^2\lambda^2}{4} + qm\lambda = a_m^2 + q^2$

or $\qquad a_m^2 = \dfrac{m^2\lambda^2}{4} + qm\lambda$ (neglecting higher power of m)

then $\qquad a_n^2 = qm\lambda \qquad$ or $\qquad a_m = \sqrt{qm\lambda}$

So, the area of the mth zone = $\pi a_m^2 - \pi a_{m-1}^2$

$$= \pi(a_m^2 - a_{m-1}^2) = \pi\{qm\lambda + [q(m-1)\lambda]\}$$

$$= \pi(qm\lambda - qm\lambda + q\lambda) = \pi q\lambda.$$

The area of the nth zone $(A) = \pi q \lambda$
So, the area of the different zones on the zone plate is same.
Focal length of the zone plate (q)

$$a_m^2 = qm\lambda$$

$$q = \frac{a_m^2}{m\lambda} \qquad (9.11)$$

Source and screen are at finite distance

We conceive that a portion of a given zone plate held vertically occupy same position with half-period zone in serial order as shown in Figure 9.10.

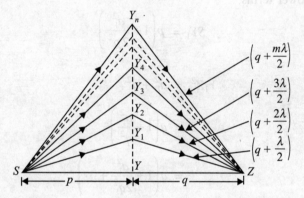

Figure 9.10 Illustration for the theory of a zone plate (when source and screen are at finite distance).

Again, we consider that projection from the source of light S on the zone plate axis and equal distance from $Y_1, Y_2, Y_3, ..., Y_m$ i.e.

$$SY = SY_1 = SY_2 = \cdots = SY_m = p$$

and

$$Y_m Z = q + \frac{m\lambda}{2}$$

For the above values, we can write

$$SY_1 + Y_1 Z = p + q + \frac{\lambda}{2}$$

$$SY_2 + Y_2 Z = p + q + \frac{2\lambda}{2}$$

$$\vdots \qquad \vdots \qquad \vdots$$

$$SY_m + Y_m Z = p + q + \frac{m\lambda}{2} \qquad (9.12)$$

Let us find SY_1 and Y_mZ from the diffeent methods

$$SY_1^2 = p^2 + a_1^2 = p^2\left(1 + \frac{a_1^2}{p^2}\right)$$

or
$$SY_1 = p\left(1 + \frac{a_1^2}{p^2}\right)^{1/2}$$

$$= p\left(1 + \frac{1}{2}\frac{a_1^2}{p^2} + \cdots\right)$$

Neglecting higher power terms,

$\therefore$
$$SY_1 = p\left(1 + \frac{a_1^2}{2p^2}\right) \tag{9.13}$$

Similarly in ΔY_1YZ

$$Y_1Z^2 = a_1^2 + q^2$$

or
$$Y_1Z = q\left(1 + \frac{a_1^2}{q^2}\right)^{1/2}$$

$$= q\left(1 + \frac{a_1^2}{2q^2} + \cdots\right)$$

Again neglecting higher power terms, we get

$$Y_1Z = q\left(1 + \frac{a_1^2}{2q^2}\right) \tag{9.14}$$

Proceeding in the same manner, we get

$$SY_m = p\left(1 + \frac{1}{2}\frac{a_m^2}{p^2}\right) \tag{9.15}$$

and
$$Y_mZ = q\left(1 + \frac{1}{2}\frac{a_m^2}{q^2}\right) \tag{9.16}$$

From Eqs. (9.15) and (9.16), we get

$$SY_m + Y_mZ = p + q + \frac{m\lambda}{2}$$

or
$$p\left(1 + \frac{1}{2}\frac{a_m^2}{p^2}\right) + q\left(1 + \frac{1}{2}\frac{a_m^2}{q^2}\right) = p + q + \frac{m\lambda}{2}$$

or
$$\frac{a_m^2}{2}\left(\frac{1}{p^2} + \frac{1}{q^2}\right) = \frac{m\lambda}{2}$$

or
$$\frac{1}{p^2} + \frac{1}{q^2} = \frac{m\lambda}{a_m^2} \quad (9.17)$$

We know that for convex lens
$$\frac{1}{v} - \frac{1}{u} = \frac{1}{f} \quad (9.18)$$

Compare Eqs. (9.17) and (9.18), i.e. zone plate with a convex lens

$$\frac{1}{f} = \frac{m\lambda}{2}$$

$$f = \frac{2}{m\lambda} \quad (9.19)$$

So, the zone plate is a kind of lens with multiple focal lengths.

EXAMPLE 9.2 The diameter of the first ring of a zone plate is 1 mm. If plane wave (λ = 500 nm) fall on the plate, where should the screen be placed so that light is focused on a brightest spot?

Solution Given r_1 = 0.5 mm = 0.05 cm and l = 500 nm = 5 × 10^{-5} cm, m = 1. The plane wave falling on the plate will be focused to a brightest spot at the first focus of the plate.

Now, the first focal length of the zone plate is given by

$$f = \frac{r_m^2}{n\lambda} = \frac{r_1^2}{\lambda}$$

$$= \frac{(0.05 \times 0.05)}{5.0 \times 10^{-5}} = 50 \text{ cm}$$

Hence, the screen should be placed at a distance of 50 cm from the plate.

9.6 COMPARISON BETWEEN THE ACTION OF A ZONE PLATE AND THAT OF A CONVEX LENS

S.No.	Zone plate	Convex lens
1.	The focal length of a zone plate depends upon wavelength.	The focal length of a convex lens depends upon wavelength also.
2.	The zone plate has multiple focal lengths and makes a series of point images of decreasing amplitude for a given point source.	The convex lens has a single focal length and makes a single image, if it is without any aberration.
3.	In case of a zone plate, all the rays reaching on image point have a path difference between the rays from the successive zone is λ.	In case of a convex lens, all the rays reaching on image point has same path.
4.	In case of a zone plate, the focal length of red colour is less than that of violet colour, i.e. $f_R < f_V$	In case of a convex lens, the focal length of violet colour is less than that of red colour, i.e. $f_V < f_R$

9.7 FRESNEL'S DIFFRACTION DUE TO A STRAIGHT EDGE

Suppose there is a TT' cylindrical wavefront diverging from a narrow slit S. This narrow slit is perpendicular to the plane of paper and is illuminated with wavelength λ. Suppose Z is a point in illuminated region. Here we have taken MN, a straight edge, which is parallel to the slit. Screen EF is parallel to straight edge as shown in Figure 9.11.

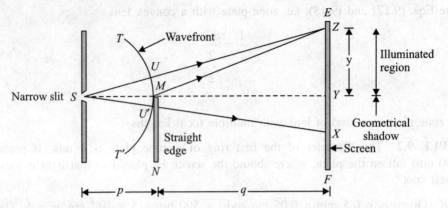

Figure 9.11 Fresnel's diffraction at straight edge (i.e. at razor blade).

From this, we can observe that:

1. If no diffraction of light wave is at the straight edge, a uniform illumination above Y has been achieved and there is complete darkness below Y.
2. But in actual practice. There have been observed a few unequally spaced bright and dark fringes are parallel to the length of the slit in the illuminated region (near and above Y).
3. The intensity falls rapidly and becomes zero at a small yet finite distance from Y in the geometrical shadow. This diffraction pattern is of the straight edge.

Figure 9.12 depicts the intensity distribution curve for the straight edge diffraction. In the illuminated region, we observed unequally spaced bright and dark fringes, while the intensity falls rapidly in gometrical shadow.

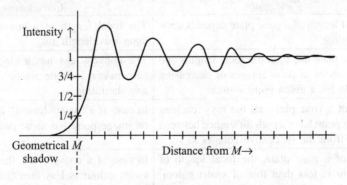

Figure 9.12 Intensity distribution curve for straight edge diffraction.

9.7.1 Mathematical Treatment

Recalling Figure 9.11, suppose p is the distance between the narrow slit and straight edge and q is the distance between the straight edge and the screen. Now, join Z to M as ZM. Here Z is a maximum or minimum, depends on MU contains an odd and even numbers of half-period strips. Then path difference is $(MZ - UZ)$.

To get bright fringes (ands)

$$MZ - UZ = (2m+1)\frac{\lambda}{2} \quad \text{[for maxima]} \tag{9.20}$$

$$MZ - UZ = m\lambda \quad \text{[for minima]} \tag{9.21}$$

where $m = 0, 1, 2, 3, 4, \ldots$

In Figure 9.11, $SM = p$, $MZ = q$ and $YZ = y$
Then

$$\begin{aligned}
AZ &= \sqrt{(YZ)^2 + (MZ)^2} \\
&= \sqrt{y^2 + q^2} = (y^2 + q^2)^{1/2} \\
&= q\left(1 + \frac{y^2}{q^2}\right)^{1/2} \\
&= q\left(1 + \frac{1}{2}\frac{y^2}{q^2} + \cdots\right) \quad \text{[on application of binomial expansion]} \\
&= q + \frac{y^2}{2q} \quad \text{(on neglection of higher order terms)}
\end{aligned} \tag{9.22}$$

and

$$UZ = SZ - SU = SZ - SM$$
$$SM = p$$
$$\begin{aligned}
SZ &= \sqrt{(YZ)^2 + (YS)^2} = \sqrt{y^2 + (p+q)^2} \\
&= (p+q)\left[1 + \frac{y^2}{(p+q)^2}\right]^{1/2} \\
&= (p+q)\left[1 + \frac{1}{2}\frac{y^2}{(p+q)^2} + \cdots\right] \quad \text{[on application of binomial expansion]} \\
&= (p+q) + \frac{y^2}{2(p+q)} \quad \text{(on neglection of higher order terms)}
\end{aligned} \tag{9.23}$$

Then

$$\begin{aligned}
UZ &= (p+q) + \frac{y^2}{2(p+q)} - p \\
&= q + \frac{y^2}{2(p+q)}
\end{aligned} \tag{9.24}$$

Then path difference

$$\Delta = MZ - UZ$$

$$= q + \frac{y^2}{2q} - q - \frac{y^2}{2(p+q)}$$

$$= \frac{py^2}{2q(p+q)} \tag{9.25}$$

Now, for bright band or to get maxima

$$\Delta = (2m+1)\frac{\lambda}{2}$$

$$\frac{py^2}{2q(p+q)} = (2m+1)\frac{\lambda}{2}$$

or

$$y = \sqrt{\frac{q(p+q)}{p}(2m+1)\lambda}$$

or

$$y = C\sqrt{(2m+1)} \tag{9.26}$$

where

$$C = \sqrt{\frac{q(p+q)}{p}\lambda}$$

where C is a constant, being p, q and λ constants. This expression provides the position of the nth maxima in the diffraction pattern.

The distance of the successive maxima for Y are [for the values of $m = 0, 1, 2, 3, 4, \ldots$ in Eq. (9.26)]

$$\left. \begin{array}{l} y_1 = C \\ y_2 = \sqrt{3}C \\ y_3 = \sqrt{5}C \\ y_4 = \sqrt{7}C \\ \vdots \quad \vdots \quad \vdots \end{array} \right\} \tag{9.27}$$

The distance between successive maxima be

$$\left. \begin{array}{l} y_2 - y_1 = C(\sqrt{3}-1) = 0.73\ C \\ y_3 - y_2 = C(\sqrt{5}-\sqrt{3}) = 0.50\ C \\ y_4 - y_3 = C(\sqrt{7}-\sqrt{5}) = 0.43\ C \\ \vdots \quad \vdots \quad \vdots \end{array} \right\} \tag{9.28}$$

It is clear from Eq. (9.28) that the separation of successive maxima are in decreasing order. Thus, as we go above Y on the screen, the fringes come closer and closer.

9.7.2 Wavelength of Monochromatic Light

Recalling Eq. (9.26) and rearranging it, we get for nth maxima

$$y_m^2 = \frac{q(p+q)\lambda}{p}(2m+1) \tag{9.29}$$

and for $(m + m_1)$th maxima

$$y_{m+m_1}^2 = \frac{q(p+q)\lambda}{p}[2(m+m_1)+1] \tag{9.30}$$

$$\therefore \quad y_{m+m_1}^2 - y_m^2 = \frac{q(p+q)\lambda}{p} \cdot 2m_1 \tag{9.31}$$

and

$$\lambda = \frac{p(y_{m+m_1}^2 - y_m^2)}{2q(p+q)m_1} \tag{9.32}$$

EXAMPLE 9.3 In an experiment for observing diffraction pattern due to a straight edge, the distance between the slit source ($\lambda = 600$ nm) and the straight edge is 6 m and that between the straight edge and eyepiece is 4 m Calculate the position of first three maxima and their separation.

Solution If p be the distance between the slit and the straight edge and q the distance between the straight edge and eyepiece, the position of the mth maxima (bright band) is given by

$$y = \sqrt{\frac{q(p+q)}{p}(2m_1+1)\lambda}$$

Here $p = 600$ cm, $q = 400$ cm, $\lambda = 600$ nm $= 6.00 \times 10^{-5}$ cm

$$\therefore \quad y = \sqrt{\frac{400 \times (600+400)}{600}(2m_1+1) \times 6.0 \times 10^{-5}}$$

$$= \frac{\sqrt{2m_1+1}}{5} \text{ cm}$$

where $m_1 = 0, 1, 2, 3, \ldots$

The position of the first three maxima (puting $m_1 = 0, 1, 2, \ldots$)

$$y_1 = \frac{1}{5} = 0.20 \text{ cm}, \quad y_2 = \frac{\sqrt{3}}{5} = 0.346 \text{ cm}, \quad y_3 = \frac{\sqrt{3}}{5} = 0.447 \text{ cm}$$

The separation we

$y_2 - y_1 = 0.346 - 0.200 = 0.146$ cm

$y_3 - y_1 = 0.447 - 0.346 = 0.101$ cm

9.8 FRESNEL'S DIFFRACTION AT A CIRCULAR DISC

Let a small circular opaque disc *AB* (Figure 9.13) be placed in the path of light from a point source *S*, and the diffraction pattern obtained on a screen *XY*. The pattern consists of a 'bright' spot at the centre *C* of the geometrical shadow *A'B'*. Surrounding the bright spot, there are a number of faint rings within the shadow *A'B'*. Outside the shadow, brighter and broader rings are obtained.

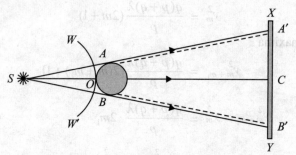

Figure 9.13 Fresnel's diffraction at a circular disc.

Explanation

Let *WW'* be the trace of the spherical wavefront starting from *S* and incident on the disc. *O* is the pole of the wavefront with respect to *C*. To find the intensity at *C*. Let us divide the wavefront into half-period zones. Some of the central zones are obstructed by the disc *AB*, and the intensity at *C* is due to the exposed zones on the wavefront. Let R_1, R_2, R_3, ..., etc. be the amplitudes due to the first, second, third, ... zones respectively.

When the disc is so small that it obstructs only the first zone, then the resultant amplitude at *C* due to remaining exposed zones is

$$R = R_2 - R_3 + R_4 - R_5 \cdots = \frac{R_2}{2} \tag{9.33}$$

i.e., equal to the half of the amplitude due to the second zone. The intensity is proportional to $R_2^2/4$. This is practically the same as the intensity, when the disc is absent, the intensity then being proportional to $R_1^2/4$. Hence *C* is almost as bright as when the disc is absent.

If the area of the disc is increased, it covers first two Fresnel's zones (or if the screen is moved forward to a position, the disc covers first two zones). As screen is gradually moved towards the disc, the area of each half-period zone decreases and the disc gradually covers first two, three, four zones. Then, the resultant amplitude at *C* is given by

$$R = R_3 - R_4 + R_5 - R_6 + \cdots = \frac{R_3}{2} \tag{9.34}$$

and the intensity is proportional to $\dfrac{R_3^2}{4}$.

Thus the intensity at *C* is proportional to $R_2^2/4$, $R_3^2/4$, $R_4^2/4$, ..., etc., according as the disc covers one, two, three zones, etc. Hence *C* is always bright but the intensity of illumination gradually decreases as the size of the disc increases or the screen is gradually moved towards the disc. Because R_2, R_3, R_4, ..., etc. are in decreasing order of magnitude.

The concentric bright and dark rings inside the geometrical shadow are formed due to interference between the exposed parts of the wavefront above A and below B, like the bands in the geometrical shadow of a narrow wire.

The rings outside the shadow are the unequally spaced 'diffraction' rings formed in the same way as the unequally spaced bands in the illuminated region of a straight edge.

9.9 FRESNEL'S DIFFRACTION AT A CIRCULAR APERTURE

Let a narrow circular aperture AB (Figure 9.14) be placed in the path of light from a point source S, and the diffraction pattern be obtained on a screen XY. The centre of the pattern C is bright or dark and is surrounded by alternate dark and bright diffraction rings in the illuminated region $A'B'$. The intensity in the geometrical shadow (above A' and below B') rapidly falls to zero.

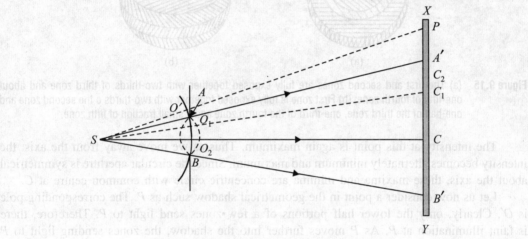

Figure 9.14 Fresnel's diffraction at a circular aperture.

Explanation

Let us first consider all the intensity at the axial point C. Let the wavefront be divided into half-period zones with respect to C. If the size of the hole is such that it allows only the first zone, the amplitude at C will be R_1. When the entire wavefront is unobstructed the resultant amplitude at C is only R_2. Thus the first zone along gives twice the amplitude or four times the intensity due to the entire wavefront. This means that at a certain point on the axis the intensity of light is greater (four times) than that without the aperture.

If the aperture is wider so that it allows two half-period zones, the resultant amplitude at C is $(R_1 - R_2)$, which is almost zero. If three zones are allowed, the resultant amplitude is $R_1 - R_2 + R_3$, which is again large, Hence the illumination at an axial point is maximum or minimum according as the aperture allows an odd or an even number of Fresnel zones.

If the screen gradually moves towards the aperture, the width of the aperture being fixed, the number of zones contained in the aperture gradually increases and becomes alternately odd and even. Hence the point C will be alternatively bright and dark.

Let us consider the intensity at a non-axial point such as C_1. Suppose the aperture allows three zones corresponding to the point C which is, therefore, bright.

As we move from C to C_1, the pole of the wavefront moves from O to O_1. Suppose in this position the first and second zones are fully exposed together with two-thirds of the third and about one-fifth of the fourth [Figure 9.15(a)]. The first two zones approximately cancel, leaving some light due to the effect of third and fourth zones. The intensity at C_1, is thus, less than that at C.

For a still higher point C_2, the pole is O_2. Now first zone is fully exposed together with about two-thirds of the second zone, one-half of the third zone, one-third of the fourth zone, and a small fraction of the fifth zone [Figure 9.15(b)].

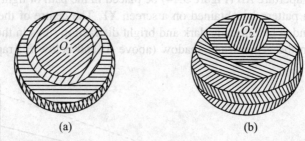

Figure 9.15 (a) The first and second zones are fully exposed together with two-thirds of third zone and about one-fifth of fourth zone; (b) First zone is fully exposed together with two-thirds o the second zone and one-half of the third zone, one-third of the fourth zone and a small fraction of fifth zone.

The intensity at this point is again maximum. Thus as we move away from the axis, the intensity becomes alternately minimum and maximum. Since the circular aperture is symmetrical about the axis, these maxima and minima are concentric circle with common centre at C.

Let us now consider a point in the geometrical shadow such as P. The corresponding pole is O'. Clearly, only the lower half portions of a few zones send light to P. Therefore, there is faint illumination at P. As P moves further into the shadow, the zones sending light to P of higher and higher order. Hence the intensity decreases rapidly in the geometrical shadow.

9.10 FRAUNHOFER'S DIFFRACTION AT A SINGLE SLIT

A slit is a rectangular aperture, whose length is large as compared to its breadth. Let a parallel beam of monochromatic light of wavelength λ be incident normally upon a narrow slit of $AB = a$, as shown in Figure 9.16.

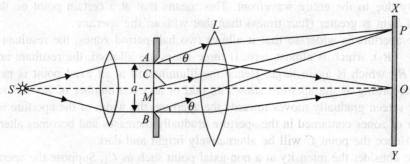

Figure 9.16 Fraunhofer's diffraction at a single slit.

Let the diffracted light be focused by a convex lens *L*. The diffraction pattern obtained on the screen consists of a central bright band, having alternate dark and weak bright bands of decreasing intensity on either side of central bright band.

According to Huygen's theory, a plane wavefront is incident normally on the slit *AB* and each point in *AB* sends out secondary wavelets in all directions. The rays proceeding in the same direction as the incident rays are focused at *O*; while those diffracted through an angle θ are focused at *P*.

Method I

Let the disturbance caused at *P* by the wavelet from unit width of slit at *M* be

$$Y_0 = A \cos \omega t \qquad (9.35)$$

Then the wavelet from width dx at *C* when it reaches at *P* has the amplitude $A dx$ and phase $\left(\omega t + \dfrac{2\pi}{\lambda} x \sin\theta\right)$.

Let this small disturbance be dy, we have

$$dy = A dx \cos\left\{\omega t + \frac{2\pi}{\lambda} x \sin\theta\right\} \qquad (9.36)$$

For the total disturbance at the point of observation at an angle θ, we get

$$y = \int_{-a/2}^{a/2} dy = \int_{-a/2}^{a/2} A \cos\left(\omega t + \frac{2\pi x \sin\theta}{\lambda}\right) dx$$

$$= A \cos\omega t \int_{-a/2}^{a/2} \cos\frac{2\pi x \sin\theta}{\lambda} dx - A \sin\omega t \int_{-a/2}^{a/2} \sin\frac{2\pi x \sin\theta}{\lambda} dx$$

$$= \left\{ A \frac{\sin\dfrac{\pi a \sin\theta}{\lambda}}{\dfrac{\pi \sin\theta}{\lambda}} \right\} \cos\omega t$$

$$\Rightarrow \quad y = Aa \left\{ \frac{\sin\dfrac{\pi a \sin\theta}{\lambda}}{\dfrac{\pi a \sin\theta}{\lambda}} \right\} \cos\omega t$$

$$y = A_0 \left\{ \frac{\sin\dfrac{\pi a \sin\theta}{\lambda}}{\dfrac{\pi a \sin\theta}{\lambda}} \right\} \cos\omega t \qquad (9.37)$$

where $A_0 = Aa$ is the amplitude for $\theta = 0$. Put

$$\frac{\pi a \sin \theta}{\lambda} = \alpha$$

Therefore, Eq. (9.37) becomes

$$y = A_0 \frac{\sin \alpha}{\alpha} \cos \omega t \tag{9.38}$$

∴ Resultant amplitude

$$R = A_0 \frac{\sin \alpha}{\alpha} \tag{9.39}$$

The resultant intensity at P is given by $I = R^2$ (as constant of proportionality is unity)

$$I = R^2 = A_0^2 \left(\frac{\sin \alpha}{\alpha}\right)^2$$

$$\Rightarrow \qquad I = I_0 \frac{\sin^2 \alpha}{\alpha^2} \tag{9.40}$$

where $I_0 = A^2$ represents the intensity at $\theta = 0$.

As $\alpha = \dfrac{\pi a \sin \theta}{\lambda}$, it is clear that α depends on the angle of diffraction θ and $\dfrac{\sin^2 \alpha}{\alpha^2}$ gives the intensity at different value of θ.

Method II

We can consider infinite point sources of secondary wavelets on the wavefront between A and B. Let the slit AB be divided into n equal parts, each part being the source of secondary wavelets. The amplitude at the point P due to the waves obtained from each part will be equal (since the width of each part is same and the screen in effectively at infinite distance from the slit, hence the amplitude remains independent of the distance and the angle of inclination); but the phase difference will be different for different parts. (The phase diference increases from 0 to $\dfrac{2\pi}{\lambda} a \sin \theta$ from A to B.) The phase difference between the waves obtained at point P from an two consecutive parts is

$$\psi = \frac{1}{m} \times \frac{2\pi}{\lambda} a \sin \theta \tag{9.41}$$

Let the amplitude at the point P due to the waves obtained from each part be A. Thus we get m waves, each of amplitude A and has phase difference ψ between the two consecutive waves. We can find their resultant amplitude at the point P by the vector method.

For this, we draw vectors MP_1, P_1P_2, P_2P_3, ... such that the magnitude of each vector is A and the angle between the two consecutive vector is ψ. These vectors form the sides of polygon.

The vector MP_1 joins the origin of the first vector M and the terminus of last vector P_m gives the resulant vector (Figure 9.17).

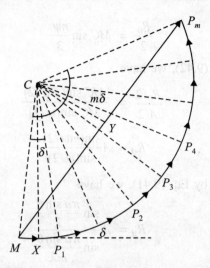

Figure 9.17 Vector polygon to find intensity in case of Fraunhofer's diffraction at a single slit.

Let the magnitude of the resultant vector MP_m be R_θ. If the centre of polygon is C, by simple geometry, we can see that each vector MP_1, P_1P_2, ... subtends an angle ψ at the centre C and the angle subtended by the resultant vector MP_m at centre C is $m\psi$. Let CX be the normal drawn from the point C on MP_1 and CY be normal drawn on MP_m from the point C.

From $\triangle CXM$

$$\frac{MX}{MC} = \sin\frac{\psi}{2}$$

or

$$MX = MC \sin\frac{\psi}{2}$$

But

$$MX = \frac{1}{2}MP_1$$

$$= \frac{A}{2} \tag{9.42}$$

∴

$$\frac{A}{2} = MC \sin\frac{\psi}{2}$$

Similarly, from $\triangle CYM$

$$\frac{MY}{MC} = \sin\frac{m\psi}{2}$$

or

$$MY = MC \sin\frac{m\psi}{2}$$

But,

$$MY = \frac{1}{2}MP_m = \frac{1}{2}R_\theta$$

$$\therefore \qquad \frac{R_\theta}{2} = MC \sin \frac{n\psi}{2} \qquad (9.43)$$

Dividing Eq. (9.43) by Eq. (9.42), we have

$$\frac{R_\theta/2}{A/2} = \frac{MC \sin(m\psi/2)}{MC \sin(\psi/2)}$$

or

$$R_\theta = A \frac{\sin(m\psi/2)}{\sin(\psi/2)}$$

Substituting the value of ψ by Eq. (9.41), we have

$$R_\theta = A \frac{\sin \dfrac{\pi a \sin \theta}{\lambda}}{\sin \dfrac{\pi a \sin \theta}{m\lambda}}$$

Let $\dfrac{\pi a \sin \theta}{\lambda} = \alpha$, then

$$R_\theta = A \frac{\sin \alpha}{\sin(\alpha/m)} = \frac{A \sin \alpha}{\alpha/m}$$

[$\because$ n is very large, $\therefore$ α/m is very small and then $\sin(\alpha/m) = \alpha/m$]

or

$$R_\theta = mA \frac{\sin \alpha}{\alpha} \qquad (9.44)$$

Now if $\theta = 0$ or $\psi = 0$ or $\alpha = 0$ (i.e. if all the waves are in the same phase), then

$$\therefore \qquad \lim_{\alpha \to 0} \left(\frac{\sin \alpha}{\alpha} \right) = 1 \qquad \therefore R_0 = mA$$

Hence from Eq. (9.44),

$$R_\theta = R_0 \left(\frac{\sin \alpha}{\alpha} \right) \qquad (9.45)$$

This expression gives the resultant amplitude at diffraction angle θ. Hence the resultant intensity at the point P_m on the screen corresponding to the angle of diffraction θ is

$$I \propto R_\theta^2$$

or

$$I = k R_\theta^2$$

$$= k R_0^2 \left(\frac{\sin \alpha}{\alpha} \right)^2$$

At $\theta = 0°$,

$$I_0 = k R_0^2$$

$$\therefore \qquad I = I_0 \frac{\sin^2 \alpha}{\alpha^2} \qquad (9.46)$$

Eq. (9.46) is similar as Eq. (9.40).

Position of central or principal maxima: For the central point O on the screen (Figure 9.16).

$$\alpha = 0$$

$$\therefore \quad \lim_{\alpha \to 0} \frac{\sin \alpha}{\alpha} = 1$$

Hence intensity at O

$$I = I_0 \frac{\sin^2 \alpha}{\alpha^2} = I_0$$

This is maximum as all waves reach at O in phase
   Again $\alpha = 0$

$$\therefore \quad \frac{\pi a \sin \theta}{\lambda} = 0$$

or $$\theta = 0$$

This shows that the waves are travelling normal to the slit and O gives the position of central maximum.

Position of minima: The intensity is minimum (0), when

$$\frac{\sin \alpha}{\alpha} = 0$$

or $$\sin \alpha = 0 \quad \text{(but } \alpha \neq 0)$$

$$\alpha = \pm m_1 \pi$$

where $m_1 = 1, 2, 3, 4, ...,$ except zero.
As

$$\alpha = \frac{\pi a \sin \theta}{\lambda}$$

$$\therefore \quad \frac{\pi a \sin \theta}{\lambda} = \pm m_1 \pi$$

or $$a \sin \theta = \pm m_1 \lambda \qquad (9.47)$$

where $m_1 = 1, 2, 3, ...$ gives the direction of first, second, third, ... minima respectively.

Secondary maxima: The direction of m_1th seconday maxima is given by

$$a \sin \theta = \pm \left(m_1 + \frac{1}{2} \right) \lambda \qquad (9.48)$$

As $\alpha = \dfrac{\pi a \sin \theta}{\lambda} = \pm \dfrac{\pi}{\lambda} \left(m_1 + \dfrac{1}{2} \right) \lambda = \pi \left(m_1 + \dfrac{1}{2} \right)$

For various values of $m_1 = 1, 2, 3, ...,$ we get

$$\alpha = \pm \frac{3\pi}{2}, \pm \frac{5\pi}{2}, \pm \frac{7\pi}{2}, ...$$

(a) *The intensity of the first secondary (subsidiary) maximum:*

$$I_1 = I_0 \left(\frac{\sin \alpha}{\alpha} \right)^2$$

$$= I_0 \left(\frac{\sin \frac{3\pi}{2}}{\frac{3\pi}{2}} \right)^2$$

$$= I_0 \left(\frac{-1}{3\pi/2} \right)^2$$

$$= \frac{4}{9\pi^2} I_0$$

$$= \frac{I_0}{22}$$

(b) *The intensity of second secondary maximum:*

$$I_2 = I_0 \left(\frac{\sin \frac{5\pi}{2}}{\frac{5\pi}{2}} \right)^2$$

$$= I_0 \left(\frac{2}{5\pi} \right)^2$$

$$= \frac{4}{25\pi^2} I_0$$

$$= \frac{I_0}{61}$$

Thus the intensity of secondary maxima falls off rapidly. Hence we find that secondary maxima of decreasing intensity occur, on either side of the central maximum.

Thus the relative intensities of the successive maxima are

$$1 : \frac{4}{9\pi^2} : \frac{4}{25\pi^2} : \frac{4}{49\pi^2} : \ldots$$

or
$$1 : \frac{1}{22} : \frac{1}{61} : \frac{1}{121} : \ldots \tag{9.49}$$

To determine the position of secondary maxima, differentiate the equation of intensity with respect to α an equate to zero.

$$\frac{dI}{d\alpha} = 0$$

$$\frac{d}{d\alpha} I_0 \left(\frac{\sin\alpha}{\alpha}\right)^2 = 0$$

$$I_0 \left(\frac{2\sin\alpha}{\alpha}\right) \frac{\alpha\cos\alpha - \sin\alpha}{\alpha^2} = 0$$

or
$$\frac{\alpha\cos\alpha - \sin\alpha}{\alpha^2} = 0$$

or
$$\alpha\cos\alpha - \sin\alpha = 0$$

or
$$\alpha = \tan\alpha \tag{9.50}$$

This is the condition for secondary maxima. This equation can be solved by plotting the graphs for $y = \alpha$ and $y = \tan\alpha$.

The curve $y = \alpha$ is a straight line inclined to x-axis at $45°$. The curve $y = \tan\alpha$ is shown in Figure 9.18. The point of intersection of two curves gives the position of secondary maxima. The positions are $\alpha_1 = 0$, $\alpha_2 = 1.43\pi$, $\alpha_3 = 2.46\pi$, $\alpha_4 = 3.47\pi$, ...

Thus, it is clear that secondary maxima do not fall half way between two minima, but are displaced towards the centre of the system by an amount which decreases with increase of m.

The intensity distribution curve of Fraunhofer's diffraction at a single slit is shown in Figure 9.19.

The principal (central) maxima occur at $\alpha = 0$, secondary minima occur at $\alpha = +\pi$, $+2\pi$, ... Secondary maxima occur at $\alpha = 1.43\pi$, 2.46π, ...

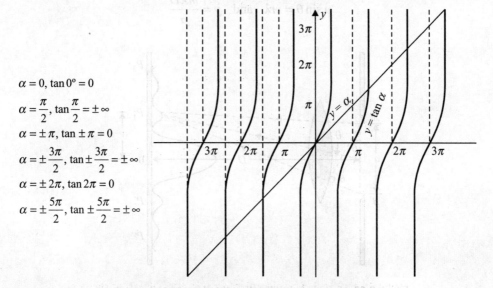

$\alpha = 0, \tan 0° = 0$

$\alpha = \dfrac{\pi}{2}, \tan\dfrac{\pi}{2} = \pm\infty$

$\alpha = \pm\pi, \tan\pm\pi = 0$

$\alpha = \pm\dfrac{3\pi}{2}, \tan\pm\dfrac{3\pi}{2} = \pm\infty$

$\alpha = \pm 2\pi, \tan 2\pi = 0$

$\alpha = \pm\dfrac{5\pi}{2}, \tan\pm\dfrac{5\pi}{2} = \pm\infty$

Figure 9.18 Graphical solution for single slit pattern.

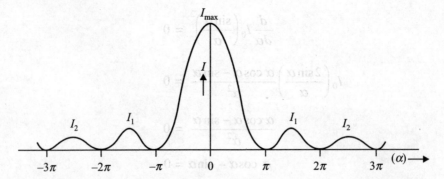

Figure 9.19 Intensity distribution curve for Fraunhofer's diffraction data single slit.

Width of central maxima: Let the distance of the first secondary minimum from the centre of the principal maximum be y

$$\therefore \quad \text{Width of central maximum} = 2y \quad (9.51)$$

If the lens L lies very close to the slit or screen in very far from the lens, then distance between the slit and screen is very large. Let this distance be D [Figure 9.20]. For $D \gg a$, this approaches a right angle and $\theta' = \theta$

$$\tan\theta = \sin\theta = \frac{y}{D} \quad (9.52)$$

Condition for minimum intensity

$$a\sin\theta = m\lambda \quad \text{and} \quad y = \frac{m\lambda D}{a}$$

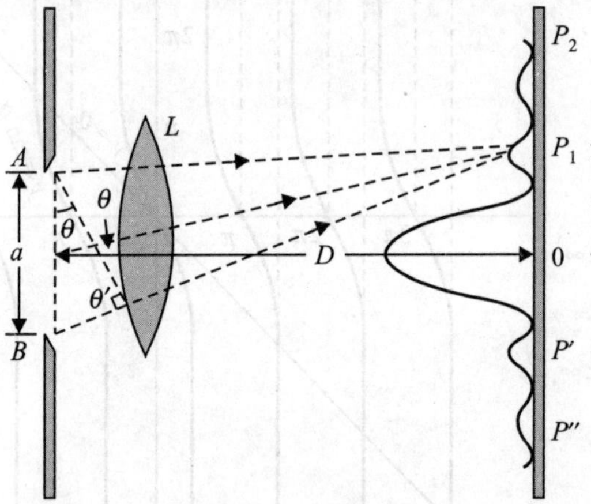

Figure 9.20 Fraunhofer's diffraction of a plane wave at a single slit.

For the first minima

$$\sin\theta = \pm\frac{\lambda}{a}$$

$$\pm\frac{\lambda}{a} = \frac{y}{D}$$

∴
$$y = \pm\frac{\lambda D}{a} \tag{9.53}$$

If f is the focal length of lens which is very close to the slit, then $D = f$

∴
$$y = \pm\frac{\lambda f}{a} \tag{9.54}$$

± sign indicates, on either side to central maxima.
∴ Width of central maxima

$$= y - (-y) = 2y$$

$$= \frac{2\lambda f}{a} \tag{9.55}$$

This shows that the width of the central maximum is directly proportional to the wavelength of light (λ) and inversely proportional to the slit width (a).

Effect of slit width: As $\sin\theta = \pm\lambda/a$

If the slit width a is large, then for a given wavelength of light, $\sin\theta$ is small and, hence, θ is small. This means that the maxima and minima be very close to the central maximum. If the slit width a is narrow, θ is large. Hence diffraction maxima and minima are quite distinct and clear.

EXAMPLE 9.4 Monochromatic light is incident on a slit of width 0.012 mm. The angular position of first bright line is 5.2°. Calculate the wavelength of incident light.

Solution Given $a = 0.012$ mm $= 1.2 \times 10^{-5}$ m, $\theta = 5.2° = \dfrac{5.2 \times \pi}{180}$ radian, $\lambda = ?$

We know the condition for maxima (to get bright fringes)

$$a\sin\theta = \left(m_1 + \frac{1}{2}\right)\lambda = \frac{3}{2}\lambda$$

or
$$\lambda = \frac{2a\sin\theta}{3} = \frac{2a\theta}{3} \quad [\because \sin\theta = \theta, \theta \text{ is small}]$$

$$= \frac{2 \times 1.2 \times 10^{-5} \times 5.2 \times 3.14}{3 \times 180}$$

$$= 7.257 \times 10^{-7} \text{ m}$$

$$= 7257 \text{ Å}$$

EXAMPLE 9.5 A parallel beam of light of wavelength $\lambda = 5890$ Å, is perpendicularly on a slit of width 0.1 mm. Calculate angular width and linear width of the central maximum formed on a screen 100 cm away.

Solution $\lambda = 5890$Å $= 5.89 \times 10^{-7}$ m, $a = 0.1$ mm $= 1.0 \times 10^{-4}$ m, $D = 100$ cm $= 1$ m.
We know that
$$a \sin \theta = m\lambda$$
$$\sin \theta = \frac{m\lambda}{a} = \frac{5.89 \times 10^{-7}}{1.0 \times 10^{-4}} = 5.89 \times 10^{-3} \text{ radian}$$

∴   Angle is very small, then $\sin\theta \cong \theta$
i.e.,
$$\theta = 5.89 \times 10^{-3} \text{ radian}$$

Total angular width of central maxima $= 2\theta = 2 \times 5.89 \times 10^{-3}$ radian $= 11.78 \times 10^{-3}$ radian.
Let y is the linear half width and D is the distance of the screen from the slit, then
$$y = D\theta \text{ [for small } \theta]$$
⇒
$$y = 1 \times 5.89 \times 10^{-3} \text{ m} = 0.589 \text{ cm}$$

The linear width of the central maximum on the screen
$$= 2y = 2 \times 0.589 = 1.178 \text{ cm}.$$

9.11 FRAUNHOFER'S DIFFRACTION AT DOUBLE SLIT

For the two fine slits, the same graphical procedure can be adopted as for the single slit. The two rectangular slits AB and CD are shown in Figure 9.21. AB and CD have equal width a and are at a distance b, i.e. the width of opaque BC. Hence separation between the corresponding points of there slits is $(a + b)$.

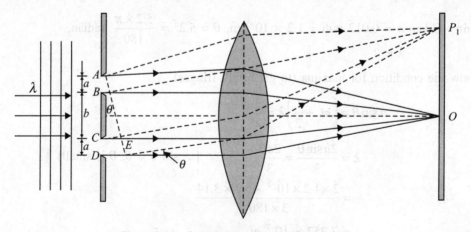

Figure 9.21 Fraunhofer's diffraction at two slits.

Suppose a parallel beam of given monochromatic light be incident upon these two slits. This beam is of wavelength λ. Again, let us consider each slit diffraction the incident beam at an angle θ.

Hence, the optical path difference for the points B and C is given by

$$CE = AC \sin\theta$$
$$= (a + b) \sin\theta \quad (\text{as } AC = a + b)$$

Then, corresponding phase difference

$$2\beta = \frac{2\pi}{\lambda} \times CE$$

$$= \frac{2\pi}{\lambda}(a + b)\sin\theta \qquad (9.56)$$

Here both slits, separately can be treated as an independent source of light.

If the monochromatic beam from these two slits be focused by a given convex lens on the screen in the focal plane of the lens as visualized in Figure 9.21. The resultant pattern will be the same as that due to the interference of two light sources placed at the corresponding points.

Let 2α be phase change between the extreme rays from the first slit, then we have

$$2\alpha = \frac{2\pi}{\lambda} a \sin\theta$$

or
$$\alpha = \frac{\pi}{\lambda} a \sin\theta \qquad (9.57)$$

On the basis of analysis, the resultant amplitude y_1 due to the extreme rays from the first slit is given by

$$y_1 = A \sin\omega t \qquad (9.58)$$

Here,

$$A = A_0 \frac{\sin\alpha}{\alpha}$$

where A_0 is the resultant amplitude of the direct rays, and A is the resultant amplitude of the diffracted rays at an angle θ from the first slit.

In the similar manner, the resultant amplitude y_2 due to the extreme rays from the second slit is given by

$$y_2 = A \sin(\omega t + 2\beta) \qquad (9.59)$$

Therefore, the resultant displacement Y due to the extreme rays from the double slits at angle θ as per superposition principle is given by

$$Y = y_1 + y_2$$
$$\therefore \quad Y = A \sin\omega t + A \sin(\omega t + 2\beta)$$
$$\therefore \quad Y = 2A \cos\beta \sin(\omega t + \beta) \qquad (9.60)$$

Hence, we get resultant amplitude as

$$R = 2A\cos\beta \qquad (9.61)$$

Putting the value of $A = A_0 \dfrac{\sin \alpha}{\alpha}$ in Eq. (9.61), we have

$$R = 2A_0 \frac{\sin \alpha}{\alpha} \cos \beta$$

and we know that intensity (I) is directly proportional to square of the amplitude (R^2), i.e.

$$I \propto R^2$$

Then

$$I \propto 4A_0^2 \frac{\sin^2 \alpha}{\alpha^2} \cos^2 \beta$$

Suppose constant of proportionality is one, then

$$I = 4A_0^2 \left(\frac{\sin \alpha}{\alpha}\right)^2 \cos^2 \beta \tag{9.62}$$

Hence the resultant intensity depends on the following two factors:

(i) $\dfrac{\sin^2 \alpha}{\alpha^2}$, and

(ii) $\cos^2 \beta$

$\dfrac{\sin^2 \alpha}{\alpha^2}$ gives diffraction pattern due to each individual slit and $\cos^2 \beta$ gives interference pattern due to diffracted light waves from the two slits.

$\dfrac{\sin^2 \alpha}{\alpha^2}$ gives a central maximum in the direction $\theta = 0$, having alternately minima and subsidiary maxima of decreasing intensity on either side, as shown in Figure 9.22(a). The minima are obtained in the directions given by

$$\sin \alpha = 0$$

or
$$\alpha = \pm m_1 \pi$$

as
$$\alpha = \frac{\pi a \sin \theta}{\lambda}$$

$\therefore$
$$\pi a \sin \alpha = \pm m_1 \pi \lambda$$

$$\alpha \sin \theta = \pm m_1 \lambda \tag{9.63}$$

where $m_1 = 1, 2, 3, \ldots$ except zero.

The term $\cos^2 \beta$ is the intensity pattern gives a set of equidistant dark and bright fringes as shown in Figure 9.22(b). The bright fringes are obtained in the directions given by

$$\cos^2 \beta = 1$$

$$\beta = \pm m\pi$$

or
$$\frac{\pi}{\lambda}(a + b) \sin \theta = \pm m\pi$$

or
$$(a + b)\sin\theta = \pm m\pi \qquad (9.64)$$

where $m = 0, 1, 2, ...$, corresponds to zero order, first order, second order, ... maxima.

The resultant intensity distribution curve is shown in Figure 9.22(c).

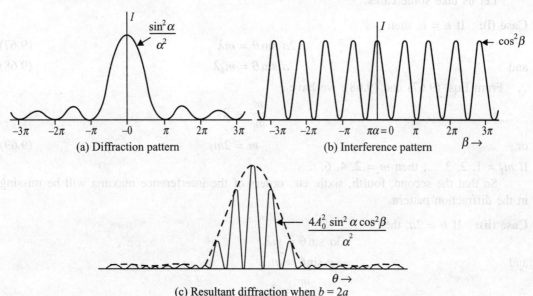

(a) Diffraction pattern (b) Interference pattern

(c) Resultant diffraction when $b = 2a$

Figure 9.22

Effect of increasing the slit width

On increasing the slit width a, the central peak will become sharper, but fringe spacing remains unchanged. Hence interference maxima fall within the central diffraction maximum.

Effect of increasing the distance between slits

On increasing the separation between slits b, keeping the slit width constant, the fringes becomes closer together, but the envelope of the pattern remains unchanged. Hence more interference maxima fall within the central envelope.

Missing order in a double diffraction pattern

We have taken the slit width as a and double slit separation b. If a is kept constant, we will observe the same diffraction pattern. However, if a is kept constant and b is varied, the spacing between interference maxima changes, depending upon the relative values of a and b. Some order of interference maxima will be missing in the resultant pattern.

We have direction for interference maxima as

$$(a + b)\sin\theta = m\lambda \qquad (9.65)$$

and the direction for diffraction minima as

$$a\sin\theta = m_1\lambda \qquad (9.66)$$

where m_1 and m are integers.

If the values of a and b are such that both Eqs. (9.65) and (9.66) are satisfied simultaneously for some value of θ, in that case position of interference maxima corresponds to that of diffraction minima.

Let us take some cases.

Case (i): If $a = b$, then

$$2a \sin\theta = m\lambda \qquad (9.67)$$

and

$$a \sin\theta = m_1\lambda \qquad (9.68)$$

∴ From Eqs. (9.67) and (9.68), we have

$$\frac{m}{m_1} = 2$$

or

$$m = 2m_1 \qquad (9.69)$$

If $m_1 = 1, 2, 3, ...$, then $m = 2, 4, 6, ...$.

So that the second, fourth, sixth, etc. orders of the interference maxima will be missing in the diffraction pattern.

Case (ii): If $b = 2a$, then

$$3a \sin\theta = m\lambda$$

and

$$a \sin\theta = m_1\lambda$$

∴

$$\frac{m}{m_1} = 3$$

or

$$m = 3m_1 \qquad (9.70)$$

If $m_1 = 1, 2, 3, ...$ then $m = 3, 6, 9, ...$, etc.

So that third, sixth, ninth, etc. orders of the interference maxima will be missing in the diffraction pattern.

Case (iii): If $a + b = a$, then $b = 0$.

In this case, we will have a single slit, so all the interference patterns will be missing. In this case, diffraction pattern observed on the screen is similar to that due to single slit with width $2a$.

EXAMPLE 9.6 A diffraction phenomenon is observed using a double slit with light of $\lambda = 5000$ Å slit width $a = 0.02$ mm and the spacing between the two slits $b = 0.10$ mm. The distance of the screen from the slit is 1 ms. Calculate (i) the distance between the central maximum and the first minimum of the fringe envelope, and (ii) distance between any two consecutive double slit dark fringes.

Solution Given $a = 0.002$ cm, $b = 0.01$ cm and $\lambda = 5 \times 10^{-5}$ cm

First minimum of the fringe envelope occurs at $\alpha = \pi$

Since

$$\alpha = \frac{\pi}{\lambda} a \sin\theta$$

∴

$$\frac{\pi}{\lambda} a \sin\theta = \pi \quad \text{or} \quad a \sin\theta = \lambda$$

or $$\sin\theta = \frac{\lambda}{a} = \frac{5 \times 10^{-5}\,\text{cm}}{0.002\,\text{cm}} = 2.5 \times 10^{-2}$$

Since θ is small
$$\sin\theta = \theta = 2.5 \times 10^{-2} \text{ radian}$$

(i) Since distance of screen from slits = 100 cm, the distance y of the first minimum from central maximum (which is at $\theta = 0$) is given by

$$\tan\theta = \theta = \frac{y}{100} = 2.5 \times 10^{-2}$$

$$y = 2.5 \text{ cm}.$$

(ii) Since $\beta = \frac{\pi}{\lambda}(a+b)\sin\theta$ and double slit dark fringes occur at

$$\beta = \pm(2m+1)\frac{\pi}{2} \quad \text{i.e.} \quad \beta = \frac{\pi}{2}, \frac{3\pi}{2}, \frac{5\pi}{2}, \ldots$$

Let θ_1 be the angle corresponding to the first dark fringe and θ_2 be the angle corresponding to the second dark fringe.

∴ For first dark fringe

$$\frac{\pi}{2} = \frac{\pi}{\lambda}(a+b)\sin\theta_1$$

or $$(a+b)\sin\theta_1 = \frac{\lambda}{2}$$

$$\sin\theta_1 = \frac{\lambda}{2(a+b)} = \frac{5 \times 10^{-5}\,\text{cm}}{2 \times (0.002 + 0.01)\,\text{cm}} = \frac{5 \times 10^{-5}}{0.024}$$

$$\theta_1 = \frac{5 \times 10^{-5}}{0.024} \quad [\text{Here } \sin\theta_1 = \theta_1]$$

and for second dark fringe

$$\frac{3\pi}{2} = \frac{\pi}{\lambda}(a+b)\sin\theta_2$$

or $$(a+b)\sin\theta_2 = \frac{3\lambda}{2}$$

or $$\sin\theta_2 = \frac{3\lambda}{2(a+b)} = \frac{3 \times 5 \times 10^{-5}}{2 \times (0.002 + 0.01)}$$

$$= \frac{3 \times 5 \times 10^{-5}}{0.024}$$

$$\theta_2 = \frac{3 \times 5 \times 10^{-5}}{0.024} \quad [\text{Here } \sin\theta_2 = \theta_2]$$

If x_1 and x_2 be respectively the distances of the first and second minima from the centre, then

$$\theta_1 = \frac{x_1}{100} = \frac{5 \times 10^{-5}}{0.024} \Rightarrow x_1 = \frac{5 \times 10^{-5} \times 100}{0.024}$$

and

$$\theta_2 = \frac{x_2}{100} = \frac{3 \times 5 \times 10^{-5}}{0.024} \Rightarrow x_2 = \frac{3 \times 5 \times 10^{-5} \times 100}{0.024}$$

∴ The linear separation between two minima

$$x = x_2 - x_1 = \frac{2 \times 5 \times 10^{-5}}{0.024} \times 100 \text{ cm} = 0.42 \text{ cm}$$

EXAMPLE 9.7 Deduce the missing order for a double slit Fraunhofer diffreation pattern, at the slit width is 0.25 mm and they are 0.5 mm apart.

Solution Given $a = 0.25$ mm and $b = 0.5$ mm. If a be the slit width and b be the separation between slits is the condition of missing order speetra is given by

$$\frac{a+b}{a} = \frac{m}{m_1}$$

$$\frac{0.25 + 0.5}{0.25} = \frac{m}{m_1}$$

$$\frac{m}{m_1} = 3$$

$$m = 3m_1$$

for $m_1 = 1, 2, 3, \ldots\ldots$

$m = 3, 6, 7, \ldots\ldots$

Thus, 3rd, 6th, 9th ... orders will be missing.

9.12 PLANE TRANSMISSION DIFFRACTION GRATING: FRAUNHOFER'S DIFFRACTION AT N PARALLEL SLITS

We know that interference of waves diffracted by individual slits determines the intensity distribution in the double slit pattern. Let us now consider the diffraction pattern produced by N slits. We used experimental arrangement in Figure 9.23 for N slits. For simplicity, we assume that

(i) each slit is of width a and has the same wavelength,
(ii) all slits are parallel to each other, and
(iii) the intervening opaque space between two successive slits is the same, equal to b.

Therefore, the distance between any two equivalent points in two consecutive slits is $(a + b)$. Let us denote it by d, which we call the *grating element*. As before, we take the source of light in the form of a slit and adjust the length of this source slit vertical and parallel to the length of N slits. An arrangement which consists of a large number of parallel, equidistant narrow rectangular slits of the same width is called *diffraction grating*.

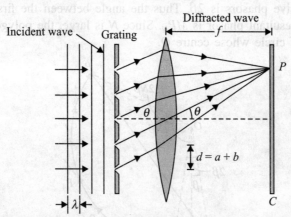

Figure 9.23 Fraunhofer's diffraction of a plane wave incident normally on a multiple slit aperture.

9.12.1 Intensity Distribution

Consider a beam of monochromatic light of wavelength λ incident on a plane transmission grating has N parallel lines. If this has N parallel slits, each of width a and opaque space in between two slits each of width b. The light waves diffracted from all these slits reach the screen with the same amplitude, but with different phases. The superposition of N diffracted waves give rise to the intensity on the screen.

Method I

Consider point P on the screen where waves diffracted at an angle θ superimpose. Now from the theory of diffraction at single slit, we know that the disturbance originating from all points of the slit can be summed up into single wavelet of amplitude A originating from the centre of the slit, is given by

$$A_0 = A \frac{\sin \alpha}{\alpha} \tag{9.71}$$

where
$$\alpha = \frac{\pi}{\lambda} a \sin \theta \tag{9.72}$$

As seen from Figure 9.23, the path difference between the diffracted light disturbances from the two nearby slits is given by

$$\Delta = (a+b)\sin \theta \tag{9.73}$$

The corresponding phase difference is

$$\frac{2\pi}{\lambda}(a+b)\sin \theta = 2\beta \tag{9.74}$$

where
$$\beta = \frac{\pi}{\lambda}(a+b)\sin \theta \tag{9.75}$$

To find the intensity (I), we have to superimpose N waves, each of amplitude A_0 differing in phase with nearby wave by 2β.

Figure 9.24 shows the phasor addition, MP_1, P_1P_2, ..., $P_{N-1}P_N$ are N phasors. The angle between two successive phasors is 2β. Thus the angle between the first phasor and the last phasor is $2N\beta$. The resultant phasor is MP_N. Since N is large, the polygon of phasors may be assumed to an arc of circle whose centre O.

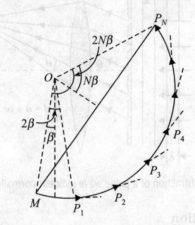

Figure 9.24 Phasor addition diagram for and N slits.

Obviously,
$$MP_N = 2\ OM \sin N\beta \qquad (9.76)$$
where OM is the radius of the arc of circle.

For a single phasor MP_1, we have (Figure 9.24)
$$MP_1 = 2\ OM \sin\beta \qquad (9.77)$$
Dividing Eq. (9.76) with Eq. (9.77), we have
$$\frac{MP_N}{MP_1} = \frac{\sin N\beta}{\sin \beta}$$
Now
$$MP_N = \text{resultant disturbance phasor } A$$
where MP_1 is single slit disturbance phasor A_0.

Hence
$$A = A_0 \frac{\sin N\beta}{\sin \beta}$$
which on using value of A_θ from Eq. (9.71), reduces to
$$A = A_0 \left(\frac{\sin \alpha}{\alpha}\right)\left(\frac{\sin N\beta}{\sin \beta}\right) \qquad (9.78)$$

The resultant intensity is square of resultant amplitude

$$I = A_0^2 \left(\frac{\sin\alpha}{\alpha}\right)^2 \left(\frac{\sin N\beta}{\sin\beta}\right)^2$$

$$\Rightarrow \qquad I = I_0 \left(\frac{\sin\alpha}{\alpha}\right)^2 \left(\frac{\sin N\beta}{\sin\beta}\right)^2 \qquad (9.79)$$

The first term in Eq. (9.79), $\left(\dfrac{\sin\alpha}{\alpha}\right)^2$ is the intensity due to single slit diffraction and the second term $\left(\dfrac{\sin N\beta}{\sin\beta}\right)^2$ may be interpreted as interference term. Both effects combined together give intensity pattern of light diffracted by plane transmission grating.

Method II

Let a parallel beam of monochromatic light of wavelength λ be incident normally on grating from the left. At an angle θ on the right, we have N light waves each of amplitude $A_\theta = A\dfrac{\sin\alpha}{\alpha}$ and successive phase difference $\psi = \dfrac{2\pi}{\lambda}(a+b)\sin\theta = 2\beta$.

The resultant displacement Y is given by

$$Y = A_\theta[\cos\omega t + \cos(\omega t + \psi) + \cos(\omega t + 2\psi) + \cdots N \text{ terms}]$$

$$= \text{Real part of } A_\theta\, e^{i\omega t}\,[1 + e^{i\psi} + e^{2i\psi} + \cdots + N \text{ terms}]$$

$$= A_\theta\, e^{i\omega t} \left[\frac{1 - e^{iN\psi}}{1 - e^{i\psi}}\right] \qquad (9.80)$$

$\therefore$ Intensity $= YY^*$
where Y^* is the complex conjugate of Y

$$I = A_\theta^2 \left[\frac{(1 - e^{iN\psi})(1 - e^{-iN\psi})}{(1 - e^{i\psi})(1 - e^{-i\psi})}\right]$$

$$= A_\theta^2 \left[\frac{1 - \cos N\psi}{1 - \cos\psi}\right]$$

$$= A_\theta^2 \left[\frac{2\sin^2\dfrac{N\psi}{2}}{2\sin^2\dfrac{\psi}{2}}\right]$$

$$= A_\theta^2 \,\frac{\sin^2\left[\dfrac{N\pi(a+b)\sin\theta}{\lambda}\right]}{\sin^2\left[\dfrac{\pi(a+b)\sin\theta}{\lambda}\right]} \qquad (9.81)$$

394 Fundamentals of Optics

$$A_\theta = A \frac{\sin \alpha}{\alpha}$$

$$I = A^2 \frac{\sin^2 \alpha}{\alpha^2} \frac{\sin^2 \left[\frac{N\pi(a+b)\sin\theta}{\lambda} \right]}{\sin^2 \left[\frac{\pi(a+b)\sin\theta}{\lambda} \right]}$$

or

$$I = A^2 \frac{\sin^2 \alpha}{\alpha^2} \frac{\sin^2 N\beta}{\sin^2 \beta} \qquad (9.82)$$

where

$$\beta = \frac{\pi(a+b)\sin\theta}{\lambda}$$

Hence the intensity distribution is the product of two terms. The first term $A^2 \frac{\sin^2 \theta}{\alpha^2}$ represents the diffraction pattern due to single slit. The second term $\frac{\sin^2 N\beta}{\sin^2 \beta}$ represents the interference pattern due to N slits.

For $N = 1$:

$$I = A^2 \frac{\sin^2 \alpha}{\alpha^2} \frac{\sin^2 \beta}{\sin^2 \beta}$$

$$= A^2 \frac{\sin^2 \alpha}{\alpha^2} \quad \text{[for single slit diffraction]} \qquad (9.83)$$

For $N = 2$:

$$I = A^2 \frac{\sin^2 \alpha}{\alpha^2} \times \frac{\sin^2 2\beta}{\sin^2 \beta}$$

$$= A^2 \frac{\sin^2 \alpha}{\alpha^2} \times \frac{4 \sin^2 \beta \cos^2 \beta}{\sin^2 \beta}$$

$$= A^2 \frac{\sin^2 \alpha}{\alpha^2} \times 4 \cos^2 \beta$$

$$= 4A^2 \frac{\sin^2 \alpha}{\alpha^2} \cos^2 \beta \quad \text{[for double slit diffraction]} \qquad (9.84)$$

For N slits:

$$I = A^2 \frac{\sin^2 \alpha}{\alpha} \frac{\sin^2 N\beta}{\sin^2 \beta} \quad \text{[Generalized expression for } N \text{ slits]} \qquad (9.85)$$

Principal maxima: Intensity would be maximum, when

$$\sin \beta = 0$$

or

$$\beta = \pm m\pi$$

where $m = 0, 1, 2, 3, 4, \ldots$.

Also $\sin N\beta = 0$. Thus

$$\frac{\sin N\beta}{\sin \beta} = \frac{0}{0} \quad \text{(Undefined)}$$

So

$$\lim_{\beta \to \pm n\pi} \frac{\sin N\beta}{\sin \beta} = \lim_{\beta \to \pm n\pi} \frac{N \cos N\beta}{\cos \beta}$$

$$= \frac{N \cos(\pm m\pi)}{\cos \pm m\pi} = \pm N$$

So intensity at principal maxima

$$I = A^2 \frac{\sin^2 \alpha}{\alpha^2} N^2 \tag{9.86}$$

$$\therefore \quad \beta = \pm m\pi$$

$$\frac{\pi}{\lambda}(a+b)\sin\theta = \pm m\pi$$

or $\qquad (a+b)\sin\theta = \pm m\lambda \qquad (9.87)$

For $m = 0$, maximum is zero order maximum and for $m = \pm 1, \pm 2, ...,$ etc. are called first, second, ... etc. order of principal maxima respectively.

Secondary minima: When $\sin N\beta = 0$, but $\sin \beta \neq 0$, then from Eq. (9.85) $I = 0$, which is minimum
or $\qquad N\beta = \pm m_1 \pi$

$$N\frac{\pi}{\lambda}(a+b)\sin\theta = \pm m_1 \pi$$

or $\qquad N(a+b)\sin\theta = \pm m_1 \lambda \qquad (9.88)$

where $m_1 = 1, 2, 3, ... (N-1)$.

If $m_1 = 0$ gives principal maxima and $m_1 = N$ also gives principal maxima, the $m_1 = 1, 2, 3, ... (N-1)$ gives minima. There is $(N-1)$ minima between two maxima.

Secondary maxima: Since there are $(N-1)$ minima between two maxima, there must be $(N-2)$ maxima between two principal maxima. To find the position of these secondary maxima, we differentiate Eq. (9.85) with respect to β and equating to zero.

$$\frac{dI}{d\beta} = \frac{A^2 \sin^2 \alpha}{\alpha^2} \frac{\sin N\beta}{\sin \beta} \left[\frac{N \sin \beta \cos N\beta - \sin N\beta \cos \beta}{\sin^2 \beta} \right]$$

$\Rightarrow \qquad \dfrac{dI}{d\beta} = 0 \quad \text{[condition for maxima]}$

$$N \sin \beta \cos N\beta - \sin N\beta \cos \beta = 0$$

$$\tan N\beta = N \tan \beta,$$

but we need $\sin N\beta$, so we make a triangle as shown in Figure 9.25.

Then from Figure 9.25

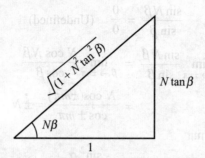

Figure 9.25 Illustration for finding intensity at secondary maxima.

$$\sin N\beta = \frac{N \tan \beta}{\sqrt{1 + N^2 \tan^2 \beta}}$$

or

$$\frac{\sin^2 N\beta}{\sin^2 \beta} = \frac{(N^2 \tan^2 \beta)/(1 + N^2 \tan^2 \beta)}{\sin^2 \beta}$$

$$= \frac{N^2}{\cos^2 \beta + N^2 \sin^2 \beta}$$

$$= \frac{N^2}{1 + (N^2 - 1)\sin^2 \beta}$$

Intensity at secondary maxima is given by

$$I' = A^2 \frac{\sin^2 \alpha}{\alpha^2} \frac{N^2}{1 + (N^2 - 1)\sin^2 \beta} \tag{9.89}$$

From Eq. (9.85), the intensity of principal maxima is directly proportional to N^2.
Then

$$\frac{\text{Intensity of secondary maxima } (I')}{\text{Intensity of principal maxima } (I)} = \frac{1}{1 + (N^2 - 1)\sin^2 \beta} \tag{9.90}$$

If N is large then the intensity of the secondary maxima is less. Since, the number of slits is very large in the grating, the secondary maxima are not visible in grating spectrum.

9.12.2 Grating Spectra

Figure 9.26 shows the intensity distribution arising from factors $\left(A \frac{\sin \alpha}{\alpha}\right)^2$ and $\frac{\sin^2 N\beta}{\sin^2 \beta}$ respectively. The intensity and angular spacing of secondary maxima and minima are so small in comparison to principal maxima, that they cannot be observed. Hence, there is uniform darkness between any two principal maxima.

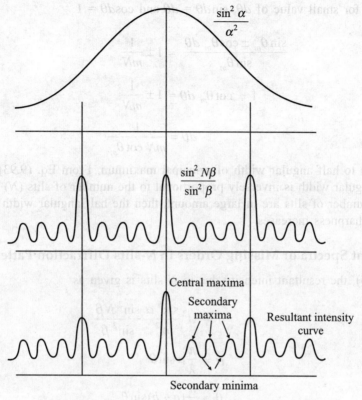

Figure 9.26 Intensity distribution curve for diffraction grating.

9.12.3 Angular Half Width of Principal Maxima

Since, we know that the condition for nth principal maxima in diffraction grating is

$$(a + b)\sin\theta_m = m\lambda \tag{9.91}$$

Let $(\theta_m + d\theta)$ gives the direction of the first secondary maxima on the two sides of the n primary maxima as per positive and negative signs, then

$$(a + b)\sin(\theta_m + d\theta) = m\lambda \pm \frac{\lambda}{N} \tag{9.92}$$

Now, we divide Eq. (9.92) by Eq. (9.91), then

$$\frac{(a+b)\sin(\theta_m + d\theta)}{(a+b)\sin\theta_m} = \frac{m\lambda \pm \dfrac{\lambda}{N}}{m\lambda}$$

or

$$\frac{\sin(\theta_m + d\theta)}{\sin\theta_m} = 1 \pm \frac{1}{mN}$$

or

$$\frac{\sin\theta_m \cos d\theta \pm \cos\theta_m \sin d\theta}{\sin\theta_m} = 1 \pm \frac{1}{mN}$$

We know that, for small value of $d\theta$, $\sin d\theta = d\theta$ and $\cos d\theta = 1$
Then

$$\frac{\sin\theta_m \pm \cos\theta_m \, d\theta}{\sin\theta_m} = 1 \pm \frac{1}{mN}$$

or

$$1 \pm \cot\theta_m \, d\theta = 1 \pm \frac{1}{mN}$$

or

$$d\theta = \frac{1}{mN \cot\theta_m} \tag{9.93}$$

Here $d\theta$ relates to half angular width of principal maximum. From Eq. (9.93) it is depicted that the half angular width is inversely proportional to the number of slits (N) or in the other words, if the number of slits are in large amount, then the half angular width will be small, therefore, the sharpness increases.

9.12.4 Absent Spectra or Missing Orders in N-slits Diffraction Pattern

From Eq. (9.85), the resultant intensity due to N slits is given as

Here

$$I = A^2 \frac{\sin^2\alpha}{\alpha^2} \frac{\sin^2 N\beta}{\sin^2\beta}$$

$$\alpha = \frac{\pi}{\lambda} a \sin\theta$$

and

$$\beta = \frac{\pi}{\lambda}(a+b)\sin\theta$$

The direction of minima for a single slit pattern is given by

$$a \sin\theta = m_1 \lambda \tag{9.94}$$

and the direction of principal maxima is grating spectra is given as

$$(a+b)\sin\theta = m\lambda \tag{9.95}$$

If the two conditions given by Eqs. (9.94) and (9.95) are simultaneously satisfied, then a particular maximum of mth order will be missed or absent in the N slits spectrum, these spectra are termed *absent spectra* or *missing order spectra*.

If we divide Eq. (9.95) by Eq. (9.94), then we have

$$\frac{(a+b)}{a} = \frac{m}{m_1} \tag{9.96}$$

This is the condition for the missing order spectra in the N slits diffraction pattern.

The following are the cases of this condition:

1. If the width of transparencies is equal to the width of opacities, i.e. $b = a$, then Eq. (9.96) becomes

$$\frac{(a+b)}{a} = \frac{m}{m_1} \Rightarrow m = 2m_1 \tag{9.97}$$

for $m_1 = 1, 2, 3, 4, ...$ and $m = 2, 4, 6, 8, ...$

So the second, fourth, sixth, eight, etc. orders of spectra will be absent corresponding to minima due to the single slit given by $m_1 = 1, 2, 3, 4, ...$.

2. If the width of transparencies is double of the width of opacities, i.e. $b = 2a$, then Eq. (9.97) becomes

$$\frac{(a+b)}{a} = \frac{m}{m_1} \Rightarrow m = 3m_1 \qquad (9.98)$$

for $m_1 = 1, 2, 3, 4, ...$ and $m = 3, 6, 9, 12, ...$

So the third, sixth, ninth, twelfth, etc. orders of the spectra will be absent corresponding to minima due to the single slit given by $m_1 = 1, 2, 3, 4, ...$.

9.12.5 Dispersive Power of Grating

The dispersive power of a plane transmission diffraction grating is the rate of change of angle of diffraction with wavelength of given monochromatic light, i.e.

$$\frac{d\theta}{d\lambda}$$

We know that the grating equation is given by

$$(a + b) \sin\theta = m\lambda$$

Now, we differentiate above equation with respect to wavelength λ, we have

$$(a + b) \cos\theta \frac{d\theta}{d\lambda} = m$$

$\therefore$
$$\frac{d\theta}{d\lambda} = \frac{m}{(a+b)\cos\theta} \qquad (9.99)$$

Equation (9.99) is an expression for the dispersive power of a grating. From this expression, we find that:

1. Dispersive power is directly proportional to the spectrum order, i.e.

$$\frac{d\theta}{d\lambda} \propto m$$

2. Dispersive power is inversely proportional to grating element, i.e.

$$\frac{d\theta}{d\lambda} \propto \frac{1}{(a+b)}$$

3. Dispersive power is inversely proportional to cosine of diffraction angle, i.e.

$$\frac{d\theta}{d\lambda} \propto \frac{1}{\cos\theta}$$

Note:
1. If the value of a diffraction angle is large, then the cosine of this angle will be smaller. In this case, dispersive power will be high.

The diffraction angle in red colour is higher than the diffraction angle in violet colour. We can say that the dispersion in red region is greater than that of in violet region.

2. If the value of a diffraction angle is very small, then the cosine of this angle will be approximately equal to one, i.e.

$$d\theta \propto d\lambda$$

This type of spectrum is known as *normal spectrum* and so on.

9.12.6 Grating at Oblique Incidence

Let us now consider a parallel beam of light incident obliquely on the grating surface at an angle of incidence, i.e., θ be the corresponding angle of diffraction (Figure 9.27).

The path difference between the secondary waves from the corresponding points A and C is given by

$$MC + CN = AC \sin i + AC \sin\theta$$
$$= (AB + BC)(\sin i + \sin\theta)$$
$$= (a + b)(\sin i + \sin\theta) \quad (9.100)$$

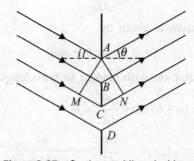

Figure 9.27 Grating at oblique incidence.

where $AC = AB + BC = (a + b)$ is the grating element, a being the width of the slit and b the width of opaque space. Equation (9.100) holds good for the upward diffracted beam. If, however, we consider the beam diffracted downwards, the path difference

$$= (a + b)(\sin i - \sin\theta)$$

For nth order principal maximum

$$(a + b)(\sin i + \sin\theta) = m\lambda$$

or
$$(a+b)\left[2\sin\frac{i+\theta}{2}\cos\frac{i-\theta}{2}\right] = m\lambda \quad (9.101)$$

$\therefore$
$$\sin\frac{i+\theta}{2} = \frac{m\lambda}{2(a+b)\cos\frac{i-\theta}{2}} \quad (9.102)$$

Here the deviation produced in the path of the incident ray is given by

$$\delta = i + \theta$$

For the deviation $\delta = i + \theta$ to be minimum, so $\sin\dfrac{i+\theta}{2}$ must be minimum in Eq. (9.102) for it $\cos\dfrac{i-\theta}{2}$ must be maximum, i.e.

$$\frac{i-\theta}{2} = 0$$

or
$$i = \theta$$

Thus the rays forming spectrum in a grating suffer minimum deviation when the angle of incidence equals angle of diffraction. Hence the angle of minimum deviation

$$\delta_m = i + \theta = 2i = 2\theta$$

$\Rightarrow$
$$\theta = \frac{\delta_m}{2}$$

and
$$i = \frac{\delta_m}{2}$$

Then the condition for nth order principal maximum from Eq. (9.101) modifies to

$$(a+b)\left(\sin\frac{\delta_m}{2} + \sin\frac{\delta_m}{2}\right) = n\lambda$$

or
$$2(a+b)\sin\frac{\delta_m}{2} = n\lambda \qquad (9.103)$$

Using the grating in the minimum deviation position is more advantageous as the definition of diffracted image considerably improves.

9.12.7 Wavelength of Incident Light by Means of Diffraction Grating

Wavelength of light can be determined using equation

$$(a+b)\sin\theta = m\lambda \qquad (9.104)$$

This equation is valid when light is incident normally on the grating. The process of determining or measuring wavelength of light involves the following steps:

(i) Setting the grating for normal incidence

Spectrometer is set for parallel rays. The collimator and telescope are arranged in line with the slit to obtain the image of slit on the crosswire. The telescope is now rotated by 90°. Keeping telescope in position, the grating is oriented on its stand so that reflected light is seen at the crosswire of the telescope. This shall happen when grating makes an angle 45° with incident light. The grating is now rotated by 45° or 135°. This sets grating normal to incident light. This is shown in Figure 9.28.

402 Fundamentals of Optics

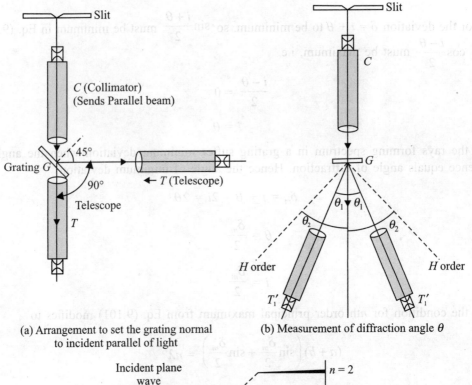

(a) Arrangement to set the grating normal to incident parallel of light

(b) Measurement of diffraction angle θ

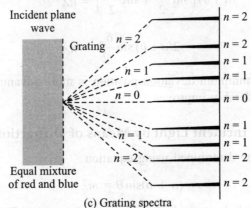

(c) Grating spectra

Figure 9.28(a to c) Setting the grating for normal incidence.

(ii) Determination of grating element
On grating, a number of lines per inch are marked, say N lines per inch. Then grating element is given by

$$(a+b) = \frac{2.54}{N} \text{cm} \qquad (9.105)$$

(iii) Determination of diffraction angle
With grating locked in position and exposed to normal parallel beam of light, the telescope is rotated to find the spectral line exactly on crosswire and reading θ_1 is noted. The telescope

then rotated on the otherside and the same spectral line is seen on the cross-wire again. The reading of θ_2 is again noted. The difference $(\theta_2 - \theta_1)$ gives twice the diffraction angle. Knowing $(a + b)$, θ and n, λ may be calculated

$$\lambda = \frac{(a+b)\sin\theta}{m} \quad (9.106)$$

EXAMPLE 9.8 What is the highest order spectrum which may be seen with monochromatic light of wavelength 6000Å by means of a diffraction grating with 5000 lines per cm?

Solution Given $\lambda = 6000Å = 6.0 \times 10^{-5}$ cm, $N = 5000$ lines/cm

Then
$$(a+b) = \frac{1}{N} = \frac{1}{5000}$$

For highest order spectrum $\theta = 90°$
We know the grating formula

$$(a + b) \sin\theta = m\lambda$$

or
$$(a + b) = m\lambda$$

or
$$m = \frac{(a+b)}{\lambda}$$

$$= \frac{10^5}{5000 \times 6} = \frac{10}{3} \cong 3 \text{ (approximately)}$$

EXAMPLE 9.9 How many orders will be visible if the wavelength of an incident radiation is 5000Å and number of lines on the grating is 2620 per inch?

Solution Given $\lambda = 5000$ Å $= 5.0 \times 10^{-7}$ cm

$N = 2620$ LPI, then grating element $(a + b) = \dfrac{2.54}{2620}$ cm

We know grating formula $(a + b) \sin\theta = m\lambda$ (for highest order $\theta = 90°$), then $(a + b) = m\lambda$

or
$$m = \frac{(a+b)}{\lambda}$$

$$= \frac{2.54}{2620} \times \frac{1}{5.0 \times 10^{-5}} = 19.38 \cong 19$$

EXAMPLE 9.10 Show that only first order spectra is possible if the width of grating element is less than twice the wavelength of the light.

Solution Given $(a + b) < 2\lambda$, suppose $(a + b) = (2\lambda - x)$, then grating formula

$$(a + b) \sin\theta = m\lambda$$

$\theta = 90°$ for highest order

$$(2\lambda - x) = m\lambda$$

$$m = \frac{2\lambda - x}{\lambda} = 2 - \frac{x}{\lambda}$$

which is less than 2 or it is first order spectra.

EXAMPLE 9.11 A parallel beam of light is made to incident on a plane transmission diffraction grating of 15000 lines per inch and angle of 2nd order diffraction is found to be 45°. Calculate the wavelength of light used:

Solution N = 15000 lines/inch = $\frac{15000}{2.54}$ lines/cm

$$m = 2, \; \theta = 45°, \; \lambda = ?$$

We know the grating formula

$$(a + b) \sin\theta = m\lambda$$

$$\lambda = \frac{(a+b)\sin\theta}{m} \tag{i}$$

$$(a + b) = \frac{1}{N} = \frac{2.54}{15000} \text{ cm} \tag{ii}$$

Putting Eq. (ii) in Eq. (i), we get

$$\lambda = \frac{2.54 \sin 45°}{15000 \times 2} = 5.987 \times 10^{-5} \text{ cm} = 5987 \text{Å}$$

EXAMPLE 9.12 A plane transmission grating has 15000 lines per inch. What is the highest order of spectra, which can be observed for wavelength 6000 Å? If opaque spaces are exactly two times the transparent spaces, which order of spectra will be absent?

Solution N = 15000 lines/inch, $(a + b) = \frac{2.54}{15000}$ cm

$$\lambda = 6000 \text{ Å} = 6.0 \times 10^{-5} \text{ cm}$$

We know the grating formula

$$(a + b) \sin\theta = m\lambda$$

For highest order, $\sin\theta = 1$

$$m = \frac{a+b}{\lambda}$$

$$= \frac{2.54 \times 10^5}{15000 \times 6} = 2.8 \cong 3 \text{ approximately}$$

The condition for the spectrum of order m to be absent

$$\frac{a+b}{a} = \frac{m}{m_1}, \text{ where } m_1 = 1, 2, 3, \ldots$$

and

$$b = 2a \Rightarrow \frac{3a}{a} = \frac{m}{m_1}$$

$$m = 3m_1$$

Therefore, 3rd, 6th, 9th, etc. order of spectra will be absent.

EXAMPLE 9.13 A grating is made of 200 wires per cm placed at equal distances apart. The diameter of each wire is 0.025 mm. Calculate the angle of diffraction for third order spectrum and also find the absent spectra, if any. The wavelength of the light used is 6000Å.

Solution Given $\lambda = 6000\text{Å} = 6.0 \times 10^{-7}$ m, $m = 3$, $a = 0.025$ mm $= 2.5 \times 10^{-5}$ m

$$N = \frac{1}{(a+b)} = 200 \text{ wires/cm} = 200000 \text{ wires/m}$$

or
$$(a+b) = \frac{1}{200000}$$
$$= 5.0 \times 10^{-5} \text{ m}$$
$$(a+b) = 5.0 \times 10^{-5} \text{ m}$$
$$a = 2.5 \times 10^{-5} \text{ m}$$

Then $a = b = 2.5 \times 10^{-5}$ m

(i) Angle of diffraction

We know that the grating equation

$$(a+b) \sin\theta = m\lambda$$

or
$$\sin\theta = \frac{m\lambda}{(a+b)}$$
$$= \frac{3 \times 6.0 \times 10^{-5}}{5 \times 10^{-3}}$$
$$\theta = \sin^{-1}(0.036)$$
$$= 2.06°$$

(ii) Condition for absent spectra

$$\frac{a+b}{a} = \frac{m}{m_1}$$

if $a = b$

Then $m = 2m_1$

for $m_1 = 1, 2, 3, \ldots$

$$m = 2, 4, 6, \ldots$$

Then second, fourth, sixth, ... order spectrum will be absent.

EXAMPLE 9.14 In a diffraction grating the width of opacities and transparencies are in the ratio 1:3. Find out the absent spectra.

Solution Given $\dfrac{b}{a} = 3$ or $b = 3a$

Then, we know condition of absent spectra

$$\frac{a+b}{a} = \frac{m}{m_1}$$

or $\dfrac{m}{m_1} = \dfrac{a+3a}{a}$

or $m = 4\ m$

For $m_1 = 1, 2, 3, \ldots$

$m = 4, 8, 12, \ldots$

Thus, fourth, eighth, twelfth ... orders will be missing.

9.13 RESOLVING POWER OF AN OPTICAL INSTRUMENT

When two objects are very close together, they may appear as one object and it may difficult for the nacked eye to see them as separate. Similarly, there are two point sources very close together, the two diffraction patterns produced by each of them may overlap and, hence, it may be difficult to distinguish them as separate. To see the two objects or two spectral lines which are very close together, optical instruments like telescopes, microscope prism, gratings, etc. are employed.

The capacity of an optical instrument to show two close objects separately is called *Resolution*.

The ability of an optical instrument to resolve the images of two close point objects is called its *Resolving power*.

The eye can see two objects as separate only if the angle subtended by them at the eye is greater than one minute, which is the resolving limit of the normal eye.

9.14 RAYLEIGH CRITERION FOR THE LIMIT OF RESOLUTION

To get a exact measure of the resolving power, Lord Rayleigh on the basis of detailed study of the resultant intensity distribution in the diffraction pattern of closely spaced point sources, laid down the following criterion for resolution.

When the central maximum of one image falls on the first minimum of another images are said to be just resolved. This limiting condition of resolution is known as *Rayleigh criterion*.

The criterion is applicable for the evaluation of resolving power of a telescope, microscope, grating, prism, etc.

To illustrate the criterion, consider the intensity distribution curve of two wavelength λ and $(\lambda + d\lambda)$. The separation between their central maxima will depend upon the wavelength difference $d\lambda$. As depicted in Figure 9.29(a), $d\lambda$ is sufficiently large so that the central maxima due to both wavelengths are quite separated and the two spectral lines are distinctly resolved. If, however, the difference in the wavelengths is smaller and has a limiting value for which the angular separation between their principal maxima is such that the principal maximum of one coincide with first minimum of the other and vice versa [as shown in Figure 9.29(b)]. In this case, the curve shows a distinct dip in the middle, indicating the presence of two spectral lines corresponding to wavelengths λ and $(\lambda + d\lambda)$. The two spectral lines under this condition are said to be just resolved. If the wavelength difference $(d\lambda)$ is smaller than the limiting value, the two principal maxima show considerable overlapping. The resultant intensity curve shown in Figure 9.29(c) indicates no dip and appears as it is a diffraction pattern of only one spectral line. Hence under this condition, the two spectral lines are no resolved.

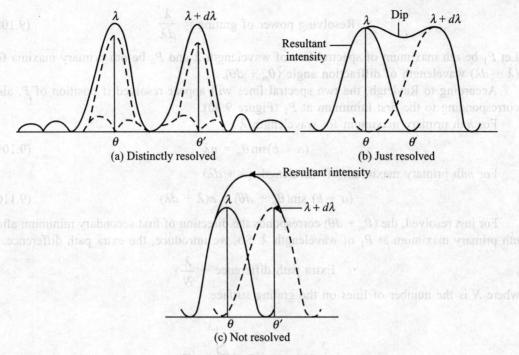

Figure 9.29 Two spectral lines.

The distribution of intensity in grating or prism spectra is of the form

$$I = I_{max} \frac{\sin^2 \alpha}{\alpha^2} \tag{9.107}$$

For the first minimum, I is minimum if $\alpha = \pi$. At the mid-point of the two maxima, the intensity due to each wavelength is obtained by substituting $\alpha = \frac{\pi}{2}$ in Eq. (9.107) [Figure 9.29(a)]. Thus the total intensity at the mid-point of the maxima is given by

$$I_{min} = 2I_{max} \frac{\sin^2(\pi/2)}{(\pi/2)^2} = \frac{8}{\pi^2} I_{max}$$

$$\therefore \quad \frac{I_{min}}{I_{max}} = \frac{8}{\pi^2}$$

Thus the Rayleigh criterion may also be stated such that the two spectral lines are just resolved if the intensity at the dip in the middle is $\frac{8}{\pi^2}$ times the intensity at either of the maxima.

9.15 RESOLVING POWER OF A PLANE DIFFRACTION GRATING

The resolving power of grating is the ratio of wavelength of any spectral line to the difference in wavelengths between this line and neighbouring line such that the two lines appear to be just resolved, i.e.

$$\text{Resolving power of grating} = \frac{\lambda}{d\lambda} \qquad (9.108)$$

Let P_1 be nth maximum of spectral line of wavelength λ and P_2 be nth primary maxima for $(\lambda + d\lambda)$ wavelength of diffraction angle $(\theta_m + d\theta)$.

According to Rayleigh, the two spectral lines will appear resolved if position of P, also corresponding to the first minimum at P_1 (Figure 9.30).

For mth primary maximum for wavelength λ

$$(a + b)\sin\theta_m = n\lambda \qquad (9.109)$$

For mth primary maximum of wavelength $(\lambda + d\lambda)$

$$(a + b)\sin(\theta_m + d\theta) = n(\lambda + d\lambda) \qquad (9.110)$$

For just resolved, the $(\theta_m + d\theta)$ corresponds the direction of first secondary minimum after nth primary maximum at P_1 of wavelength λ. So, we introduce, the extra path difference.

$$\text{Extra path difference} = \frac{\lambda}{N}$$

where N is the number of lines on the grating surface.

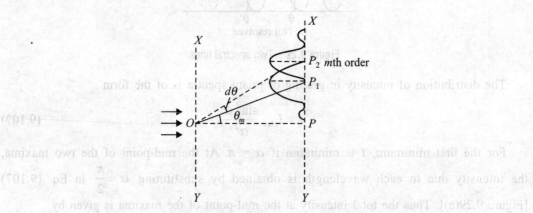

Figure 9.30 Resolution of spectral lines by plane diffraction grating.

So,

$$(a + b)\sin(\theta_m + d\theta) = m\lambda + \frac{\lambda}{N} \qquad (9.111)$$

From Eqs. (9.110) and (9.111),

$$m(\lambda + d\lambda) = m\lambda + \frac{\lambda}{N}$$

$$\Rightarrow \quad m\lambda + md\lambda = m\lambda + \frac{\lambda}{N}$$

$$\Rightarrow \quad \frac{\lambda}{d\lambda} = mN \qquad (9.112)$$

For central maxima $m = 0$, hence resolving power is zero. As the dispersive power of a grating is given by

$$\frac{d\theta}{d\lambda} = \frac{m}{(a+b)\cos\theta} \qquad (9.113)$$

Therefore, the resolving power of a diffraction grating may be expressed as

$$\frac{\lambda}{d\lambda} = mN = N(a+b)\cos\theta \frac{d\theta}{d\lambda} \qquad (9.114)$$

∴ Resolving power of grating = Total aperture × dispersive power. (9.115)

EXAMPLE 9.15 What is the least separation between wavelengths that can be resolved near 640 nm in the second order, using diffraction grating that is 5 cm wide and ruled with 32 lines per mm.

Solution $\lambda = 640$ nm $= 6.40 \times 10^{-7}$ m, $N = 32 \times 50 = 1600$, $d\lambda = ?$

We know the resolving power of grating is given by

$$\frac{\lambda}{d\lambda} = mN$$

$$d\lambda = \frac{\lambda}{mN}$$

$$= \frac{640 \times 10^{-9} \text{ m}}{2 \times 1600} = \frac{6400}{3200} \times 10^{-10} \text{ m}$$

$$= 2 \times 10^{-10} \text{ m} = 2 \text{ Å}$$

EXAMPLE 9.16 A plane transmission grating has 40000 lines per inch. If length of diffraction grating is 2 inch, then determine the resolving power in third order ($n = 3$) for a wavelength 5000 Å.

Solution The resolving power of the grating = mN
 = order of spectrum × number of lines on the grating
Length of grating = 2 inches
Number of lines per inch on the grating = 40000
Total number of lines on a grating = 40000 × 2 = 80000

So, resolving power = $\frac{\lambda}{d\lambda} = mN = 3 \times 80000 = 240000$

EXAMPLE 9.17 A plane transmission grating has 40000 lines in all with grating element 12.5×10^{-5} cm. Calculate the maximum resolving power for which it can be used in the range of wavelength 5000 Å.

Solution We know that $\frac{\lambda}{d\lambda} = mN$

Also $(a+b)\sin\theta = m\lambda$

$$\therefore \quad \frac{\lambda}{d\lambda} = N \frac{(a+b)\sin\theta}{\lambda}$$

$$\text{Maximum resolving power} = \frac{N(a+b)}{\lambda}$$

Here $N = 40000$; $(a+b) = 12.5 \times 10^{-5}$ cm; $\lambda = 5000\text{Å} = 5.0 \times 10^{-5}$ cm

Maximum resolving power

$$= \frac{40000 \times 12.5 \times 10^{-5}}{5 \times 10^{-5}} = 100000$$

Again

$$m_{\max} = \frac{(a+b)}{\lambda} = \frac{12.5 \times 10^{-5}}{5 \times 10^{-5}} = 2.5 \cong 2$$

Hence, the maximum number of order available with grating is 2. Therefore, maximum resolving power of the grating

$$N(m_{\max}) = 40000 \times 2 = 80000$$

EXAMPLE 9.18 Two spectral lines with average wavelength 6000Å are resolved in second order by a grating having 500 lines per cm. The least width of grating is 2 cm. Find the difference in the wavelength of the lines.

Solution Given $N = 500$ lines/cm, than $N = 2 \times 500$, $n = 2$, $\lambda = 6000\text{Å} = 6.0 \times 10^{-5}$ cm, $d\lambda = ?$

We know resolving power of grating $\dfrac{\lambda}{d\lambda} = mN$

or

$$d\lambda = \frac{\lambda}{mN}$$

$$= \frac{6000 \times 10^{-8}}{2 \times 1000} = 3 \times 10^{-8} \text{ cm} = 3\text{Å}$$

9.16 RESOLVING POWER OF A PRISM

As in the case of a grating, the resolving power of a prism is given by the ratio $\lambda/d\lambda$, where $d\lambda$ is the smallest wavelength difference that can be just resolved at the wavelength λ.

Figure 9.31 shows schematically the resolution of close spectral lines by a prism. A parallel beam of light consisting of slightly different wavelengths λ and $(\lambda + d\lambda)$ is incident on the prism ABC. DB is the incident wavefront. We know that the refractive index of the material of a prism depends on the wavelength of light and also that the refractive index decreases with increase of wavelength. Let n and $(n - dn)$ be the refractive indices for wavelengths λ and $(\lambda + d\lambda)$ respectively. Since the deviation of a ray by a prism depends on its refractive index, being more if the refractive index is more, the rays corresponding to λ and $(\lambda + d\lambda)$ are refracted by different amounts. The emergent wavefronts are

CE for wavelength λ and CF for wavelength $(\lambda + d\lambda)$. The converging lens L_2 focuses the wavefront CE at P_1 and CF at P_2. Thus P_1 and P_2 are the principal maxima of the diffraction pattern of λ and $(\lambda + d\lambda)$ respectively. So, the spectral lines of wavelengths λ and $(\lambda + d\lambda)$ will be observed at P_1 and P_2 respectively. According to Rayleigh's criterion, the two spectral lines will be just resolved if the first minimum of λ falls exactly at P_2, or equivalently the first minimum of $(\lambda + d\lambda)$ falls exactly at P_1.

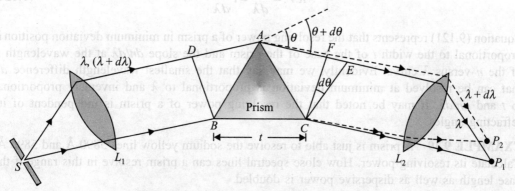

Figure 9.31   Resolution of spectral lines by a prism.

Since wavelength difference $d\lambda$ is small, the same setting of the prism is at the minimum deviation for both the wavelengths. Therefore, the rays of both the wavelengths travel through the prism parallel to the base BC. Let t be the thickness of the base.

According to Fermat's principle, the total path from the source to the image is the same for each wavelength. Therefore, for wavelength λ

$$DA + AE = nBC = nt \tag{9.116}$$

Similarly, for wavelength $(\lambda + d\lambda)$

$$DA + AF = (n - dn)t \tag{9.117}$$

Subtracting Eq. (9.117) from Eq. (9.116), we have

$$AE - AF = tdn$$

or

$$EF = tdn \tag{9.118}$$

Now, from ΔEFC

$$d\theta = \frac{EF}{EC}$$

or

$$EF = EC\, d\theta$$

Further, from the theory of diffraction, we know that for the first minimum of λ to fall at P_2

$$d\theta = \frac{\lambda}{a} \tag{9.119}$$

where a is the width of the beam which is equal to EC here. Thus,

$$EF = a\left(\frac{\lambda}{a}\right) = \lambda \qquad (9.120)$$

Combining Eqs. (9.118) and (9.120)

$$\lambda = t\, d\lambda$$

Therefore, the resolving power is

$$R.P. = \frac{\lambda}{d\lambda} = t\frac{dn}{d\lambda} \qquad (9.121)$$

Equation (9.121) represents that the resolving power of a prism in minimum deviation position is proportional to the width t of the base of the prism and the slope $dn/d\lambda$ at the wavelength λ of the n versus λ curve. Evidently, we may say that the smallest wavelength difference $d\lambda$ that can be resolved at minimum deviation is proportional to λ and inversely proportional to t and $dn/d\lambda$. It may be noted that the resolving power of a prism is independent of its refracting angles.

EXAMPLE 9.19 A prism is just able to resolve the sodium yellow lines 5890 Å and 5896 Å. Calculate its resolving power. How close spectral lines can a prism resolve in this range if the base length as well as dispersive power is doubled.

Solution Given $\lambda_1 = 5890$Å and $\lambda_2 = 5896$Å;

$$\lambda = \frac{\lambda_1 + \lambda_2}{2} = 5893\text{Å}$$

$$d\lambda = \lambda_1 \sim \lambda_2 = 6\text{Å}$$

We know the resolving power $\dfrac{\lambda}{d\lambda} = \dfrac{5893\text{Å}}{6\text{Å}} = 982$

Now, if t = base length of prism and dispersive power $\dfrac{dn}{d\lambda}$, than

$$\frac{\lambda}{d\lambda} = t \times \frac{dn}{d\lambda}$$

The smallent resolvable wavelength difference is

$$d\lambda = \frac{\lambda}{t(dn/d\lambda)}$$

If t and $\dfrac{dn}{d\lambda}$ are both doubled, then $d\lambda$ is reduced to one-fourth.

Thus, new prism can resolve a wavelength difference = $\dfrac{6}{4}\text{Å} = 1.5$ Å.

9.17 THEORY OF CONCAVE GRATING

A concave reflection grating consists of polished spherical surface. The surface is ruled with fine parallel line equally placed along the chord of the arc joining the extreme rulings. When light is incident as such a grating, it is diffracted and focused without the use of lenses. The concave grating is shown in Figure 9.32.

Let WW' be the concave grating with centre of curvature C. P and P' be the two corresponding points on the grating, such that PP' is equal to $(a + b)$, i.e. grating element. S is a narrow vertical source illuminate with monochromatic light of wavelength λ. SP and SP' be the two rays, incident on grating at the angles i and $(i + di)$ respectively. $P'S'$ and PS'

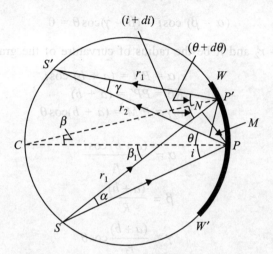

Figure 9.32 Concave grating.

be the diffracted rays with angle of diffraction θ and $(\theta + d\theta)$ respectively. Now we drawn PN and $P'M$ perpendiculars from P and P' respectively. So, the path difference between the rays $SP'S'$ and SPS'

$$\Delta = (SP' - SP) - (PS' - P'S')$$
$$= P'N - PM$$
$$= PP' \sin i - PP' \sin\theta$$
$$= PP'(\sin i - \sin\theta)$$
$$= (a + b)(\sin i - \sin\theta)$$

So, for maxima,
$$(a + b)(\sin i - \sin\theta) = n\lambda \tag{9.122}$$

In order that all the diffracted rays of given wavelength may focused at S'. The path difference for any such pair of corresponding rays should be same.

$$(a + b)(\sin i - \sin\theta) = \text{constant}$$

Differentiating above equation, we get
$$\cos i \, di - \cos\theta \, d\theta = 0 \tag{9.123}$$

Let α, β and γ are the angles in Figure 9.32. From the Figure,
$$\beta_1 = \alpha + i \quad \text{and} \quad \beta_1 = \beta + i + di$$

Then
$$\alpha + i = \beta + i + di$$
or
$$di = \alpha - \beta \tag{9.124}$$

Similarly,
$$\beta + \theta = \gamma + \theta + d\theta$$
$$d\theta = \beta - \gamma \tag{9.125}$$

Substituting the values of di and $d\theta$ in Eq. (9.123), we get

$$(\alpha - \beta)\cos i - (\beta - \gamma)\cos\theta = 0 \tag{9.126}$$

Let $P'S = r_1$ and $PS' = r_2$ and R be the radius of curvature of the grating, then

$$r_1\alpha = PN = (a + b)\cos i$$
$$R\beta = PP' = (a + b)$$
$$r_2\gamma = P'M = (a + b)\cos\theta$$

So
$$\alpha = \frac{(a+b)\cos i}{r_1}$$

$$\beta = \frac{(a+b)}{R}$$

and
$$\gamma = \frac{(a+b)}{r_2}\cos\theta$$

Substituting the values of α, β and γ in Eq. (9.126)

$$\left[\frac{(a+b)}{r_1}\cos i - \frac{(a+b)}{R}\right]\cos i - \left(\frac{a+b}{R} - \frac{a+b}{r_2}\cos\theta\right)\cos\theta = 0$$

$$\Rightarrow \left(\frac{\cos i}{r_1} - \frac{1}{R}\right)\cos i - \left(\frac{1}{R} - \frac{\cos\theta}{r_2}\right)\cos\theta = 0 \tag{9.127}$$

This is general equation for the position of S'. Since $i \neq 0$

$$\frac{\cos i}{r_1} - \frac{1}{R} = 0$$

or
$$r_1 = R\cos i \tag{9.128}$$

Similarly,
$$r_2 = R\cos\theta \tag{9.129}$$

These are polar equations of circle. If S lies on a circle of radius R, then S' also lies on the same circle. Thus if the slit and the concave grating are placed at the circumference of a circle whose diameter is equal to the radius of curvature of the grating, the spectra are focused on the circumference of circle. The circle is known as *Rowland circle*.

9.18 MOUNTINGS OF CONCAVE GRATING

The concave grating can be used to obtain spectra in a number of ways, called 'mountings'. In all mountings, the grating is tangent to Rowland circle, and the slit and the eyepiece (or photographic plate) are placed on the same circle. There are three kinds of mountings used for the concave grating. These are Rowland mounting, Paschen mounting and Eagle mounting. Here we shall study them briefly.

Rowland mounting

The geometrical fact that three vertices of a right-angled triangle be on a circle with the hypotenuse as its diameter has been utilized in the mounting by Rowland and has been named after him. The principle of the mounting has been illustrated in Figure 9.33.

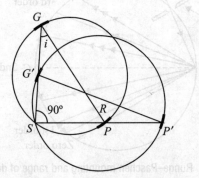

Figure 9.33 Rowland mounting.

The concave grating G and photographic plate P are mounted on opposite ends of a beam of length R. The ends of this beam rest on two movable rails SG and SP which are at right angle to each other. A slit S is placed at the junction of two rails SG and SP. S, G and P all lie on a circle of radius R, which is Rowland circle. Various regions of the spectrum are focused on the plate by sliding the beam which changes the angle of incidence. GP and $G'P'$ represent two positions of the beam.

Since plate P lies always at the centre of curvature of the grating G, hence the angle of diffraction θ is always nearly zero, and general relation for diffraction is

$$(a + b)(\sin i + \sin \theta) = m\lambda \tag{9.130}$$

Since $\theta \sim 0$, we get the position of spectral as

$$(a + b)\sin i = m\lambda \tag{9.131}$$

Now, from ΔSGP

$$\sin i = \frac{SP}{PG} = \frac{SP}{R}$$

This makes Eq. (9.131) as

$$(a+b)\frac{SP}{R} = m\lambda \quad \text{or} \quad \lambda = \left(\frac{a+b}{mR}\right)SP \tag{9.132}$$

Since $(a + b)$ and R constant for a particular concave grating, hence for a given order of spectrum SP is proportional to λ. As explained earlier, this type of spectrum is called normal spectrum. The railing SP can be graduated to given wavelength directly. This method of mountings is good and easier for the determination of wavelength. However the defect of this mounting is that it shows astigmatism with higher order spectra and is now rarely used.

Runge–Paschen mounting

This is the most popular term of concave grating mountings. It was devised by Runge and Paschen and is shown in Figure 9.34.

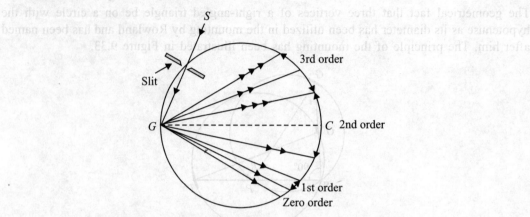

Figure 9.34 Runge–Paschen mounting and range of different spectra.

In this mounting, the slit is held on the Rowland circle and the various order of spectra are obtained on the photographic plate placed along the circumference of the circle. The different orders of spectra obtained in the various regions can be photographed simultaneously. This is its chief advantage. It is not possible with the other mountings. The chief drawback of this mounting is that spectrum is not normal and there astigmatism is of every order of spectrum, except zero order.

Eagle mounting

In this type of mounting, the grating GG' is mounted on a stand as shown in Figure 9.35.

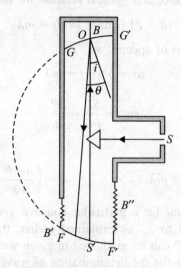

Figure 9.35 Eagle mounting.

The grating can be moved along the stand and can also be turned around the vertical axis. Slit S and right-angled prism are placed as shown in the Figure. The arrangement becomes equal to an arrangement having slit at S' such that $PS = PS'$. Bellows BB' are provided to allow setting of photographic plate FF' at a suitable tilt.

To obtain spectra, first the grating GG' is set at suitable inclination. Then it is moved along the stand so as to get some spectral lines in focus at S'. Finally, the tilt of the plane FF' is adjusted so that all spectral lines obtained along FF' are well focused. The Rowland circle passes through effective slit S and the plate FF' as shown by dotted circle in Figure 9.35.

Merits and demerits of Rowland mounting Runge-Paschen mounting and eagle mounting

Merits	Rowland mounting	Runge-Paschen mounting	Eagle mounting
Merits	1. The spectrum is normal. 2. The rail SP can be graduated to read wavelengths directly. 3. This method of mounting is good. 4. It is easier to find out the wavelength.	1. The arrangement is very rigid. 2. The plate may be placed anywhere along the frame and any portion of the spectrum can be photographed without altering the adjustment. 3. By using a large photographic film, curved to the curvature of Rowland circle, the spectra of all orders may be simultaneously photographed. 4. Two or more slits can be placed at different points on the Rowland circle to work at different angles of incidence.	1. This mounting is most compact and requires much less space than other mountings. Hence, it is widely used in vacuum spectrographs for the study of extreme ultraviolet region (because a much smaller volume of space has to be evacuated). 2. Temperature control is easier. 3. High order spectra can be attained. 4. The astigmatism is much smaller than in other mountings.
Demerits	1. It needs large space. 2. It is expensive and involves a considerable amount of mechanism. 3. The adjustment is disturbed when the spectral region is changed and, hence, refocusing is to be done. 4. It shows some astigmatism.	1. The spectrum is normal only in the region which falls near the grating normal, i.e. near C (Figure 9.34). 2. Except where the spectrum is normal, there is a strong astigmatism. 3. The mounting requires a large space and a solid steady floor insulated from the vibrations of the building. 4. It is difficult to avoid temperature changes which may cause shift in the spectral lines during long exposures.	1. The spectrum is not normal. 2. The instrument is not simultaneously in focus for all spectral regions. 3. It involves both translation and rotation of the grating, together with rotation of the plate. 4. The arrangement of this mounting is not very simple.

9.19 ECHELON GRATING

The resolving power of a grating is given by

$$R.P. = N \times m$$

where N is the total number of lines on the grating and m is the order of spectra. Naturally, to increase the resolving power, we have to increase N and m. In practice, the diamond point ruling the grating fails when N approaches 100,000. It, thus, sets a limit to the size of the surface which can be ruled. On the other hand, with a ruled grating, we cannot use an order of spectrum greater than third, because the intensity is very poor in the higher orders.

Michelson in 1898, devised a special type of grating in which N is kept fairly small (say 20 to 4), whereas the order of spectrum m is made very high, but with no loss in intensity. This is called an *Echelon grating* and is shown in Figure 9.36.

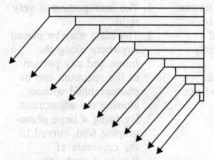

Figure 9.36 Echelon grating.

Construction

It consists of a few plane, parallel and equally thick plates of optically worked glass or quartz. The plates are arranged in a step formation such that each plate projects the same distance beyond the one below it. This distance (*i.e.*, the width of each step) is usually 1 mm. When a parallel beam of light falls normally on the grating, each point on the steps acts as a secondary source of disturbance sending out rays in all possible directions. A telescope (not shown in Figure 9.36) collects these diffracted rays, since the width of each diffracting element (step) is very large compared with the wavelength of light.

Working equation

Let w be the width of each step, t the thickness of each plate and n the refractive index of glass (Figure 9.37).

Let us consider two parallel rays diffracted through an angle θ from corresponding points A and B. The path difference between these rays

$$= nBD - AC$$
$$= nBD - (CN - AN)$$
$$= nBD - (BM - AN)$$
$$= nBD - (BD \cos\theta - AD \sin\theta)$$

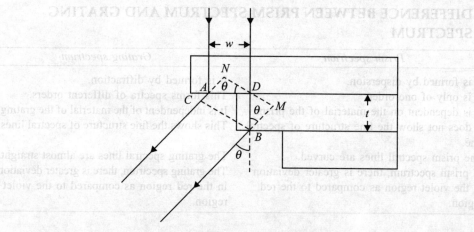

Figure 9.37 Illustration for working equation of Echelon grating.

$$= nt - (t\cos\theta - w\sin\theta)$$
$$= nt - t + w\theta \quad [\text{As } \theta \text{ is small, } \cos\theta = 1, \sin\theta = 0]$$
$$= (n-1)t + w\theta$$

The grating equation giving the principal maximum is then

$$(n-1)t + w\theta = m\lambda$$

where m is order of spectrum.

$$m = \frac{(n-1)t + w\theta}{\lambda} \tag{9.133}$$

Taking $n = 1.5$, $t = 1$ cm, $\theta = 0$ and $\lambda = 5 \times 10^{-5}$ cm, we have

$$m = \frac{(1.5-1) \times 1\,\text{cm}}{5 \times 10^{-5}\,\text{cm}} = 10{,}000$$

This is the order of the first spectrum observed.

If the number of plates (N) is 40, the resolving power is

$$N \times m = 40 \times 10{,}000 = 400{,}000 \tag{9.134}$$

Thus a very high resolving power is achieved. Further as the light is diffracted through very narrow angles, practically all the light is thrown into one or two orders which result in a high intensity.

The disadvantage of the Echelon grating is that the spectra of successive orders, being very high, overlap.

Its advantage is in the analysis of the hyperfine structure of an individual spectral line, which is actually composed of several components very close together. To avoid overlapping, an auxiliary prism is used to isolate the individual line under examination.

9.20 DIFFERENCE BETWEEN PRISM SPECTRUM AND GRATING SPECTRUM

S.No.	Prism spectrum	Grating spectrum
1.	It is formed by dispersion.	It is formed by diffraction.
2.	It is only of one order.	This forms spectra of different orders.
3.	It is dependent on the material of the prism.	It is independent of the material of the grating.
4.	It does not show the fine structure of spectral line.	This shows the fine structure of spectral lines.
5.	The prism spectral lines are curved.	The grating spectral lines are almost straight.
6..	In prism spectrum, there is greater deviation in the violet region as compared to the red region.	The grating spectrum, there is greater deviation in the red region as compared to the violet region.

9.21 DIFFERENCE BETWEEN DISPERSIVE POWER AND RESOLVING POWER OF A GRATING

S.No.	Dispersive power (D.P.)	Resolving power (R.P.)
1.	It is the rate of change of angle of diffraction with the wavelength of light used, i.e. $d\theta/d\lambda$	It is the ratio of the wavelength of any spectral line to the smallest wavelength difference between the neighbouring lines for which the spectral line can be just resolved.
2.	$\text{D.P.} \Rightarrow \dfrac{d\theta}{d\lambda} = \dfrac{m}{(a+b)\cos\theta}$	$\text{R.P.} \Rightarrow \dfrac{\lambda}{d\lambda} = mN$
3.	It is independent of the number of lines on grating surface.	R.P. increases as the number of lines on the grating surface increases.
4.	It is dependent on the grating element.	It is independent of the grating element.
5.	This gives an idea of the angular separation of two spectral lines.	Resolving power tells us about the closeness of the two spectral lines.
6.	It is independent on the number of lines per centimetre on the grating.	It is dependent on the total number of lines on the ruled surface.

FORMULAE AT A GLANCE

9.1 Fresnel's half-period zone

Consider a plane wavefront falling on aperture. We divide the plane wavefront into a large number of circles (zones) around a point such that

(a) Average distances of zones from screen

$$= p + (2m \pm 1)\dfrac{\lambda}{4}$$

(b) Area of half-period zone $= \pi p \lambda$

(c) Obliquity factor $(-\theta_m) = \pi \lambda \theta_m$

9.2 Zone plate consists of large number of concentric circles whose radii are proportional to the square root of natural numbers and alternate annular zones are opaque for light. This a convex lens which produces image of light source on the screen placed at a finite distance.

(a) $r_m^2 = bm\lambda$

$r_m \propto \sqrt{m}$

(b) $f = \dfrac{r_m^2}{m\lambda}$

f = focal length of particular zone plate.

9.3 Fraunhofer type of diffraction due to single slit

(a) **Resultant intensity**

$$I = R^2 = A^2 \dfrac{\sin^2 \alpha}{\alpha^2}$$

and $\alpha = \dfrac{\pi a \sin \theta}{\lambda}$ where a is slit width.

(b) **Direction of minima** $a \sin \theta = \pm m\theta$

(c) **Ratio of principal maxima's intensities**

$$I_0 : I_1 : I_2 : I_3 := 1 : \dfrac{4}{9\pi^2} : \dfrac{4}{25\pi^2} : \dfrac{4}{49\pi^2} \cdots = 1 : \dfrac{1}{22} : \dfrac{1}{61} : \dfrac{1}{121} : \cdots$$

(d) **Width of central maxima**

$$\theta = \pm \dfrac{\lambda}{a} = \dfrac{2\lambda}{a}$$

(e) **Principal maximum**

$\theta = 0$

$\alpha = \dfrac{\pi a \sin \theta}{\lambda} = 0$ or $\sin \theta = 0$

$\alpha = 0$

$I \propto A^2 \propto R^2$

(f) **Minimum intensity**

$\sin \alpha = 0$; $\alpha = \pm\pi, \pm 2\pi, \pm 3\pi, \pm 4\pi, \ldots,$ etc.

$\dfrac{\pi a \sin \theta}{\lambda} = \pm m_1 \pi$

$a \sin \theta = \pm m_1 \lambda$

where $m_1 = 1, 2, 3, \ldots,$ etc.

a = width of the single slit

θ = angle of diffraction

m_1 = order of diffraction

so different θ's correspond to position of minimas on both sides of screen.

(g) Secondary maxima

$$R = \frac{ma \sin \alpha}{\alpha}$$

for R_{max}

$$\frac{d\left(\frac{\sin \alpha}{\alpha}\right)}{\delta \alpha} = 0$$

$$\Rightarrow \tan \alpha = \alpha$$

(Refer to Figure 9.18)

$\tan \alpha = \alpha$ (maxima)

The root $\alpha = 0$ corresponds to the central maxima and other roots can be obtained by $\alpha = \tan \alpha$ conditions. The roots occur at $\alpha = 1.43\pi, 2.46\pi$, etc. known as 1st maxima, 2nd maxima.

9.4 Fraunhofer diffraction at two slits

(a) Resultant intensity

$$I = 4A_0^2 \left(\frac{\sin \alpha}{a}\right)^2 \cos^2 \beta$$

(b) For maxima

$$\beta = \pm m\pi$$
$$(a + b) \sin \theta = \pm m\lambda$$

(c) For minima

$$(a + b) \sin \theta = (2m \pm 1)\frac{\lambda}{2}$$

(d) For diffracted maxima and minima

(i) For maxima

$$\sin \alpha = 1 = \sin \frac{m\pi}{2}$$

$$\frac{\pi}{\lambda} a \sin \theta = \frac{m\pi}{2} \quad \text{or} \quad a \sin \theta = \frac{m\lambda}{2}$$

(ii) For maxima

$$\sin \alpha = 0 = \sin m_1 \pi$$
$$\alpha = m_1 \pi$$

$$\frac{\pi}{\lambda} a \sin \theta = m_1 \pi \quad \text{or} \quad a \sin \theta = m_1 \lambda$$

a = width of opaque region

b = width of transparent region

$(a + b)$ = distance between two consecutives slits.

m = order of diffraction pattern.

Diffraction pattern will be symmetrical on both sides of central maxima.

9.5 *Diffraction grating:* A diffraction grating consists of large number of equally spaced identical slits. It can be constructed by ruling large number of parallel, equidistant lines on a optically flat glass or plastic plate with the help of diamond suffer. Ruled or scratch lines act as opaque region to light while remaining region are transparent to light and act as slits. For example a grating ruled with 5,000 lines/cm has slit spacing $(a + b)$ equal to inverse of this number. Hence,

$$(a+b) = \frac{1}{5000}\left(\frac{cm}{lines}\right) = 2 \times 10^{-4} \, cm$$

Here
$$\beta = \frac{\pi(a+b)\sin\theta}{\lambda}$$

$$\alpha = \frac{\pi b \sin\theta}{\lambda}$$

where b = width of single slit.
  a = width of opaque region.
  $(a + b)$ = distance between two consecutive slits grating element.

(a) **Resultant intensity**
$$I = R^2 = A^2 \frac{\sin^2\alpha}{\alpha^2} \frac{\sin^2 N\beta}{\sin^2\beta}$$

(b) **Principal maxima**
$$\beta = \pm m\pi \quad \text{where } n = 0, 1, 2, 3, \ldots$$
$$(a+b)\sin\theta = \pm m\lambda$$

(c) **Secondary minima**
$$\sin N\beta = 0$$
$$N\beta = \pm m_1 a$$
$$N(a+b)\sin\theta = \pm m_1\lambda \quad \text{where } n = 1, 2, 3, \ldots (N-1)$$

(d) **Secondary maxima**
$$I' = A^2 \frac{\sin^2\alpha}{\alpha^2} \frac{N^2}{1+(N^2-1)\sin^2\beta}$$

$$\frac{\text{Intensity of secondary maxima}}{\text{Intensity of principal maxima}} = \frac{1}{1+(N^2-1)\sin^2\beta}$$

9.6 Angular half width of principal maxima
$$d\theta = \frac{1}{mN \cot\theta_n}$$

9.7 Absent spectra or missing order
$$\frac{a+b}{a} = \frac{m}{m_1}$$

m is missing order, which depends on a, b and m_1.

9.8 Dispersive power of a grating

$$\frac{d\theta}{d\lambda} = \frac{m}{(a+b)\cos\theta}$$

9.9 Resolving power of prism

$$\frac{\lambda}{d\lambda} = t\frac{dn}{d\lambda}$$

9.10 Resolving power of grating

$$\frac{\lambda}{d\lambda} = mN$$

9.11 Rowland mountings

$$\lambda = \left(\frac{a+b}{mR}\right)SP$$

where SP is a distance.

9.12 Echelon grating

$$m = \frac{(n-1)t + w\theta}{\lambda}$$

where m = order of spectrum, n = refractive index of glass, t = thickness of each plate, w = width of each step.

SOLVED NUMERICAL PROBLEMS

PROBLEM 9.1 An object is placed at 20 cm from a zone plate and the brightest image is situated at 20 cm from the zone plate with light of λ = 4000 Å. Calculate the number of Fresnel's zones in a radius of 1 cm of that plate.

Solution Given

$$u = 20$$
$$v = 20$$
$$a_m = 1 \text{ cm}$$
$$f = ?$$
$$m = ?$$

Since we know that for a zone plate

$$\frac{1}{u} + \frac{1}{v} = \frac{m\lambda}{a_m^2} = \frac{1}{f}$$

we have

$$\frac{1}{20} + \frac{1}{20} = \frac{1}{f}$$

or

$$f = 20 \text{ cm}$$

Now

$$\frac{m\lambda}{a_m^2} = \frac{1}{f}$$

$$m = \frac{a_m^2}{f\lambda}$$

$$= \frac{(1\text{ cm})^2}{10\text{ cm} \times (4 \times 10^{-5}\text{ cm})}$$

$$= 2500$$

PROBLEM 9.2 In Fraunhofer's diffraction due to narrow slit, a screen is placed 1.5 m away from the lens to obtain pattern. If wavelength of light is 500 nm and the first minima lie 5 mm on either side of the central maxima, compute the slit width.

Solution In Fraunhofer's diffraction due to a narrow slit (Figure 9.20), given that

$$D = 1.5 \text{ m} = 150 \text{ cm}$$
$$\lambda = 500 \text{ nm} = 5 \times 10^{-5} \text{ cm}$$
$$m = 1$$
$$y = 5 \text{ mm} = 0.5 \text{ cm}$$
$$a = ?$$

We know that
$$a \sin\theta = m\lambda$$

or
$$\sin\theta = \frac{\lambda}{a}$$

and
$$\sin\theta = \frac{y}{D}$$

$\therefore$
$$a = \frac{\lambda D}{y}$$

$$= \frac{5 \times 10^{-5} \times 150}{0.5}$$

$$= 1500 \times 10^{-5}$$
$$= 1.5 \times 10^{-2}$$
$$= 0.015 \text{ cm}$$

PROBLEM 9.3 A single slit is illuminated by light composed of two wavelengths λ_1 and λ_2. One observes that due to Fraunhofer's diffraction, the first minima obtained for λ_1 coincides with the second diffraction minima of λ_2. What is the relation between λ_1 and λ_2?

Solution The angular position of nth minimum is given by

$$a \sin\theta = m\lambda$$

The angular position of the first minimum for wavelength λ_1 is given by

$$a \sin\theta_1 = 1 \times \lambda_1 \tag{i}$$

The angular position of the second minimum for wavelength λ_2 is given by

$$a \sin\theta_2 = 2 \times \lambda_2$$

426 Fundamentals of Optics

But,
$$\theta_1 = \theta_2$$
$$\therefore \quad a \sin\theta_1 = 2\lambda_2 \qquad \text{(ii)}$$

From Eqs. (i) and (ii), we have
$$\lambda_1 = 2\lambda_2$$

PROBLEM 9.4 In a double slit Fraunhofer's diffraction pattern, the screen is placed 170 cm away from the slits. The width of the slit is 0.08 mm and the slits are 0.4 mm apart. Calculate the wavelength of light, if the fringe width is 0.25 cm. Also find the missing order.

Solution Given
$$D = 170 \text{ cm} = 1.7 \text{ m}$$
$$\beta = 0.25 \text{ cm} = 2.5 \times 10^{-3} \text{ m}$$
$$a = 0.08 \text{ mm} = 8 \times 10^{-5} \text{ m}$$
$$b = 0.4 \text{ mm} = 4 \times 10^{-4} \text{ m}$$
$$d = b = 4 \times 10^{-4} \text{ m}$$

In double slit Fraunhofer's diffraction pattern, the fringe width is given by
$$\beta = \frac{\lambda D}{d}$$
or
$$\lambda = \frac{\beta d}{D}$$
$$= \frac{2.5 \times 10^{-3} \times 4 \times 10^{-4}}{1.7}$$
$$= 0.5882 \times 10^{-6}$$
$$= 588.2 \text{ nm}$$

The condition for missing order
$$\frac{(a+b)}{a} = \frac{m}{m_1}$$
or
$$m = \left(\frac{a+b}{a}\right) m_1$$
$$= \left(\frac{8 \times 10^{-5} + 4 \times 10^{-4}}{8 \times 10^{-5}}\right) m_1$$
$$= \frac{48}{8} m_1$$
$$= 6 m_1$$

Hence the missing orders are 6, 12, 18, 24, 30, ...

PROBLEM 9.5 Light of wavelength 500 nm falls normally on a plane transmission grating having 15,000 lines in 3 cm. Find the angle of diffraction for maximum intensity in first order (Given $\sin^{-1} 0.25 = 14°29'$).

Solution Given

$$\lambda = 500 \text{ nm} = 5.0 \times 10^{-7} \text{ m}$$
$$N = 15000$$
$$\text{Grating length} = 3 \text{ cm}$$
$$m = 1$$

When light falls normally on a grating, the directions θ of the maximum intensity in the diffracted light are given by

$$(a + b)\sin\theta = \pm m\lambda, \quad \text{where } n = 0, 1, 2, 3, \ldots$$

where $(a + b)$ is the grating element, n is the order of the maximum.

For $n = 1$,

$$(a+b)\sin\theta = \lambda \Rightarrow \sin\theta = \frac{\lambda}{(a+b)}$$

Here

$$(a+b) = \frac{3}{15000} \text{ cm}$$
$$= 2 \times 10^{-4} \text{ cm}$$
$$= 2 \times 10^{-6} \text{ cm}$$

$$\therefore \qquad \sin\theta = \frac{5 \times 10^{-7} \text{ cm}}{2 \times 10^{-6} \text{ cm}}$$
$$= 0.25$$

or $\qquad \theta = \sin^{-1}(0.25) = 14°29'$

EXAMPLE 9.6 How many orders will be observed by a grating having 4000 lines/cm, if it is illuminated by a visible light of wavelength in the range of 4000 Å to 7500 Å?

Solution Given

$$(a+b) = \frac{1}{4000} \text{ cm}$$
$$= \frac{1}{400000} \text{ m}$$
$$\lambda_1 = 4000 \text{ Å} = 4 \times 10^{-7} \text{ m}$$
$$\lambda_2 = 7500 \text{ Å} = 7.5 \times 10^{-7} \text{ m}$$
$$\theta = 90° \quad \text{(for maximum order)}$$

(a) $\qquad (a+b)\sin\theta = m_1\lambda_1$

or $\qquad m_1 = \frac{(a+b)\sin 90°}{\lambda_1}$

$$= \frac{1 \times 1}{400000 \times 4 \times 10^{-3}}$$
$$= 6.25 \cong 6$$

(b)
$$(a+b)\sin\theta = m_2\lambda_2$$

or
$$m_2 = \frac{(a+b)\sin 90°}{\lambda_2}$$

$$= \frac{1\times 1}{400000 \times 7.5 \times 10^{-7}}$$

$$= 3.33$$
$$\cong 3$$

Hence, the order of spectrum varies from 3 to 6.

PROBLEM 9.7 Light of $\lambda = 500$ nm falls on a grating normally. Two adjacent principal maxima occur at $\sin\theta = 0.2$ and $\sin\theta = 0.3$ respectively. Calculate grating element.

Solution The principal maxima of order n is given by

$$(a+b)\sin\theta = m\lambda$$

Two adjacent orders means n and $(n + 1)$, they are occurred at $\sin\theta = 0.2$ and $\sin\theta = 0.3$ respectively.
Thus,
$$(a+b)0.2 = m\lambda \qquad \text{(i)}$$
$$(a+b)0.3 = (m+1)\lambda \qquad \text{(ii)}$$

Subtracting Eq. (i) by Eq. (ii), we get

$$(a+b)0.1 = \lambda = 500 \text{ nm}$$

$$(a+b) = 5 \times 10^{-6} \text{ m}$$

PROBLEM 9.8 What is maximum number of lines of a grating which will resolve the third order spectrum of two lines having wavelengths 5890 Å and 5896 Å?

Solution Given

$$m = 3$$
$$\lambda = 5890 \text{ Å}$$
$$d\lambda = 6 \text{ Å}$$

We know that the resolving power of grating is

$$\frac{\lambda}{d\lambda} = mN$$

or
$$N = \frac{1}{m}\frac{\lambda}{d\lambda}$$

$$= \frac{1}{3} \cdot \frac{5890 \text{ Å}}{6 \text{ Å}}$$

$$= 327.22 \cong 327$$

PROBLEM 9.9 A grating has 6000 lines per cm drawn on it. If its width of 10 cm, compute:
(a) The resolving power is second order.
(b) The smallest wavelength that can be resolved in the third order in 6000 Å wavelength region.

Solution Given
$$N = 6000 \times 10 = 6 \times 10^4$$
(a) $m = 2$
$$\text{Resolving power of grating} = mN$$
$$= 2 \times 6 \times 10^4$$
$$= 1.2 \times 10^5$$
(b) Resolving power for $\lambda = 6000$ Å and $m = 2$
$$\frac{\lambda}{d\lambda} = mN$$
or
$$d\lambda = \frac{\lambda}{mN}$$
$$= \frac{6000}{3 \times 6 \times 10^4}$$
$$= 0.033 \text{ Å}$$

PROBLEM 9.10 What other spectral lines in the visible range 400 nm to 700 nm will coincide with the 5th order line of 600 nm in the grating spectrum.

Solution We know that grating equation
$$(a + b)\sin\theta = m\lambda$$
Suppose the values of wavelength λ_2 of the other spectral lines coincide with the nth order line of λ_1 for mth order. Then,
$$(a + b)\sin\theta = m\lambda_1$$
$$(a + b)\sin\theta = m_1\lambda_2$$
and
$$m\lambda_1 = m_1\lambda_2$$
Here, we have
$$\lambda_1 = 600 \text{ nm}$$
$$m = 5$$
Hence,
$$5 \times 600 = m_1\lambda_2$$
or
$$\lambda_2 = \frac{300}{m_1}\text{nm}$$
Then
$$\lambda_2 = 750 \quad \text{nm} \quad \text{for } m_1 = 4$$
$$\lambda_2 = 600 \quad \text{nm} \quad \text{for } m_1 = 5$$
$$\lambda_2 = 500 \quad \text{nm} \quad \text{for } m_1 = 6$$
$$\lambda_2 = 428.6 \text{ nm} \quad \text{for } m_1 = 7$$
$$\lambda_2 = 375 \quad \text{nm} \quad \text{for } m_1 = 8$$

Hence, the other spectral lines in the visible range of 400 nm to 700 nm are 500 nm and 428.6 nm, which will coincide with the 5th order of 600 nm.

PROBLEM 9.11 A parallel beam of light (λ = 5890 Å) is incident perpendicularly on a slit of width 0.1 mm. Calculate the angular width and the linear width of the central maximum formed on a screen 100 cm away.

Solution In single slit arrangement (Fraunhofer)
   Given
$$\lambda = 5890 \text{ Å} = 5.89 \times 10^{-7} \text{ m}$$
$$a = 0.1 \text{ mm}$$
$$= 1.0 \times 10^{-4} \text{ m}$$
$$D = 100 \text{ cm} = 1 \text{ m}$$

We know that
$$a \sin\theta = m\lambda$$
$$\sin\theta = \frac{m\lambda}{a}$$
$$= \frac{5.89 \times 10^{-7}}{1.0 \times 10^{-4}}$$
$$= 5.89 \times 10^{-3} \text{ rad}$$

Since angle is very small, then $\sin\theta \cong \theta$, i.e.
$$\theta = 5.89 \times 10^{-3} \text{ rad}$$

Total angular width of central maximum = 2θ
$$= 2 \times 5.89 \times 10^{-3} \text{ rad}$$
$$= 11.78 \times 10^{-3} \text{ rad}$$
$$= 1.178 \times 10^{-4} \text{ rad}$$

If y is the linear half width and D is the distance of the screen from the slit, then
$$y = D\theta \qquad \text{[for small } \theta\text{]}$$

∴ The linear width of the central maximum on the screen = $2y$
$$= 2 \times 0.589$$
$$= 1.178 \text{ cm}$$

PROBLEM 9.12 What is the highest order spectrum which may be seen with monochromatic light of wavelength 6000 Å by means of a diffraction grating with 5000 lines/cm?

Solution Given
$$\lambda = 6000 \text{ Å} = 6.000 \times 10^{-5} \text{ cm}$$
$$N = 5000 \text{ lines/cm}$$

Then
$$(a + b) = \frac{1}{N} = \frac{1}{5000}$$

For highest order spectrum $\theta = 90°$

$$(a+b)\sin\theta = m\lambda$$

or $$(a+b) = m\lambda$$

or $$m = \frac{(a+b)}{\lambda}$$

$$= \frac{10^5}{5000 \times 6}$$

$$= \frac{100}{30}$$

$$= \frac{10}{3} \approx 3 \quad \text{(approximately)}$$

PROBLEM 9.13 A parallel beam of light is made incident on a plane transmission diffraction grating of 15000 lines per inch and angle of 2nd order diffraction is found to be 45°. Calculate the wavelengths of light used.

Solution Given

$$N = 15000 \text{ lines/inch} = \frac{15000}{2.54} \text{ lines/cm}$$
$$m = 2$$
$$\theta = 45°$$
$$\lambda = ?$$

We know that

$$(a+b)\sin\theta = m\lambda$$

$$\lambda = \frac{(a+b)\sin\theta}{m} \quad \text{(i)}$$

$$(a+b) = \frac{1}{N} = \frac{2.54}{15000} \text{ cm} \quad \text{(ii)}$$

Putting Eq. (ii) in Eq. (i), we get

$$\lambda = \frac{2.54 \sin 45°}{15000 \times 2}$$

$$= 5.987 \times 10^{-5}$$

$$= 5987 \text{ Å}$$

PROBLEM 9.14 A plane transmission grating has 15000 per inch. What is the highest order of the spectra, which can be observed for wavelength 6000 Å? If opaque spaces are exactly two times of the transparent spaces, which order of spectra will be absent?

Solution Given

$$N = 15000 \text{ lines/inch}$$

$$(a+b) = \frac{2.54}{15000} \text{ cm}$$

$$\lambda = 6000 \text{ Å} = 6.0 \times 10^{-5} \text{ cm}$$

We know that
$$(a+b)\sin\theta = m\lambda$$
For highest order
$$\sin\theta = 1$$
$$m = \frac{(a+b)}{\lambda}$$
$$\Rightarrow \quad m = \frac{2.54 \times 10^5}{15000 \times 6}$$
$$= 2.8 \cong 3 \text{ (approximately)}$$

Hence the third order is highest order visible.

The condition for the spectrum of order m to be absent
$$\frac{(a+b)}{a} = \frac{m}{m_1}$$
where $m_1 = 1, 2, 3, \ldots$.
Here
$$b = 2a \Rightarrow \frac{3a}{a} = \frac{m}{m_1}$$
$$m = 3\,m$$

for $m = 1, 2, 3, 4, \ldots$ the value of $m_1 = 3, 6, 9, 12, \ldots$.

Therefore, third, sixth, nineth, etc. orders of spectra will be absent.

PROBLEM 9.15 A diffraction grating having 4000 lines per cm is illuminated normally by light of wavelength 5000 Å. Calculate its dispersive power in the third order spectrum.

Solution The dispersive power of grating is given by
$$\frac{d\theta}{d\lambda} = \frac{m}{(a+b)\cos\theta}$$
Also
$$(a+b)\sin\theta = m\lambda$$
Here
$$m = 3$$
$$\lambda = 5000 \text{ Å} = 5000 \times 10^{-8} \text{ cm} = 5.0 \times 10^{-5} \text{ cm}$$
$$(a+b) = \frac{1}{4000} \text{ cm}$$
$$\sin\theta = \frac{3 \times 5.0 \times 10^{-5}}{1} \times 4000 = 0.6$$
$$\cos\theta = \sqrt{1 - \sin^2\theta}$$
$$= \sqrt{1 - (0.6)^2} = 0.8$$

$$\therefore \quad \frac{d\theta}{d\lambda} = \frac{3 \times 4000}{0.8}$$
$$= 15000$$

PROBLEM 9.16 A plane transmission grating has 40000 times in all with grating element 12.5×10^{-5} cm. If the maximum resolving power of the grating is 80000, find out the range of wavelength for which it can be used.

Solution Given
$$N = 40000$$
$$(a + b) = 12.5 \times 10^{-5} \text{ cm}$$
$$(\text{Resolving power})_{max} = 80000$$

Maximum resolving power $\dfrac{\lambda}{d\lambda} = mN$

$$mN = 80000$$
$$m = \frac{80000}{40000} = 2$$

For maximum
$$(a + b)\sin 90° = m\lambda$$
$$\Rightarrow \quad \lambda = \frac{(a+b)}{2}$$
$$= \frac{12.5 \times 10^{-5} \text{ cm}}{2}$$
$$= 6.5 \times 10^{-5} \text{ cm}$$
$$= 6500 \text{ Å}$$

Smallest wavelength difference
$$\frac{\lambda}{d\lambda} = mN$$
or
$$d\lambda = \frac{\lambda}{d\lambda}$$
$$= \frac{65}{8} \times 10^{-2}$$
$$= 8.33 \times 10^{-2}$$
$$= 0.08 \text{ Å}$$

Hence $\lambda_1 = 6500$ Å and $\lambda_2 = (6500 \pm 0.08)$Å.

PROBLEM 9.17 How many orders will be visible if the wavelength of the incident radiation is 4800 Å and the number of lines on the grating is 2500 per inch.

Solution We know that
$$(a + b)\sin \theta = m\lambda$$

434 Fundamentals of Optics

The maximum possible value of $\sin \theta = 1$, hence the maximum number of orders is given by

$$(a + b) = m\lambda$$

or
$$m = \frac{(a+b)}{\lambda}$$

Here,
$$(a + b) = \frac{2.54}{2500} \text{ cm}$$

$$\lambda = 4800 \text{ Å}$$
$$= 4.8 \times 10^{-5} \text{ cm}$$

$\therefore$
$$m = \frac{2.54}{2500 \times 4.8 \times 10^{-5}}$$

$$= \frac{2.54 \times 10^5}{2500 \times 4.8}$$

$$= \frac{25.4 \times 10000}{4.8 \times 2500}$$

$$> 21 = 21 \text{ (approximately)}$$

PROBLEM 9.18 What should be the minimum number of lines per inch in a half inch width grating to resolve the $D_1(5896 \text{ Å})$ and $D_2(5890 \text{ Å})$ lines of sodium?

Solution Given

$$\lambda_1 = 5896 \text{ Å} = 5.896 \times 10^{-5} \text{ cm}$$
$$\lambda_2 = 5890 \text{ Å} = 5.89 \times 10^{-5} \text{ cm}$$
$$\lambda = \frac{\lambda_1 + \lambda_2}{2}$$
$$= \frac{5896 + 5890}{2}$$
$$= 5893 \text{ Å}$$
$$= 5.893 \times 10^{-5} \text{ cm}$$

Now, applying condition for just resolution

$$\frac{\lambda}{d\lambda} = mN$$

or
$$N = \frac{1}{m} \frac{\lambda}{d\lambda}$$

$$= \frac{5.893 \times 10^5}{1 \times 0.006 \times 10^{-5}}$$

$$= 982 \text{ lines/cm} \qquad [\text{for } m = 1]$$

Hence the two lines will be just resolved in the first order with minimum lines per inch.

PROBLEM 9.19 An angle of the second order diffraction is found to be 30°, when a beam of light (λ = 6000 Å) is made incident perpendicular on a plane transmission diffraction grating. Calculate number of lines per cm of the grating.

Solution Given

$$m = 2$$
$$\theta = 30°$$
$$\lambda = 6000 \text{ Å} = 6.0 \times 10^{-5} \text{ cm}$$
$$N = ?$$

We know that

$$(a + b) \sin \theta = m\lambda$$

$$(a + b) = \frac{m\lambda}{\sin \theta}$$

$$= \frac{2 \times 6.0 \times 10^{-5}}{\sin 30°}$$

$$= 24.000 \times 10^{-5}$$

$$= 2.4 \times 10^{-4} \text{ cm}$$

Then

$$N = \frac{1}{(a + b)}$$

$$= \frac{1}{2.4 \times 10^{-4}}$$

$$= \frac{100000}{24}$$

$$= 4170 \text{ lines/cm}$$

PROBLEM 9.20 A diffraction grating which has 4000 lines to a cm is used at normal incidence. Calculate the dispersive power of the grating in the third order spectrum in the wavelength region 5000 Å.

Solution Given

$$N = 4000 \text{ lines/cm}$$
$$m = 3$$
$$\lambda = 5000 \text{ Å}$$

The dispersive power is defined as

$$\frac{d\theta}{d\lambda} = \frac{m}{(a + b) \cos \theta}$$

$$= \frac{m}{(a + b)\sqrt{1 - \sin^2 \theta}}$$

But, we have

$$(a + b) \sin \theta = m\lambda$$

or $\sin\theta = \dfrac{m\lambda}{(a+b)}$

$$\dfrac{d\theta}{d\lambda} = \dfrac{m}{(a+b)\sqrt{1-\left(\dfrac{m\lambda}{a+b}\right)^2}}$$

$$= \dfrac{3}{\dfrac{1}{4000}\left[1-\left(\dfrac{3\times 5\times 10^{-5}}{1/4000}\right)^2\right]^{1/2}}$$

$$= \dfrac{3\times 4000}{\sqrt{1-0.36}}$$

$$= \dfrac{3\times 4000}{0.8}$$

$$= 15000 \text{ radian/cm}$$

$$= 1.5 \times 10^4 \text{ rad/cm}$$

PROBLEM 9.21 A Rowland (concave) grating has 6000 lines per cm. Find the angular separation between the two yellow lines of neon (5882 Å and 5852 Å) in the first order spectrum observed normally.

Solution For a concave grating, we have

$$(a+b)(\sin i - \sin\theta) = m\lambda$$

Differentiating above equation, we get

$$(a+b)\cos\theta \dfrac{d\theta}{d\lambda} = m$$

or $d\theta = \dfrac{m\,d\lambda}{(a+b)\cos\theta}$

When the spectrum is observed normally ($\theta = 0$), so that $\cos\theta = 1$, then we have

$$d\theta = \dfrac{m\,d\lambda}{(a+b)}$$

Here $m = 1$

$d\lambda = (5882 - 5852) \times 10^{-8} = 30 \times 10^{-8}$ cm

and $(a+b) = \dfrac{1}{6000}$ cm

∴ $d\theta = \dfrac{1\times (30\times 10^{-8}\text{cm})}{(1/6000)\text{cm}}$

$$= 1.8 \times 10^{-3} \text{ rad}$$

CONCEPTUAL QUESTIONS

9.1 What is the difference between interference and diffraction fringes?

Or

Explain the difference between interference and diffraction phenomena.

Ans

S.No.	Interference	Diffraction
1.	In the phenomenon of interference, the interaction takes place between two coherent sources.	In the phenomenon of diffraction, the interaction takes place between secondary wavelets originating from different points of exposed parts of the same wavefront.
2.	In interference pattern, the region of minimum intensity is usually almost perfect dark.	While it is not so in diffraction pattern.
3.	The width of the fringes in interference may or may not be equal or uniform.	While in diffraction pattern fringe width of various fringes are never equal.
4.	In an interference pattern, the maxima are of same intensity.	But in a diffraction pattern, they are of varying intensity.

9.2 What do you mean by diffraction of light?

Or

What is diffraction?

Ans If an opaque obstacle or aperture be placed between a source of light and a screen, a sufficiently distinct shadow is obtained on the screen. If the size of the obstacle or aperture is small, there is a departure from straight line propagation, and the light bends round the corners of the obstacle or aperture and enters the geometrical shadow. This bending of light is called 'diffraction'.

9.3 Distinguish between Fresnel and Fraunhofer class of diffraction.

Or

Distinguish between Fraunhofer and Fresnel diffraction.

Or

Differentiate between Fresnel and Fraunhofer diffraction.

Ans

Fresnel's Diffraction	Fraunhofer's Diffraction
In this class of diffraction source and screen are placed at finite distance from the aperture of obstacle having sharp edges. In this class no lens is used for making rays parallel or convergent. The incident wavefront is either spherical or cylindrical.	In this class of diffraction source and screen or telescope (through which image is viewed) are placed at infinity. In this case, the wavefront which is incident on the aperture or obstacle is plane.

9.4 What is the cause of diffraction?

Ans Diffraction occurs due to interference of secondary wavelets between different portions of a wavefront allowed to pass a small aperture or obstacle.

9.5 What two main changes in diffraction pattern of single slit will you observe when the monochromatic source of light is replaced by a source of white light?

Ans When the monochromatic source is replaced by a source of white light, the diffraction pattern show following changes:
  (i) In each diffraction order, the diffracted image of the slit gets dispersed into component colours of white light. As fringe width is directly proportional to wavelengths, so the red fringe with higher wavelength is wider than the violet fringe with smaller wavelength.
  (ii) In higher spectra, the dispersion is more and it causes overlapping of different colours.

9.6 Is it correct to say that diffraction is interference between different parts of the same wavefront?

Ans Yes, diffraction is due to the interference of secondary wavelets starting from different parts of the wavefront that passes through an aperture or that is unobstructed by the obstacle.

9.7 Sound wave can undergo diffraction at the corner of a building, but not light wave. Why?

Ans The condition that a wave can undergo diffraction is that its wavelength must be comparable with the dimension of the diffracting obstacle. Now, the wavelength of sound is 10 m that of light wave is ~ 5893×10^{-8} cm. So, the sharpness of the corner of a building satisfies the condition of diffraction for the sound wave but not for the light wave. Hence, though the corner of a building can diffract sound wave and it fails to diffract light wave.

9.8 Two students are separated by a 7 m partition in a room of 10 m high. If both sound and light waves can bend around obstacle, how is it that the students are unable to see each other, even then they can converse easily.

Ans The size of partition is very large as compared to wavelength of light and hence it is not diffracted and two students cannot see each other.

The wavelength or sound is nearly of the order of the height of the partition. It causes diffraction of sound and hence they can converse easily with each other.

9.9 Why is diffraction of sound waves more evident in our daily life than that of light wave?

Or

Interference and diffraction, which is more common and why.

Ans To obtain a well defined diffraction pattern, the size of the obstacle or aperture should be of the same order as the wavelength. The wavelength of sound is comparable to the size of most of the obstacle or aperture, we come across in daily life. The diffraction of sound is, therefore, a common experience. But wavelength of light is very small as compared to the size of the obstacles or aperture, we come across in daily life. We, therefore, do not observe diffraction of light as an everyday phenomenon.

9.10 What is the significance of diffraction phenomenon on the nature of light?

Ans The diffraction phenomenon is a very strong and convincing proof in favour of Huygen's hypothesis of the undulatory nature of light.

9.11 What should be the order of size of obstacle/aperture for diffraction of light?

Or

State essential condition for diffraction light to occur.

Ans Size of obstacle/aperture should be of the same order as that of the wavelength of light.

9.12 A single slit diffraction pattern is obtained using a beam of red light. What happens if the red light is replaced by blue light?

Ans Diffraction fringes become narrower and crowded together.

9.13 What is the effect of increasing the number of lines per cm of the grating on the diffraction grating?

Or

What are the advantages of increasing the number of rulings in a grating?

Ans On increasing the number of lines per cm, decreases the grating element $(a + b)$. As a result the angle of diffraction θ increases for a given order. This results in a less number of spectra separated by large angles or an increased dispersive power.

9.14 Why optical diffraction grating cannot be employed to study the diffraction of X-rays?

Ans Diffraction effect is possible if the width of the slit on the grating is of the order of wavelength of light. Since the width of the slits on the optical diffraction grating is quite large as compared to wavelength of X-rays and hence optical diffraction grating cannot be employed to study the diffraction of X-rays.

9.15 If we look at the sun through a piece of fine cloth, we observe coloured spectra at the site of hole in the cloth. Why?

Ans At the site of fine apertures or holes in the cloth, we get diffraction patterns. As the light from the sun is white light, consisting of colours from red to violet, we get coloured spectra or coloured streaks.

9.16 In a plane transmission grating 15000 lines/inch are taken, why?

Ans With the increase in number of lines, the secondary maxima relative to the principal maxima decreases and becomes negligible. When N becomes large the secondary maxima are not visible with a grating having about 15000 lines or more per inch.

9.17 Explain why grating of larger number of lines are preferred.

Ans A grating with larger number of lines is preferred because
  (i) with the increase of number of lines per cm, the dispersive power of grating increases.
  (ii) with increase in total number of lines in the effective part of the grating resolving power increases.

9.18 Differentiate dispersive power and resolving power of a grating.

Ans

Dispersive power (D.P.)	Resolving power (R.P.)
Dispersive power is the rate of change of angle of diffraction with the wavelength of light used i.e., $(d\theta/d\lambda)$.	Resolving power is the ratio of the wavelength of any spectral line to the smallest wavelength difference between neighbouring lines for which the spectral line can just resolved i.e., $(\lambda/d\lambda)$.
D.P. $= \dfrac{d\theta}{d\lambda} = \dfrac{n}{(a+b)\cos\theta}$	R.P. $= \dfrac{\lambda}{d\lambda} = nN$
D.P. is independent of the number of lines on grating surface.	R.P. increases as number of lines on grating surface increases.
D.P. depends on grating element.	R.P. is independent of the grating element.
D.P. gives the idea of the angular separation of two spectral lines.	R.P. tells us about the closeness of the two spectral lines.
D.P. depends upon the number of lines per centimeter on the grating.	R.P. depends on the total number of lines on the rules surface.

EXERCISES

Theoretical Questions

9.1 Write any two differences between the interference and diffraction.
   [Agra University, 2011, 2012]

9.2 Give the difference between interference and diffraction (only four differences).
   [Avadh Univesity, 2011; Bundelkhand University, 2006, 2007]

9.3 What is diffraction? Give examples of Fresnel and Fraunhofer diffraction.
   [Bundelkhand University, 2012]

9.4 Distinguish between the Fresnel and Fraunhofer class of diffraction.
   [Kanpur University, 2011; Avadh Univesity, 2010, 2011]

9.5 Explain the difference between Fresnel and Fraunhofer class of diffraction phenomenon.
   [Avadh University, 2012]

9.6 What is the difference between Fresnel and Fraunhofer class of diffraction?
   [Avadh University, 2013]

9.7 Compare the Fresnel's and Fraunhofer's diffraction. **[Kanpur University, 2010]**

9.8 Compare a zone plate with a biconvex lens for its similarities and differences (three each). **[Bundelkhand University, 2008; Agra University, 2006, 2007]**

9.9 What is zone plate? How it is constructed? Show that a zone plate behaves like a convex lens of multiple foci. Deduce expression for its focal length. **[Agra University, 2008]**

9.10 Explain multiple focal lengths of a zone plate. [Agra University, 2009]

9.11 What are Fresnel's half period zones? Prove that the area of a half period zone on a plane wavefront is independent of the order of the zone and that the amplitude due to large wavefront at a point in front of it is just half that due to the first half period zone acting alone. [Bundelkhand University, 2007]

9.12 Prove that the intensity at any point due to plane wavefront is one fourth of that due to the first half period zone. [Bundelkhand University, 2010]

9.13 Explain the half period zones. [Bundelkhand University, 2011]

9.14 What is a zone plate? Describe the construction and working of a zone plate. Show that a zone plate has multiple foci. [Bundelkhand University, 2012]

9.15 What is zone plate and how is it made? Explain how a zone plate acts like convergent lens lying multiple foci? Derive an expression for its focal length. [Avadh University, 2010]

9.16 What is diffraction? Explain Fresnel's theory of half period zones. [Avadh University, 2011]

9.17 What is zone plate? Give its theory. Show that zone plate has multiple foci. Compare the zone plate with a convex lens. What is meant by phase reversal zone plate? [Avadh University, 2012]

9.18 What is a zone plate? Describe the construction and working of a zone plate. Show that a zone plate has multiple foci. [Kanpur University, 2007]

9.19 Compare the working of a zone plate with a converging lens. [Kanpur University, 2008]

9.20 Using Fresnel's theory of half-period zone, discuss rectilinear propagation of light. [Kanpur University, 2009]

9.21 How construction of zone plate and explain how zone plate differ from convex lens. [Kanpur University, 2011]

9.22 Explain, why higher number zone do not contribute for intensity in Fresnel's half-period zone. [Kanpur University, 2011]

9.23 What is significance of half-period zone in Fresnel's class of diffraction? [Kanpur University, 2012]

9.24 What are Fresnel's half-period zones? Prove that the area of a half-period zone on a plane wavefront is independent of the order of the zone, and that the amplitude due to large wavefront at a point infront of it is just half due to the first half-period zone acting alone. Give Fresnel's explanation of the rectilinear propagation of light.

9.25 What is a zone plate. Give its theory. Show that a zone plate has multiple foci. Compare the zone plate with a convex lens.

9.26 A sharp razor blade is held parallel to a narrow slit illuminated by monochromatic source of light. Discuss intensities of light at different positions on a screen placed at a distance on the other side of source. [Kanpur University, 2009]

9.27 What is meant by diffraction of light? Explain the phenomenon of diffraction of light at straight edge, and explain the formation of alternate bright and dark bands. [Kanpur University, 2012]

9.28 Describe with necessary theory, the Fresnel type of diffraction due to a straight edge. Show the intensity distribution in the diffraction pattern and comment on its features. How would you use it to determine the wavelength of light?

9.29 A sharp razor blade is held vertically in a beam of monochromatic light coming from a narrow slit parallel to the edge of the blade. Discuss the position and intensities observed on a screen placed behind the blade.

9.30 Discuss the Fraunhofer diffraction at a circular aperture and obtain the expression for the half angular width of central bright ring. **[Agra University, 2013]**

9.31 Describe the diffraction effects due to narrow circular aperture illuminated by monochromatic light from a point source. Show that if the screen on which diffraction pattern is obtained be moved towards the aperture, the intensity at the centre would be alternately maximum and minimum.

9.32 Discuss Fresnel's diffraction at a small opaque disc illuminated by monochromatic light. Explain why the centre of the shadow of the disc is bright. What happens if the size of the disc is increased?

9.33 Clearly differentiate between Fresnel and Fraunhofer class of diffraction. Find the expression for intensity due to single slit of the width a. **[Agra University, 2006, 2007]**

9.34 Distinguish between the diffraction fringes due to single slit and the interference fringes due to two narrow slits. **[Agra University, 2013]**

9.35 Obtain expression for the intensity distribution due to diffraction at a single slit and discuss it graphically. Distinguish between the diffraction fringes to the interference fringes due to two narrow slits. **[Bundelkhand University, 2005]**

9.36 Discuss intensity distribution in single slit Fraunhofer diffraction.
[Bundelkhand University, 2011]

9.37 Discuss single slit Fraunhofer's diffraction. **[Avadh University, 2010]**

9.38 Describe single slit Fraunhofer diffraction and derive the positions for maxima and minima. **[Avadh University, 2011]**

9.39 Describe the phenomenon of Fraunhofer class of diffraction at a circular aperture.
[Avadh University, 2013]

9.40 Explain by the Fresnel's theory, the rectilinear propagation of light.
[Kanpur University, 2006]

9.41 What is meant by diffraction of light? Explain the phenomenon of diffraction of light at straight edge. **[Kanpur University, 2006]**

9.42 Obtain an expression for resolving power of grating. **[Avadh University, 2013]**

9.43 Prove that the width of central maximum for single slit in the Fraunhofer class diffraction is $2\theta = 2\lambda/a$; where a = width of slit and λ = wavelength of light used.
[Kanpur University, 2011]

9.44 In Fraunhofer class diffraction show that the width of principal maxima increases as slit width decreases. **[Kanpur University, 2012]**

9.45 Explain:
(a) The phenomenon of diffraction of light.
(b) The difference between Fresnel's and Fraunhofer's classes of diffraction phenomenon.
(c) The difference between interference and diffraction of light.

9.46 Describe Fraunhofer's diffraction due to a single slit and deduce the positions of the maxima and minima. Show that the relative intensities of successive maxima are nearly $1 : \dfrac{1}{22} : \dfrac{1}{61} : \cdots$ What will happen if the width of the slit is made equal to the wavelength of light.

9.47 Explain Fraunhofer's diffraction due to a double slit. How does its intensity distribution curve differ from the curve obtained due to single slit? What is the effect of
(i) increasing the slit width,
(ii) increasing the slit separation, and
(iii) increasing the wavelength of light?
What are missing orders?

9.48 Give the construction and theory of a plane diffraction grating of the transmission type, and explain the formation of spectra.

9.49 Explain and obtain the condition of absent spectra in a plane transmission grating. If the width a of the opaque space (ruling) is equal to the width b of transparent space, which order will be absent? Which if $a = 2b$?

9.50 Obtain the condition of absent spectra in plane transmission grating. If the width a of opaque space (ruling) is equal to the width b of transparent space, which orders will be absent? Is $b = 2a$? **[Kanpur University, 2007]**

9.51 What do you mean by an optical transmission grating? **[Avadh University, 2013]**

9.52 (a) Find the maximum number of orders available with a grating. Show that if the width of the grating elements is less than twice the wavelength of light, then only first order is available.
(b) Define dispersive power of a grating and obtain and expression for it.
(c) Deduce an expression for a half angular width of a principal maximum in the diffraction pattern of a grating having N slits.

9.53 How would you use it to find the wavelength of light?

9.54 Show that the rays forming spectrum in a grating suffers minimum deviation when the angle of incidence equals the angle of diffraction.

9.55 In what respects does
(a) a prism spectrum differ from a grating spectrum, and
(b) resolving power differ from dispersive power of a grating?

9.56 Explain Rayleigh's criteria of just resolution. **[Agra University, 2008]**

9.57 What do you understand by the resolving power of optical instrument? What is Rayleigh's criterion for just resolution? Find an expression for the resolving power of plane grating.
[Agra University, 2011]

9.58 What is Rayleigh's criterion for the limit resolution? **[Bundelkhand University, 2007]**

9.59 What is difference between resolving power and dispersive power?
[Bundelkhand University, 2012]

9.60 Deduce an expression for the resolving power of grating. **[Avadh University, 2010]**

9.61 Explain the Rayleigh criterion of resolution. Find an expression for the resolving power of telescope. **[Avadh University, 2011]**

9.62 What is meant by resolving power of a plane transmission grating? Derive an expression for it. **[Avadh University, 2012]**

9.63 Explain clearly what do you understand by the limit of resolution of telescope and obtain an expression for it. **[Avadh University, 2012]**

9.64 What do you mean by resolving criterion of Rayleigh for two wavelengths?
[Avadh University, 2013]

9.65 Describe resolving power of a telescope. **[Avadh University, 2013]**

9.66 Explain and obtain the condition of absent spectra in plane transmission grating.
[Kanpur University, 2006]

9.67 What is meant by the resolving power of an optical instrument? Describe the Rayleigh criterion for resolution. **[Kanpur University, 2006]**

9.68 Deduce an expression for the resolving power of a plane transmission grating.
[Kanpur University, 2006]

9.69 What is Rayleigh criterion of resolving power? **[Kanpur University, 2007]**

9.70 Explain the difference between magnifying power and resolving power of a telescope.
[Kanpur University, 2010]

9.71 Explain the Rayleigh's criterion of resolution and find out the expression for the resolving power of a grating. **[Kanpur University, 2010]**

9.72 What is Rayleigh, criterion of resolving power of two spectral lines?
[Kanpur University, 2011]

9.73 What is Rayleigh's criterion of resolving power of two spectral lines? Find the resolving power of the grating. **[Kanpur University, 2012]**

9.74 What was the main difficulty in plane transmission grating and how it was removed by concave grating? **[Bundelkhand University, 2006]**

9.75 In what respect does a prism spectrum differ from grating spectrum.
[Bundelkhand University, 2008]

9.76 State advantages of concave grating over plane grating.
[Bundelkhand University, 2011]

9.77 Discuss the properties of Cornu's spiral and explain its relationship with Fresnel's half period zones. Show that how the spiral can be used to obtain the intensity distribution in Fresnel diffraction pattern of a straight edge. **[Avadh University, 2008]**

9.78 Give the theory of concave reflection grating and deduce the conditions of focusing the spectra. Describe the working of Rowland's mounting. What is normal spectrum? Why are concave grating preferable to plane grating? **[Avadh University, 2008]**

9.79 Describe the working of Rowland's mounting. What are the advantages of concave grating over to plane grating? What do you understand by normal spectrum?
[Avadh University, 2009]

9.80 State the advantage of concave grating over a plane grating.
[Kanpur University, 2007]

9.81 Prove that the half angular width of the principal maximum in any order m is given by $d\theta = 1/(N\,m\cot\theta_m)$, where N is the number of lines in the grating and θ_m is the angle of diffraction. [Kanpur University, 2008, 2011]

9.82 Discuss merits and demerits of different mountings in a concave grating.
[Kanpur University, 2009]

9.83 Describe Fraunhofer diffraction due to a narrow slit and deduce the positions of minima and maxima. Also show that the relative intensities of successive maxima are approximately

$$I_0 : I_1 : I_2 : I_3 : \cdots = 1 : \frac{4}{9\pi^2} : \frac{4}{25\pi^2} : \frac{4}{49\pi^2} \cdots = 1 : \frac{1}{22} : \frac{1}{61} : \frac{1}{121} : \cdots$$

Show the intensity distribution graphically. [Kanpur University, 2008, 2011, 2012]

9.84 Given the theory of concave reflection grating and deduce conditions for focusing the spectra. Describe the working of Rowland's mounting.

9.85 Give the theory of concave reflection grating. Describe various methods of mounting the same and point out then relative merits and demerits.

9.86 How does the resolution improved in Echelon grating? Illustrate your answer by Michelson's echelon grating.

Numerical Problems

9.1 The diameter of the first ring of a zone plate is 1 mm. If the plane wave ($\lambda = 5000$ Å) falls on the plate, where should the screen be placed so that light is focused on a brightest spot. [**Ans.** $f = 50$ cm]

9.2 If the distance of screen from zone plate is 50 cm and the wavelength of light used is 5×10^{-4} mm then find out the area of any zone. [**Ans.** 7.85×10^{-7} m^2]
[Avadh University, 2011]

9.3 The innermost zone of a zone plate has a diameter of 0.425 mm.
 (a) Find the focal length of the plate when it is used with parallel incident light of wavelength 4471 Å from a helium lamp.
 (b) Find its first subsidiary focal length. [**Ans.** (a) 40.40 cm (b) 13.47 cm]

9.4 A zone plate is found to give a series of images of a point source on its axis. If the strongest and second strongest images are at distances of 30 cm and 6 cm respectively from the zone plate, both on the same side remote from the source, calculate the distance of the source from the zone plate, the principal focal length and the radius of the first zone. [Wavelength of light 5×10^{-5} cm] [**Ans.** $d = 30$ cm, $f = 15$ cm, $R = 2.74 \times 10^{-2}$ cm]

9.5 A zone plate is illuminated by parallel white light. A lens of crown glass is to be placed in contact with it so that the resultant image is free from chromatic aberration. Design a suitable lens and find the position of the image, if the zone plate has 'principal' focal length of 100 cm for red light of wavelength 8×10^{-5} cm. [Refractive index of crown glass = 1.500 for wavelength 8×10^{-5} and 1.535 for wavelength 4×10^{-5} cm]

[**Ans.** 12.3 cm]

9.6 A screen is placed at a distance of 100 cm from a circular hole illuminated by a parallel beam of light of wavelength 6400 Å. Compute the radius of the fourth half-period zone.

[**Ans.** $r_4 = 0.16$ cm]

9.7 A concave grating has 15,000 lines to an inch. Calculate the angular separation of two mercury lines 5770 Å and 5461 Å in the first order spectrum in Rowland mounting.

[**Ans.** $d\theta = 1.82 \times 10^{-2}$ rad]

9.8 In an experiment for observing diffraction pattern due to a straight edge, the distance between the slit source ($\lambda = 6000$ Å) and the straight edge is 6 m and that between the straight edge and eyepiece is 4 m. Calculate the position of first-three maxima and their separations. [**Ans.** $x_1 = 0.20$ cm, $x_2 = 0.346$ cm, $x_3 = 0.447$ cm,

$x_2 - x_1 = 0.146$ cm and $x_3 - x_2 = 0.101$]

9.9 A parallel beam of monochromatic light of wavelength 6000 Å falls perpendicularly on an opaque screen in which there is a circular hole of radius 0.12 cm. The emergent light is received on another screen which is at a distance of 80 cm from the opaque screen. Show that the centre of the diffraction, pattern is a bright spot. [**Ans.** $m = 3$]

9.10 In a straight edge diffraction pattern, one observes that the most intense maximum occurs at a distance of 1 mm from the edge of the geometrical shadow. Calculate the wavelength of light, if the distance between the screen and the straight edge is 300 cm.

[**Ans.** $\lambda \approx 4480$ Å]

9.11 A convex lens of focal length 20 cm is placed after a slit of width 0.6 mm. If the centre of the second dark band is 1.6 cm from the middle of central bright band, deduce the wavelength of light. [**Ans.** 0.04 cm]

9.12 A single slit of width 0.14 mm is illuminated normally by monochromatic light and diffraction hands are observed on a screen 2 m away. If the centre of the second dark band is 1.6 cm from the middle of central bright band, deduce the wavelength of light.

[**Ans.** $\lambda = 5600$ Å]

9.13 In a double slit Fraunhofer's diffraction pattern, the screen is 160 cm away from the slit width of 0.08 cm and they are 0.4 mm apart. Calculate the wavelength of light of the fringe spacing is 0.25 cm. [**Ans.** $\lambda = 6250$ Å]

9.14 Show that for a double slit Fraunhofer pattern if $a = mb$, the number of bright fringes (or parts thereof) within the central diffraction maximum will be equal to 2 m.

9.15 A single slit of width 0.14 mm is illuminated normally by monochromatic light and diffraction bands are observed on a screen 2 m away. If the centre of the second dark band is 1.6 cm from the middle of central bright band, deduce the wavelength of light.

[**Ans.** $\lambda = 5600$ Å]

9.16 A grating having 15000 lines per inch produces spectra of mercury arc. The green line of mercury spectrum has a wavelength of 5461 Å. What is the angular separation between the first order green line and the second order green line? [**Ans.** $\theta_1 \sim \theta_2 = 21.36°$]

9.17 A diffraction grating is just able to resolve two lines of wavelengths 5140.34 Å and 5140.85 Å in the first order. Will it resolve the lines 8037.2 Å and 8037.5 Å in second order? **[Ans. Will not resolve]**

9.18 Calculate the number of lines per cm in a plane transmission grating which gives first order line of the light of the wavelength 6×10^{-5} cm at angle of a diffraction.
[Ans. $N = 8333$]

9.19 A plane transmission grating has 16000 lines to an inch over a length of 5 inches. The wavelength region of 6000 Å is found in the second order. Find:
(a) The resolving power of the grating, and
(b) The smallest wavelength difference that can be resolved.
[Ans. (a) 16000, (b) 0.0375 Å]

9.20 A diffraction grating used at normal incidence gives a yellow line ($\lambda = 6000$ Å) in a certain spectral order superimposed on a blue line ($\lambda = 4800$ Å) of next higher order. If the angle of diffraction is $\sin^{-1}\left(\dfrac{3}{4}\right)$, calculate grating element.
[Ans. 3.2×10^{-6} m]

9.21 In the second order spectrum of a diffraction grating, a certain spectral line appears at an angle of 10°, while another line of wavelength 0.5 Å higher appears at an angle 3″ greater. Find the wavelengths of the lines and the minimum grating width required to resolve them. **[Ans. $\lambda_1 = 6063.8$ Å, $\lambda_2 = 6064.3$ Å, $W = 4.2$ cm]**

9.22 Microwaves of 6000 MHz frequency are incident on a slit of variable width and large length. A receiver is placed at a large distance on the other side at angle 30° with normal. For what width of slit, would receiver show zero intensity?
$$\left[\textbf{Ans.}\quad a = \frac{m}{10},\ \text{i.e.}\ 0.1\ \text{m}, 0.2\ \text{m}, \cdots\right]$$

9.23 A grating has 1000 lines ruled over it in the region of wavelength 6000 Å. Find the separation between two wavelengths that can be just resolved in the first order spectrum and the resolving power in the second order. **[Ans. RP = 2000]**

9.24 A plane diffraction grating has 40,000 lines. Determine its resolving power in the second order for a wavelength of 6000 Å. **[Ans. 80000]**

9.25 What is the longest wavelength that can be observed in the fourth order for a transmission grating having 5000 lines/cm? Assume normal incidence. **[Ans. 5000 Å]**
[Agra University, 2013]

9.26 Light of wavelength 5000 Å is incident normally on a plane transmission grating having 15000 lines in 3 cm. Find the angle of diffraction for maximum intensity in first order. Given $\sin^{-1} 0.25 = 14°29'$. **[Ans. 14°29′]**
[Kanpur University, 2008]

9.27 What should be number of lines in a grating, which will just resolve in the third order for the sodium doublet ($\lambda = 5890$ Å and 5896 Å). **[Ans. 327] [Kanpur University, 2009]**

9.28 How many orders will be visible if the wavelength of light is 5000Å and the number of lines per inch on the grating is 2650? **[Ans. 2] [Kanpur University, 2010]**

9.29 A diffraction grating is ruled with 100,000 lines in a distance of 8.0 cm and used in the first order to study the structure of a spectrum line at $\lambda = 4230$ Å. How does the chromatic resolving power compare with that of 60° glass prism with a base of 8.0 cm and refractive indices 1.5608 at $\lambda = 4010$ Å and 1.5462 at $\lambda = 4450$ Å ?

[**Ans.** Grating resolving power = 100,000; prism resolving power = 26,550]

9.30 Light is incident at 70° on a plane reflection grating which has 6000 lines per cm ruled over a width of 3 cm. What is the maximum resolving power available at a wavelength of 5000 Å ? Find the corresponding angle of diffraction and the smallest resolvable difference in wavelength. [**Ans.** $\dfrac{\lambda}{\delta\lambda} = 1.08 \times 10^5$; $\theta = 59°21'$; $\delta\lambda = 0.046$ Å]

9.31 White light falls normally on a transmission grating that contains 1000 lines per cm. At what angle red light ($\lambda_0 = 650$ nm) emerge in the first order spectrum. [**Ans.** $\Theta_1 = 3.73°$]

9.32 What is the total number of lines a grating must have in order just to separate the sodium doublet ($\lambda_1 = 5895.9$ Å, $\lambda_2 = 5890.0$ Å) in the third order. [**Ans.** 333]

9.33 A diffraction grating is just able to resolve two lines of wavelengths 5140.34 Å and 5140.85 Å in the first order. Will it resolve the lines 8037.2 Å and 8037.5 Å in second order? [**Ans.** Will not resolved]

9.34 Calculate the number of lines per centimeter in a plane transmission grating which gives first order line of the light of the wavelength 6×10^{-5} cm at angle of diffraction 30°.

[**Ans.** $N = 8333$]

9.35 A diffraction grating used at normal incidence gives a line (5400 Å) in a certain order superimposed on the violet line (4050 Å) of the next higher order. How many lines per cm are there in the grating if angle of diffraction is 30°? [**Ans.** 3086]

9.36 In a diffraction grating which spectral line in 4th order will overlap with 3rd order line of the wavelength 5461 Å? [**Ans.** $\lambda = 4095.75$ Å]

9.37 Give the formula for the resolving power of transmission grating. What is the ratio of resolving power of two gratings, which have 15000 lines in 2 cm and 10000 lines in 1 cm in 1st order? Each grating has lines in its 2.5 cm width. [**Ans.** 3 : 4]

9.38 A plane transmission diffraction grating has 15000 lines/inch. Determine the angle of separation of the 5048 Å and 5016 Å lines of helium in the 2nd order spectrum.

[**Ans.** $\Delta\theta = 1.5'$]

9.39 Sodium light is incident normally on a plane transmission grating having 3000 lines per cm. Find the direction of the first order for the D-lines and the width of the grating necessary to resolve them. [Wavelength of D-lines: 5890 Å and 5896 Å]

[**Ans.** $\Theta = 10°11'$ and $d = 3.27$ mm]

9.40 Light is incident normally on a grating 0.5 cm wide with 2500 lines. Find the angle of diffraction of two sodium lines in the first order spectrum. Are two lines resolved?

[**Ans.** $\theta_1 = 17.13°$, $\theta_2 = 17.15°$, Yes]

9.41 D_1 and D_2 lines of sodium are 6Å apart. What should be the minimum number of lines in a diffraction grating to resolve them. [**Ans.** $N = 982$ for $m = 1$]

9.42 Two spectral lines have wavelength λ and $(\lambda + d\lambda)$ respectively. Show that if $d\lambda \ll \lambda$, their angular separation $d\theta$ in a grating spectrometer is given by

$$d\theta = \frac{d\lambda}{\left[\left(\frac{a+b}{m}\right)^2 - \lambda^2\right]^{1/2}}$$

where $(a + b)$ is the grating element and n is the order at which lines are observed.

9.43 Light composed of two spectral lines with wavelengths 6000 Å and 6000.5 Å falls normally on a diffraction grating 10 mm wide. At a certain diffraction angle θ these lines are close being resolved (according to Rayleigh criterion). Find θ. [Ans. $\theta = 46°$]

9.44 Two spectrum lines at $\lambda = 6200$ Å have a separation of 0.652 Å. Find the minimum number of lines a diffraction grating must have to just resolve this doublet in the second order spectrum. [Ans. $N = 4755$]

9.45 If the reflection coefficient of plates in Fabry–Perot etalon is 0.9 and the wavelength of light used is 6000 Å. Then calculate:
  (i) the coefficient of sharpness of fringes [Ans. 0.10]
  (ii) resolving power of etalon for maxima of 20000th order. [Ans. 1050]

[Agra University, 2013]

9.46 A concave grating has 5000 lines per cm. Calculate the angular separation between two lines of wavelengths 5882 Å and 5852 Å in the first order spectrum when seen normally. [Ans. 1.5×10^{-3} radian]

[Bundelkhand University, 2005]

9.47 In a concave grating with Rowland mounting two lines of known wavelengths 4886.24 Å and 488.7.37 Å fall at linear positions 12.372 mm and 14.268 mm. Calculate the wavelength of the line whose reading is 11.638 mm. [Ans. $\lambda = 4885.8$ Å]

Multiple Choice Questions

9.1 A zone plate behaves like a convex lens of focal length 50 cm for a light of wavelength 5000 Å. The radius of the first-half period zone is
  (a) 5 mm (b) 0.5 mm
  (c) 1 mm (d) 1.5 mm

9.2 Fresnel's half-period zone differs from each other by a phase difference of
  (a) 2π (b) π
  (c) $\frac{\pi}{2}$ (d) $\frac{\pi}{2}$

9.3 The principal focal length of a zone plate is given by
  (a) $f = \frac{a_n^2}{n\lambda}$ (b) $f = \frac{a_n}{n\lambda}$
  (c) $f = \frac{a_n}{n\lambda^2}$ (d) $f = \frac{1.22\, a^2}{n\lambda}$

9.4 The aperture of the objective of a microscope has diameter of 0.8 cm. The object is illuminated with visible light of wavelength 4040 Å. The maximum limit of resolution for the microscope is
   (a) 5.05×10^{-5} rad
   (b) 6.16×10^{-5} rad
   (c) 4.1×10^{-5} rad
   (d) 3.2×10^{-5} rad

9.5 In a single slit diffraction pattern, for a slit width (d) and wavelength (λ), the separation between the central maximum and the first minimum is
   (a) $\theta = \dfrac{\lambda}{d}$
   (b) $\theta = \dfrac{\lambda}{2d}$
   (c) $\theta = \dfrac{\lambda}{4d}$
   (d) $\theta = \dfrac{\pi}{2}$

9.6 In a plane transmission grating the angle of diffraction for the second order principal maxima for the wavelength 5×10^{-5} cm is 30°. The number of lines in one cm of the grating surface will be
   (a) 100
   (b) 10
   (c) 1000
   (d) 5000

9.7 A grating resolves a given doublet in the first order and the other grating of same width resolves the same doublet in the second order. The ratio of number of lines on the two grating is
   (a) 1:2
   (b) 1:4
   (c) 2:1
   (d) 4:1

9.8 A plane diffraction grating is 5 cm wide and has 5000 lines per cm. The resolving power of the grating in the third order spectrum is
   (a) 5,000
   (b) 10,000
   (c) 25,000
   (d) 75,000

9.9 The power of an optical instrument by which it can form separate images of two close objects is called
   (a) dispersive power
   (b) magnifying power
   (c) resolving power
   (d) diopter

9.10 The grating formula is given by
   (a) $(a + b) \sin\theta = n\lambda$
   (b) $(a + b) \cos\theta = n\lambda$
   (c) $(a + b) \sin\theta = \left(n + \dfrac{1}{2}\right)\lambda$
   (d) $(a + b) \sin\theta = 0$

   where letters have their usual meanings.

9.11 A plane transmission grating having 5000 lines per cm is being used under normal incidence of light. The highest order spectrum that can be seen for the light of wavelength 4800 Å is
   (a) 1
   (b) 2
   (c) 3
   (d) 4

9.12 A plane transmission grating is 5 cm wide and has 5000 lines per cm. The resolving power of the grating in the second order spectrum is
   (a) 5,000
   (b) 10,000
   (c) 25,000
   (d) 50,000

9.13 In the spectrum formed by a diffraction grating, the ratio of intensity of a secondary maximum adjacent to a principal maximum is
(a) between 1 and 0.01
(b) between 0.1 and 0.01
(c) between 0.01 and 0.001
(d) less than 0.001

9.14 The grating element of a 2.0 cm wide plane grating, which resolves two lines in the second order, differing in wavelength by 6Å, of mean wavelength 6000 Å, will be
(a) 4×10^{-3} m
(b) 3×10^{-3} m
(c) 2×10^{-3} m
(d) 1×10^{-3} m

9.15 Two spectral lines of a light source are coincident in the second and third orders of the grating. If the wavelength of the first line is λ, then there wavelength of the second line is
(a) $\dfrac{\lambda}{2}$
(b) λ
(c) $\dfrac{2}{3}\lambda$
(d) $\dfrac{1}{3}\lambda$

9.16 A monochromatic beam of light of wavelength 5460 Å falls on the grating normally and gives a second order image at an angle of 45°. The grating element is of the order of
(a) 10^{-2} cm
(b) 10^{-3} cm
(c) 10^{-4} cm
(d) 10^{-5} cm

9.17 The angular separation, between the two lines of neon 5882 Å and 5852 Å, in the first order spectrum of a plane transmission grating, having 6000 lines per cm, illuminated normally, will be
(a) 1.7×10^{-3} rad
(b) 1.8×10^{-3} rad
(c) 1.9×10^{-3} rad
(d) 2.0×10^{-3} rad

9.18 The resolving power of a prism can be expressed as
(a) $\dfrac{\lambda}{d\lambda} = nN$
(b) $\dfrac{\lambda}{d\lambda} = t\dfrac{d\mu}{d\lambda}$
(c) $\dfrac{1}{\theta} = \dfrac{2\mu \sin\theta}{1.22\lambda}$
(d) $\dfrac{1}{\theta} = \dfrac{a}{1.22\lambda}$

9.19 The mounting is most compact and requires much less space than other mountings, is known as
(a) Rowland mounting
(b) Runge–Paschen mounting
(c) Eagle mounting
(d) None of these

9.20 Echelon grating was devised in 1898 by
(a) Einstein
(b) Newton
(c) Fraunhofer
(d) Michelson

Answers

9.1 (b) 9.2 (b) 9.3 (a) 9.4 (b) 9.5 (a) 9.6 (d) 9.7 (c) 9.8 (d)
9.9 (c) 9.10 (a) 9.11 (d) 9.12 (d) 9.13 (b) 9.14 (c) 9.15 (c) 9.16 (d)
9.17 (b) 9.18 (b) 9.19 (c) 9.20 (d)

CHAPTER 10

Polarization of Light Waves

"The alignment of the transverse electric vibration of electromagnetic radiation. Such waves of aligned vibrations are said to be polarised."

—Paul G. Hewitt

IN THIS CHAPTER

- Polarization of Light by Tourmaline Crystals Experiments
- Plane of Vibration and Plane of Polarization
- Pictorial Representation of Light Vibrations
- Methods of Producing Plane-Polarized Light
- Matrix Representation of Plane-Polarized Light Waves
- Optical Activity
- Polarimeter
- Photoelasticity

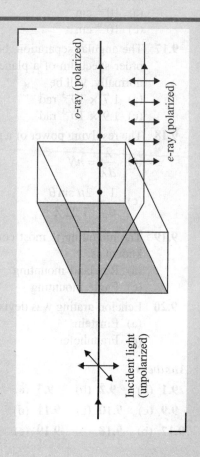

10.1 INTRODUCTION

Interference and diffraction phenomena provide the best evidence that light is wave-like. As we know that waves can be either longitudinal or transverse. Sound waves are longitudinal, which means that the vibratory motion is along the direction of wave travel. But when we shake a taut rope, the vibratory motion travelling along the rope is perpendicular or transverse to the rope. Both longitudinal and transverse waves exhibits interference and diffraction effects. Are light waves, then longitudinal or transverse? Polarization of the light waves demonstrates that the light waves are *transverse*.

The phenomenon in which the vibrations of the particles of the medium produced by light waves are restricted to one particular direction or plane is called *polarization of light waves*. The light wave having one particular direction of vibration is called *polarized light*. Ordinary light (from any source) in which the particles of the medium can vibrate in any number of directions all perpendicular to the direction of propagation is called *unpolarized light*.

The meaning of polarization can be more easily illustrated by taking simple mechanical experiment:

Let us consider, a long string PQ is passing through two rectangular slits A and B as visualized in Figure 10.1. The end Q of the string is tied to a hook in a wall and the free end A is jerked in all possible directions perpendicular to the length of the string so as to generate transverse waves in it. The portion PA of the string has vibrations in all directions perpendicular to PQ, so that the wave is unpolarized. The first slit A will permit only those vibrations to pass through it which are parallel to slit A and will cut off all other vibrations. Thus the wave emerging from and will cut off all other vibrations. Thus the wave emerging from the slit A is *plane polarized*. The slit A is called the *polarizer*. If the second slit B called the *analyzer* is held parallel to A the wave from A will pass through B unchanged. If B is held perpendicular to A, no vibrations will emerge from the slit S_2. This indicates that the slit A has polarized the incoming wave in the vertical plane.

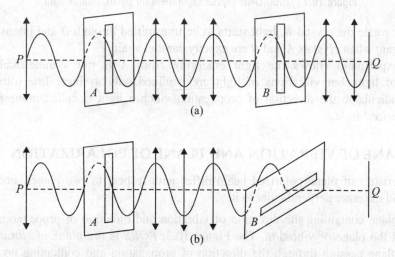

Figure 10.1 Experiment to demonstrate polarization of a wave through string.

10.2 POLARIZATION OF LIGHT BY TOURMALINE CRYSTALS EXPERIMENTS

Suppose we have a tourmaline crystal A, which cut parallel to the crystallographic axis and allow a beam of unpolarized (ordinary), light to pass normally through it. On rotating the crystal about the ray as the axis, the character of the light transmitted through it remains same. Now let us take another tourmaline crystal B and put it parallel to previous crystal A. We observed that, light is approximately transmitted to the second crystal so the intensity is maximum in this position of crystal pair as shown in Figure 10.2(a).

If now crystal B is rotated about an axis parallel to the incident beam of light (i.e., the axes of crystals A and B are inclined each other), the transmitted beam decreases in intensity and become zero, when the axes of both crystals are at 90° to each other. This shows no light is transmitted through the crystal B [Figure 10.2(b)].

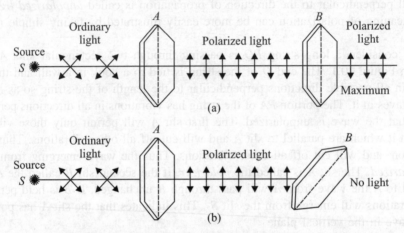

Figure 10.2 Tourmaline crystal experiment for polarization of light.

If we further rotate the crystal B, light starts to be transmitted through B and intensity becomes maximum again when crystals A and B are exactly parallel again.

This experiment shows the light emerging from A is not symmetrical about the propagation of light, but vibrations of light are confined only a single line (direction) in a plane perpendicular to the direction of propagation. Such a light is called *plane polarized* or *linearly polarized light*.

10.3 PLANE OF VIBRATION AND PLANE OF POLARIZATION

The characteristics of plane polarized beam differ with respect to two planes, one containing vibrations and the other perpendicular to it.

- The plane containing the direction of vibration and direction of propagation of light is called the plane of vibration. The Figure 10.3, PQRS is the *plane of vibration*.
- The plane passing through the direction of propagation and containing no vibration is called *plane of polarization*. The plane TUVW is the plane of polarization as shown in the Figure 10.3.

The plane of polarization is always at the right angle to the plane of vibration as shown in the Figure 10.3.

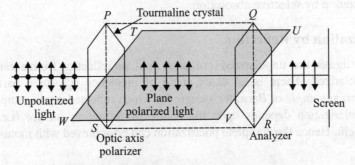

Figure 10.3 Production of plane polarized light.

10.4 PICTORIAL REPRESENTATION OF LIGHT VIBRATIONS

In an unpolarized beam of light, all directions of vibration at right angles to that of propagation of light are possible, hence it is pictorially represented by a star(✻) as shown in Figure 10.4(a) and partially polarized light is represented by arrows (↕) as shown in Figure 10.4(b).

Figure 10.4 Pictorial view of unpolarized light.

In plane polarized beam of light, the vibrations are along a single straight line. When the plane polarized light has got vibrations in the plane of paper, they are represented by (↕) arrows as shown in Figure 10.5(a), while that perpendicular to the paper are represented by dots (•) as shown in Figure 10.5(b).

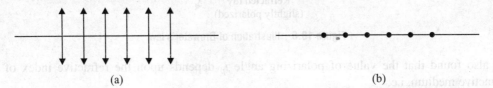

Figure 10.5 Pictorial view of polarized light.

10.5 METHODS OF PRODUCING PLANE POLARIZED LIGHT

There are several methods to polarize any ordinary light beam. Some of them are:
1. Polarization by reflection,
2. Polarization by refraction,

3. Polarization by double refraction,
4. Polarization by scattering,
5. Polarization by selective absorption.

10.5.1 Polarization by Reflection

E.L. Malus investigated that unpolarized (ordinary) light, on reflection from a transparent surface, gets partially polarized. There is an exact angle of incidence (i_p) for a particular material, is known as *polarizing angle* or *Brewster's angle* for which reflected ray is completely polarized. The angle of polarization depends upon the nature of refracting surface (i.e., material) and wavelength of light. Hence the complete polarization can be achieved with monochromatic light.

Brewster's law

The Scottish physicist David Brewster discovered that for a certain angle of incidence, monochromatic light was 100% polarized upon reflection. The refracted beam is partially polarized, while the reflected beam is completely polarized parallel to the reflecting surface. Furthermore, the noticed that this angle of incidence, the reflected and refracted beam are perpendicular as shown in Figure 10.6.

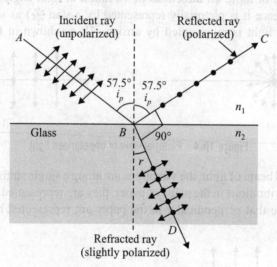

Figure 10.6 Illustration of Brewster's law.

He also found that the value of polarizing angle i_p depends upon the refractive index of the refractive medium, i.e.

$$n = \tan i_p \tag{10.1}$$

Proof: Suppose the light beam is travelling in a medium with refractive index n_1 and being partially reflected at the boundary with a medium of refractive index n_2. The angle of incidence and reflection are i_p, i.e. polarizing angle. The angle of refraction is r. Then Snell's law for incident nd refracted beams is

$$n_1 \sin i_p = n_2 \sin r$$

According to Brewster's law
$$i_p + r = 90°$$
$$\therefore \quad r = 90° - i_p$$

We substitute the value of r in the Snell's law equation,
$$n_1 \sin i_p = n_2 \sin r$$
$$\therefore \quad n_1 \sin i_p = n_2 \sin(90° - i_p)$$
$$\therefore \quad n_1 \sin i_p = n_2 \cos i_p$$
$$\therefore \quad \frac{\sin i_p}{\cos i_p} = \frac{n_2}{n_1}$$
$$\therefore \quad \tan i_p = \frac{n_2}{n_1}$$
$$\therefore \quad \tan i_p = {}_1 n_2 \qquad (10.2)$$

This equation is known as *Brewster's law*. Usually, the incident beam is travelling in air, so $n_1 \approx 1.00$, and the equation becomes $\tan i_p = n_2$. The polarizing angle is sometime referred as Brewster's angle of the material.

Note: The refractive index of material varies slightly with the wavelength of incident light. The polarizing angle therefore also depends on wavelength, so a beam of white light does not have a unique polarizing angle.

EXAMPLE 10.1 When the angle of incidence on a certain material is 60°, the reflected light is completely polarized. Find the refractive index for the material and also the angle of refraction.

Solution By using Brewster's law
$$n = \tan i_p$$
Here $\quad i_p = 60°$

$\therefore$ Refractive index of the material
$$n = \tan 60° = \sqrt{3} = 1.732$$
Angle of refraction = $180° - (90° + 60°) = 30°$

EXAMPLE 10.2 Find the polarizing angle for light incident from
   (i) air to glass   (ii) glass to air
  (iii) water to glass   (iv) glass to water
   (v) air to water, and   (vi) water to air
[Given n for glass = 1.54; n for water = 1.33; n for air = 1.00]

Solution (i) When light is incident from air to glass
$$n = \frac{\text{refractive index of glass}}{\text{refractive index of air}} = \frac{n_2}{n_1} = \frac{1.54}{1} = 1.54$$

Now using Brewster's law $\tan i_p = n$

or polarizing angle $i_p = \tan^{-1}(n) = \tan^{-1}(1.54) = 57°$

(ii) When light is incident from glass to air

$$n = \frac{\text{refractive index of air}}{\text{refractive index of glass}} = \frac{1}{1.54} = 0.6494$$

$\therefore \quad i_p = \tan^{-1}(n) = \tan^{-1}(0.6494) = 33°$

(iii) When light is incident from water to glass

$$n = \frac{\text{refractive index of glass}}{\text{refractive index of water}} = \frac{1.54}{1.33} = 1.158$$

$\therefore \quad i_p = \tan^{-1}(n) = \tan^{-1}(1.158) = 49°12'$

(iv) When light is incident from glass to water

$$n = \frac{\text{refractive index of water}}{\text{refractive index of glass}} = \frac{1.33}{1.54} = 0.8636$$

$\therefore \quad i_p = \tan^{-1}(n) = \tan^{-1}(0.8636) = 40°49'$

(v) When light is incident from air to water

$$n = \frac{\text{refractive index of water}}{\text{refractive index of air}} = \frac{1.33}{1} = 1.33$$

$\therefore \quad i_p = \tan^{-1}(1.33) = 53°4'$

(vi) Similarly when light is incident from water to air

$$n = \frac{\text{refractive index of air}}{\text{refractive index of water}} = \frac{1}{1.33} = 0.7519$$

$\therefore \quad i_p = \tan^{-1}(0.7519) = 36°57'$

EXAMPLE 10.3 At what angle the light should be incident on a glass plate ($n = 1.5697$) to get a plane polarized light by reflection?

Solution According to Brewster's law for the polarization of glass plate.

$n = \tan i_p$

here, $n = 1.5697$

Then $i_p = \tan^{-1}(1.5697) = 57.5°$

10.5.2 Polarization by Refraction (A Pile of Plates)

When an unpolarized light is incident on the upper surface of a glass slab at the polarizing angle i_p (Figure 10.7), a small fraction is reflected, which is completely plane polarized with vibrations perpendicular to the plane of incidence, while the rest is refracted and is only partially polarized having vibrations both in the plane of incidence as well as perpendicular to the plane of incidence.

The refracted light is incident at the lower surface at angle r, where r is the angle of refraction at upper face.

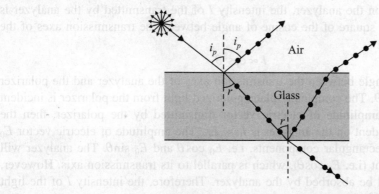

Figure 10.7 Polarization by refraction.

Now
$$\tan r = \frac{\sin r}{\cos r} = \frac{\sin r}{\sin(90° - r)} = \frac{\sin r}{\sin i_p} \quad [\text{as } i_p + r = 90°]$$
$$= {_g}n_a \tag{10.3}$$

Thus by Brewster's law, r is the polarizing angle for the reflection at the lower surface of the plate. Hence the light reflected at the lower face is completely plane polarized, while that refracted into the air is partially plane polarized.

Hence, if a beam of unpolarized light be incident at the polarizing angle on a pile of plates as shown in Figure 10.8, the sequential refracted beams become poorer and poorer in the perpendicular component. If a good number of plates be used, the refracted beam will contain mostly the parallel component, i.e. it will almost be plane polarized. The greater the number of the plates, more pure is the refracted polarized light.

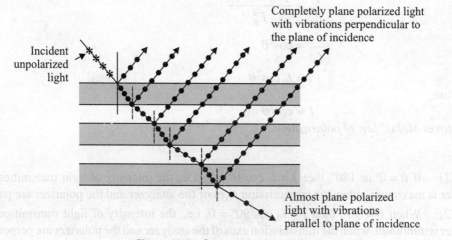

Figure 10.8 Sequential refraction.

Malus' law

According to Malus' law, (which is named after Etienne-Louis Malus), when completely plane polarized light is incident on the analyzer, the intensity I of the transmitted by the analyzer is directly proportional to the square of the cosine of angle between the transmission axes of the analyzer and polarizer, i.e.

$$I \propto \cos^2 \theta$$

Let us consider the angle between the transmission axes of the analyzer and the polarizer is θ as shown in Figure 10.9. The completely plane polarized light from the polarizer is incident on analyzer. If E_0 is the amplitude of electric vector transmitted by the polarizer, then the intensity I_0 of the light incident on the analyzer is $I_0 \propto E_0^2$. The amplitude of electric vector E_0 can be resolved into two rectangular components, i.e. $E_0 \cos\theta$ and $E_0 \sin\theta$. The analyzer will transmit only the component (i.e. $E_0 \cos\theta$), which is parallel to its transmission axis. However, the component $E_0 \sin\theta$ will be absorbed by the analyzer. Therefore, the intensity I of the light transmitted by the analyzer is

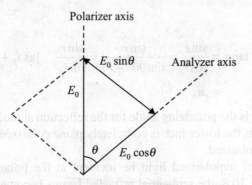

Figure 10.9 Malus' law.

$$I \propto (E_0 \cos\theta)^2$$

or

$$\frac{I}{I_0} = \frac{(E_0 \cos\theta)^2}{E_0^2}$$

∴

$$\frac{I}{I_0} = \cos^2\theta$$

∴

$$I = I_0 \cos^2\theta \qquad (10.4)$$

Therefore,

$$I \propto \cos^2\theta$$

This proves *Malus' law of polarization*.

Cases:

Case (1): If $\theta = 0°$ or $180°$, then $I = I_0 \cos^2 0° = I_0$, i.e., the intensity of light transmitted by the analyzer is maximum, when the transmission axes of the analyzer and the polarizer are parallel.

Case (2): When $\theta = 90°$, then $I = I_0 \cos^2 90° = 0$, i.e., the intensity of light transmitted by the analyzer is minimum, when the transmission axes of the analyzer and the polarizer are perpendicular to each other.

EXAMPLE 10.4 If the plane of vibration of the incident beam makes an angle of 30° with the optic axis, compare the intensities of extraordinary and ordinary light.

Solution For the division of *e*-ray and *o*-ray given that
$\theta = 30°$ = angle between vibration of electric vector and optic axis.
$\dfrac{I_e}{I_o} = ?$, where I_e and I_o are the intensity of *e*-ray and *o*-ray respectively.

$$\frac{I_e}{I_o} = \frac{\cos^2\theta}{\sin^2\theta} = \frac{3}{1}$$

EXAMPLE 10.5 Intensity of light through a polarizer and analyzer is maximum, when their principal planes are parallel. Through what angle the analyzer must be rotated so that the intensity gets reduced to ¼ of the maximum value?

Solution Given that
$$I = I_0 - \frac{I_0}{4} = \frac{3I_0}{4}$$

Then from Malus' law,
$$I = I_0 \cos^2\theta$$
Hence
$$\cos^2\theta = \frac{3I_0/4}{I_0} = \frac{3}{4}$$
$$\Rightarrow \cos\theta = \frac{\sqrt{3}}{2} \Rightarrow \theta = 30°$$

10.5.3 Polarization by Double Refraction (Birefringence)

An optical property in which a single ray of unpolarized light entering an anisotropic medium is split into two rays, each travelling in a different direction. One ray (called the extraordinary ray) is bent or refracted, at an angle as it travels through the medium, the other ray (called the ordinary ray) passes through the medium unchanged. This phenomenon of splitting the ray into two components is known as *double refraction*.

- The ordinary ray (*o*-ray) obeys the laws of refraction, i.e. it always lies in plane of incidence and its velocity in the crystal is the same in all directions.
- The extraordinary ray (*e*-ray) does not obey the laws of refraction. It travels in the crystal with different speeds in different orientations.

The velocities of *e*-ray and *o*-rays are same along the optic axis of the anisotropic medium like calcite crystal, the phenomena of double refraction can be explained by a simple experiment. Mark an ink dot on a piece of paper and place the calcite crystal over this dot. Two images of ink mark will be observed. On rotating the crystal either clockwise or anticlockwise, it is found that one image rotates with the rotation of crystal and the other image remains stationary as shown in Figure 10.10. The stationary image is known as *ordinary image* and the rotating image is known as *extraordinary image*. The refracted ray which produces ordinary image is known as ordinary (*o*) ray and obeys ordinary laws of refraction and the refracted ray which produces extraordinary image is known as extraordinary ray (*e*-ray) and does not obey ordinary law of refraction.

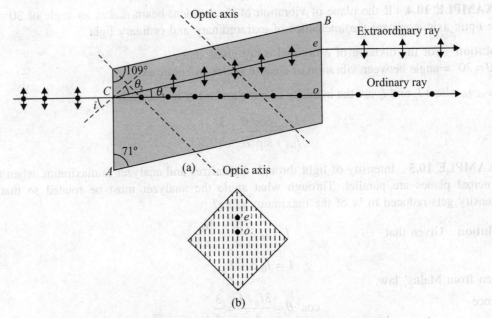

Figure 10.10 An illustration for double refraction (extraordinary and ordinary rays).

Points to remember

- In case of *ordinary ray*, the refractive index $n_o = \dfrac{\sin i}{\sin \theta_1}$ is constant.
- In case of *extraordinary ray*, the refractive index $n_e = \dfrac{\sin i}{\sin \theta_2}$ is not constant, but varies with i, i.e. angle of incidence.
- For *quartz crystal*, $\theta_2 < \theta_1$, therefore, $n_e > n_o$, hence $v_o > v_e$, i.e. inside the quartz crystal the ordinary ray travels faster than the extraordinary ray. Such type of crystals are said to be *uniaxial positive crystals*.
- For *calcite crystal*, $\theta_2 > \theta_1$, therefore, $n_o > n_e$, hence $v_e > v_o$, i.e. inside the calcite crystal the extraordinary ray travels faster than the ordinary ray. Such type of crystals are said to be *uniaxial negative crystals*.

Geometry of Calcite Crystal

The calcite crystal is a colourless transparent crystal to visible as well as ultraviolet light is also known as *Iceland spar*. Its chemical name is calcium carbonate ($CaCO_3$) and occurs in nature in various forms, all of which give rhombohedron on cleavage as shown in Figure 10.11. Each face of the crystal is a parallelogram having angles 102° and 78°. At two diametrically opposite corners, the angles P and W of three faces meet there, are obtuse (>90°), while at the rest six corners, two angles are acute and the remaining one is obtuse. These corners P and W are said to be blunt corners.

Optic axis: A line passing through one of the blunt corners and equally inclined to all the three edges meeting over there gives the direction of optic axis. Any line parallel to this line is also known as optic axis as shown in Figure 10.11(b). A line joining two opposite blunt corners is not an optic axis, but only in case of cubic crystal where all the three edges are equal, a line joining two opposite blunt corners will be an optic axis.

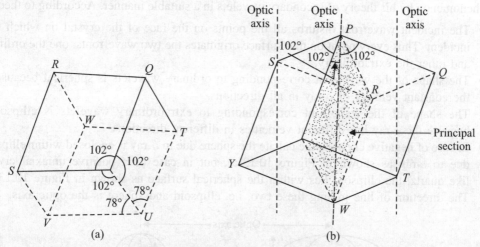

Figure 10.11 Geometry of calcite crystal.

Principal section: A plane containing optic axis and perpendicular to the opposite faces of the crystal is called *principal section* of the crystal as shown in Figure 10.12(a). As there are six faces in a crystal, one corresponding to each pair of opposite faces. The principal section always cuts the surface of calcite crystal in parallelogram with angles 109° and 71° as visualized in Figure 10.12(b).

Principal plane of the crystal: The principal plane or principal section of the ordinary ray (o-ray) is defined as a plane which contains the ordinary ray and the optic axis, whereas the principal plane of the extraordinary ray (e-ray) is a plane which contains the optic axis and extraordinary ray. Ordinarily, the ordinary and extraordinary planes do not coincide except in a case when the plane of incidence and principal section are same.

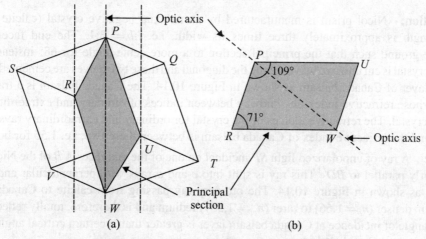

Figure 10.12 Principal section of the calcite crystal.

Huygen's theory of double refraction in uniaxial crystals

Huygen studied the polarization of light by double refraction in calcite crystal and explained this phenomenon by his theory of secondary wavelets in a suitable manner. According to theory:

- The incident wavefront disturbs all the points on the face of the crystal on which it is incident. Thus every point on that surface originates the two wavefronts, one the ordinary and other the extraordinary.
- The shape of the wavefront corresponding to ordinary wavelets is spherical because of the constant velocity of o-ray in all directions.

  The shape of the wavefront corresponding to extraordinary wavelets is ellipsoidal because the e-ray has different velocities in different directions.
- In case of negative crystals like calcite the sphere due to o-ray is enclosed within ellipsoid due to e-ray as shown in Figure 10.13(a), but in case of a positive uniaxial crystal like quartz, the ellipsoid lies within the spherical surface as shown in Figure 10.13(b). The direction of line joining these two, i.e. ellipsoid and sphere is the optic axis.

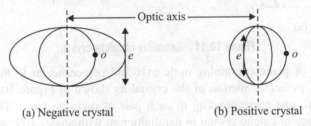

(a) Negative crystal   (b) Positive crystal

Figure 10.13   Double refracting crystals.

Nicol prism

Nicol prism is an optical device. It is made from calcite crystal and is used in many instruments for producing and analyzing plane polarized light. Actually, William Nicol in 1826 invented this device, so it is named after his name.

The principle of Nicol prism is based on the phenomenon of polarization by double refraction.

Construction: Nicol prism is manufactured by an uniaxial negative crystal (calcite crystal). Whose length is approximately three times its width, i.e. $l:b = 3:1$. The end faces of this crystal are ground such that the principal section to a more acute angle for 68° instead of 71°. Then, the crystal is cut into two halves along the diagonal and these two halves are cemented together by a thin layer of Canada balsam as shown in Figure 10.14. The Canada balsam is a transparent material whose refractive index lies midway between indices of ordinary and extraordinary rays of calcite crystal. The refractive indices of the crystal for ordinary and extraordinary rays are 1.66 and 1.49 and the refractive index of Canada balsam is between these two, i.e. 1.55 for both rays.

Working: A ray of unpolarized light SP incident on one of the end faces $A'B$ of the Nicol prism and is nearly parallel to BD'. This ray is split into e- and o-rays with perpendicular and parallel vibrations as shown in Figure 10.14. The ordinary ray passing from calcite to Canada balsam travels from denser ($n_o = 1.66$) to rarer ($n_{cb} = 1.55$) medium and is therefore, totally reflected only when the angle of incidence at Canada balsam layer is greater than a certain critical angle for two media (calcite and Canada balsam), i.e. the critical angle

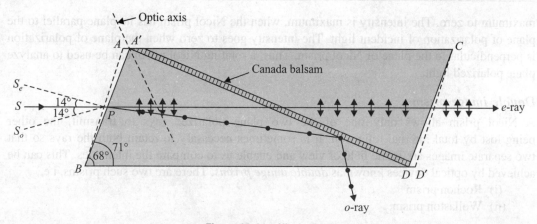

Figure 10.14 Nicol prism.

$$\phi_c = \sin^{-1}\left(\frac{n_{cb}}{n_o}\right) = \sin^{-1}\left(\frac{1.55}{1.66}\right)$$
$$= \sin^{-1}(0.933) = 69° \tag{10.5}$$

This total internally reflected ray is, as a result, absorbed by the side BD' of the prism and only extraordinary ray can be transmitted through this prism as it is travelling from the rarer ($n_e = 1.49$) to the denser ($n_{cb} = 1.55$) medium.

Hence, in Nicol prism, the ordinary ray is eliminated by total internal reflection (TIR) and only extraordinary ray with vibrations parallel to principal section passes through and the light beam emerging from crystal is plane polarized. Therefore, Nicol prism is used as polarizer.

Drawbacks

- Nicol prism can act as polarizer only if the incident beam is slightly convergent or slightly divergent and fails, if the incident beam is highly convergent or divergent.
- If the angle of incidence of incident ray S_oP at the crystal surface is increased, the angle of incidence at Canada balsam surface decreases. If $\angle SPS_o$ is greater than 14°, the angle of incidence at Canada balsam surface is less than 69° and ordinary ray is also transmitted through Nicol prism. Hence emergent ray from Nicol prism will be mixture of o-ray and e-ray, i.e. will not be plane polarized.
- The refractive index of calcite crystal is different for different directions of e-ray, being minimum when it is travelling at right angle to optic and minimum when it travelling along optic axis.
- Being along optic axis e-ray and o-ray travel with same speed for intermediate angles, it is between 1.49 and 1.66. For particular value of angle of incidence of ray S_eP, $n_e > n_{cb}$ and e-ray will also be totally internally reflected and no light emerges from the Nicol.

Nicol prism as an analyzer: Nicol prism produces plane polarized light by the process of double refraction. It can also be used as an analyzer in addition to being used as a polarizer. Plane polarized light when allowed to pass through a rotating Nicol prism indicates a variation in maximum to zero. The intensity is maximum, when the Nicol prism has its plane parallel to the plane of polarization of incident light. The intensity goes to zero when the plane of polarization is perpendicular to the plane of Nicol prism. Thus, a rotating Nicol prism can be used to analyze plane polarized light.

Double image prisms

A Nicol prism allows only one of the two plane polarized rays to transmit, the other being lost by total internal reflection. It is sometimes necessary to retain both the rays so that two separate images are in the field of view and enable us to compare the intensities. This can be achieved by optical devices known as *double image prisms*. There are two such prisms, i.e.,
   (i) Rochon prism
   (ii) Wollaston prism

(i) Rochon prism: It consists of two right-angled prisms *CAB* and *BCD* cemented by castor oil as shown in Figure 10.15.

The optic axis of one of these prisms is parallel to base *AB* and perpendicular to base plane of the paper when incident light falls normal to *AC*, it travels in the direction axis, and hence, is neither deviated nor split. Thus it does not suffer double refraction. After crossing the first prism, it enters the second prism in the direction perpendicular to optic axis, and thus, splits into *e*- and *o*-rays. The *o*-ray remains undeviated but *e*-ray is deviated towards the apex of prism *ABC*. Thus we get split plane polarized *e*- and *o*-rays. One has to take a precaution in using this prism. This is that in using this prism, it should always kept in such a way that light travels first along the optic axis, i.e. on the face *AC*.

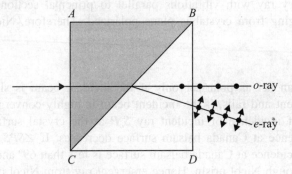

Figure 10.15 Rochon prism.

(ii) Wollaston prism: This is similar in construction to the Rochon prism except that the first prism *ABC* (Figure 10.16) has its optic axis parallel to the face *AB*.

In this case the incident ray entering normally to the surface *AB* travels perpendicular to the optic axis. Therefore, the *o*- and *e*-rays travel in the first prism in the same direction, but with different velocities. Since the second prism is cut with its optic axis perpendicular to that of the first, the *o*-ray in the first prism will become the *e*-ray in the second and vice-versa. Hence both the rays are deviated and dispersed in the second prism resulting into greater and greater separation of two rays. This prism is especially useful in experiments in which intensities of polarized lights are involved.

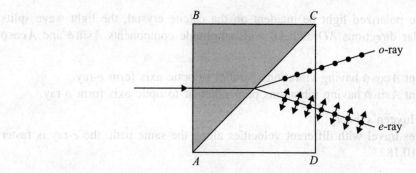

Figure 10.16 Wollaston prism.

The double image prisms have the additional advantage that they can be used in ultraviolet region where a Nicol prism cannot be used because Canada balsam absorbs ultraviolet light.

Plane, circularly and elliptically polarized light

Plane-polarized light: When light travels along a certain direction, the vibrations take place in the direction at right angle to the direction of propagation. If the vibrations of the ether particles are linear and take place parallel to a plane through the axis of the beam or the direction of propagation, light is said to be plane polarized.

Elliptically polarized light: When the vibrations of the ether particles are elliptical, having a constant period and take place in a plane perpendicular to the direction of propagation, the light is said to be elliptically polarized. In elliptically polarized light, the amplitude of vibrations changes in magnitude as well as in direction.

Circularly polarized light: When the vibrations of the ether particles are circular having constant period and take place in the transverse plane, light is said to be circularly polarized. In circularly polarized light, the amplitude of vibrations remains constant, but changes only the direction.

Superposition of two plane polarized waves having perpendicular vibrations

Let us consider a beam of plane polarized light be incident normally on a uniaxial crystal plate like calcite plate cut with its optic axis parallel to its faces. Let the linear vibration in the incident light along AB be making an angle ϕ with optic axis. Let A be the amplitude of vibration in the incident light as shown in Figure 10.17.

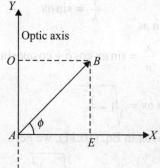

Figure 10.17 Geometry of linearly polarized light.

When the plane polarized light be incident on the calcite crystal, the light wave splits into two perpendicular directions AO and AE with amplitude components $A\sin\phi$ and $A\cos\phi$ respectively.

- The component $A\cos\phi$ having vibrations parallel to optic axis form e-ray.
- The component $A\sin\phi$ having vibrations perpendicular to optic axis form o-ray.

We know that, with Huygen's theory:

If the two waves travel with different velocities along the same path, the e-ray is faster as shown in Figure 10.18.

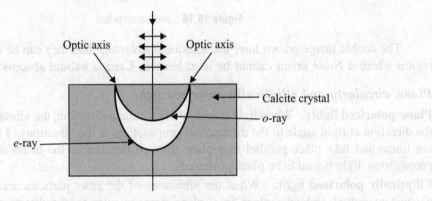

Figure 10.18 Velocities of elliptically and circularly polarized light.

Theory: Let the incident wave is $A\sin\omega t$, then the two components will be represented by

$$x = A\cos\phi \sin(\omega t + \delta) \quad \text{(for } e\text{-ray)} \tag{10.6}$$

and

$$y = A\sin\phi \sin\omega t \quad \text{(for } o\text{-ray)} \tag{10.7}$$

Here δ is the phase difference between e-ray and o-ray

Let $A\cos\phi = a$ and $A\sin\phi = b$

Then, we have

$$\frac{x}{a} = \sin(\omega t + \delta) \tag{10.8}$$

and

$$\frac{y}{b} = \sin\omega t \tag{10.9}$$

Equation (10.8) can also be written as

$$\frac{x}{a} = \sin\omega t \cos\delta + \cos\omega t \sin\delta \tag{10.10}$$

From Eq. (10.9), we have

$$\cos\omega t = \sqrt{1 - \frac{y^2}{b^2}} \tag{10.11}$$

Putting the values of $\sin\omega t$ and $\cos\omega t$ in Eq. (10.11), we get

$$\frac{x}{a} = \frac{y}{b}\cos\delta + \sqrt{1 - \frac{y^2}{b^2}}\sin\delta$$

or
$$\frac{x}{a} - \frac{y}{b}\cos\delta = \sqrt{1 - \frac{y^2}{b^2}}\sin\delta$$

Squaring and arranging again, we get

$$\frac{x^2}{a^2} - \frac{2xy}{ab}\cos\delta + \frac{y^2}{b^2} = \sin^2\delta \tag{10.12}$$

This is general equation of an ellipse.

Special cases

Case (1): If $\delta = 0$, $\sin\delta = 0$ and $\cos\delta = 1$, then from Eq. (10.12), we get

$$\frac{x^2}{a^2} + \frac{y^2}{b^2} - \frac{2xy}{ab} = 0$$

or
$$\left(\frac{x}{a} - \frac{y}{b}\right)^2 = 0$$

or
$$y = \frac{b}{a}x \tag{10.13}$$

This is an equation of straight line as shown in Figure 10.19(a). Thus the emergent is plane polarized with the same direction of vibrations as the incident light.

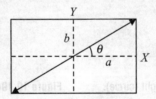

Figure 10.19(a) Plane polarized light (straight line).

Case (2): If $\delta = \pi, 3\pi, 5\pi, \ldots$, then $\cos\delta = -1$ and $\sin\delta = 0$. Then Eq. (10.12) reduces as

$$\frac{x^2}{a^2} + \frac{y^2}{b^2} + \frac{2xy}{ab} = 0$$

or
$$y = -\frac{b}{a}x \tag{10.14}$$

This equation is also for straight line with negative slope as shown in Figure 10.19(b).

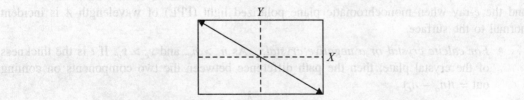

Figure 10.19(b) Plane polarized light (straight line with negative slope).

This represents a plane polarized light with vibrtion direction making an angle $2\delta \left(= 2\tan^{-1}\dfrac{b}{a}\right)$ with that of incident light.

Case (3): If $\delta = \dfrac{\pi}{2}$, then $\cos\delta = 0$ and $\sin^2\delta = 1$, and $a = b$. Then Eq. (10.12) reduces as

$$x^2 + y^2 = a^2 \qquad (10.15)$$

Equation (10.15) is an equation of a circle. The emergent ray is circularly polarized light as shown in Figure 10.19(c).

Hence the resultant of two plane polarized beams with equal amplitudes and a phase difference of $\dfrac{\pi}{2}$ is circularly polarized light.

Case (4): If $\delta = \dfrac{\pi}{2}$ and $a \ne b$, $\cos\delta = 0$, $\sin\delta = 1$, then Eq. (10.12) becomes

$$\dfrac{x^2}{a^2} + \dfrac{y^2}{b^2} = 1 \qquad (10.16)$$

This is an equation of ellipse. The emergent ray is elliptically polarized light as shown in Figure 10.19(d).

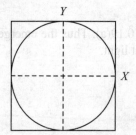

Figure 10.19(c) Plane polarized light (circle).

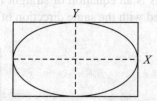

Figure 10.19(d) Plane polarized light (ellipse).

Hence the resultant of two waves becomes with unequal amplitude and a phase difference of $\pi/2$ is an elliptical polarized light.

Retardation plates

The crystal plate of doubly refracting crystal that retards the motion of one of the refracted beams (*o*- and *e*-rays) are known as *retardation plates*. These are of two kinds:

(i) Quarter wave plate (QWP): It is a crystal plate of a double refracting material of a uniaxial crystal cut with its optic axis parallel to the refracting faces. The thickness of the plate is such as to produce a path difference of $\lambda/4$ (or phase difference of $\pi/2$) between the *o*-ray and the *e*-ray when monochromatic plane polarized light (PPL) of wavelength λ is incident normal to the surface.

- *For calcite crystal or a negative crystal:* As $n_o > n_e$, and $v_e > v_o$. If t is the thickness of the crystal plate, then the path difference between the two components on coming out $= t(n_o - n_e)$

  $\therefore \qquad (n_o - n_e)t = \dfrac{\lambda}{4}$

$$t = \frac{\lambda}{4(n_o - n_e)} \tag{10.17}$$

- *For quartz crystal or a positive crystal:* As $n_e > n_o$, and $v_o > v_e$, then

$$(n_e - n_o)t = \frac{\lambda}{4}$$

$$t = \frac{\lambda}{4(n_e - n_o)} \tag{10.18}$$

Note: QWP is applicable to produce circularly and elliptically polarized light.

- If the angle of incidence of PPL is equal to 45° with optic axis, then we get circularly polarized light.
- If the angle of incidence of PPL is not equal to 45°, then we get elliptically polarized light.

(ii) Half wave plate (HWP): It is crystal plate of a doubly refracting material of a uniaxial crystal cut with its optic axis parallel to the refracting faces. The thickness of the plate is such as to produce a path difference of $\lambda/2$ (or phase difference of π) between o-ray and e-ray, when monochromatic plane polarized light (PPL) of wavelength λ is incident normal to the surface.

- *For calcite crystal or a negative crystal:* As $n_o > n_e$, and $v_e > v_o$. If t be the thickness of the crystal plate, then the path difference between the two components on coming out = $t(n_o - n_e)$

$$\therefore \quad (n_o - n_e)t = \frac{\lambda}{2}$$

$$\therefore \quad t = \frac{\lambda}{2(n_o - n_e)} \tag{10.19}$$

- *For quartz crystal or a positive crystal:* As $n_e > n_o$, and $v_o > v_e$, then

$$\therefore \quad (n_e - n_o)t = \frac{\lambda}{2}$$

$$t = \frac{\lambda}{2(n_e - n_o)} \tag{10.20}$$

Note:

- If PPL is incident upon HWP at a particular angle with direction of optic axis, we get plane polarized vibrations at double of that angle.
- HWP is applied to make Laurent's half shade device in a polarimeter.

EXAMPLE 10.6 Compute the minimum thickness of a quarter wave plate of calcite for $\lambda = 5460$ Å. The principal indices of calcite are 1.652 and 1.488.

Solution The minimum thickness of a quarter wave plate of calcite (for which $n_o > n_e$) is given by

$$t = \frac{\lambda}{4(n_o - n_e)}$$

Given $\lambda = 5460 \times 10^{-10}$ m, $n_o = 1.652$ and $n_e = 1.488$

$$\therefore \quad t = \frac{5460 \times 10^{-10}}{4(1.652 - 1.488)} = \frac{5460 \times 10^{-10}}{4 \times 0.161} = 8.32 \times 10^{-7} \text{ m}$$

EXAMPLE 10.7 A beam of linearly polarized light is changed into circularly polarized light by passing it through a slice of crystal 0.003 cm thick. Calculate the difference in the refractive indices for the two rays in the crystal assuming this to be of minimum thickness that will produce the effect and the wavelength is 6×10^{-7} m.

Solution If the vibrations in the incident plane polarized light make an angle $\theta = 45°$ with the optic axis of the crystal plate and the thickness of the plate in such that the least phase difference between the emergent o-ray and e-rays is $\pi/2$, then these rays combine to form circularly polarized light. A phase difference of $\pi/2$ is equivalent to the path difference of $\pi/4$ and thus linearly polarized light is converted into circularly polarized light by means of a $\pi/4$ plate. Hence, the given crystal slice is quarter wave plate.

Now for a quarter wave plate,

$$t = \frac{\lambda}{4(n_o - n_e)}$$

Given that $t = 0.003$ cm $= 3 \times 10^{-5}$ m and $\lambda = 6 \times 10^{-7}$ m

$$\therefore \quad (n_o - n_e) = \frac{\lambda}{4t} = \frac{6 \times 10^{-7}}{4 \times 3 \times 10^{-5}}$$

$$= 5 \times 10^{-3} = 0.005$$

EXAMPLE 10.8 A given calcite plate behaves as a half wave plate for a particular wavelength λ. Assuming variation of refractive index with λ to be negligible, how would the above plate behave in another light of wavelength 2λ.

Solution The thickness of a half wave plate for a negative calcite crystal ($n_o > n_e$) is given by

$$t = \frac{\lambda}{2(n_o - n_e)} \quad \text{or} \quad t = \frac{2\lambda}{4(n_o - n_e)} = \frac{\lambda'}{4(n_o - n_e)}$$

which is expression for thickness of a quarter wave plate for wavelength λ'. Hence, the half wave plate for λ will behave as a quarter wave plate for $\lambda' = 2\lambda$ provided the variation of n with λ is neglected.

EXAMPLE 10.9 A beam of linearly polarized light is changed into circularly polarized light by passing it through a sliced crystal of thickness 0.03 cm. Calculate the difference in refractive indices of the two rays in the crystal and using their minimum thickness that will produce the effect. The wavelength of light used is 6×10^{-7} m.

Solution For a quarter wave plate, given that

$t = 0.03$ cm $= 3 \times 10^{-4}$ m, $\lambda = 6 \times 10^{-7}$ m, $(n_e - n_o) = ?$

For quarter wave plate $t = \dfrac{\lambda}{4(n_e - n_o)}$.

So $(n_e > n_o) = \dfrac{\lambda}{4t}$

$$= \dfrac{6 \times 10^{-7}}{4 \times 3 \times 10^{-4}} = 5.0 \times 10^{-4}$$

EXAMPLE 10.10 Plane polarized light ($\lambda = 5 \times 10^{-7}$ m) is incident on a quartz plate cut parallel to the optic axis. Find the least thickness of the plate for which the ordinary and extraordinary rays combine to form plane polarized light on emergence. What multiples of this thickness would give the same result? The indices of refraction of quartz are $n_e = 1.5533$ and $n_0 = 1.5442$.

Solution In this case the quartz plate must act as half wave $\left(\dfrac{\lambda}{2}\right)$ plate. Thus, if t be the required thickness then we have

$$(n_e - n_o)\,t = \dfrac{\lambda}{2} \quad \text{or} \quad t = \dfrac{\lambda}{2(n_e - n_o)}$$

Putting the given values, we get

$$t = \dfrac{5 \times 10^{-7} \text{ m}}{2(1.5533 - 1.5442)}$$

$$= \dfrac{5 \times 10^{-7} \text{ m}}{2 \times 0.0091} = 2.75 \times 10^{-5} \text{ m} = 2.75 \times 10^{-3} \text{ cm}$$

The thickness which would give the same results are t, $3t$, $5t$, ...
$$= (2.75 \times 10^{-3} \text{ cm}) \times 1, 3, 5, ...$$

Production of polarized light

(i) Plane polarized light (PPL): The unpolarized light is allowed to pass through a Nicol prism. The Nicol prism converts unpolarized light into plane polarized light as shown in Figure 10.20.

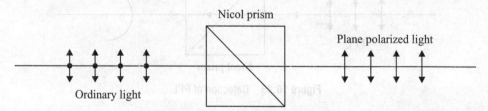

Figure 10.20 Production of PPL.

(ii) Circularly polarized light (CPL): Ordinary light (or unpolarized light) is first allowed to pass through a Nicol prism to convert it to PPL. This PPL is then allowed to fall on a QWP at an angle 45°. The emergent beam is obtained as a result of two polarized beams with a phase difference of $\pi/2$ and of equal amplitude. This is circularly polarized light (Figure 10.21).

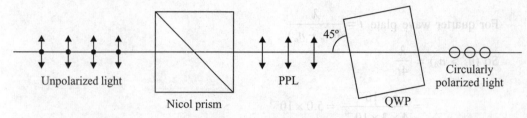

Figure 10.21 Production of CPL.

(iii) Elliptically polarized light (EPL): Unpolarized light is first passed through a Nicol prism to convert it PPL. This PPL is allowed to fall on QWP at an angle which is not equal to 45°. The emergent beam is achieved as a result of two polarized beams with a phase difference of $\pi/2$ and unequal amplitude. This is elliptically polarized light (Figure 10.22).

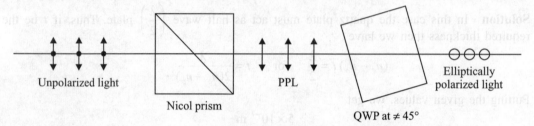

Figure 10.22 Production of EPL.

Detection of polarized light

(i) Plane-polarized light (PPL): The incident beam is allowed to pass through a Nicol prism, which acts as an analyzer. The Nicol prism is rotated by 360°. The intensity of the incident varies from maximum to zero minimum, as shown in Figure 10.23, the incident beam is then known as plane polarized.

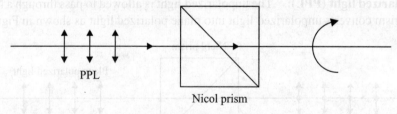

Figure 10.23 Detection of PPL.

(ii) Circularly polarized light (CPL): The beam of CPL is allowed to pass through a Nicol prism (which acts as analyzer). As this Nicol prism is rotated, its intensity does not vary at all. This indicates that the incident beam is either CPL or unpolarized. For confirmation, this beam is allowed to pass through QWP and then analyzed through Nicol prism. On rotating the Nicol prism, the intensity varies from maximum to zero (minimum) as shown in Figure 10.24. This confirms availability of polarized light.

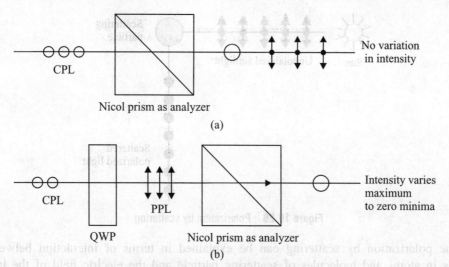

Figure 10.24 Detection of CPL.

(iii) Elliptically polarized light (EPL): Incident beam of EPL is first analyzed through a rotating Nicol prism. If the variation is intensity is from maximum to non-zero minima, the incident light is either elliptically polarized or partially polarized. To confirm this further, light is allowed to pass through QWP and then analyzed through a Nicol prism. If the variation of intensity is from maximum to zero minimum, the incident light is said to be elliptically polarized, otherwise, it is partially polarized (Figure 10.25).

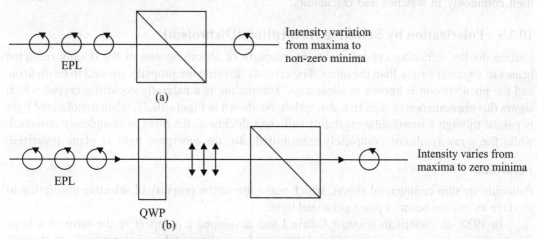

Figure 10.25 Detection of EPL.

10.5.4 Polarization by Scattering

On viewing the clear blue portion of the sky through a polarizer or calcite crystal, the intensity of transmitted light varies as the polarizer (or calcite crystal) is rotated. This shows that the blue light coming from sky is polarized. This polarization is due to scattering of sun light by particles in earth's atmosphere. This is shown symbolically in Figure 10.26.

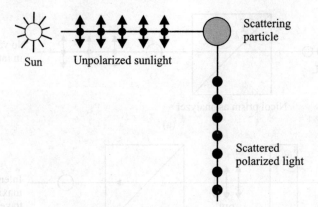

Figure 10.26 Polarization by scattering.

The polarization by scattering can be explained in terms of interaction between the electrons in atoms and molecules of scattering particle and the electric field of the incident radiations. The electrons are set into oscillations by the electric field and re-emit electromagnetic waves according to the laws of electromagnetic theory of radiation. A liquid crystal display, abbreviated as LCD, uses the process mentioned above. LCD consists of long molecules whose direction is controlled by applying electric fields.

These molecules are used to modify plane polarized light falling on them. The polarization of scattered light is adjusted to be perpendicular to the axis of analyzer, which cuts it off. These dark regions can be controlled and are used to form different letters and numbers. LCDs are used commonly in watches and calculators.

10.5.5 Polarization by Selective Absorption (Dichroism)

Certain doubly refracting crystals have the property of absorbing one of the doubly refracted beams to a greater extent than the other. The crystals showing the property are said to be *dichroic* and the phenomenon is known as *dichroism*. Tourmaline is a naturally occurring crystal which shows this phenomenon of selective absorption. As shown in Figure 10.27, when unpolarized light is passed through a tourmaline crystal of sufficient thickness, the *o*-ray is completely absorbed, while the *e*-ray is almost completely transmitted. So, the emergent light is plane polarized.

Polaroids

Polaroids are thin commercial sheets, which make use of the property of selective absorption to produce an intense beam of plane polarized light.

In 1932, an American scientist Edwin Land developed a polarizer in the form of a large sheets. When a paste of quinine iodosulphate made in nitrocellulose is squeezed out through a fine slit, the needle-shaped crystals of quinine iodosulphate align themselves parallel to their optic axis. These crystals are highly dichroic. They absorb one of the doubly refracted beams completely. The thin polarizing sheet so obtained is enclosed between two thin glass plates for mechanical support and we get a *polaroid*. Each polaroid has a characteristic direction called polaride axis (shown by parallel lines). A polaroid transmits only those vibrations which are parallel to its polaroid axis.

Polarization of Light Waves **477**

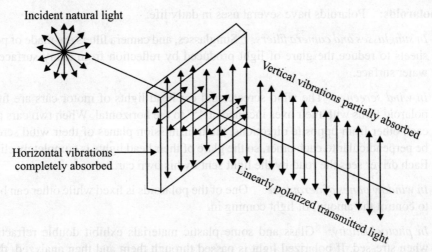

Figure 10.27 Polarization by selective absorption.

As shown in Figure 10.28, when a beam of unpolarized light falls on a polaroid P_1, it transmits only those vibrations which are parallel to its polaroid axis. It absorbs the vibrations in the perpendicular direction.

Thus the transmitted light is plane polarized. It can be examined by using a second polaroid P_2. When the polaroid axes of the two polaroids are parallel to each other [Figures 10.28(a) and (b)], the plane-polarized light transmitted by P_1 is also transmitted by P_2. When the second polaroid is rotated through (cross polaroids), no light is transmitted by P_2 [Figures 10.28(c) and (d)].

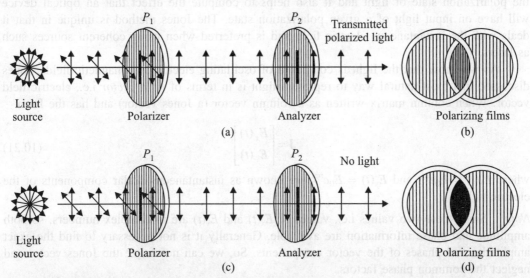

Figure 10.28 Polaroid films. When the film sheets are oriented with the same polarization direction, the transmitted light is polarized (a and b). When one of the sheets is rotated 90° (crossed polaroids) no light is transmitted (c and d).

Uses of polaroids: Polaroids have several uses in daily life:

1. *In sunglasses and camera filters:* Sunglasses, and camera filters are made of polarizing sheets to reduce the glare of light produced by reflection from shiny surface such as water surface.

2. *In wind screens:* The wind screens and car head lights of motor cars are fitted with polaroid films with their axes inclined at 45° to the horizontal. When two cars approach each other from opposite directions, the transmission planes of their wind screens will be perpendicular to each other, so the glare of their head lights is completely eliminated. Each driver sees the road by the light sent by his own car.

3. *In window panes of aeroplane:* One of the polaroids is fixed while other can be rotated to control the amount of light coming in.

4. *In photoelasticity:* Glass and some plastic materials exhibit double refraction only when stressed. If polarized light is passed through them and then analyzed, the bright coloured lines indicate the existence of strains. In engineering work, plastic models of structures are constructed and weaknesses are examined in this way.

10.6 MATRIX REPRESENTATION OF PLANE-POLARIZED LIGHT WAVES

In 1941, R. Clark Jones invented a broadly useful representation for polarized light. Although these are several methods are available to represent the polarized light, but the Jones method is very good because of its simplicity. The Jones method provides the mathematical detail of the polarization state of light and it also helps to compute the effect that an optical device will have on input light of a given polarization state. The Jones method is unique in that it deals with the instantaneous electric field and is preferred when using coherent sources such as lasers.

As we know that the light is composed of oscillating electric and magnetic fields, Jones discussed that most natural way to represent light is in terms of *light vector* i.e., electric field vector. When column matrix written as a column vector (a Jones vector) and has the form

$$\mathbf{J} = \begin{bmatrix} E_x(t) \\ E_y(t) \end{bmatrix} \qquad (10.21)$$

where $E_x(t) = E_0 e^{i\phi_1}$ and $E_y(t) = E_0 e^{i\phi_2}$ are known as instantaneous scalar components of the electric field.

Note: Since these two values i.e., values of $E_x(t)$ and $E_y(t)$ are in complex numbers, so both amplitude and phase information are available. Generally it is not necessary to find the exact amplitudes and phases of the vector components. So, we can normalize the Jones vector and neglect the common phase factors.

Although some information is missed out, but these results are very simplified expressions. For example, the following vectors contains varying degrees of information but are all Jones vector representations for the same polarization state

$$\begin{bmatrix} E_0 e^{i\phi_1} \\ E_0 e^{i\phi_2} \end{bmatrix} \rightarrow \begin{bmatrix} e^{i\phi_1} \\ e^{i\phi_2} \end{bmatrix} \rightarrow \begin{bmatrix} 1 \\ e^{i(\phi_1-\phi_2)} \end{bmatrix} \quad (10.22)$$

If the dot product of the vector with its complex conjugate yields a value of unity, then a complex vector is known as normalized.

The Jones vector may be represented for horizontal and vertical linear polarization states i.e.,

$$\mathbf{E}_h = \begin{bmatrix} E_x(t) \\ 0 \end{bmatrix} \text{ and } \mathbf{E}_v = \begin{bmatrix} 0 \\ E_y(t) \end{bmatrix} \quad (10.23)$$

Equation (10.23) may be represented in normalized form as

$$\mathbf{E}_h = \begin{bmatrix} 1 \\ 0 \end{bmatrix} \text{ and } \mathbf{E}_v = \begin{bmatrix} 0 \\ 1 \end{bmatrix} \quad (10.24)$$

where $\mathbf{E}_h$ and $\mathbf{E}_v$ are horizontally and vertically polarized light respectively. The sum of two coherent light beams is given by the sum of their corresponding Jones vector components, so the sum of $\mathbf{E}_h$ and $\mathbf{E}_v$, when $E_y(t) = E_x(t)$ is given by

$$\mathbf{E}_{45°} = \begin{bmatrix} E_x(t) \\ E_x(t) \end{bmatrix} \quad (10.25)$$

Equation (10.25) in normalized form may be written as

$$\mathbf{E}_{45°} = \frac{1}{\sqrt{2}} [1] \quad (10.26)$$

Expression (10.26) shows the representation of the polarization state in which the electric field is oriented at 45° with respect to the basis state.

The right circular and left circular are two common polarization states. Two components have equal amplitude in both cases but for right circular, the phase of the y-component leads the x-component by $\frac{\pi}{2}$, while for left circular is at the x-component that leads. Thus, the Jones vector representation for right circular is

$$\mathbf{E}_R = \begin{bmatrix} E_0 e^{i\phi_1} \\ E_0 e^{i(\phi_1-\frac{\pi}{2})} \end{bmatrix} \quad (10.27)$$

Equation (10.27) in normalized form, when factorial out a constant phase of $e^{i\phi_1}$ yields.

$$\mathbf{E}_R = \frac{1}{\sqrt{2}} \begin{bmatrix} 1 \\ e^{i\frac{\pi}{2}} \end{bmatrix} = \frac{1}{\sqrt{2}} \begin{bmatrix} 1 \\ -i \end{bmatrix} \quad (10.28)$$

In similar fashion, the normalized representation for left-circular light is

$$\mathbf{E}_L = \frac{1}{\sqrt{2}} \begin{bmatrix} 1 \\ i \end{bmatrix} \qquad (10.29)$$

Plane Polarized Light

Suppose a beam of light represented by the Jones vector is incident on an optical device

$$\mathbf{E}_i = \begin{bmatrix} E_{ix} \\ E_{iy} \end{bmatrix} \qquad (10.30)$$

The light interacts with particular device and a new polarization state of light upon the existing device will be

$$\mathbf{E}_t = \begin{bmatrix} E_{tx} \\ E_{ty} \end{bmatrix} \qquad (10.31)$$

The coupling between these two vectors can be fully described by a set of four coefficients according the following pair of linear equations

$$E_{tx} = aE_{ix} + bE_{iy} \qquad (10.32)$$

and

$$E_{ty} = cE_{ix} + dE_{iy} \qquad (10.33)$$

Equations (10.32) and (10.33) may be written as

$$\mathbf{E}_t = \mathbf{J}\mathbf{E}_i \qquad (10.34)$$

where

$$\mathbf{J} = \begin{bmatrix} a & b \\ c & d \end{bmatrix} \qquad (10.35)$$

Equation (10.34) is known as Jones matrix of the optical device.

The Jones matrices for common optical devices are given in the following table.

Optical devices	Jones matrix
Vertical linear	$\begin{bmatrix} 0 & 0 \\ 0 & 1 \end{bmatrix}$
Horizontal linear polarizer	$\begin{bmatrix} 1 & 0 \\ 0 & 0 \end{bmatrix}$
Linear polarizer at 45°	$\frac{1}{2}\begin{bmatrix} 1 & 1 \\ 1 & 1 \end{bmatrix}$

Lossless fibre transmission	$\begin{bmatrix} e^{i\phi_1}\cos\theta & -e^{i\phi_2}\sin\theta \\ e^{i\phi_2}\sin\theta & -e^{-i\phi_2}\cos\theta \end{bmatrix}$
Right circular polarizer	$\dfrac{1}{2}\begin{bmatrix} 1 & i \\ -i & 1 \end{bmatrix}$
Left circular polarizer	$\dfrac{1}{2}\begin{bmatrix} 1 & -i \\ i & 1 \end{bmatrix}$
Quarter-wave plate, fast axis vertical	$e^{i\frac{\pi}{4}}\begin{bmatrix} 1 & 0 \\ 0 & -i \end{bmatrix}$
Quarter-wave plate, fast axis horizontal	$e^{i\frac{\pi}{4}}\begin{bmatrix} 1 & 0 \\ 0 & i \end{bmatrix}$

10.7 OPTICAL ACTIVITY

If a plane-polarized light is to pass through certain materials or crystals, it is found that the plane of polarization of light is rotated about the direction of propagation of light through a certain angle. This property of that particular material is known as its *optical activity* and the substance is said to be *optically active*. This phenomenon is known as *rotatory polarization* or *optical rotation*.

The optically active materials are of two types:

1. *Right-handed or dextrorotatory:* The materials or crystals, that rotate the plane of vibration (or plane of polarization) in the clockwise direction as seen by an observer facing the emergent light are said to be *right-handed* or *dextrorotatory*.

2. *Left-handed or laevorotatory:* The substances that rotate the plane of vibration (or plane of polarization) in the anticlockwise direction as seen by an observer facing the emergent light are said to be *left-handed* or *laevorotatory*.

10.7.1 Biot's Laws of Optical Activity: Specific Rotation

In 1815, Biot investigated the phenomenon of optical rotation and proposed the following laws for this:

1. The optical rotation θ of its plane of polarization is directly proportional to length of the material, i.e.
$$\theta \propto l$$

2. The optical rotation θ produced by a solution of an optically active substance is directly proportional to its concentration (C), i.e.
$$\theta \propto C$$

3. The optical rotation θ produced by a particular length of an optically active substance is inversely proportional to the square of wavelength of light passing through it, i.e.

$$\theta \propto \frac{1}{\lambda^2}$$

4. The optical rotation produced by the various substances is the algebraic sum of the optical rotations produced by them separately, i.e.

$$\theta = \theta_1 + \theta_2 + \theta_3 + \cdots$$

5. The optical rotation θ depends on temperature (T) also, i.e.

$$\theta = f(T)$$

The optical activity of any material can be measured in terms of its specific rotation [S] or specific rotatory power.

The specific rotation or specific rotatory power for a given wavelength of light at a particular temperature is defined as the rotation of plane of polarization in degrees produced by a path of one decimetre length of the solution of concentration 1.0 g/cc. Mathematically, it is represented by

$$[S]_T^\lambda = \frac{\theta}{l \times C} \tag{10.36}$$

or
$$\text{Specific rotation} = \frac{\text{Rotation in degrees}}{\text{Length in dm} \times \text{density in g/cc}}$$

Note:
1. Specific rotation is related to molecular rotation as

   Molecular rotation = Specific rotation × Molecular weight

2. Specific rotation is different for different materials.

EXAMPLE 10.11 A tube of sugar solution 20 cm long is placed between crossed Nicols and illuminated with light of wavelength 6×10^{-5} cm. If the optical rotation produced is 13° and specific rotation is 65°/dm/g/cc. Determine the strength of the solution.

Solution The specific rotation S of a solution is given by

$$[S]_T^\lambda = \frac{\theta}{l \times C}$$

Hence $\theta = 13°$, $l = 2.0$ dm and $S = 65/\text{dm}/\text{g/cc}$

$$C = \frac{13}{2.0 \times 65} = 0.1 \text{ g/cc} = 10\%$$

EXAMPLE 10.12 Calculate the specific rotation for sugar solution using the following data:

(i) Length of the tube = 20 cm

(ii) Volume of the tube = 120 cm³
(iii) Quantity of sugar dissolved = 6 g
(iv) Angle of rotation for the analyzer for reforming equal intensity = 6.6°.

Solution For specific rotation through sugar solution given that

$l = 20$ cm $= 2$ dm, volume of the tube $(V) = 120$ cm³

Quantity of sugar dissolved = 6 g, $\theta = 6.6°$.

$$C = \frac{m}{V} = \frac{1}{20} \text{ g/cc}; \text{ specific rotation}(S) = ?$$

$$S = \frac{\theta}{lC} = \frac{6.6 \times 20}{2 \times 1} = 6.6°/\text{dm/g/cc}$$

EXAMPLE 10.13 A 200 mm long tube containing 48 cm³ of sugar solution produces an optical rotation of 11° when placed in the Laurent's half shade polarimeter. If the specific rotation of sugar solution 66°, calculate the quantity of sugar contained in the tube in the form of a solution.

Solution For specific rotation through sugar solution given that

$l = 20$ cm $= 2$ dm

Volume of the tube $(V) = 48$ cm³

Quantity of sugar dissolved $(m) = ?$

$\theta = 11°$

$$C = \frac{m}{V}$$

$$S = 66°$$

$$S = \frac{\theta}{lC} = \frac{\theta V}{lm}$$

$$\Rightarrow \quad m = \frac{\theta V}{lS} = \frac{11 \times 48}{2 \times 66} = 4 \text{ g}$$

EXAMPLE 10.14 A certain length of 5% solution causes the optical rotation of 20°. How much length of 10% solution of the same substance will cause 35° rotations?

Solution

$$S = \frac{\theta_1}{l_1 \times C_1} = \frac{\theta_2}{l_2 \times C_2}$$

$$\Rightarrow \quad \frac{20°}{l_1 \times 5\%} = \frac{35°}{l_2 \times 10\%}$$

$$\Rightarrow \quad l_2 \times 10\% \times 20° = l_1 \times 5\% \times 35°$$

$$\Rightarrow \quad l_2 = \frac{5\% \times 35°}{10\% \times 20°} l_1 \quad \Rightarrow \quad l_2 = \frac{7}{8} l_1$$

EXAMPLE 10.15 80 g of impure sugar when dissolved in a litre of water gives an optical rotation of 9.9° when placed in a tube of length 20 cm. If the specific rotation of sugar is 66°, find the percentage purity of the sugar sample.

Solution The strength of the solution is given by

$$C = \frac{\theta}{l \times S}$$

Here $\theta = 9.9°$, $l = 20$ cm $= 2.0$ dm and $S = 66°$/dm/g/cc

$$\therefore \quad C = \frac{9.9°}{2.0 \times 66°} = 0.075 \text{ g/cc} = 75 \text{ g/litre} \quad [1 \text{ litre} = 10^3 \text{ cc}]$$

The sugar sample dissolved in a litre of water in 80 g in which 75 g is pure sugar. Therefore, purity is

$$\frac{75}{80} \times 100 = 93.75\%$$

10.7.2 Fresnel's Explanation of Optical Rotation

French physicist Fresnel explained the optical rotation mechanism using the following assumptions:

1. The incident plane polarized light, on entering an optically active substance/material, it is broken up into two circular polarized waves, one clockwise and the other anticlockwise.
2. In an optically in active substance, the two waves (*e*- and *o*-rays) travel with same velocity, but in optically active substance, they travel with different velocities. Hence a phase difference between these is developed.
3. In a dextrorotary substance the clockwise wave travels faster, while in the laevorotatory substances the anticlockwise wave travels faster.
4. On emergence, the two circular components recombine to form plane polarized light whose plane of polarization is rotated with respect to that of the incident light by an angle depending on the phase difference between them.

Mathematical treatment

Let us consider a plane polarized wave travelling inside an optical active material along *x*-direction. As we known that the light has transverse nature, the vibrations of two circularly polarized will be in *yz*-plane, Then two circularly polarized waves can be represented as

$$\left.\begin{array}{c} y_1 = \dfrac{a}{2}\cos \omega t \quad \text{and} \quad z_1 = \dfrac{a}{2}\sin \omega t \\[6pt] y_2 = -\dfrac{a}{2}\cos \omega t \quad \text{and} \quad z_2 = \dfrac{a}{2}\sin \omega t \end{array}\right\} \qquad (10.37)$$

and

Since these circularly polarized waves travel through the optical substance at different velocities. Therefore, when they come out of the calcite plate, a phase difference of δ is introduced between them. Then, the emergent vibrations can be given as

$$y_1 = \frac{a}{2}\cos\omega t \quad \text{and} \quad z_1 = \frac{a}{2}\sin\omega t$$

and
$$y_2 = -\frac{a}{2}\cos(\omega t + \delta) \quad \text{and} \quad z_2 = \frac{a}{2}\sin(\omega t + \delta) \qquad (10.38)$$

The resultant vibrations along two axes are, then, given by

$$Y = y_1 + y_2 = \frac{a}{2}\cos\omega t - \frac{a}{2}\cos(\omega t + \delta)$$

$$= a\sin\frac{\delta}{2}\sin\left(\omega t + \frac{\delta}{2}\right) \qquad (10.39)$$

and
$$Z = z_1 + z_2 = \frac{a}{2}\sin\omega t - \frac{a}{2}\sin(\omega t + \delta)$$

$$= a\cos\frac{\delta}{2}\sin\left(\omega t + \frac{\delta}{2}\right) \qquad (10.40)$$

Dividing Eq. (10.39) by Eq. (10.40), we get

$$\frac{Y}{Z} = \tan\frac{\delta}{2}$$

or
$$Y = Z\tan\frac{\delta}{2} \qquad (10.41)$$

If n_L and n_R be the refractive indices of calcite along optic axis for left handed and right handed circularly polarized waves, then path difference between them is

$$\Delta = (n_L - n_R)t \qquad (10.42)$$

where t is the thickness of the material.
The corresponding phase difference is given by

$$\delta = \frac{2\pi}{\lambda}(n_L - n_R)t \qquad (10.43)$$

Then the angle of rotation will be

$$\theta = \frac{\delta}{2} = \frac{\pi}{\lambda}(n_L - n_R)t$$

$$\theta = \frac{\pi}{\lambda}\left(\frac{c}{v_L} - \frac{c}{v_R}\right)t$$

$$\therefore \quad \theta = \frac{\pi c}{\lambda}\left(\frac{1}{v_L} - \frac{1}{v_R}\right)t$$

$$\therefore \quad \theta = \frac{\pi}{T}\left(\frac{1}{v_L} - \frac{1}{v_R}\right)t \qquad \left[\text{Here } \frac{c}{\lambda} = T\right] \qquad (10.44)$$

where c is the speed of light, v_L is the velocity of left-handed wave, v_R is the velocity of right-handed wave and T is the time period.

There are the following cases:
  (i) when $v_L > v_R$, θ is negative, the rotation is clockwise (dextrorotatory substance).
  (ii) when $v_R > v_L$, θ is positive, the rotation is anticlockwise (laevorotatory substance).
  (iii) when $v_L = v_R$, $\theta = 0$, no rotation at all (inactive substance).

EXAMPLE 10.16 The indices of quartz for right handed and left handed circularly polarized light of wavelength 7620 Å are 1.53914 and 1.53920 respectively. Calculate the rotation of the plane polarization of the light in degrees produced by a plate 0.5 mm thick.

Solution The rotation of plane of polarization is given by

$$\theta = \frac{\pi t}{\lambda}(n_L - n_R)$$

Here $t = 0.5$ mm, $n_L - n_R = 1.53920 - 1.53914 = 0.00006 = 6 \times 10^{-5}$
and $\lambda = 7620$ Å $= 7.620 \times 10^{-4}$ mm

$$\theta = \frac{180 \times 0.5 \text{ mm}}{7.620 \times 10^{-4} \text{ mm}} \times (6 \times 10^{-5}) = 7.1°$$

EXAMPLE 10.17 The specific rotation of quartz at 5086 Å is 29.73°/mm, calculate the difference in the refractive indices.

Solution The optical rotation θ is given by

$$\theta = \frac{\lambda l}{\lambda}(n_L - n_R)$$

Thus
$$n_L - n_R = \frac{\lambda}{\pi}\frac{\theta}{l}$$

The specific rotation $\dfrac{\theta}{l} = 29.73°$/mm

and $\lambda = 5086$ Å $= 5.086 \times 10^{-4}$ mm

∴ $$n_L - n_R = \frac{5.086 \times 10^{-4}}{180} \times 29.73 = 8.4 \times 10^{-5}$$

10.8 POLARIMETER

- Polarimeter is an optical device which is used to measure the rotation of the plane of vibration of the polarized light.
- Polarimeter is an instrument which is used to measure the angle through which the plane of polarization is rotated by the optically active substance.

When the polarimeter is used to determine the quantity in sugar solution, it is said to be Saccharimeter.

The following are the types of polarimeter:
  1. Laurent's half shade polarimeter

2. Biquartz polarimeter
3. Lippich's polarimeter
4. Soleil's compensated biquartz polarimeter

10.8.1 Laurent's Half Shade Polarimeter

This instrument is used to measure the optical rotation of certain solutions. For known specific rotation the concentration of sugar solution can be determined.

Construction

The Laurent's half shade polarimeter is shown in Figure 10.29. It consists of two Nicol prisms, one serves the purpose of polarizer, while other works as analyzer. These polarizer and analyzer are capable to rotate about an axis. A glass tube is provided, which is filled with the solution of an optically active substance. A half shade device is placed between the polarizer and glass tube. The ends of this glass tube are covered firmly with two glass plates. This tube is placed between polarizer and analyzer.

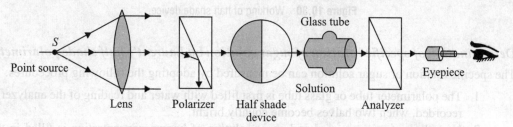

Figure 10.29 Experimental arrangement for determination of specific rotation.

Function

The half shade device is a circular plate. Its half portion is made up of quartz, while the other half is made up of glass. The thickness of quartz half is taken such that it produces a path difference of $\lambda/2$ (or phase difference of π) between ordinary and extraordinary rays, i.e. it serves the purpose of half wave plate. The thickness of the glass plate is so chosen as to absorb the same light to that of quartz half.

Working

Suppose the plane polarized light emerging from the polarizer has the vibrations along *OA*. Now from the glass plate the emergent light will have same direction of vibration along *OA* (Figure 10.30).

In case of quartz plate, the incident light splits up into *o*-and *e*-rays, *e*-rays travel faster. They gain a phase difference of π(or path difference of $\lambda/2$) on emergence from the quartz plate. Therefore, the emergent light from the quartz plate will have the vibrations along *OX'* (*o*-ray) and along *OX*(*e*-ray). Thus from quartz plate, resultant will be along *OB*. Now when the principal plane of analyzer is parallel to *BOB'*, the light from the quartz plate will pass through completely, while from glass plate will transmit partially. Therefore, the half part at the left side will be brighter as compared to right half. When the principal plane is parallel to

AOA', the right half will be brighter. But, when the principal plane of analyzer is set parallel to optic axis of half shade device, i.e. along XOY, the two halves will be equally bright. This position of analyzer is recorded. With the help of this recorded value of specific rotation of the solution can be determined by adopting procedure given below:

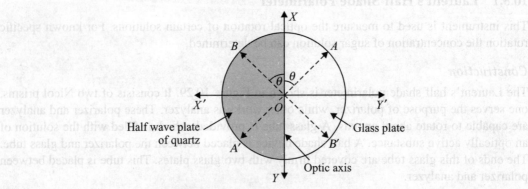

Figure 10.30 Working of half shade device.

Determination of specific rotation of sugar solution by Laurent's half shade polarimeter

The specific rotation of sugar solution can be measured by adopting the following procedures:

1. The polarimeter tube or glass tube is first filled with water and reading of the analyzer is recorded, when two halves become equally bright.
2. Now the water is removed and sugar solution of known concentration is filled in the tube. The analyzer is again rotated till the two halves are equally bright and reading is noted.
3. The difference in reading obtained by above two steps gives the angle of rotation of the plane of polarization for that concentration of the solution.
4. Now the same procedure is repeated with the solutions of different concentrations and the corresponding values of θ are determined.
5. A graph (Figure 10.31) is plotted between the angle of rotation (θ) and the concentration of solution (C). This graph is a straight line.

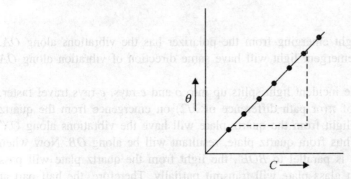

Figure 10.31 Concentration versus angle of rotation of sugar.

6. The slope of the graph gives the value of θ/C and, therefore, the specific rotation of sugar is calculated with the help of expression

$$[S]_T^\lambda = \frac{\theta}{lC}$$

where l is the length of tube in decimetre and temperature and wavelength are constant throughout the whole experiment.

10.8.2 Biquartz Polarimeter

The optical arrangement in biquartz polarimeter is almost same as that of Laurent's half shade polarimeter. However, the following are the *two* differences between them:

1. In place of Laurent's half shade plate, it consists of a biquartz plate.
2. A white light source is used in biquartz polarimeter in place of monochromatic light in half shade polarimeter.

Function

The biquartz plate is made up of two semicircular quartz plates. One of them is of left-handed quartz, while the other is of right-handed quartz. Both the plates are cut perpendicular to optic axis and joined together, so as to form a complete circle as shown in Figure 10.32.

The dextrorotatory (right-handed) and laevorotatory (left-handed) semicircular quartz plates rotate the plane of polarization in clockwise and anticlockwise directions respectively. The thickness of each semiconductor quartz plate is chosen that it rotates the plane of polarization of yellow light by 90°.

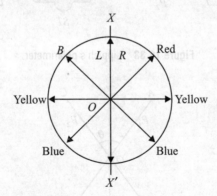

Figure 10.32 Function of biquartz plate.

Working

When a beam of light is allowed to incident on a polarizing Nicol, it becomes plane polarized. When this polarized light is incident normally on a biquartz plate, it rotates the various constituent wavelengths by different angles. The left handed plate rotates the plane of polarization in anticlockwise direction, whereas the right handed plate rotates in clockwise direction. Since the rotation of plane of polarization is inversely proportional to the wavelength of light, the red component of white light is rotated least, while the violet component is rotated by maximum angle.

Since the thickness of the plate is so chosen as to rotate the plane of polarization of yellow light by 90°, the yellow vibrations will be along the line yellow-O-yellow. If the principal plane of analyzer is set along the line XOX', the yellow vibrations may be stopped completely. In this portions, both the halves look grey coloured (a combination of red and blue). If the analyzer is now rotated slightly from this position, each half appears to be having different colours either blue or red, i.e. if one half is red, the other will be blue and vice versa. The position of analyzer, when both the halves are grey, is recorded and with the help of this reading the angle of rotation is determined.

10.8.3 Lippich's Polarimeter

Lippich's polarimeter is an improved polarizer. In this, half shade plate used in Laurent's half shade polarimeter is replaced by a small Nicol prism P_2. This replacement enables the Lippich's polarimeter to be used for light of any wavelength. The schematic diagram of this polarimeter is shown in Figure 10.33.

In this diagram, S is a source of light and P_1 is a polarizer. In front of it is mounted Nicol prism P_2 such that its lower surface touches the common central line of polarizing Nicol P_1 and the analyzer A. This enables focusing the edge of the Nicol P_2 with the telescope. The sample tube (ST) containing liquid is placed between the two diaphragms facing P_2 and A. The principal planes of P_1 and P_2 are inclined to each other by a small angle θ as shown in Figure 10.34.

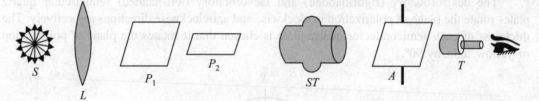

Figure 10.33 Lippich's polarimeter.

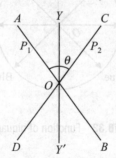

Figure 10.34 Principal plane of P_1AB inclined at angle θ with principal plane of P_2CD.

In Figure 10.34, angle θ is called half shadow angle. When the analyzer A is rotated such that its principal plane is parallel to AB, the left half of the circle will appear more bright as compared to right half. However, if the analyzer is rotated such that its principal plane becomes parallel to CD, then the right half will appear more bright compared to left half. However, if its principal plane of analyzer is parallel to YY', a bisector of angle AOC, then the field of view of two halves will be equally illuminated. A slight rotation of analyzer on the either side YY' will lead to unequal brightness of two halves. Therefore, by rotating the analyzer A, the position of equal brightness of two halves can be accurately obtained.

To get angle of rotation and, thus, specific rotation of a solvent, two observations are needed. First, fill the sample tube with solvent and obtain the position of analyzer for equal brightness of two halves. The difference of reading on the circular scale attached to the analyzer gives the angle of rotation for the solution. The specific rotation (θ) can be determined knowing the concentration (C) of the solution and accurate length of the sample tube.

10.8.4 Soleil's Compensated Biquartz Polarimeter

This polarimeter is better and more sensitive than the others. The schematic diagram of this polarimeter is shown in Figure 10.35.

It consists of a polarizer P_1 biquartz (BQ), sample tube (ST), right-handed quartz plate Q, whose optic axis is perpendicular to its end face, double wedge L called compensator, analyzer A and telescope T.

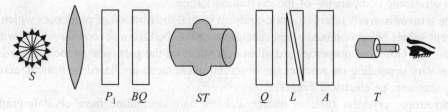

Figure 10.35 A schematic diagram of Soleil's polarimeter.

A plane polarized light from polarizer P_1 passes through biquartz (BQ) and then passing through the solution in sample (ST) falls on right-handed quartz tube Q, whose optic axis is perpendicular to its end face. On emergence from it, passes through a double wedge, each wedge being made of left handed quartz of variable thickness. The optic axis of each wedge is perpendicular to its face. The emergent light from the wedge is received by the analyzer and then by telescope.

The wedges act as a compensator and any amount of either left-hand or right-hand rotation can be introduced by it to neutralize the rotation produced by the optically active material in the sample tube. This introduced rotation by the wedges can be read off from the scale attached to the wedges. The procedure for obtaining the angle of rotation is simple. First, introduce the solvent in the tube and obtain the position of wedge by adjusting it for the position of the *tint of passage*. Read it from the scale attached to the wedge. Now introduce solution in the tube and again adjust the wedge for the tint of passage and read the position on the attached scale. The difference of two readings gives the angle of rotation. This polarimeter is specially suitable for the sugar work, because the rotatory dispersion of sugar and quartz is same.

10.9 PHOTOELASTICITY

Photoelasticity is the change in optical properties of a transparent material when it is subjected to mechanical stress. An example of such properties is *birefringence*. The mechanical birefringence of certain materials enables the determination of stress and strains from the interference fringe patterns they produce.

10.9.1 Birefringence or Double Refraction[1]

Birefringence is defined as the double refraction of light in a transparent, molecularly ordered material, which is manifested by the existence of orientation-dependent difference in refractive index.

Many transparent solids are optically isotropic, meaning that the index of refraction is equal in all directions throughout the crystalling lattice. Examples of isotropic solids are glass, sodium chloride and many polymers.

Crystals are classified as isotropic and anisotropic depending upon their optical behaviours and whether or not their crystallographic axes are equivalent. All isotropic crystals have equivalent axes that interact with light in a similar manner, regardless of the crystal orientation with respect to incident light waves. Light entering an isotropic crystal is refracted at a constant angle and passes through the crystal at a single velocity without being polarized by interaction with the electronic components of the crystalline lattice.

The term *anisotropy* refers to a non-uniform spatial distribution of properties which results in different values obtained when specimens are probed from several directions within the same material. Observed properties are often dependent on the particular probe being employed and often vary depending on whether the observed phenomena are based on optical, acoustical, thermal, magnetic or electrical events.

Anisotropic crystals, such as quartz, calcite, and tourmaline have crystallographically distinct axes and interact with light by a mechanism that is dependent upon the orientation of the crystalline lattice with respect to the incident light angle. When light enters the optical axes of anisotropic crystals, it behaves in a manner similar to the interactions with isotropic crystals, and passes through at a single velocity. However, when light enters or non-equivalent axis, it is refracted into two rays, each polarized with vibration directions oriented at right angles (mutually perpendicular) to one another and travelling at different velocities. This phenomenon is termed *double refraction* or *birefringence* (Figure 10.36).

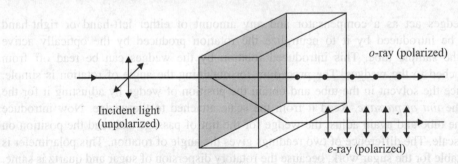

Figure 10.36   Double refraction.

The phenomenon of double refraction is based on the laws of electromagnetism, first proposed by British Mathematician James Clerk Maxwell in 1860. One of the most

---
[1] Double refraction has also been discussed in section 10.5.3.

dramatic demonstrations of double refraction occurs with calcium carbonate (calcite) crystals. When anisotropic calcite crystals refract light, they split the incoming rays into two components that take different paths during their journey through the crystal and emerge as separate light rays. This unusual behaviour is attributed to the arrangement of atoms in the crystalline lattice. Because the precise geometrical ordering of the atoms is not symmetrical with respect to the crystalline axes, light rays passing through the crystal can experience different refractive indices, depending upon the direction of propagation.

One of the rays passing through an anisotropic crystal obeys the laws of normal refraction, and travels with the same velocity in every direction through the crystal. This light is termed the ordinary ray (*o*-ray). The other ray travels with a velocity that is dependent upon the propagation direction within the crystal, and is termed as extraordinary ray (*e*-ray).

The difference in refractive indices or birefringence, between the extraordinary and ordinary rays travelling through an anisotropic crystal is a measurable quality, and can be expressed as an absolute value by the equation

$$\text{Birefringence } (B) = [n_e - n_o] \qquad (10.45)$$

where n_e and n_o are the refractive indices experienced by the extraordinary and ordinary rays respectively.

The optical path difference is defined by the relative phase shift between the ordinary and extraordinary rays as they emerge from an anisotropic material. The optical path difference is computed by multiplying the specimen thickness (*t*) by the refractive index, but only when the medium is homogeneous and does not contain significant refractive index deviations or gradients. For a system with two refractive index values (n_1 and n_2), the optical path difference (Δ) is determined from the equations:

$$\text{Optical path difference } (\Delta) = (n_1 - n_2) \times t \qquad (10.46)$$

10.9.2 Stress Optic Law

The stress optic law (also known as Brewster's law) states that the relative change in the index of refraction is proportional to the difference of principal stresses

$$n_2 - n_1 = C_b(\sigma_1 - \sigma_2) \qquad (10.47)$$

where σ_1 and σ_2 are the principal stresses at some point within the solid, n_1 and n_2 are the principal indices of refraction, and C_b is the relative strain optic coefficient, which is a property of the photoelastic material used.

10.9.3 Theory of Photoelasticity

The stress field at any point in a photoelastic specimen can be related to its index of refraction through (Maxwell's stress optic laws). Photoelastic materials are birefringent, i.e. they act as temporary waveplates, refracting light differently for different light amplitude orientations, depending upon the state of stress in the material. In the unloaded state, the material exhibits and index of refraction n_0 i.e., independent of orientation. Therefore, light of all orientation propagating along all axes through the material propagate with the same speed, namely v/n_0. In the loaded state, however, the orientation of a given light amplitude vector with respect to the principal stress axes, and the magnitudes of the principal stresses, determine the index of refraction for the light wave.

Hence, when the plane polarized light arrives at the specimen, it is refracted and split into two separate waves, one wave vibrating parallel to one principal vibration direction and the other wave parallel to the other orthogonal principal vibration direction and the other wave parallel to the other orthogonal principal vibration direction. The velocities of these waves will be determined by the relevant refractive indices along the given directions. For an anisotropic material the refractive indices will be different in two directions and, therefore, the waves will become progressively out of phase upon passing through the material. Upon emerging, the two wave recombine, however, the exact way they recombine will depend on the phase difference between them and how they interfere.

Consider a photoelastic model of uniform thickness is under a load as shown in Figure 10.37. The stress experienced by a model along the principal axis can be given as σ_x and σ_y or σ_1 and σ_2 respectively. Let n_0 be the refractive index of the model when no load is applied. When the load is applied, the model behaves as birefringent or double refracting.

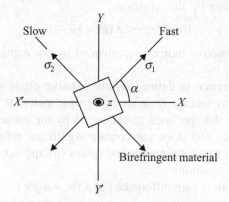

Figure 10.37 Principal stress model.

Let n_1 and n_2 be the refractive indices of the directions of vibration, then according to Brewster's law

$$n_1 - n_0 = C_1\sigma_1 + C_2\sigma_2 \quad (10.48)$$

and

$$n_2 - n_0 = C_1\sigma_2 + C_2\sigma_1 \quad (10.49)$$

where C_1 and C_2 are the stress coefficients for the direct and transverse modes. When the linearly polarized light is incident at point z of the photoelastic model, they resolved into two components as σ_1 and σ_2. These two components of light travel with different velocities in the material model. Let

$$\Delta = \text{phase difference}$$

Then, phase difference between these two components will be

$$\Delta = \frac{2\pi t}{\lambda}(n_2 - n_1) \quad (10.50)$$

where λ is the wavelength and t is the thickness of the specimen.

The value of $(n_2 - n_1)$ can be computed from Eqs. (10.48) and (10.49)

$$(n_2 - n_0) - (n_1 - n_0) = (C_1\sigma_2 + C_2\sigma_1) - (C_1\sigma_1 + C_2\sigma_2)$$

$$\Rightarrow \qquad (n_2 - n_1) = (C_2 - C_1)(\sigma_1 - \sigma_2) \qquad (10.51)$$

$$\therefore \qquad \Delta = \frac{2\pi t}{\lambda}(C_2 - C_1)(\sigma_1 - \sigma_2) \qquad (10.52)$$

$$\therefore \qquad \Delta = \frac{2\pi t}{\lambda} C(\sigma_1 - \sigma_2) \qquad (10.53)$$

where $C = (C_2 - C_1)$ is the relative stress optic coefficient. Therefore, for a transparent and isotropic material, the phase difference between the two components is directly proportional to the difference of the principal stress σ_1 and σ_2. From these values, the wavelength of relative path difference can be calculated as

$$N = \frac{\Delta}{2\pi}$$

$$= \frac{\frac{2\pi t}{\lambda} C(\sigma_1 - \sigma_2)}{2\pi}$$

$$N = \frac{t}{\lambda} C(\sigma_1 - \sigma_2) \qquad (10.54)$$

Equation (10.54) is known as *stress optic law*. Using this equation, materials fringe can be calculated as

$$(\sigma_1 - \sigma_2) = \frac{N\lambda}{tC}$$

$$\Rightarrow \qquad (\sigma_1 - \sigma_2) = \frac{N}{t} f \qquad (10.55)$$

where $f = \lambda/C$ is called the *material fringe value*. For $\sigma_1 = \sigma_2$, the fringe order is zero and the black dots appear. These points are called *isotropic points*. If $2\theta = 0$ or $90°$, the transmitted intensity is zero and one of the permitted vibration direction lies parallel to the polarizer direction. These settings are known as *extinction positions* and produce isoclinic fringes. These isochromatics fringes are defined as dark lines corresponding to a constant value of principal stress difference.

10.9.4 Applications of Photoelasticity

The photoelasticity is applicable in many important areas. Some important applications are given below.
1. **Aircraft industry:** To analyze stresses in aeroplane wings and window frames under both static and flight conditions. Stresses in the landing gear, compressor, blades and aircraft body can be determined.
2. **Automobile industry:** For design changes and optimization of shape of the components.
3. **Mining and civil engineering:** To analyze stresses in rocks, bridges and dam locks, etc.
4. **Ordinance:** To analyze stresses due to impact such as those caused by shocks, explosions and bullets and explosive forming of metals.
5. **Biomechanics:** To analyze stresses in human skull, femur, legs and hips, etc. and other mechanical medical aids.

FORMULAE AT A GLANCE

10.1 Brewster's law (reflection method): When an unpolarized light beam is reflected from a surface, the reflected beam may be completely polarized, partially polarized or unpolarized, depending on the angle of incidence. Reflected beam will be polarized when angle of incidence (Brewster angle i_p) becomes polarization angle. Thus,

$$\tan i_p = n$$

n = Refractive index of the material

For example, Brewster angle for crown glass ($n = 1.52$) has the value $i_p = 56.7°$. Since refractive index is wavelength dependent, i_p is also function of wavelength.

10.2 At polarization angle, reflected and refracted beams are perpendicular to each other.

10.3 Malus' law: It measures the intensity of transmitted light coming out of the analyzer, the incoming light should be polarized (i.e., output of the polarizer).

Thus, $I = I_0 \cos^2\theta = a^2 \cos^2\theta$

Where, I_0 = intensity of the incoming light

θ = angle between planes of polarizer and analyzer

10.4 Quarter wave plate (QWP): When a plane polarized beam of light incident on doubly refracting crystal (quartz, calcite) it split into two components such as ordinary (o) and extraordinary (e) beams and both components travel with different velocities. This difference of velocities is maximum in the direction perpendicular to the optic axis. If thickness t of the crystal is such that it introduces a phase change of $\pi/2$ or path difference of $\lambda/4$ between o- and e-beams then plate is called QWP. It is used to produce circularly and elliptically polarized light.

$$t = \frac{\lambda}{4|(n_o - n_e)|}$$

where n_o = refractive index for ordinary beam

n_e = refractive index for extraordinary beam

10.5 Half wave plate (HWP): Thickness of plate $t = \dfrac{\lambda}{2|(n_o - n_e)|}$ half wave plate produces phase of π or path change of $\lambda/2$. It is used to produce plane polarized light.

10.6 General equation of an ellipse $\dfrac{x^2}{a} + \dfrac{y^2}{b} - \dfrac{2xy}{ab}\cos\phi = \sin^2\phi$ is used to describe the condition for production of plane, circularly or elliptically polarized light. Since phase difference (ϕ) between o- and e-beams is deciding parameters, which can be manipulated with the help of thickness t of the doubly refracting plate. Here incident light beam is plane polarized, which slit into o- and e-inside the plate.

Case 1: If $\phi = 0, 2\pi, 4\pi$, $\sin\phi = 0$; $\cos\phi = 1$ then equation reduces to $y = b/a \times x$ (straight line i.e., light coming out of the plate is plane polarized)

Case 2: If $\phi = \pi, 3\pi, 5\pi$ then $y = (-b/a \times x)$. It again provides plane polarized light at the output.

Case 3: If (QWP) $\phi = \pi/2$ and $a = b$; then $x^2 + y^2 = a^2$ (circle). Emergent light would be circularly polarized.

Case 4: (QWP) $\phi = \dfrac{\pi}{2}$ and $a \neq b$; then $\dfrac{x^2}{a^2} + \dfrac{y^2}{b^2} = 1$ (ellipse). Emergent light would be elliptically polarized.

10.7 Jones matrices of common optical devices

Optical devices	Jones matrix
Vertical linear	$\begin{bmatrix} 0 & 0 \\ 0 & 1 \end{bmatrix}$
Horizontal linear polarizer	$\begin{bmatrix} 1 & 0 \\ 0 & 0 \end{bmatrix}$
Linear polarizer at 45°	$\dfrac{1}{2}\begin{bmatrix} 1 & 1 \\ 1 & 1 \end{bmatrix}$
Lossless fibre transmission	$\begin{bmatrix} e^{i\phi_1}\cos\theta & -e^{i\phi_2}\sin\theta \\ e^{i\phi_2}\sin\theta & -e^{-i\phi_2}\cos\theta \end{bmatrix}$
Right circular polarizer	$\dfrac{1}{2}\begin{bmatrix} 1 & i \\ -i & 1 \end{bmatrix}$
Left circular polarizer	$\dfrac{1}{2}\begin{bmatrix} 1 & -i \\ i & 1 \end{bmatrix}$
Quarter-wave plate, fast axis vertical	$e^{i\frac{\pi}{4}}\begin{bmatrix} 1 & 0 \\ 0 & -i \end{bmatrix}$
Quarter-wave plate, fast axis horizontal	$e^{i\frac{\pi}{4}}\begin{bmatrix} 1 & 0 \\ 0 & i \end{bmatrix}$

10.8 *Optical rotation:* There are certain optically active substances like sugar solution, sodium chlorate, cinnabar etc. which rotate the plane of polarization of light when light is passed through it. This phenomenon is known as optical activity by optically active solutions is directed the angle of rotation.

10.9 Optical rotation depends on (i) length of substance, (l) (ii) concentration of solution, (C) and indirectly depends on wavelength and temperature.

$$\theta_T^\lambda \propto lC$$

$$\theta_T^\lambda = SlC$$

$\Rightarrow$ Specific rotation $[S]_T^\lambda = \dfrac{\theta}{l.C}$

where θ = in degree, l = in decimetre and C = in g/cc

10.10 Angle of rotation (Fresnel's method)

$$\theta = \frac{\pi}{\lambda}(n_L - n_R)t = \frac{\pi}{T}\left(\frac{1}{v_L} - \frac{1}{v_R}\right)$$

Where n_L and n_R are the refractive indices of calcite crystal along optic axis for left handed and right handed circularly polarized waves and v_L and v_R are velocities for left handed and right handed circularly polarized waves.

10.11 Birefringence $(B) = |n_e - n_o|$

10.12 Optical path $(\Delta) = (n_1 - n_2)t = \dfrac{2\pi t}{\lambda}(n_1 - n_2) = \dfrac{2\pi t}{\lambda}C(\sigma_1 - \sigma_2)$

SOLVED NUMERICAL PROBLEMS

PROBLEM 10.1 The polarizing angle for air-glass interface is 56.3°. What is the angle of refraction if light is incident from air on a glass slab at the polarizing angle?

Solution We know that $r + i_p = \dfrac{\pi}{2}$

Hence, $r = \dfrac{\pi}{2} - i_p = 90° - 56.3° = 33.7°$

PROBLEM 10.2 A glass plate is to be used as a polarizer. Find the angle of polarization and the angle of refraction, the refractive index of glass is 1.54.

Solution Given

$$n = 1.54$$

We know that

$$\tan i_p = n$$
$$i_p = \tan^{-1} n$$
$$= \tan^{-1} 1.54 = 57°$$

If r be the angle refraction, then

$$i_p + r = 90°$$

and

$$r = 90° - 57° = 33°$$

PROBLEM 10.3 What is the polarizing angle for a beam of light travelling in air, when it is reflected by a pool of water ($n = 1.33$)?

Solution Using Brewster's law

$$\tan i_p = n$$

$\therefore$ $$\tan i_p = 1.33$$

or $$i_p = \tan^{-1}(1.33) = 53.1°$$

PROBLEM 10.4 Two polarizing sheets have their directions so that the intensity of transmitted light is maximum. At what angle should either sheet be turned so that the intensity becomes one half of the initial?

Solution Given

$$I = \frac{I_0}{2}$$

From Malus' law

$$I = I_0 \cos^2 \theta$$

$\therefore$ $$\frac{I_0}{2} = I_0 \cos^2 \theta$$

$\therefore$ $$\cos \theta = \pm \frac{1}{\sqrt{2}} \Rightarrow \theta = 45° \text{ or } 135°$$

PROBLEM 10.5 Four polarizing sheets are placed one above the other that the direction of polarization of each sheet makes an angle of 30° with the direction of polarization of its preceding sheet. If an unpolarized light is incident on this system, what fraction of it will emerge out of it?

Solution Let us consider I_1, I_2, I_3 and I_4 be the intensities of plane-polarized light transmitted through the first, second, third and fourth polarizing sheets respectively. Then

$$I_1 = \frac{I_0}{2}$$

$$I_2 = I_1 \cos^2 30° = \frac{3}{8} I_0$$

$$I_3 = I_2 \cos^2 30° = \frac{3}{8} I_0 \times \frac{3}{4} = \frac{9}{32} I_0$$

and $$I_4 = \left(\frac{9}{32} I_0\right) \cos^2 30° = \frac{9}{24} I_0 \times \frac{3}{4}$$

$$= \frac{27}{128} I_0$$

PROBLEM 10.6 Unpolarized light falls on two polarizing sheets placed one over another. What will be the angle between their transmission axes if the intensity of light transmitted finally is one-third of the intensity of the incident light? Assume that each polarizing sheet acts as an ideal polarizer.

Solution Since both sheets are ideal, the intensity of the incident unpolarized beam I will reduce to half after passing through one of them as shown in Figure 10.38. After passing through the second polarizing sheet, the intensity reduces to one-third of the original value.

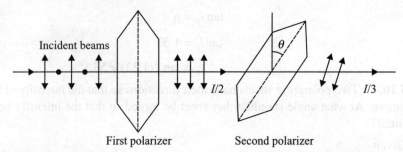

Figure 10.38 Unpolarized light beam of intensity I passing through two-polarizing sheets.

According to Malus' law

$$I = I_0 \cos^2\theta$$

Here

$$I = \frac{I}{3}$$

and

$$I_0 = \frac{I}{2}$$

Then

$$\frac{I}{3} = \frac{I}{2}\cos^2\theta$$

or

$$\cos^2\theta = \frac{2}{3} = 0.666$$

or

$$\cos\theta = \sqrt{0.666} \implies \theta = \cos^{-1}\sqrt{0.666} = 35.3°$$

PROBLEM 10.7 Calculate the thickness of a quarter wave plate for light of wavelength 589 nm. The refractive indices for o- and e-rays are 1.55 and 1.50 respectively.

Solution Given

$$n_e = 1.55, \text{ and}$$
$$n_o = 1.50$$

We know that thickness of QWP

$$t = \frac{\lambda(n_o - n_e)}{4}$$

$$= \frac{589 \text{ nm}(1.55 - 1.50)}{4}$$

$$= 7.363 \text{ nm} = 7.4 \text{ nm}$$

PROBLEM 10.8 Compute the thickness of a half wave plate for sodium light, given $n_o = 1.54$ and the ratio of the velocity of o-component and e-component is 1.007. Is the crystal positive or negative?

Solution Given $n_o = 1.54$

$$\frac{v_o}{v_e} = 1.007$$

We know that

$$\frac{v_o}{v_e} = \frac{n_e}{n_o} = 1.007$$

or
$$n_e = 1.007 \times n_o$$
$$= 1.007 \times 1.54 = 1.551$$

∵ $n_e > n_o$, the crystal is positive

∴ Thickness of HWP,

$$t = \frac{\lambda}{2(n_e - n_o)}$$

$$= \frac{589.3 \text{ nm}}{2(1.551 - 1.54)} = 0.2679 \times 10^{-4} \text{ m}$$

PROBLEM 10.9 What is the optical thickness of a quarter waveplate for the light of wavelength 600 nm, the birefringence of the plate $(n_e - n_o)$ being 0.172?

Solution Given $\lambda = 600 \text{ nm} = 6000 \times 10^{-10} \text{ m}$
$(n_e - n_o) = 0.172$

The optical thickness of QWP

$$t = \frac{\lambda}{4(n_e - n_o)}$$

$$= \frac{6000 \times 10^{-10}}{4 \times 0.172} \text{ m}$$

$$= 8.72 \times 10^{-7} \text{ m}$$

PROBLEM 10.10 A 20 cm tube contains sugar solution made by dissolving 15 g of sugar in 100 cc of water. What is the angle of rotation of the plane of vibration of a plane polarized light passing through the tube? Specific rotation of sugar = 66.5/dm/g/cc.

Solution Given $l = 20 \text{ cm} = 2 \text{ m}$

$$C = \frac{15}{100} \text{ g/cc}$$

$[S]_T^\lambda = 66.5/\text{dm/g/cc}$

∵ We know that the specific rotation

$$[S]_T^\lambda = \frac{\theta}{l \times C}$$

or
$$\theta = [S]_T^\lambda [l \times C]$$

$$= \frac{66.5 \times 2 \times 15}{100} = 19.95°$$

PROBLEM 10.11 The refractive indices for right handed and left handed circularly polarized light of wavelength 762 nm for quartz are 1.53914 and 1.53920 respectively. Compute the rotation of the plane of polarization of light in degrees produced by a plate of 0.5 mm thickness.

Solution Given
$$n_L = 1.53914$$
$$n_R = 1.53920$$
$$\lambda = 762 \text{ nm} = 7.62 \times 10^{-7} \text{ m}$$
$$t = 0.5 \text{ mm} = 5.0 \times 10^{-4} \text{ m}$$

We know that the rotation of the plane of polarization of light
$$\theta = \frac{\pi t}{\lambda}(n_L - n_R)$$

$$\therefore \theta = \frac{\pi \times 0.5 \times 10^{-3} \times (1.53914 - 1.53920)}{7.62 \times 10^{-7}}$$

$$= \frac{\pi \times 30}{762} \text{ rad} = \frac{\pi \times 30}{762} \times \frac{180°}{\pi}$$

$$= 7.1°$$

PROBLEM 10.12 Deduce the speeds of o- and e-rays in calcite: (i) In a plane perpendicular to the optic axis. (ii) Along the optic axis. Hence find the relative phase difference between the rays for light of wavelength 6000 Å in travelling a distance 3×10^{-3} cm at right angles to the optic axis. [Given that $n_o = 1.658$ and $n_e = 1.486$]

Solution (i) Velocity of ordinary ray
$$v_o = \frac{c}{n_o} = \frac{3 \times 10^8}{1.658} = 1.8 \times 10^8 \text{ m/s}$$

Velocity of e-ray
$$v_e = \frac{c}{n_e} = \frac{3 \times 10^8}{1.486} = 2.02 \times 10^3 \text{ m/s}$$

Along the optic axis, the two rays travel with the same velocity and its value is equal to the o-ray.
Hence,
$$v_e = v_o = 1.8 \times 10^8 \text{ m/s}$$

Now, the time difference in travelling a distance x
$$\Delta t = \frac{x}{v_o} - \frac{x}{v_e} = \frac{x}{c}\left[\frac{x}{v_o} - \frac{x}{v_e}\right] = \frac{x}{c}[n_o - n_e]$$

Hence path difference $= c \times \Delta t = x[n_o - n_e]$

$$\therefore \text{ Phase difference} = \frac{2\pi}{\lambda} \times \text{path difference} = \frac{2\pi}{\lambda} \times x[n_o - n_e]$$

$$= \frac{2 \times 3.14}{6000 \times 10^{-10}} \times 3 \times 10^{-3} [1.658 - 1.486] = 54 \text{ rad.}$$

CONCEPTUAL QUESTIONS

10.1 What do you mean polarization of light?

Ans Maxwell predicted that transverse nature of light has been established by the polarization experiments. In transverse wave, the particles of the medium execute periodic oscillations in a direction perpendicular to the direction of propagation of light wave. In electromagnetic wave, the light-emitting atoms oscillate at random and emit individual wave trains in all possible directions. The corresponding light wave is called natural light or unpolarized.

If, however, the oscillations are confined to only one direction, then we get a plane polarized light. If electric vector of transversed electromagnetic wave is identified as light vector. Hence, in plane polarized light, the light vectors (electric vector) vibrate in a single plane called the plane of vibration and the plane perpendicular to it is the plane of polarization.

Polarized light is of three types, viz.

(i) plane polarized light (ii) circularly polarized light and (iii) elliptically polarized light.

10.2 Light is generally characterized by electric vector, although it also possesses magnetic vector. Explain.

Ans Light is form of electromagnetic wave motion. The electromagnetic wave has electric and magnetic fields vectors oscillating in phase but perpendicular to each other. The interaction of electromagnetic waves with matter can be explained either with the help of electric vector or with magnetic vector. But peak value of electric vector is much higher than that of magnetic vector. Moreover, our eyes are more sensitive to electric vector than to magnetic vector. Hence, polarization and other optical phenomenon are preferably described in terms of electric vector. This is why light is characterized by electric vector although it also possess magnetic vector.

10.3 Why light waves can be polarized, while sound waves cannot be polarized?

Ans Sound waves are longitudinal in nature, whereas light waves are transverse in nature. Only transverse waves can be polarized.

10.4 Prove that the maximum transmission through polarizer is half of the maximum even if the polarizer is ideal.

Ans Let the intensity of the incident on the polarizer is I_0. Let θ be the angle between the polarizing direction of the polarizer and the direction of vibration of the electric vector in the incident light.

Therefore, applying Malus' law, the intensity of the transmitted light through the polarizer (say I) will be $I = I_0 \cos^2 \theta$.

The incident light is unpolarized one. So, it can have any direction of vibration perpendicular to the direction of propagation of light. So, to compute the intensity I, we have to take the average value of I. Now, average value of $\cos^2 \theta$ is $\dfrac{1}{2}$.

Therefore, $I_{max} = I_0 \cos^2 \theta$ is $\dfrac{1}{2}$.

An ideal polarizer will transmit this maximum quantity which is equal to the half of the intensity of incident one. This implies that even an ideal polarizer can, at best, have a maximum transmission of 50%.

10.5 If unpolarized light falls on a system of two crossed polarized sheets, no light is transmitted. If third polarizing sheet is placed between them, will light be transmitted? Explain.

Ans Let A and B are two crossed polarizing sheets and I_0 be the intensity of the light incident on A. Then the intensity of light transmitted through A is

$$I_A = \frac{1}{2} I_0$$

when the third sheet C is inserted between A and B (but not parallel to either), then the intensity of light transmitted by C is

$$I_C = I_A \cos^2 \theta = \frac{1}{2} I_0 \cos^2 \theta$$

Since A and B are crossed the angle between C and B is $(90° - \theta)$, so the intensity of light transmitted by B will be

$$I_B = I_C \cos^2(90° - \theta) = I_C \sin^2 \theta = \frac{1}{2} I_0 \cos^2 \theta \sin^2 \theta$$

Now if $\theta = 90°$, $I_B = 0$ and if $\theta = 0°$, $I_0 = 0$, but when $\theta = 45°$

$$I_B = \frac{1}{2} I_0 \left(\frac{1}{2}\right)\left(\frac{1}{2}\right) = \frac{I_0}{8}$$

So, introduction of third sheet can give transmitted part of the incident light even if the two polarizing sheets are perpendicular to each other.

10.6 What are uniaxial and biaxial crystals?

Ans The crystals having one direction (optic axis) along which the two refracted rays travel with the same velocity are called as uniaxial crystals. In biaxial crystals, there are two optic axes.

10.7 What is difference between 'ordinary' and 'extraordinary' rays?

Ans While passing through a double refracting crystal, the ray which obeys the laws of refraction is the ordinary ray while the other, which does not obey the laws of refraction, is extraordinary ray.

10.8 Define the plane of vibration.

Ans The plane in which the vibrations take place (i.e., the plane containing the direction of vibration and the direction of propagation) is called plane of vibration.

10.9 What do you mean by plane of polarization?

Ans A plane perpendicular to the plane of vibration is called plane of polarization. Thus, the plane of polarization is the plane passing through the direction propagation and containing no vibrations.

10.10 How does the polarization of light afford a convincing evidence of transverse nature of light?

Ans If light waves were not transverse in nature then the vibrations would have passed through two tourmaline crystals with their optic axes perpendicular to each other, i.e., in crossed position.

10.11 What is phenomenon of double refraction?

Ans When the ordinary light is incident on a certain crystal and it splits in two refracted rays, the phenomenon is called the double refraction.

10.12 A calcite crystal is placed over a dot on a piece of paper and rotated. What will you observe through the calcite crystal?

Ans One dot rotating about the other.

10.13 What is the 'principle section' of crystal?

Ans The plane having optic axis and perpendicular to opposite sides of the crystal is principal section.

10.14 What do you mean by optic axis?

Ans Line passing through any one of the blunt corners and making equal angles with three faces is called optic axis. It is the direction along which o- and e-rays travel with same velocity.

10.15 Differentiate between plane polarized, circularly polarized and elliptically polarized light.

Ans In polarized light,
  (i) If the electric vector vibrates along a straight line, it is plane polarized.
  (ii) If the electric vector rotates along a circle i.e., does not change magnitude while rotating, it is circularly polarized, and
  (iii) If the electric vector rotates along an ellipse i.e., changes magnitude while rotating, it is elliptically polarized.

10.16 What is the direction of electric vector of polarized reflected light for light incident at Brewster's angle at an air-glass interface?

Ans The electric vector is perpendicular to the plane of incidence i.e., parallel to interface.

10.17 Can you define polarimeter?

Ans Yes, this is an instrument used for measuring the angle of rotation of plane of polarization produced by an optically active substance.

10.18 What do you mean by half shade device or Laurent plate?

Ans Laurent's half shade device consists of two semicircular plates, one of quartz and another of glass. Quartz plate serves the purpose of half wave plate, i.e., its thickness is such that it introduces a phase change of π between extraordinary and ordinary rays. The thickness of glass plate is chosen such that it absorbs the same amount of light as quartz plate. The quartz plate is cut parallel to optic axis and connected along diameter with the glass plate.

10.19 What is a wave plate? What are its types?

Ans A wave plate is a transparent plate made from uniaxial crystal.

The wave plates are of two types:
  (i) *Quarter wave plate:* It is a thin plate made from a doubly refracting uniaxial crystal such as calcite cut with its refracting faces parallel to the direction of optic axis. The thickness of the plate is so chosen such that it introduces a path difference of $\lambda/4$ between o-ray and e-ray during propagation through it.

(ii) *Half wave plate:* It is a thin plate made from a doubly refractive uniaxial crystal cut its refracting faces parallel to direction of optic axis. The thickness of the plate is so chosen such that it introduces a path difference of $\lambda/2$ between o-ray and e-ray during the propagation through the crystal.

10.20 What will be the state of polarization of the emerging light when:

(i) A beam of circularly polarized light is passed through a quarter wave plate (QWP).

(ii) A beam of plane polarized light is passed through a quarter wave plate.

(iii) A beam of elliptically polarized light is passed through quarter wave plate.

(iv) A beam of plane-polarized light is passed through a QWP such that the plane-polarized light falling on QWP makes an angle of 45° with optic axis.

Ans (i) Plane/linearly polarized light.

(ii) Circularly or elliptically or plane-polarized light.

(iii) Plane/linearly polarized light.

(iv) Circularly polarized light.

10.21 A transparent plate is given. Using the Nicol prisms how would you find whether the given plate is a quarter wave plate or a half wave plate or a simple glass plate.

Ans For recognition of the given plate (quarter wave plate/half wave plate/simple glass plate). Light transmitted through N_1 will be plane-polarized light:

(i) If there is no light after N_2 then plate is half wave plate (HWP).

(ii) If on rotating N_2, there is variation in light intensity after N_2 but not zero then plate is quarter wave plate (QWP).

(iii) If on rotating N_2, there is variation in light intensity after N_2 from zero to maximum then plate is a simple glass plate.

10.22 What is meant by photoelasticity?

Ans It is a phenomenon by which an optically isotropic material such as glass or plastic becomes a doubly refracting material under the application of an external mechanical stress.

10.23 What is meant by photoelastic effect?

Ans Certain materials like glass when subjected to mechanical stress, they exhibit temporary doubly refraction. This effect is known as photoelastic effect.

10.24 What is induce birefringence or artificial double refraction?

Ans The process of inducing optical anisotropy in the isotropic material is called birefringence or artificial double refraction. On removing the external force, the induce anisotropy disappears.

EXERCISES

Theoretical Questions

10.1 What is polarization? Explain the methods of production of plane-polarized light from reflected, and refracted light.

10.2 Explain the terms:
  (a) Polarization of light
  (b) Plane of polarization
  (c) Plane of vibration

10.3 What is polarization? Write the name of different type of polarization.

10.4 Explain Brewster's law. Use this law to show that when light is incident on a transparent material at the polarizing angle, the reflected and refracted rays are at right angle.

10.5 Explain Brewster's law. **[Lucknow University, 2009, 2010]**

10.6 Give Brewster's law for polarization. **[Agra University, 2012]**

10.7 State and prove law of Malus. **[Lucknow University, 2013]**

10.8 What are retardation plates. **[Lucknow University, 2010]**

10.9 State law of Malus. **[Lucknow University, 2014]**

10.10 Explain the phenomenon of double refraction with neat and clean diagram. Clearly explaining e-ray, o-ray and optic axis. Under what condition we can have single image of a point object through a uniaxial crystal when light falls normally on a uniaxial crystal.
[Agra University, 2006]

10.11 Give Huygen's theory of double refraction in uniaxial crystals.
[Agra University, 2007, 2012]

10.12 Write Jones matrices for quarter wave plate and half wave plate. **[Agra University, 2013]**

10.13 What are negative and positive crystals? Give one example of each.
[Agra University, 2008; Bundelkhand University, 2009]

10.14 What is meant by double refraction? Describe the media in which such phenomenon occurs. Explain double refraction on the basis of electromagnetic theory.
[Bundelkhand University, 2005]

10.15 What do you understand by double refractoin? Explain on the base of Huygen's theory of propagation of light in double refracting uniaxial crystal. **[Lucknow University, 2008]**

10.16 Distinguish between 0 and e-rays. **[Lucknow University, 2008, 2012]**

10.17 What do you understand by double refraction? Describe a Nicol prism and explain how it acts as a polarizer and analyzer. **[Lucknow University, 2010]**

10.18 Show that at polarizing angle reflected and refracted rays are mutually perpendicular.
[Lucknow University, 2011]

10.19 Explain the phenomenon of double refraction. **[Lucknow University, 2011]**

10.20 What is a plane polarized light? How do you obtain plane-polarized light?
[Bundelkhand University, 2012]

10.21 Describe the phenomenon of double refraction in uniaxial crystals. Distinguish between negative and positive crystals. How is double refraction explained by Huygen's theory.
[Avadh University, 2008]

10.22 Explain the following terms
  (a) optic axis of crystal
  (b) ordinary and extraordinary rays **[Lucknow University, 2014]**

10.23 Explain the phenomenon of double refraction in uniaxial crystal.
[Kanpur University, 2006]

10.24 What is optic axis and principal section in calcite crystal? Explain the phenomenon of double refraction in calcite crystal. **[Kanpur University, 2011]**

10.25 Explain the action of a pile of plates in producing plane-polarized light.

10.26 Give a short account of doubly refracting crystals with special reference of calcite.

10.27 What do you understand by double refraction? What are ordinary and extraordinary rays in a uniaxial crystal? How can you show that they are plane polarized?

10.28 Give an account of Huygen's theory of double refraction.

10.29 Explain the propagation of plane waves in uniaxial crystals.
[Bundelkhand University, 2011]

10.30 What is Nicol prism? Write down limitations of Nicol prism.
[Lucknow University, 2011]

10.31 Explain what do you understand by principal refractive indices of crystal?
[Kanpur University, 2009; Avadh University, 2012]

10.32 Describe any one double image prism indicating their advantage over a Nicol prism.
[Kanpur University, 2006]

10.33 Explain the geometry of the calcite crystal. **[Kanpur University, 2009]**

10.34 Explain working of double image prism. **[Kanpur University, 2009]**

10.35 Describe the construction of Nicol prism. Explain how it can be used as a polarizer and as an analyzer. Would a similar prism prepared from quartz serve a similar purpose?

10.36 Write a short note on double image prisms, indicating their advantages over a Nicol prism.

10.37 Describe Nicol prism and explain how it acts as a polarizer and as an analyzer.
[Lucknow University, 2012]

10.38 Describe how, with the help of a Nicol prism and a quarter wave plate, plane polarized light, circularly polarized light and elliptically polarized light are produce and detected.

10.39 What do you meant by optical rotation. Given an outline of Fresnel's theory of optical rotation.

10.40 Explain with figure can we produce and analyse plane, circular and elliptically polarized light from the unpolarized light. **[Agra University, 2006]**

10.41 Explain how can you transform an elliptically polarized light into plane-polarized light.
[Agra University, 2010]

10.42 Explain Fresnel's theory of plane of polarization. **[Kanpur University, 2006]**

10.43 What are plane polarized, circularly polarized and elliptically polarized light? How will you combine two plane polarized waves to produce circularly polarized, elliptically polarized and plane polarized wave? **[Avadh University, 2009]**

10.44 Explain plane polarized, circularly polarized light and elliptically polarized light.
[Avadh University, 2011]

10.45 What is meant by plane-polarized light, circularly polarized light and elliptically polarized light? Show that plane polarized and circularly polarized light are special cases of elliptically polarized light. **[Avadh University, 2012]**

10.46 Define the polarization phenomenon. **[Avadh University, 2013]**

10.47 How would you change a left handed circularly polarized light beam into a right handed circularly polarized light beam? **[Lucknow University, 2013]**

10.48 Explain how will you produce circularly polarized light from unpolarized light and how will you detect the circularly polarized light. **[Kanpur University, 2011]**

10.49 Explain how will you detect the elliptically polarized light from unpolarized light. **[Kanpur University, 2012]**

10.50 Explain construction and working of quarter wave plate and half wave plate. **[Agra University, 2010]**

10.51 Give the construction and working of a quarter wave plate. **[Bundelkhand University, 2007]**

10.52 Describe the method to detect plane, circularly and elliptically polarized light with the help of Nicol prism and quarter wave plate. **[Lucknow University, 2010]**

10.53 Explain the principle of construction of quarter wave plate and half wave plate. How will you distinguish between these two? **[Avadh University, 2010]**

10.54 Describe the phenomenon of double refraction in uniaxial crystal. **[Kanpur University, 2008; Avadh University, 2010]**

10.55 How plane polarized light is obtained using Nicol prism? **[Avadh University, 2010]**

10.56 What are quarter wave and half wave plates? How are these used in the study of different types of polarized light? **[Avadh University, 2011]**

10.57 What are quarter wave and half wave plates? Explain their use in the study of different types of polarized light? **[Avadh University, 2012]**

10.58 Describe the construction and working of retardation plates. Explain the limitations of retardation plates. **[Lucknow University, 2012]**

10.59 What are quarter wave and half wave plates? **[Kanpur University, 2006]**

10.60 What are quarter wave and half wave plates? Explain their uses in the study of different types of polarized light.

10.61 What is polarised? Explain its use. **[Lucknow University, 2011, 2012]**

10.62 What do you understand by dichroism? **[Lucknow University, 2013]**

10.63 Explain what is meant by 'dichroism'. Give a brief account of 'polaroids' and their uses.

10.64 What do you mean by
 (a) Plane polarized light
 (b) Circularly polarized light
 (c) Elliptically polarized light
 (d) Superposition of two linearly polarized light

10.65 Write Jones Matrix for linear polarizer with axis horizontal. **[Lucknow University, 2014]**

10.66 Define specific rotation. Describe construction and working of a Laurent's half shade polarimeter, explaining fully the action of the half shade device. How would you use it to determine the specific rotation of sugar solution?

10.67 Explain working of a biquartz polarimeter.

10.68 Define specific rotation. Describe the construction and working of Laurent half shade polarimeter. Give Fresnel's explanation of optical rotaton. **[Lucknow University, 2014]**

10.69 Using Jones matrix for right and left circularly polarised light, show that when they are superimposed emergent light is plane polarised. **[Lucknow University, 2014]**

10.70 Write Jones matrix for retardation plates. **[Lucknow University, 2014]**

10.71 What do you understand by rotatory dispersion? **[Lucknow University, 2014]**

10.72 What do you understand by rotatory polarization and rotatory dispersion? How can specific rotation of sugar be determined using Biquartz polarimeter?
[Lucknow University, 2012]

10.73 What do you mean by optically active compounds? Give examples.
[Avadh University, 2013]

10.74 Give construction and working of Nicol prism. **[Agra University, 2007]**

10.75 Discuss the relative merits of biquartz polarimeter and half shade polarimeter.
[Avadh University, 2012]

10.76 Define specific rotation. Describe the construction and working of a Laurent's half shade polarimeter, explaining fully the action of the half shade device.
[Avadh University, 2008]

10.77 Write Jones matrix of quarter wave plate. **[Lucknow University, 2008]**

10.78 Define specific rotation. **[Lucknow University, 2008, 2010]**

10.79 What do you understand by rotatory polarization and rotatory dispersion? What is biquartz device and why it is used in polarimeters? **[Lucknow University, 2008]**

10.80 What is meant by optical rotation? Define specific rotation.
[Kanpur University, 2007; Bundelkhand University, 2010]

10.81 Define optical rotation. Explain the Fresnel's theory of optical rotation.
[Bundelkhand University, 2006, 2008]

10.82 Explain optical rotation. Give Fresnel's theory of optical rotation. Give experimental evidence of this principle. **[Agra University, 2013]**

10.83 What is meant by optical rotation? State the laws of optical rotation.
[Agra University, 2008]

10.84 Define specific rotation. Describe the construction and working of Biquartz polarimeter. How will you use it to find specific rotation of sugar solution?
[Lucknow University, 2007]

10.85 Define specific rotatoin. Describe the construction and working of Laurent half shade polarimeter to determine the specific rotation of sugar. **[Lucknow University, 2013]**

10.86 What is specific rotation? How to find specific rotation of sugar solution using Laurent half shade polarimeter? **[Lucknow University, 2011]**

10.87 What is specific rotation? [Lucknow University, 2012]
10.88 Explain (i) polarizing angle and (ii) Brewster's law. [Lucknow University, 2012]
10.89 What is birefringence? Discuss the theory and applications of photoelasticity.
10.90 Mention any two applications of photoelasticity.
10.91 Write short note on the following:
  (a) Half shade polarimeter [Kanpur University, 2007; Agra University, 2008]
  (b) Quarter wave plate [Kanpur University, 2007]
  (c) Double image prism indicating their advantage over a Nicol prism
   [Kanpur University, 2007]
  (d) Retardation plates [Lucknow University, 2007, 2008, 2013]
  (e) Jones matrix of quarter wave plate
   [Lucknow University, 2007, 2008, 2010, 2011, 2012]
  (f) Fresnel's explanation of optical rotation [Lucknow University, 2008]
  (g) Rotatory polarization [Lucknow University, 2010]
  (h) Biquartz polarimeter [Lucknow University, 2011]
  (i) Laurent half shade polarimeter [Lucknow University, 2012]
  (j) Rotatory dispersoin and rotatory polarization [Lucknow University, 2013]
10.92 Discuss Fresnel's theory of rotatory polarization. [Avadh University, 2010]
10.93 Define specific rotation. Discuss the working of a polarimeter. [Avadh University, 2011]
10.94 Describe the function of a half shade polarimeter with the help of a ray diagram.
 [Avadh University, 2013]
10.95 What do you understand by rotatory polarization? [Kanpur University, 2010]
10.96 Describe the Fresnel's theory of polarization. [Avadh University, 2013]
10.97 Explain what do you mean by circularly polarized light. Describe how it can be produced. How will you distinguish between elliptically and circularly polarized light?
 [Lucknow University, 2007]
10.98 Distinguish between elliptically polarized and circularly polarized light.
 [Lucknow University, 2008, 2014]

Numerical Problems

10.1 What will be the Brewster's angle for a glass slab ($n = 1.5$) immersed in water.
 [Ans 56.7°]
10.2 Find the intensity of transmitted light when unpolarized light of intensity I_0 falls on the two polarizing sheets, when angle of transmission axis, between the two polarizing sheets is 30°. [Kanpur University, 2012]
10.3 The ratio of intensities of ordinary and extraordinary light is 0.65. Determine the angle made by the plane of vibration of the incident light with the optic axis.
 [Ans $\theta = 38°52'$]

10.4 A partially polarized light beam is passed through Nicol. After 90° rotation of Nicol, the intensity changes from I_m to $0.3\, I_m$. Find the degree of polarization. [Ans $\frac{7}{13}$]

10.5 When the angle of incidence on a certain material is $\frac{\pi}{3}$, the reflected light is completely polarized. What is the refractive index of material? [Ans $n = 1.732$]

10.6 Critical angle for refraction for glass to air is 40°. Compute the polarizing angle for glass. [Ans 57.3°]

10.7 When light is incident at an angle of 60° to the normal, the reflected light is plane polarized
(a) what is the refractive index of the transparent medium?
(b) what is the angle of refraction corresponding to the angle of incidence of 60°?
(c) what is the angle between the reflected and refracted components?
[Ans (a) 1.732, (b) 30°, (c) 90°]

10.8 For sodium light of wavelength 5893Å, calculate thickness of quarter wave plate. Given $n_e = 1.553$ and $n_o = 1.544$. [Ans. 1.64×10^{-5} m] **[Agra University, 2009]**

10.9 Calculate the thickness of a double refracting crystal plate required to introduce a path difference of $\lambda/2$ between the ordinary and extraordinary rays, when $\lambda = 6000$ Å; $n_o = 1.55$ and $n_e = 1.54$. [Ans. 3×10^{-5} m] **[Kanpur University, 2007]**

10.10 Calculate the thickness of a doubly refracting crystal to make it a quarter wave plate for $\lambda = 6000$ Å. Given that $n_o = 1.53$ and $n_e = 1.54$ for the crystal. [Ans 15 μm]

10.11 Two polarizers are oriented with their plane at angle of 30°. What percentage of incident unpolarized light of wavelength 6000 Å shall be pass through the system. [Ans 86.8%]

10.12 Find the thickness of quarter wave plate for wavelength 5890 Å. Given $n_0 = 1.55$ and $n_e = 1.54$. [Ans 1.47×10^{-5} m] **[Lucknow University, 2010]**

10.13 Calculate the thickness of quarter wave plate given that: $n_e = 1.553$, $n_0 = 1.544$, $\lambda = 6 \times 10^{-5}$ cm. [Ans 1.66×10^{-5} m] **[Lucknow University, 2011]**

10.14 Calculate the minimum thickness of a plate which converts polarized light into circularly polarized light. Given $n_0 = 1.53$, $n_e = 1.54$, $\lambda = 5890$ Å. [Ans 1.47×10^{-5} m]
[Lucknow University, 2014]

10.15 The specific rotation of quartz at 5086 Å is 29.73 deg/mm. Calculate the difference in the refractive indices. [Ans $\Delta n = 8.4 \times 10^{-5}$]

10.16 Two Nicols are oriented with their planes making an angle of 60°. What percentage of incident unpolarized light will pass through the system. [Ans 37.8%]

10.17 Calculate the thickness of such a plate made from calcite, for which the principal refractive indices are $n_o = 1.658$ and $n_e = 1.486$ at the wavelength of light used is 5890 Å.
[Ans 8.56×10^{-4}]

10.18 Two Nicols are crossed to each other. Now, one of them rotated through 60°. What percentage of incident unpolarized light will pass through the system. [Ans 37.8%]

10.19 If the plane of vibration of the incident beam makes an angle 30° with the optic axis, compare the intensities of ordinary and extraordinary rays. [Ans 3]

10.20 Plane polarized light of wavelength 5×10^{-5} cm is incident on a piece of quartz cut parallel axis. Find the least thickness for which the o- and e-rays combine to form plane polarized light. Find the least thickness for which o- and e-rays combine to form plane polarized light. **[Ans** $t = 2.75 \times 10^{-2}$ cm**]**

10.21 The values of n_e and n_o for quarts are 1.5508 and 1.5418 respectively. Calculate the phase retardation for $\lambda = 5000$ Å when the plate thickness is 0.032 mm. **[Ans** 1.15 π radian**]**

10.22 Calculate the thickness of half wave plate a material for which refractive indices or ordinary and extraordinary rays are $n_o = 1.65$ and $n_e = 1.60$, with sodium light of wavelength $\lambda = 5893$ Å. **[Ans** $T = 589.3$ nm**]**

10.23 A cellophane sheet behaves as a half wave plate of light of wavelength 4000 Å. How would this sheet behave for light of wavelength 8000 Å? It may be assumed that refractive indices do not change with wavelength. **[Ans** QWP**]**

10.24 Two polaroids are adjusted so as to obtain maximum intensity. At what angle should one polaroid be rotated to reduce the intensity to
(a) half, and
(b) one-fourth **[Ans** (a) 45°, (b) 60°**]**

10.25 Two Nicol prisms have their planes parallel to each other. If one of the prisms is rotated such that it makes an angle of 45° with the other. What is the percentage of light transmitted by the second prism? **[Ans** 50%**]**

10.26 Compute the minimum thickness of a quarter wave plate of calcite for $\lambda = 546$ nm. The principal indices of calcite are 1.652 and 1.488. **[Ans** 8.32×10^{-7} m**]**

10.27 A given calcite plate behaves as a half wave plate for a particular wavelength λ. Assuming variation of refractive index with λ to be negligible, how would the above plate behave for another light of wavelength 2λ?

$$\left[\text{Hint: } t = \frac{\lambda}{2(n_o - n_e)} = \frac{\lambda'}{4(n_o - n_e)} \text{ The HWP for } \lambda \text{ will behave as QWP for } \lambda' = 2\lambda \right]$$

10.28 For a certain crystal, $n_o = 1.5442$ and $n_e = 1.5533$ for light of wavelength 6500 nm. Calculate the least thickness for a quarter wave plate made from the crystal for use with light of this wavelength. **[Ans** $t = 1.65 \times 10^{-5}$ m**]**

10.29 The indices of quartz for right-handed and left-handed circularly polarized light of wavelength 762 nm are 1.53914 and 1.5392 respectively. Compute the rotation of the plane polarization of the light in degree produced by a plate of 0.5 mm thick.

[Ans 7.1°**]**

10.30 A tube of sugar solution of 20 cm long is placed between crossed Nicols and illuminated with light of wavelength 6×10^{-5} cm. If optical rotation produced is 13° and specific rotation is 65°, determine the strength of solution. **[Ans.** 0.1 g/cc**]**

[Agra University, 2011]

10.31 A certain length of 5% solution causes the optical rotation of 20°. How much length of 10% solution of the same substance will cause 30° rotations? **[Ans.** (3/4)l**]**

[Avadh University, 2008]

10.32 Determine the specific rotation of sugar solution if the plane of polarization is turned through 13.2°. The length of the tube containing 10% sugar solution is 20 cm.

[Ans. 66°/dm/g/cc**] [Avadh University, 2009, 2011]**

10.33 A tube 20 cm long filled with solution of 15 g of cane sugar in 100 cc of water produces an optical rotation of 19.8°. Find the specific rotation of cane sugar.

[**Ans** 66.5 dm^{-1} (g/cc)$^{-1}$]

[**Kanpur University, 2011**]

10.34 A 20 cm long tube containing 48 cc of sugar solution produces an optical rotation of 11° when placed in a saccharimeter. If the specific rotation of sugar is 66°, calculate the quantity of sugar contained in the tube in the form of solution. [**Ans** 4 g]

[**Lucknow University, 2013**]

10.35 A beam of linearly polarized light is converted into circularly polarized light by passing it through a crystal of thickness 0.0028 cm. Assuming this it to be minimum thickness which produces the desired effect, calculate the difference in refractive indices of the two rays if the wavelength of light used is 550 nm. [**Ans** 4.91 × 10^{-3}]

10.36 A solution of a glucose of specific rotation 52° is kept in a polarimeter tube 2 decimetres long. If the rotation produce is 4°, calculate the strength of the solution.

[**Ans** 4% (approx.)]

10.37 A tube of length 20 cm containing sugar solution rotates the plane of polarization by 11°. If the specific rotation of sugar is 66°, calculate the strength of the rotation.

[**Ans** 0.0833 gm/cc]

10.38 A tube of sugar solution 20 cm long is placed between crossed Nicols and illuminated with light of wavelength 6 × 10^{-5} cm. If the optical rotation is produced is 13° and the specific rotation is 65°/dm/g/cc, determine the strength of the solution. [**Ans** 93.75%]

10.39 A tube of sugar solution 20 cm long is placed between crossed Nicols and illuminated with light of wavelength 600 nm. The optical rotation produced is 13° and specific rotation is 65°/dm/g/cm^3, determine the strength of the solution. [**Ans** C = 0.1 g/cc = 10%]

10.40 A certain length of 5% solution cause, the optical rotation of 20°. How much length of 10% solution of the same substance will cause 35° rotation?

$$\left[\textbf{Ans } l_2 = \frac{7}{8}n\right]$$

10.41 The specific rotation of quartz at 508.6 nm is 29.73 deg/mm, calculate the difference in refractive indices. [**Ans** $n_L - n_R$ = 8.4 × 10^{-5}]

10.42 The indices of refraction of quartz for right handed and left handed circularly polarized lights of wavelength 6500 Å travelling in the direction of optic axis have the following values n_R = 1.53914 and n_L = 1.53920. Calculate the rotation of the plane of polarization of light in degrees produced by the plate 0.2 mm thick. [**Ans** 3°20′]

Multiple Choice Questions

10.1 Two Nicol prisms are first crossed and then one of them is rotated through 60°. The percentage of incident light transmitted is
(a) 12.5 (b) 25.0
(c) 37.5 (d) 50.0

Polarization of Light Waves

10.2 The thickness of a $\frac{\lambda}{4}$ plate for light of wavelength 6000 Å will be (in cm)
 (a) 1.5×10^{-4} (b) 1.54×10^{-4}
 (c) 1.55×10^{-4} (d) 1.5×10^{-3}
 [For a plate $n_o = 1.54$ and $n_e = 1.55$]

10.3 A quarter wave plate is designated for 6000 Å. If change in refractive index with wavelength is negligible, for 4500 Å the phase retardation will be
 (a) $\frac{\pi}{2}$ (b) $\frac{2\pi}{3}$
 (c) π (d) $\frac{\pi}{3}$

10.4 When an unpolarized light is incident on a calcite crystal, it splits into two refracted rays. The phenomenon is known as
 (a) scattering (b) dispersion
 (c) double refraction (d) diffraction

10.5 The phenomenon of rotating the plane of vibration of a polarized light is known as
 (a) polarization (b) optical activity
 (c) double refraction (d) Kerr effect

10.6 When light is incident on a plane of a transparent material at the angle of polarization, the reflected and refracted beams are
 (a) parallel (b) inclined by 45°
 (c) inclined by 60° (d) perpendicular to each other

10.7 A plane polarized beam of light is passed through a quarter wave plate. The transmitted light beam is analysed using a Nicol shows no variation in intensity as the Nicol is turned through 360°. It means that the transmitted light is
 (a) circularly polarized
 (b) elliptically polarized
 (c) unpolarized
 (d) mixture of plane polarized and unpolarized

10.8 When a beam of light is incident on a glass plate at polarizing angle, then the angle between reflected and refracted beam is
 (a) 0° (b) 45°
 (c) 60° (d) 90°

10.9 A plane polarized beam of light is passed through a quarter wave plate. The transmitted beam is analysed using a Nicol shows no variation in intensity as Nicol is turned through 360°. The quarter wave plate is turned through 90° and the transmitted light is again analysed similarly. Now, we will observe
 (a) two zero intensity minima (b) no change in intensity
 (c) two non-zero minima (d) four zero intensity minima

10.10 An unpolarized beam is incident at an angle of 60° on a glass surface and after reflection it is linearly polarized. The approximate refractive index of the glass is
 (a) 1.4 (b) 1.5
 (c) 1.7 (d) 1.6

10.11 In a Nicol prism, the refractive index of calcite crystal for ordinary ray is 1.66 and that for extraordinary ray is 1.49. The refractive index of Canada balsam is
  (a) 1.49
  (b) more than 1.66
  (c) between 1.49 and 1.66
  (d) less than 1.49

10.12 Plane polarized light falls normally on a quarter wave plate and the plane of polarization of incident light makes an angle of 45° with the principal plane. The emergent will be
  (a) unpolarized
  (b) linearly polarized
  (c) circularly polarized
  (d) elliptically polarized

10.13 A beam of transverse waves whose vibrations occur in all directions perpendicular to their direction of motion is
  (a) polarized
  (b) resolved
  (c) unpolarized
  (d) diffracted

10.14 Which one or more of the following cannot be polarized?
  (a) sound
  (b) white light
  (c) x-rays
  (d) none of these

10.15 A solution of dextrose (specific rotation 52.5°) causes a rotation of 12° in a column 10 cm long; then the concentration of the solution is
  (a) 0.528 g/cc
  (b) 0.628 g/cc
  (c) 0.228 g/cc
  (d) 0.828 g/cc

10.16 Elliptically polarized light is produced if the amplitudes of the ordinary and extraordinary rays are unequal and there is a phase difference of
  (a) π
  (b) $\dfrac{\pi}{2}$
  (c) $\dfrac{\pi}{4}$
  (d) zero

10.17 Polarization of light proves the
  (a) corpuscular nature of light
  (b) quantum nature of light
  (c) longitudinal nature of light
  (d) transverse nature of light

10.18 Polarization cannot occur in
  (a) light waves
  (b) x-rays
  (c) radio waves
  (d) sound waves

10.19 The plane of polarization is at .... to the plane of vibration
  (a) 45°
  (b) 90°
  (c) 60°
  (d) 180°

10.20 Which of the following methods produce polarized light?
  (a) selective absorption
  (b) double refraction
  (c) reflection
  (d) all of the above

10.21 The correct relation between the Brewster's angle i_p and the refractive index n is
  (a) $\tan i_p = n$
  (b) $\cot i_p = n$
  (c) $\sin i_p = n$
  (d) $\cos i_p = n$

10.22 A beam of unpolarized light passes through a tourmaline crystal A, then it passes through a second tourmaline crystal B oriented, so that its principal plane is parallel to that of A. The intensity of the emergent light is I_0. Now B is rotated by 45° about the ray. The emergent light will have intensity

(a) $\dfrac{I_0}{2}$ (b) $\dfrac{I_0}{\sqrt{2}}$
(c) $I_0\sqrt{2}$ (d) $2I_0$

10.23 A calcite crystal is placed over a dot on a piece of paper and then rotated. On viewing through calcite, we observe
(a) a single dot (b) two stationary dots
(c) two rotating dots (d) one dot rotating about the other

10.24 A Nicol prism can be used
(a) only to analyze polarized light (b) for producing and analysing polarized light
(c) only to produce polarized light (d) none of the above

10.25 In elliptically polarized light
(a) amplitude of vibrations changes in direction only
(b) amplitude of vibrations changes in both magnitude and direction
(c) amplitude of vibrations changes in magnitude only
(d) none of these

10.26 In a quarter plate, the path difference between ordinary and extraordinary rays is
(a) one-fourth of the incident wavelength (b) zero
(c) half of the incident wavelength (d) none of these

10.27 Polaroid sunglasses decrease glare on a sunny day because they
(a) block a portion of light (b) have a special colour
(c) completely absorb the light (d) refract the light

10.28 The phenomenon of rotation of plane-polarized light is called
(a) double refraction (b) refraction
(c) optical activity (d) dichroism

10.29 Laurent's half shade polarimeter requires a
(a) monochromatic source of light (b) coherent source of light
(c) reflection (d) all of the above

10.30 Birefringence (B) is
(a) $|(n_e - n_o)|$ (b) $(n_0 - n_1)|$
(c) $|(n_1 - n_2)|$ (d) none of these

Answers

10.1 (b)   10.2 (d)   10.3 (b)   10.4 (c)   10.5 (b)   10.6 (d)   10.7 (a)   10.8 (d)
10.9 (a)   10.10 (c)   10.11 (c)   10.12 (c)   10.13 (c)   10.14 (a)   10.15 (c)   10.16 (b)
10.17 (d)   10.18 (d)   10.19 (b)   10.20 (d)   10.21 (a)   10.22 (a)   10.23 (d)   10.24 (b)
10.25 (b)   10.26 (a)   10.27 (a)   10.28 (c)   10.29 (a)   10.30 (a)

PART IV

Electromagnetic Waves

Chapter 11   Electromagnetic Waves

Chapter 11

Electromagnetic Waves

"Even in the vast and mysterious reaches of sea we are brought back to the fundamental truth that nothing lives to itself."

—Rachel Carson

IN THIS CHAPTER

- Production of Electromagnetic Waves by an Antenna
- Wave Equation for Waves in Space
- Wave Propagation in Lossy Dielectric Medium
- Conductors and Dielectrics
- Wave Propagation in Good Dielectrics
- Wave Propagation in a Good Conductor
- Depth of Penetration: Skin Depth
- Poynting Vector and Poynting Theorem
- Polarization
- Reflection of Uniform Plane Waves by Perfect Dielectric—Normal Incidence

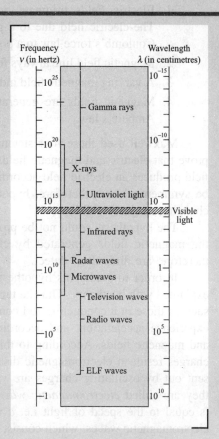

11.1 INTRODUCTION

The time-varying electric and magnetic fields give rise to phenomenon of electromagnetic wave propagation. Changing electric field produces changing magnetic field which, in turn, generates an electric field and so on, with a resulting propagation of energy. The uniform plane wave, based on Maxwell's equations, illustrates the principle behind the propagation of energy. The behaviour of a wave depends on its own and medium's characteristics. Some of the quantities of interest are velocity of propagation, wavelength, wave impedance, the phase and attenuation constant, and Poynting vector for finding the power density. A wave when striking a boundary between two media is partly transmitted and partly reflected, depending on the nature of the boundary. When a wave strikes a pure conductor, it produces standing waves. This chapter deals with the fundamentals of electromagnetic uniform plane waves.

During the early stages of their study and development, electric and magnetic phenomena were thought to be unrelated. In 1865, however, James Clerk Maxwell provided a mathematical theory that showed a close relationship between all electrical and magnetic phenomena. Additionally, his theory predicted that electric and magnetic fields can move through space as waves. The theory developed by Maxwell is based on the following statements:

1. Electric fields originate on positive charges and terminate on negative charges. The electric field due to a point charge can be determined at a location by applying Coulomb's force law to a positive test charge placed at that location.
2. Magnetic field line always form close loops, i.e. they do not begin or end any where.
3. A varying magnetic field induces an e.m.f. and, hence, an electric field (Faraday's law).
4. Magnetic fields are generated by moving charges (or currents), as summarized in Ampere's law.

Maxwell used these four statements within a corresponding mathematical framework to prove that electric and magnetic fields play a symmetrical role in nature. A changing magnetic field produces an electric field according to Faraday's law. Maxwell suspected that nature must be symmetrical and, therefore, hypothesized that a changing electric field should produce a changing magnetic field.

The hypothesis could not be proven experimentally at the time it was developed, because the magnetic fields generated by changing electrical fields are generally very weak and, therefore, are difficult to detect.

In order to justify his hypothesis, Maxwell searched for other phenomena that might be explained with his theory. Thus he turned his attention to the motion rapidly oscillating charges, such as those in a conducting rod connected to an alternating voltage. Such oscillating charges experience acceleration, and according to Maxwell's predictions, generate changing electric and magnetic fields. According to this theory, the changing fields produced by the oscillating charges result in electromagnetic disturbances that travel through space as waves. The waves sent out by oscillating charges are viewed as fluctuating electric and magnetic fields, hence, they are called *electromagnetic waves*. Maxwell computed the speed of these waves, which is equal to the speed of light i.e. $c = 3 \times 10^8$ m/s. Thus, he concluded that light waves are electromagnetic waves, which consist of fluctuating electric and magnetic fields travelling space with a velocity of 3×10^8 m/s.

11.2 PRODUCTION OF ELECTROMAGNETIC WAVES BY AN ANTENNA

Whenever a charged particle undergoes an acceleration, it must radiate energy. This is fundamental mechanism. An alternating voltage applied to the wires of an antenna forces an electric charge in the antenna to oscillate. This is common technique for accelerating charged particles and is the source of radio waves emitted by the antenna of a radio station.

Figure 11.1 illustrates the production of an electromagnetic wave by oscillating electric charges in an antenna. Two metal rods are connected to an a.c. generator, which causes charge to oscillate between the two rods. The output voltage of the generator is sinusoidal. At $t = 0$, the upper rod is given the maximum positive charge and the bottom rod, given an equal negative charge, as shown in Figure 11.1(a). The electric field near the antenna at this instant is also shown in the Figure. As the charge oscillates, the rod becomes less charged, the field near the rod decreases in strength, and the downward-directed maximum electric field produced at $t = 0$ moves away from the rod. When the charges are neutralized, as shown in Figure 11.1(b), the electric field has dropped to zero. This occurs at a time equal to one quarter of the period of oscillation. Continuing in this fashion, the upper rod soon obtains a maximum negative charge and the lower rod becomes positive, as shown in Figure 11.1(c), resulting in an electric field directed upward. This occurs after a time equal to one-half period of oscillation. The oscillation continues as indicated in Figure 11.1(d). Note that the electric field near the antenna oscillates in phase with charge distribution, i.e. the field points down when the upper rod is positive and up when the upper rod is negative. Furthermore, the magnitude of the field at any instant depends on the amount of charge on the rods at that instant.

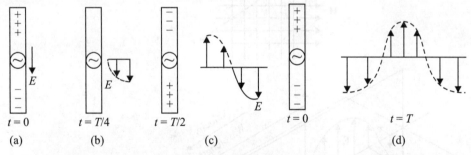

Figure 11.1 Production of electromagnetic wave by oscillating electric charges.

As the charge continues to oscillate (and accelerate) between rods, the electric field set up by the charges moves away from the antenna at the speed of light. Figure 11.1 shows the electric field pattern at many times during oscillation cycle. One cycle of charge oscillation produces one full wavelength in the electric field pattern.

Since the oscillating charges create a current in the rods, a magnetic field is also generated when the current in the rods is upwards, as shown in Figure 11.2.

The magnetic field lines circle the antenna and are perpendicular to the electric field at all points. As the current changes with time, the magnetic field lines spread out from the antenna. At large distances from the antenna, the strengths of electric and magnetic fields become very weak. However, at these large distances it is necessary to take into account the fact that,

524 Fundamentals of Optics

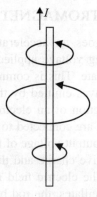

Figure 11.2 Magnetic field associated with a current element.

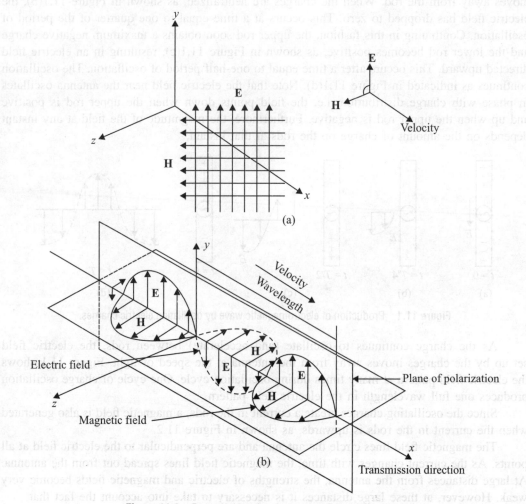

Figure 11.3 Propagation of a plane wave, **E** and **H** vectors.

(i) a changing magnetic field produces a changing electric field, and
(ii) a changing electric field produces a changing magnetic field, as predicted by Maxwell. These induced electric and magnetic fields are in phase, at any point the two fields reach their maximum at the same instant. It is shown in Figure 11.3 at one instant of time.

It may be noted that
(i) the two fields are perpendicular to each other, and
(ii) both the fields are perpendicular to the direction of motion of the wave.

The second fact is a property characteristic of transverse wave, an electromagnetic wave is a transverse wave.

The following properties are associated with an electromagnetic wave travelling through free space:

1. Electromagnetic waves travel with speed of light.
2. Electromagnetic waves are transverse waves.
3. The ratio of electric (**E**) to magnetic field (**B**) in an electromagnetic wave equals the speed of light $\left(\dfrac{E}{B} = c\right)$.
4. Electromagnetic waves carry both energy and momentum, which can be delivered to a surface. Figure 11.4 gives the spectrum of electromagnetic radiation just for general information.

A detailed analysis of electromagnetic waves will be given in the following sections.

11.3 WAVE EQUATION FOR WAVES IN SPACE

We know that space variations of electric and magnetic field components are related to time variations of the magnetic and electric field components, respectively. This interdependence gives rise to the phenomenon of electromagnetic wave propagation. We shall derive the general wave equation using Maxwell's equations.

In the solution of any electromagnetic problem, the following fundamental relations and correlations are to be satisfied.

Relations:

$$\nabla \cdot \mathbf{D} = \rho \tag{11.1}$$

$$\nabla \cdot \mathbf{B} = 0 \tag{11.2}$$

$$\nabla \times \mathbf{E} = -\dfrac{\partial \mathbf{B}}{\partial t} \tag{11.3}$$

$$\nabla \times \mathbf{H} = \mathbf{J} + \dfrac{\partial \mathbf{D}}{\partial t} \tag{11.4}$$

Correlations:

$$\mathbf{D} = \varepsilon \mathbf{E} \tag{11.5}$$

$$\mathbf{B} = \mu \mathbf{H} \tag{11.6}$$

$$\mathbf{J} = \sigma \mathbf{E} \tag{11.7}$$

526 Fundamentals of Optics

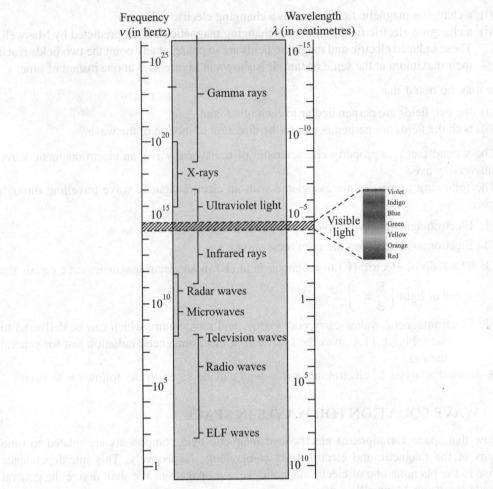

Figure 11.4 Spectrum of electromagnetic radiation.

When Eqs. (11.5)–(11.7) are inserted in Eqs. (11.3)–(11.4), Maxwell's equations become differential equations relating the electric and magnetic field strengths **E** and **H**. If they are, then solved as simultaneous equations, they will determine the laws which both **E** and **H** must obey.

Solution for perfect dielectric conditions

Let us consider a simple case of electromagnetic phenomenon in a perfect dielectric (in a homogeneous dielectric with no charges and no conduction current, i.e. $\rho = 0$, $\sigma = 0$, $J = 0$). For this case, the field equations become.

$$\nabla \cdot \mathbf{D} = 0, \qquad \nabla \cdot \mathbf{E} = 0 \qquad (11.8)$$

$$\nabla \cdot \mathbf{B} = 0, \qquad \nabla \cdot \mathbf{H} = 0 \qquad (11.9)$$

$$\nabla \times \mathbf{E} = -\frac{\partial \mathbf{B}}{\partial t} = -\mu \frac{\partial \mathbf{H}}{\partial t} \qquad (11.10)$$

$$\nabla \times \mathbf{E} = -\frac{\partial \mathbf{D}}{\partial t} = \varepsilon \frac{\partial \mathbf{E}}{\partial t} \qquad (11.11)$$

Equation (11.10) states that a changing magnetic field will produce an electric field and Eq. (11.11) states that a changing electric field will produce a magnetic field.

A changing electric field produces a changing magnetic filed and that, in turn, produces an electric field which produces a magnetic field, and so on. Some kind of series of energy transfer is started whenever any electric or magnetic distribution takes place. Energy will be transferred from the electric to the magnetic field and so on indefinitely. The magnetic energy is not confined to precise the same location in space as the electric energy derived from the magnetic energy is again a little further advanced in space and so on, the energy changing from one form to another form is also being propagated.

From two equations of **E** and **H**, we shall be getting only one equation for **E** (or **H**).

Taking the curl of both sides in Eq. (11.10)

$$\nabla \times [\nabla \times \mathbf{E}] = -\mu \left[\nabla \times \frac{\partial \mathbf{H}}{\partial t} \right] \qquad (11.12)$$

and on expanding L.H.S., we get

$$\nabla(\nabla \cdot \mathbf{E}) - \nabla^2 \mathbf{E} = -\mu \left[\nabla \times \frac{\partial \mathbf{H}}{\partial t} \right]$$

$$-\nabla^2 \mathbf{E} = -\mu \frac{\partial}{\partial t}(\nabla \times \mathbf{H}) \qquad (11.13)$$

We have $\nabla \cdot (\nabla \cdot \mathbf{E}) = 0$ and on R.H.S., curl is partial derivative with respect to distance, and it operates on partial derivative with respect to time, mathematically, order of partial differential makes no difference.

Substituting Eq. (11.11) in Eq. (11.13), we get

$$\nabla^2 \mathbf{E} = -\mu \frac{\partial}{\partial t} \left(\varepsilon \frac{\partial \mathbf{E}}{\partial t} \right)$$

$$\nabla^2 \mathbf{E} = \mu \varepsilon \frac{\partial^2 \mathbf{E}}{\partial t^2} \qquad (11.14)$$

Equation (11.14) is the general form of wave equation and is the law **E** must obey. Equation (11.14) can be expanded in Cartesian coordinates to a set of three equations:

$$\left. \begin{array}{l} \dfrac{\partial^2 E_x}{\partial x^2} + \dfrac{\partial^2 E_x}{\partial y^2} + \dfrac{\partial^2 E_x}{\partial z^2} = \mu \varepsilon \dfrac{\partial^2 E_x}{\partial t^2} \\[6pt] \dfrac{\partial^2 E_y}{\partial x^2} + \dfrac{\partial^2 E_y}{\partial y^2} + \dfrac{\partial^2 E_y}{\partial z^2} = \mu \varepsilon \dfrac{\partial^2 E_y}{\partial t^2} \\[6pt] \dfrac{\partial^2 E_z}{\partial x^2} + \dfrac{\partial^2 E_z}{\partial y^2} + \dfrac{\partial^2 E_z}{\partial z^2} = \mu \varepsilon \dfrac{\partial^2 E_z}{\partial t^2} \end{array} \right\} \qquad (11.15)$$

and

Similarly, we can get the general form of wave equation of **H** by taking curl of Eq. (11.11) and then substituting Eqs. (11.8) and (11.10), we get

$$\nabla^2 \mathbf{H} = \mu\varepsilon \frac{\partial^2 \mathbf{H}}{\partial t^2} \tag{11.16}$$

In general case, electromagnetic wave propagation involves electric and magnetic fields having more than one component, each dependent on all three coordinates, in addition to time, Eq. (11.15). However, a simple and very useful type of wave that serves as a building block in the study of electromagnetic waves will consist of electric and magnetic fields that are perpendicular to each other and to the direction of propagation and are uniform in planes perpendicular to the direction of propagation. These waves are called *uniform plane waves*. We shall consider for simplicity

$$\mathbf{E} = E_y(x, t)\mathbf{j} \tag{11.17}$$

and assume that E_x and E_z do not exist, then set of Eq. (11.15) reduces to

$$\frac{\partial^2 E_y}{\partial x^2} = \mu\varepsilon \frac{\partial^2 E_y}{\partial t^2} \tag{11.18}$$

Equation (11.18) is a second order differential equation, which occurs frequently in electromagnetism.

Its solution is of the form

$$E_y = f_1(x - vt) + f_2(x + vt) \tag{11.19}$$

where $v = 1/\sqrt{\mu\varepsilon}$ is velocity and f_1 and f_2 are functions of $(x - vt)$ and $(x + vt)$ respectively. Examples of $f_1(x - vt)$ are $A \cos\beta(x - vt)$ or $Ce^k(x - vt)$.

A wave is a physical phenomenon that occurs at one place at a given time and is reproduced at another place at a later time, the time delay being proportioned to the space separation from the first location. This group of phenomenon constitutes a wave. Wave is not necessarily a repetitive phenomenon in time.

The functions $f_1(x - vt)$ and $f_2(x + vt)$ describe such a wave mathematically, the variation of wave being confined to one dimension in space.

If a fixed time is taken say t_1, then the function $f_1(x - vt)$ becomes a function of x, since vt, is constant. Such a function is represented by Figure 11.5(a). If another time t_2 is taken, another function of x is obtained, exactly the same shape as the first except that Figure 11.5(b) is displaced to the right by a distance $v(t_2 - t_1)$. On the other hand, the function $f_2(x + vt)$ corresponds to a wave in the negative x-direction. A general solution of wave equation consists of two waves, one travelling to the right and other travelling to the left (back towards the source). With no reflecting surface the second component is zero.

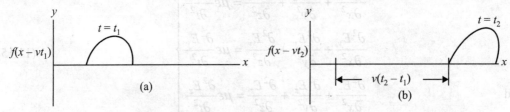

Figure 11.5 A wave travelling in time x-direction.

$$E_y = f_1(x - vt) \tag{11.20}$$

As an example, we can consider a sinusoidal wave of type

$$E_y = \sin(x - ct)$$

with $v = c = \dfrac{1}{\sqrt{\mu_0 \varepsilon_0}}$, which is velocity in free space.

Figure 11.6 shows the wave; for different values of time the wave is moving in x-direction. For some value of E_y, the parameter x will have to increase as the time increases. No physical entity is moving at velocity c. The electric field is not moving, it is merely changing in such a way that if it were visible there would appear to be waves of that velocity.

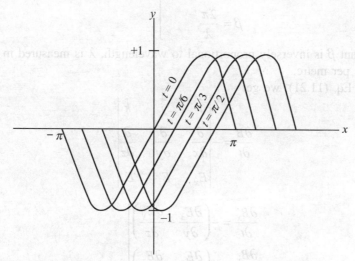

Figure 11.6 A sinusoidal wave travelling in positive x-axis.

Magnetic field

Electrical component of the wave must be accompanied by a magnetic component which can be determined from electric component by means of Maxwell's equation

$$\nabla \times \mathbf{E} = -\frac{\partial \mathbf{B}}{\partial t}$$

or
$$\frac{\partial \mathbf{B}}{\partial t} = -\nabla \times \mathbf{E} \tag{11.21}$$

Knowing $\mathbf{E}$, we can take its curl and then taking integration with respect to time to get $\mathbf{B}$.

Let us take a specific wave, the most common wave in practice is sinusoidal. Let the electric field be

$$E_x = 0, \quad E_y = E_m \cos\beta(x - vt), \quad E_z = 0 \tag{11.22}$$

where y component of field is a function of $(x - vt)$ and coefficients E_m and β are mathematical constants, but physically E_m is the maximum amplitude of the wave and β is the phase constant, which determines the frequency of its sinusoidal variation. Such a wave is shown in Figure 11.3(b).

Letting $\beta v = \omega$, we may write

$$E_x = 0, \quad E_y = E_m \cos(\omega t - \beta x), \quad E_z = 0 \tag{11.23}$$

where ω is the angular frequency, which is equal to $2\pi v$ and v is the frequency of the wave. With λ as the wavelength, and v as the velocity of wave propagation is a given medium, the following simple and useful relations are apparent:

$$\left.\begin{aligned} v &= \frac{\omega}{\beta} = 2\pi \frac{v}{\beta} \\ \lambda &= \frac{v}{v} = \frac{2\pi}{\beta} \\ \beta &= \frac{2\pi}{\lambda} \end{aligned}\right\} \tag{11.24}$$

The phase constant β is inversely proportional to wavelength, λ is measured in per cycle (Hz) and β in radians per metre.

Expanding Eq. (11.21), we get

$$\frac{\partial \mathbf{B}}{\partial t} = -\begin{vmatrix} \hat{i} & \hat{j} & \hat{k} \\ \dfrac{\partial}{\partial x} & \dfrac{\partial}{\partial y} & \dfrac{\partial}{\partial z} \\ E_x & E_y & E_z \end{vmatrix}$$

$$\left.\begin{aligned} \frac{\partial B_x}{\partial t} &= -\left(\frac{\partial E_z}{\partial y} - \frac{\partial E_y}{\partial z}\right) \\ \frac{\partial B_y}{\partial t} &= -\left(\frac{\partial E_x}{\partial z} - \frac{\partial E_z}{\partial x}\right) \\ \frac{\partial B_z}{\partial t} &= -\left(\frac{\partial E_y}{\partial x} - \frac{\partial E_x}{\partial y}\right) \end{aligned}\right\} \tag{11.25}$$

and

With the existing field component $E_y = E_m \cos(\omega t - \beta x)$, the components of Maxwell's equation become

$$\left.\begin{aligned} \frac{\partial B_x}{\partial t} &= 0 \quad \frac{\partial B_y}{\partial t} = 0 \\ \frac{\partial B_z}{\partial t} &= \frac{\partial E_y}{\partial x} = -\beta E_m \sin(\omega t - \beta x) \end{aligned}\right\} \tag{11.26}$$

The components of magnetic field are found by integration with respect to time

$$\left.\begin{aligned} B_x &= 0 \\ B_y &= 0 \\ B_z &= \frac{\beta}{\omega} E_m \cos(\omega t - \beta x) = \frac{\beta}{\omega} E_y \end{aligned}\right\} \tag{11.27}$$

ω/β being the velocity of phase propagation

$$E_y = vB_z \tag{11.28}$$

For a simple plane wave travelling in a dielectric medium, the electric and magnetic components of the wave are identical in form and are perpendicular to each other. The both are perpendicular to the direction of travel of the wave, and travel with velocity $v = 1/\sqrt{\mu\varepsilon}$. When the wave is in free space, the velocity becomes

$$c = \frac{1}{\sqrt{\mu_0\varepsilon_0}} = 3 \times 10^8 \text{ m/s}, \text{ with } B = \mu H$$

$$E_y = v\mu H_z = \frac{\mu}{\sqrt{\mu\varepsilon}} H_z = \sqrt{\frac{\mu}{\varepsilon}} H_z$$

or $\quad H_z = \sqrt{\frac{\varepsilon}{\mu}} E_m \cos(\omega t - \beta x) \tag{11.29}$

This gives relation between **E** and **H**. The ratio of **E** to **H** in a wave is denoted by η, and is called *intrinsic impedance*.

$$\eta = \frac{E}{H} = \sqrt{\frac{\mu}{\varepsilon}} \tag{11.30}$$

For free space

$$\eta = \sqrt{\frac{\mu_0}{\varepsilon_0}} \cong 377 \quad \text{or} \quad 120\pi \tag{11.31}$$

The lines of electric and magnetic fields are shown in Figure 11.3(a), in a simple way. Throughout the plane, **E** and **H** are uniform. If the plane shown in Figure is visualized as being fixed in space, **E** and **H** in that plane are constantly changing with time. If, on the other hand, the plane indicated is visualized as advancing along the x-axis in free space with speed of light, **E** and **H** in that plane are constant and unchanging. This is indeed the distinctive and defining quality of plane wave.

Figure 11.3(b) is a graphic representation of the sinusoidal wave showing **E** and **H** vectors. At any fixed point, the electrical and magnetic intensities vary sinusoidally with time.

The wave is called *uniform plane wave* because its value is uniform throughout any plane, x = constant. It represents an energy flow in the positive x-direction (as per relation $\mathbf{E} \times \mathbf{H}$). Both the electric and magnetic fields are perpendicular to the direction of propagation, or both lie in a plane that is transverse to the direction of propagation; the uniform plane wave is a transverse electromagnetic wave, or a TEM wave.

Uniform plane waves do not exist in practice because they cannot be produced by finite sized antennas. At large distances from physical antennas and ground, however, the waves can be approximated as uniform plane waves. Furthermore, the principle of guiding of electromagnetic waves along transmission lines and wave guides, and the principles of many other wave phenomena can be studied basically in terms of uniform plane waves.

Exponential notation

In the easier form, we have written

$$E_y = E_m \cos(\omega t - \beta x) \tag{11.32}$$

This may be written in exponential form as

$$E_y = \text{Real part of } E_m \, e^{j(\omega t - \beta x)}$$
$$= Re|E_m \, e^{j(\omega t - \beta x)}| \tag{11.33}$$

In Eq. (11.32), E_m is necessarily real, and the phase is indicated by an additional term is argument of cosine. For a phase difference of ϕ, Eq. (11.32) becomes

$$E_y = E_m \cos(\omega t - \beta x + \phi) \tag{11.34}$$
$$E_y = |E_m| \, e^{i(\omega t - \beta x + \phi)} \tag{11.35}$$

or
$$E_y = |E_m| \, e^{i\phi} \, e^{i(\omega t - \beta x)}$$
$$= E_m \, e^{i(\omega t - \beta x)} \tag{11.36}$$

where $E_m = |E_m| \, e^{i\phi}$. This inclusion of phase in E_m is customary.

Use of exponential function is very helpful in manipulation of equations, and it is common in alternating current circuit work as well as in field theory. Impedance is the ratio of exponential voltage to exponential current, but not the ratio of voltage to current when expressed as sine or cosine function.

Sinusoidal electromagnetic waves in free space

Electric field intensity in exponential form is written as

$$E(x, y, z, t) = E_m(x, y, z)e^{i\omega t}$$
$$= E_m(x, y, z) \cos \omega t + i \, E_m(x, y, z) \sin \omega t \tag{11.37}$$

$E(x, y, z, t)$ on left-hand side has the characteristics of both space vector or a phasor.

Real and imaginary components are considered but R_e and I_m terms are implied and not written

$$\frac{\partial \mathbf{E}}{\partial t} = E_m i \omega e^{i\omega t} = \omega \mathbf{E}$$

and
$$\frac{\partial^2 \mathbf{E}}{\partial t^2} = i^2 \omega^2 \mathbf{E} = -\omega^2 \mathbf{E} \tag{11.38}$$

In Eq. (11.14), we have derived the wave equation, which for free space becomes

$$\nabla^2 \mathbf{E} = \mu_0 \varepsilon_0 \frac{\partial^2 \mathbf{E}}{\partial t^2} \tag{11.39}$$

Combining Eqs. (11.38) and (11.39), we have

$$\nabla^2 \mathbf{E} = -\mu_0 \varepsilon_0 \omega^2 \mathbf{E} \tag{11.40}$$

Taking only the y-component of $\mathbf{E}$, variations with respect to x, we have

$$\frac{\partial^2 E_y}{\partial x^2} = -\mu_0 \varepsilon_0 \omega^2 E_y$$

or
$$\frac{\partial^2 E_y}{\partial x^2} + \mu_0 \varepsilon_0 \omega^2 E_y = 0 \tag{11.41}$$

It is a second order differential equation having complete solution.

$$E_y = Ae^{-i\omega(\mu_0\varepsilon_0)^{1/2}x} + Be^{i\omega(\mu_0\varepsilon_0)^{1/2}x} \quad (11.42)$$

where

$$\omega(\mu_0\varepsilon_0)^{1/2} = \frac{\omega}{c} = \beta, \text{ so}$$

$$E_y = A\, e^{-i\beta x} + Be^{i\beta x} \quad (11.43)$$

Showing the time variation of E_y, Eq. (11.43) becomes

$$E_y = Ae^{-i\beta x}e^{i\omega t} + Be^{i\beta x}e^{i\omega t}$$
$$= Ae^{i(\omega t - \beta x)} + Be^{i(\omega t + \beta x)} \quad (11.44)$$

Equation (11.44) represents the sum of two travelling waves in opposite directions. Real part becomes

$$E_y(t) = A\cos(\omega t - \beta x) + B\cos(\omega t + \beta x) \quad (11.45)$$

and imaginary part is

$$E_y(t) = A\sin(\omega t - \beta x) + B\sin(\omega t + \beta x) \quad (11.46)$$

There are special cases of

$$E_y = f_1(x - vt) + f_2(x + vt)$$

when sinusoidal time variation is considered.

EXAMPLE 11.1 Given $E(x, t) = 10^3 \sin(6 \times 10^8 t - \beta x)j$ V/m in free space, sketch the wave at $t = 0$ and at time t_1, when it has travelled $\lambda/4$ along the x-axis. Find t_1, β and λ.

Solution

$$E(x, t) = 10^3 \sin(6 \times 10^8 t - \beta x)j$$
$$= -10^3 \sin(\beta x - 6 \times 10^8 t)j$$
$$= -10^3 \sin\beta(x - v_0 t)j$$

$$\beta = \frac{\omega}{v_0} = \frac{6\times 10^8}{3\times 10^8} = 2 \text{ rad/m}$$

$$\lambda = \frac{2\pi}{\beta} = \frac{2\pi}{2} = \pi \text{ m} \quad \text{and} \quad \frac{\lambda}{4} = \frac{\pi}{4}$$

$$T = \frac{1}{v} = \frac{2\pi}{\omega}$$

$$t_1 = \text{time for } \frac{\lambda}{4} = \frac{T}{4} = \frac{2\pi}{4\times 6\times 10^8}$$

$$= 2.62 \times 10^{-9} \text{ s}$$

The sketch is shown in Figure 11.7.

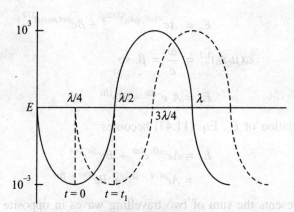

Figure 11.7 Wave propagation.

11.4 WAVE PROPAGATION IN LOSSY DIELECTRIC MEDIUM

In the earlier section, we have considered perfect dielectric medium. This is done be setting conductivity $\sigma = 0$ in Maxwell's equations. If the material is lossy, it will have finite conductivity σ considering the electric and magnetic fields vary sinusoidally, the Maxwell's equations becomes

$$\nabla \cdot \mathbf{D} = 0 \qquad (11.47)$$

$$\nabla \cdot \mathbf{E} = 0 \qquad (11.48)$$

$$\nabla \times \mathbf{E} = -i\omega\mu\mathbf{H} \qquad (11.49)$$

$$\nabla \times \mathbf{H} = (\sigma + i\omega\varepsilon)\mathbf{E} \qquad (11.50)$$

and

$$\nabla \times (\nabla \times \mathbf{E}) = \nabla \cdot (\nabla \cdot \mathbf{E}) - \nabla^2 \mathbf{E} \qquad (11.51)$$

Substituting values from Eqs. (11.48) and (11.49) in Eq. (11.51) we get

$$-\nabla^2 \mathbf{E} = -i\omega\mu(\nabla \times \mathbf{H}) \qquad (11.52)$$

Substituting Eq. (11.50) in Eq. (11.52), we have

$$\nabla^2 \mathbf{E} = i\omega\mu(\sigma + i\omega\varepsilon)\mathbf{E} \qquad (11.53)$$

or

$$\nabla^2 \mathbf{E} = \gamma^2 \mathbf{E} \qquad (11.54)$$

where

$$\gamma^2 = i\omega\mu(\sigma + i\omega\varepsilon) \qquad (11.55)$$

γ (gamma) is called *propagation constant*, and had real and imaginary parts

$$\gamma = \alpha + i\beta \qquad (11.56)$$

α is called *attenuation constant*, and has units of Neper per meter (Np/m) and β is *phase constant* in rad/m.

The solution of wave equation if we consider only E_y components is of the form [similar to Eqs. (11.43) and (11.44)]

$$E_y = E_m e^{i\omega t} e^{\pm \gamma x} \qquad (11.57)$$

or
$$E_y = E_m e^{-\alpha x} e^{i(\omega t - \beta x)} \quad (11.58)$$
where
$$-\gamma x = -\alpha x - i\beta x$$

Comparing Eq. (11.58) with Eq. (11.44), we find that a wave in lossy dielectric is similar to a wave in space or in a perfect dielectric except that it grows smaller as a result of losing energy as it travels. The factor $e^{-\alpha x}$ shows the attenuation of wave. The phase constant β for a lossy dielectric is different from phase constant of a perfect dielectric of same dielectric constant and permeability, β increases with conductivity. Hence wavelength corresponding to given frequency becomes smaller and velocity of propagation is less.

The propagation constant can be obtained from Eq. (11.55)
$$\gamma^2 = i\omega\mu(\sigma + i\omega\varepsilon)$$
or
$$\gamma = i\omega\sqrt{\mu\varepsilon}\sqrt{1 - i\frac{\sigma}{\omega\varepsilon}} \quad (11.59)$$

γ can be separated into α and β such that
$$\gamma^2 = (\alpha + i\beta)^2 = i\omega\mu(\sigma + i\omega\varepsilon) \quad (11.60)$$
$$\alpha^2 - \beta^2 = -\omega^2\mu\varepsilon \quad (11.61)$$
$$2\alpha\beta = \omega\mu\sigma \quad (11.62)$$

From Eqs. (11.51) and (11.52), we may separate α and β

$$\alpha = \omega\sqrt{\frac{\mu\varepsilon}{2}\left(\sqrt{1 + \left(\frac{\sigma}{\omega\varepsilon}\right)^2} - 1\right)} \quad (11.63)$$

$$\beta = \omega\sqrt{\frac{\mu\varepsilon}{2}\left(\sqrt{1 + \left(\frac{\sigma}{\omega\varepsilon}\right)^2} + 1\right)} \quad (11.64)$$

Considering only E_y component of **E** varying with respect to x

$$\nabla \times \mathbf{E} = \frac{\partial E_y}{\partial t}\hat{k} \quad \text{[as derived earlier in Eq. (11.35)]}$$

$$= \hat{k}\frac{\partial}{\partial x}E_m e^{-\gamma x}$$

$$= -\hat{k}\gamma E_m e^{-\gamma x} \quad (11.65)$$

Substituting Eq. (11.49) in Eq. (11.65), we get
$$-i\omega\mu\mathbf{H} = -\hat{k}\gamma E_m e^{-\gamma x}$$

H has H_z component corresponding to E_y
$$H_z = \frac{\gamma}{i\omega\mu}E_y \quad (11.66)$$

So the intrinsic impedance
$$\eta = \frac{E_y}{H_z} = \frac{i\omega\mu}{\gamma} \quad (11.67)$$

Substituting γ from Eq. (11.59)

$$\eta = \sqrt{\frac{\mu}{\varepsilon\left(1 + \dfrac{\sigma}{i\omega\varepsilon}\right)}} = \sqrt{\frac{i\omega\mu}{\sigma + i\omega\varepsilon}} \tag{11.68}$$

η is a complex quantity. With

$$E_y = E_m\, e^{-\alpha x}\, e^{i(\omega t - \beta x)} \tag{11.69}$$

the magnetic field becomes

$$H_z = \frac{E_m}{\eta}\, e^{-\alpha x}\, e^{i(\omega t - \beta x)}$$

The electric and magnetic fields are no longer in time phase.

The factor $e^{-\alpha x}$ causes an exponential decrease in amplitude with increasing value of x, η is a complex quantity in the first quadrant, so *in the lossy dielectric, the electric field leads the magnetic field in time phase.*

EXAMPLE 11.2 For a lossy dielectric material having $\mu_r = 1$, $\varepsilon_r = 48$, $\sigma = 20$ S/m, calculate the attenuation constant, phase constant and intrinsic impedance at a frequency of 16 GHz.

Solution

$$\frac{\sigma}{\omega\varepsilon} = \frac{20 \times 10^{12}}{2\pi \times 16 \times 10^9 \times 48 \times 8.856} = 0.47$$

$$\gamma = i\,\omega\sqrt{\mu\varepsilon}\,\sqrt{1 - i\frac{\sigma}{\omega\varepsilon}}$$

$$= i\,(2\pi)\,16 \times 10^9\, \sqrt{4\pi \times 10^{-7} \times 48 \times 8.854 \times 10^{-12}} \times \sqrt{1 - i(0.47)}$$

$$= i\,2323.25\sqrt{1.0966 \angle -24.23°}$$

$$= 2432.88 \angle 77.89° = 510.4 + i\,2378.7\ \text{m}^{-1}$$

$$\alpha = 510.4\ \text{Np/m}$$

$$\beta = 2378.7\ \text{rad/m}$$

The intrinsic impedance

$$\eta = \sqrt{\frac{i\omega\mu}{\sigma + i\omega\varepsilon}} = \sqrt{\frac{\mu}{\varepsilon\left(1 + \dfrac{\sigma}{i\omega\varepsilon}\right)}}$$

$$= \sqrt{\frac{4\pi \times 10^{-7}}{48 \times 8.854 \times 10^{-12}}} \times \sqrt{\frac{1}{1 - j(0.47)}}$$

$$= \frac{54.377}{\sqrt{1.0966 \angle -24.23°}}$$

$$= 51.93 \angle 12.12°\ \Omega$$

The electric field (E_y) leads the magnetic field (H_z) by 12.12° at every point.

11.5 CONDUCTORS AND DIELECTRICS

In electromagnetics, materials are divided roughly into two classes:
1. Conductors
2. Dielectrics or insulators

The dividing line between two classes is not sharp and some media are considered as conductors in one part of radio frequency range, and as dielectric (with loss) in another part of the range.

In the Maxwell's equation
$\nabla \times \mathbf{H} = \sigma \mathbf{E} + i\omega\varepsilon \mathbf{E}$, the ratio $\sigma/\omega\varepsilon$ is, therefore, just the ratio of conduction current density to displacement current density in medium. For good conductors $\sigma/\omega\varepsilon \gg 1$ over entire radio frequency range. For good dielectrics $\sigma/\omega\varepsilon \ll 1$ in the ratio frequency range.

The term $\sigma/\omega\varepsilon$ is referred to as the *loss tangent* (similar to loss tangent in case of a capacitor) or *dissipation factor*. In practice, the following observations are true:

- For good *conductors* σ and ε are nearly independent of frequency.
- For most *dielectrics* σ and ε are functions of frequency, but ratio $\sigma/\omega\varepsilon$ is often constant over the frequency range of interest.

Based on the values of $\sigma/\omega\varepsilon$, we can approximate the relations for attenuation constant, phase constant and intrinsic impedance for conductors and dielectric materials.

11.6 WAVE PROPAGATION IN GOOD DIELECTRICS

For good dielectrics,
$$\sigma \ll \omega\varepsilon$$
or
$$\frac{\sigma}{\omega\varepsilon} \ll 1$$

We may write
$$\sqrt{\left(1 + \frac{\sigma^2}{\omega^2\varepsilon^2}\right)} \approx \left(1 + \frac{\sigma^2}{2\omega^2\varepsilon^2}\right) \quad \text{[using Binomial theorem]} \quad (11.70)$$

Equations (11.63) and (11.64), then become

$$\alpha = \omega\sqrt{\frac{\mu\varepsilon}{2}\left(\sqrt{1 + \left(\frac{\sigma}{\omega\varepsilon}\right)^2} - 1\right)} \approx \omega\sqrt{\frac{\mu\varepsilon}{2}\left(1 + \frac{\sigma^2}{2\omega^2\varepsilon^2} - 1\right)}$$

$$\alpha = \frac{\sigma}{2}\sqrt{\frac{\mu}{\varepsilon}} \quad (11.71)$$

$$\beta = \omega\sqrt{\frac{\mu\varepsilon}{2}\left(\sqrt{1 + \left(\frac{\sigma}{\omega\varepsilon}\right)^2} + 1\right)} \approx \omega\sqrt{\frac{\mu\varepsilon}{2}\left(1 + \frac{\sigma^2}{2\omega^2\varepsilon^2} + 1\right)}$$

$$\beta = \omega\sqrt{\mu\varepsilon}\left(1 + \frac{\sigma^2}{8\omega^2\varepsilon^2}\right) \quad (11.72)$$

where $\omega\sqrt{\mu\varepsilon} = \dfrac{\omega}{v}$ is the phase shift for a perfect dielectric. The effect of small amount of loss (due to conductivity) is to add a second term in Eq. (11.72) and β increases.

$$\text{Velocity of wave } v = \frac{\omega}{\beta} = \frac{1}{\sqrt{\mu\varepsilon}\left(1 + \dfrac{\sigma^2}{8\omega^2\varepsilon^2}\right)}$$

$$v \cong \frac{1}{\sqrt{\mu\varepsilon}}\left(1 - \frac{\sigma^2}{8\omega^2\varepsilon^2}\right) \tag{11.73}$$

where $\dfrac{1}{\sqrt{\mu\varepsilon}}$ is the velocity of the wave in dielectric when the conductivity is zero, i.e. in perfect dielectric. The effect of a small amount of loss is to reduce slightly the velocity of propagation of the wave.

$$\text{Intrinsic impedance } \eta = \sqrt{\frac{i\omega\mu}{\sigma + i\omega\varepsilon}}$$

$$= \sqrt{\frac{\mu}{\varepsilon} \cdot \frac{1}{1 + \dfrac{\sigma}{i\omega\varepsilon}}}$$

$$\eta = \sqrt{\frac{\mu}{\varepsilon}}\left(1 + i\frac{\sigma}{2\omega\varepsilon}\right) \tag{11.74}$$

where $\sqrt{\dfrac{\mu}{\varepsilon}}$ is the intrinsic impedance of dielectric with $\sigma = 0$. *The chief effect of loss is to add a small reactive component to the intrinsic impedance.*

The above approximations may be mode in the cases where $\dfrac{\sigma}{\omega\varepsilon} < 0.1$, but for a perfect dielectric $\dfrac{\sigma}{\omega\varepsilon} \leq 0.01$.

11.7 WAVE PROPAGATION IN A GOOD CONDUCTOR

For a good conductor $\sigma \ll \omega\varepsilon$

$$\frac{\sigma}{\omega\varepsilon} \ll 1$$

The propagation constant γ can be written as

$$\gamma^2 = i\omega\mu(\sigma + i\omega\varepsilon)$$

$$= i\omega\mu\sigma\left(1 + i\frac{\omega\varepsilon}{\sigma}\right)$$

$$\cong i\omega\mu\sigma$$

$$\gamma = \sqrt{i\omega\mu\sigma} \tag{11.75}$$

$$= \sqrt{\omega\mu\sigma} \angle 45°$$
$$= \sqrt{\omega\mu\sigma}[\cos 45° + i \sin 45°]$$
$$= \sqrt{\frac{\omega\mu\sigma}{2}} + i\sqrt{\frac{\omega\mu\sigma}{2}}$$
$$\gamma = (1+i)\sqrt{\frac{\omega\mu\sigma}{2}} \qquad (11.76)$$

So
$$\alpha = \beta = \sqrt{\frac{\omega\mu\sigma}{2}} = \sqrt{\pi\nu\mu\sigma} \qquad (11.77)$$

The velocity of propagation
$$v = \frac{\omega}{\beta} = \sqrt{\frac{2\omega}{\mu\sigma}} \qquad (11.78)$$

The intrinsic impedance of the conductor is
$$\eta = \sqrt{\frac{i\omega\mu}{\sigma + i\omega\varepsilon}} \approx \sqrt{\frac{i\omega\mu}{\sigma}} \qquad (11.79)$$

$$\eta = \sqrt{\frac{\omega\mu}{\sigma}} \angle 45°$$

In a conductor, α and β are large. The wave attenuates greatly as it progresses and phase shift per unit length is also large. The velocity of the wave is small. The intrinsic impedance is small and has a reactive component with impedance angle of 45°.

For a good conductor let $\alpha = 60$ Np/m, and $\beta = 60$ rad/m, and $\eta = 30 \angle 45°$. An **E** wave is expressed as
$$\mathbf{E} = 4e^{-\alpha x} \sin(10^8 t - \beta x)\hat{k} \text{ V/m}$$
and the wave is propagating in x-direction, we want to find H. H will be represented as
$$\mathbf{H} = H_0 e^{-\alpha x} \sin\left(\omega t - \beta x - \frac{\pi}{4}\right)\hat{j}$$
$$H_0 = \frac{E_0}{\eta} = \frac{4}{30} = 0.133 \text{ A/m}$$

With propagation in x-direction and **E** being in $\hat{k}$ direction, H will be in $-\hat{j}$-direction (such that $\mathbf{E} \times \mathbf{H}$ vectors must give propagation direction).
$$\mathbf{H} = -0.133 \, e^{-60x} \sin\left(\omega t - 60x - \frac{\pi}{4}\right)\hat{j} \text{ A/m}$$

11.8 DEPTH OF PENETRATION: SKIN DEPTH

In a medium of high conductivity, the wave is attenuated as its progress due to those losses which occur in the medium. In a good conductor, the rate of attenuation is very great and the wave may penetrate only a very short distance before being reduced to a negligible small percentage of its original strength. A term that has significance is called *depth of penetration* or *skin depth*. The depth of penetration δ is defined as the depth in which the wave has been attenuated to $1/e$ or approximately 37% of the original value as shown in Figure 11.8.

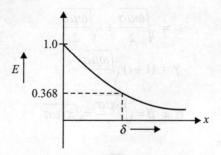

Figure 11.8 Field in a conductor, skin depth.

The amplitude of the wave decreases by a factor $e^{-\alpha x}$, where α is the attenuation constant, it is apparent that a distance x which makes $\alpha x = 1$, the amplitude is only $1/e$ times its value at $x = 0$. This distance is equal to δ, the depth of penetration we have

$$\alpha x = 1$$

or
$$\alpha \delta = 1$$

or
$$\delta = \frac{1}{\alpha}$$

$$\delta = \frac{1}{\alpha} = \frac{1}{\omega \sqrt{\dfrac{\mu \varepsilon}{2}} \sqrt{1 + \dfrac{\sigma^2}{\omega^2 \varepsilon^2} - 1}}$$

For good conductor, $\dfrac{\sigma}{\omega \varepsilon} \gg 1$, so

$$\delta = \frac{1}{\alpha} \approx \sqrt{\frac{2}{\omega \mu \sigma}} = \sqrt{\frac{1}{\pi \nu \mu \sigma}} \tag{11.80}$$

For copper with $\sigma = 5.8 \times 10^7$ S/m, $\mu = \mu_0$ at $\nu = 1$ MHz

$$\delta = \frac{1}{\sqrt{\pi \times 10^6 \times 4\pi \times 10^{-7} \times 5.8 \times 10^7}} = 0.0661 \text{ mm}$$

The same value at 50 Hz,

$$\delta = \frac{1}{\pi \times 50 \times 4\pi \times 10^{-7} \times 5.8 \times 10^7} = 9.35 \text{ mm}$$

At a microwave frequency of 10,000 MHz, $\delta = 6.61 \times 10^{-4}$ mm or about one-eight of the wavelength of visible light.

All fields in a good conductor such as copper are essentially zero at distances greater than a few skin depths from the surface. Electromagnetic energy is not transmitted in the interior of the conductor, it travels in the region surrounding the conductor, while the conductor merely guides the waves. At microwave frequencies only the surface coating of the guiding conductor is important. A silver coating on a piece of glass may be and excellent conductor at these frequencies.

EXAMPLE 11.3 A medium like copper conductor is characterised by the parameter $\sigma = 5.8 \times 10^7$ mho/m, $\varepsilon_r = 1$, $\mu_r = 1$ supports a uniform plane wave of frequency 60 Hz. Find the attenuation constant, propagation constant, intrinsic impedance, wavelength and phase velocity of the wave.

Solution Let us obtain the ratio

$$\frac{\sigma}{\omega\varepsilon} = \frac{5.8 \times 10^7}{2\pi \times 60 \times 8.854 \times 10^{-12}} = 173 \times 10^{14}$$

This is very very larger than 1, therefore, it is a very good conductor.

Attenuation constant $\alpha = \left(\frac{\omega\mu\sigma}{2}\right)^{1/2} = 117.2$ per m

Phase constant $\beta = \left(\frac{\omega\mu\sigma}{2}\right)^{1/2} = 117.2$ per m

Propagation constant $k = \alpha + j\beta = 117.2 + j117.2$ per m

Wavelength $\lambda = \frac{2\pi}{\beta} = \frac{2\pi}{117.2} = 0.0536$

Intrinsic impedance $\eta = \sqrt{\frac{j\omega\mu}{\sigma}} = (2.002 + j2.002) \times 10^{-6}$ Ω

Phase velocity of wave $v = \lambda\nu = 0.0536 \times 60 = 3.216$ m/s

EXAMPLE 11.4 Find the skin depth δ at a frequency of 1.6 MHz in aluminium, where $\sigma = 38.2$ MS/m and $\mu_r = 1$. Also find the propagation constant and wave velocity.

Solution $\delta = \sqrt{\frac{1}{\pi\nu\mu\sigma}} = \frac{1}{\sqrt{\pi \times 1.6 \times 10^6 \times 4\pi \times 10^{-7} \times 38.2 \times 10^6}}$

$= 0.06438$ mm

$\alpha = \beta = \delta^{-1}$

$\gamma = 15.53 \times 10^3 + j15.53 \times 10^3$

$= 21.96 \times 10^3 \angle 45°$ m^{-1}

$v = \frac{\omega}{\beta} = \omega\delta = 2\pi \times 1.6 \times 10^6 \times 0.06438 \times 10^{-3}$

$= 647.2$ m/s

11.9 POYNTING VECTOR AND POYNTING THEOREM

The cross or vector product of electric is field intensity **E** and magnetic field intensity **H** is known as *Poynting vector*. It is denoted by **P**. It is defined as

$$\mathbf{P} = \mathbf{E} \times \mathbf{H} \qquad (11.81)$$

It is interpreted as an instantaneous power density that is measured in W/m². Poynting vector **P** can be used to determine total power crossing the surface in an outward sense. Total power can be determined by integrating vector over a closed surface. Poynting vector ($\mathbf{P} = \mathbf{E} \times \mathbf{H}$) can be proved using Maxwell's equation

$$\nabla \times \mathbf{H} = \mathbf{J} + \frac{\partial \mathbf{D}}{\partial t} \qquad (11.82)$$

Taking the dot product of both sides of Eq. (11.82) with **E**, we obtain

$$\mathbf{E}.(\nabla \times \mathbf{H}) = \mathbf{E}.\left(\mathbf{J} + \frac{\partial \mathbf{D}}{\partial t}\right)$$

$$= \mathbf{E}.\mathbf{J} + \mathbf{E}.\frac{\partial \mathbf{D}}{\partial t}$$

or $\qquad \mathbf{E}.(\nabla \times \mathbf{H}) = \mathbf{J}.\mathbf{E} + \mathbf{E}.\frac{\partial \mathbf{D}}{\partial t} \qquad (11.83)$

Now using vector identity given as

$$\nabla.(\mathbf{E} \times \mathbf{H}) = -\mathbf{E}.(\nabla \times \mathbf{H}) + \mathbf{H}.(\nabla \times \mathbf{E}) \qquad (11.84)$$

Substituting Eq. (11.83) in Eq. (11.84), we obtain

$$\nabla.(\mathbf{E} \times \mathbf{H}) = -\mathbf{E}.(\nabla \times \mathbf{H}) + \mathbf{H}.(\nabla \times \mathbf{E})$$

$$= -\left\{\mathbf{J}.\mathbf{E} + \left(\mathbf{E}.\frac{\partial \mathbf{D}}{\partial t}\right)\right\} + \mathbf{H}.(\nabla \times \mathbf{E}) \qquad (11.85)$$

From Maxwell's third equation, we know that

$$\nabla \times \mathbf{E} = -\frac{\partial \mathbf{B}}{\partial t}$$

Substituting above value in Eq. (11.85), we obtain

$$\nabla.(\mathbf{E} \times \mathbf{H}) = -\mathbf{J}.\mathbf{E} - \mathbf{E}.\frac{\partial \mathbf{D}}{\partial t} + \mathbf{H}.\left(-\frac{\partial \mathbf{B}}{\partial t}\right)$$

$$= -\mathbf{J}.\mathbf{E} - \mathbf{E}.\frac{\partial \mathbf{D}}{\partial t} - \mathbf{H}.\frac{\partial \mathbf{B}}{\partial t}$$

$$= -\mathbf{J}.\mathbf{E} - \mathbf{E}.\frac{\partial(\varepsilon \mathbf{D})}{\partial t} - \left[\mathbf{H}.\frac{\partial(\mu \mathbf{H})}{\partial t}\right]$$

$$= -\mathbf{J}.\mathbf{E} - \mathbf{E}.\frac{\partial(\varepsilon \mathbf{D})}{\partial t} - \mu \mathbf{H}.\frac{\partial \mathbf{H}}{\partial t}$$

or
$$\nabla \cdot (\mathbf{E} \times \mathbf{H}) = -\mathbf{J} \cdot \mathbf{E} - \varepsilon \mathbf{E} \cdot \frac{\partial \mathbf{E}}{\partial t} - \mu \mathbf{H} \cdot \frac{\partial \mathbf{H}}{\partial t} \qquad (11.86)$$

Substituting

$$\mu \mathbf{H} \cdot \frac{\partial \mathbf{H}}{\partial t} = \frac{\mu}{2} \frac{\partial (H^2)}{\partial t^2} = \frac{\partial}{\partial t}\left(\frac{\mu H^2}{2}\right)$$

and

$$\varepsilon \mathbf{E} \cdot \frac{\partial \mathbf{E}}{\partial t} = \frac{\varepsilon}{2} \frac{\partial (E^2)}{\partial t^2} = \frac{\partial}{\partial t}\left(\frac{\varepsilon E^2}{2}\right)$$

From Eq. (11.86), we have

$$\nabla \cdot (\mathbf{E} \times \mathbf{H}) = -\mathbf{J} \cdot \mathbf{E} - \frac{\partial}{\partial t}(\varepsilon E^2) - \frac{\partial}{\partial t}(\mu H^2)$$

$$= -\mathbf{J} \cdot \mathbf{E} - \frac{\partial}{\partial t}\left(\frac{\varepsilon E^2}{2} + \frac{\mu H^2}{2}\right)$$

or
$$-\nabla \cdot (\mathbf{E} \times \mathbf{H}) = \mathbf{J} \cdot \mathbf{E} + \frac{\partial}{\partial t}\left(\frac{\varepsilon E^2}{2} + \frac{\mu H^2}{2}\right) \qquad (11.87)$$

Integrating both sides of Eq. (11.87) throughout a volume V,

$$-\int_V \nabla \cdot (\mathbf{E} \times \mathbf{H}) \, dV = \int_V \left[\mathbf{J} \cdot \mathbf{E} + \frac{\partial}{\partial t}\left(\frac{\varepsilon E^2}{2} + \frac{\mu H^2}{2}\right)\right] dV$$

or
$$-\int_V \nabla \cdot (\mathbf{E} \times \mathbf{H}) \, dV = \int_V \mathbf{J} \cdot \mathbf{E} \, dV + \frac{\partial}{\partial t} \int_V \left(\frac{\varepsilon E^2}{2} + \frac{\mu H^2}{2}\right) dV \qquad (11.88)$$

From Gauss divergence theorem, we know that

$$\int_V \nabla \cdot (\mathbf{E} \times \mathbf{H}) \, dV = \oint_S (\mathbf{E} \times \mathbf{H}) \cdot d\mathbf{S} \qquad (11.89)$$

Substituting Eq. (11.89) in Eq. (11.88), we get

$$-\oint_S (\mathbf{E} \times \mathbf{H}) \cdot d\mathbf{S} = \underbrace{\int_V (\mathbf{J} \cdot \mathbf{E}) \, dV}_{\text{I}} + \underbrace{\frac{\partial}{\partial t} \int_V \left(\frac{\varepsilon E^2}{2} + \frac{\mu H^2}{2}\right) dV}_{\text{II}} \qquad (11.90)$$

- Equation (11.90) is the integral form of the Poynting theorem. If we assume that there is no source within the volume V, then the first integral on right hand sides of Eq. (11.90) is that the total instantaneous ohmic power dissipated in the volume.
- The integral in the second term of right hand side of Eq. (11.90) is the total energy stored in both electric and magnetic fields. The partial derivative with respect to time represents the time rate of increase of energy stored within the volume V. Thus we can say that right hand side of Eq. (11.90) expresses the total power flowing into this volume.
- Total power flowing out of the volume V is given as

$$\oint (\mathbf{E} \times \mathbf{H}) \cdot d\mathbf{S}$$

where the integral is over the closed surface surrounding the volume.

We have already defined that the cross product of **E** and **H** is known as the Poynting vector.

$$\text{Poynting vector} = \mathbf{P} = \mathbf{E} \times \mathbf{H} \tag{11.91}$$

The Poynting vector is applied to time constant fields. The direction of the vector **P** indicates the direction of the instantaneous power flow at the point. Many of us think that Poynting vector is a pointing vector. This homogeneous is accidental, but correct. The Poynting vector **P** has the dimensions of power per unit area and its unit is W/m^2. It is a vector, because it indicates not only the magnitude of the energy flow but also its direction. It is Poynting theorem that the vector product or cross product $\mathbf{P} = \mathbf{E} \times \mathbf{H}$ at any point is a measure of the time rate of energy flow per unit area at that point.

The direction of flow is perpendicular to **E** and **H** and is the direction of Poynting vector $\mathbf{P} = \mathbf{E} \times \mathbf{H}$.

EXAMPLE 11.5 The electromagnetic wave intensity received on the surface of the earth from the sum is found to be 1.33 kW/m^2. Find amplitude of electric field vector associated with sunlight as received on earth surface. Assume sun's light to be monochromatic (λ = 6000 Å).

Solution The energy transported by an electromagnetic wave per unit area per second during propagation is represented by Poynting vector **S** as

$$\mathbf{S} = \mathbf{E} \times \mathbf{H}$$

The energy flux per unit area per second is

$$|\mathbf{S}| = |\mathbf{E} \times \mathbf{H}| = EH \sin 90° = EH$$

The energy flux per unit area per second at the earth surface.

$$|\mathbf{S}| = 1.33 \text{ kW/m}^2 = 1.33 \times 10^3 \text{ Jm}^{-2}\text{s}^{-1}$$

$$\therefore \quad |\mathbf{S}| = 1330 \text{ Jm}^{-2}\text{s}^{-1} \tag{i}$$

We know that

$$Z_0 = \left|\frac{\mathbf{E}}{\mathbf{H}}\right| = \sqrt{\frac{\mu_0}{\varepsilon_0}} = \sqrt{\frac{4\pi \times 10^{-7} \text{ Wb/A-m}}{8.854 \times 10^{-12} \text{ C}^2/\text{Nm}^2}} = 376.72 \, \Omega$$

$$\frac{E}{H} = 376.72 \, \Omega \tag{ii}$$

Multiplying Eq. (i) and Eq. (ii), we get

$$EH \times \frac{E}{H} = 1330 \times 376.72$$

or

$$E^2 = 501037.6$$

$$E = 707.8 \text{ V/m}$$

Substituting this value in Eq. (i)

$$H = \frac{1330}{707.8} = 1.879 \text{ A/m}$$

Therefore, the amplitude of electric and magnetic fields of radiation are

$$E_0 = E\sqrt{2} = 707.8\sqrt{2} = 1000.8292 = 1000 \text{ V/m}$$

and

$$H_0 = H\sqrt{2} = 1.928\sqrt{2} = 2.73 \text{ A/m}$$

EXAMPLE 11.6 If the earth receives 2 cal min^{-1} cm^{-2} solar energy, what are the amplitudes of electric and magnetic fields of radiation?

Solution As Poynting vector,

$$\mathbf{S} = \mathbf{E} \times \mathbf{H} = EH \sin 90° = EH$$

Solar energy = 2 cal min^{-1} cm^{-2}

$$= \frac{2 \times 4.18 \times 10^4}{60} \text{ J/ms}$$

Both are energy flux per unit area per second.

Hence
$$EH = \frac{2 \times 4.18 \times 10^4}{60} \approx 1400$$

But
$$\frac{E}{H} = \sqrt{\frac{\mu_0}{\varepsilon_0}} = 120\pi = 377$$

∴
$$EH \times \frac{E}{H} = 1400 \times 377$$

$$E = \sqrt{1400 \times 377} = 726.5 \text{ V/m}$$

Now,
$$H = \frac{E}{377} = 1.927 \text{ A/m}$$

Amplitudes of electric and magnetic fields of radiation are

$$E_0 = E\sqrt{2} = 1024.3 \text{ V/m}$$
$$H_0 = H\sqrt{2} = 2.717 \text{ A/m}$$

11.10 POLARIZATION

In electromagnetic fields, we say that wave is polarized in a particular way depending on the direction of **E** vector. For example, for a plane wave travelling in z-direction, $E_z = 0$ and with the components of E_x and E_y in x- and y-directions respectively as shown in Figure 11.9.

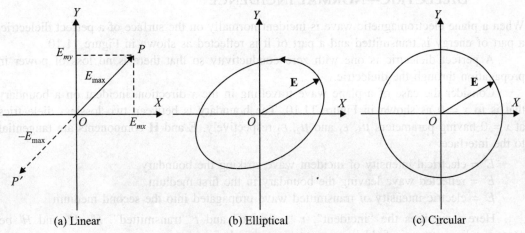

(a) Linear   (b) Elliptical   (c) Circular

Figure 11.9 Polarization of waves.

$$\mathbf{E} = i\,E_x + jE_y \qquad (11.92)$$

With $E_x = E_{mx} \sin \omega t$ and $E_y = E_{my} \sin \omega t$ at any point on a plane $z =$ constant, $E = \sqrt{E_x^2 + E_y^2}$, the length of vector will vary from 0 to E_{max} in one direction and 0 to $-E_{max}$ in other direction as shown in Figure 11.9(a), vector **E** always lies along the straight line PP' shown and the wave is said to be *linear polarized*.

If the field components E_x and E_y have time phase difference like

$$x = E_x = E_{mx} \sin \omega t \qquad (11.93)$$

and
$$y = E_y = E_{my} \sin \omega t \qquad (11.94)$$

By eliminating ωt, we can get equation representing the laws of tip of the vector **E**, which is ellipse with its axis inclined to x and y axes. **E** changes its magnitude and direction with vector **E** describing an *ellipse* as shown in Figure 11.9(b). The magnitude attains a maximum value E_{max} in the direction of major axis of ellipse, and minimum value E_{min} in the direction of minor axis. Vector **E** rotates around the origin at ω radians per second, this wave is said to be *elliptically polarized* plane wave.

In a special case of elliptical polarization, where $E_{mx} = E_{my} = E_m$, and the phase difference $\phi = \pi/2$, we get

$$x = E_x = E_m \sin \omega t \qquad (11.95)$$

$$y = E_y = E_m \sin\left(\omega t + \frac{\pi}{2}\right)$$

$$= E_m \cos \omega t \qquad (11.96)$$

and eliminating ωt, we get

$$x^2 + y^2 = E_m^2 \qquad (11.97)$$

It is the equation of a *circle* of constant radius. E_m described by the tip of vector **E** rotating at ω radians per second as shown in Figure 11.9(c). This wave is then called *circularly polarized* plane wave.

11.11 REFLECTION OF UNIFORM PLANE WAVES BY PERFECT DIELECTRIC—NORMAL INCIDENCE

When a plane electromagnetic wave is incident normally on the surface of a perfect dielectric, a part of energy is transmitted and a part of it is reflected as shown in Figure 11.10.

A perfect dielectric is one with zero conductivity so that there is no loss of power in propagation through the dielectric.

Consider the case of a plane wave travelling in the x-direction, incident on a boundary that is to $x = 0$ as shown in Figure 11.10. The boundary is between two lossless dielectrics at $x = 0$ having parameters μ_1, ε_1 and μ_2, ε_2 respectively. **E** and **H** components are tangential to the interface.

E^i = electrical intensity of incident wave striking the boundary
E^r = reflected wave leaving the boundary in the first medium
E^t = electric intensity of transmitted wave propagated into the second medium

Here i signifies the "incident", r "reflected" and t "transmitted". H^i, H^r and H^t be corresponding magnetic field components respectively.

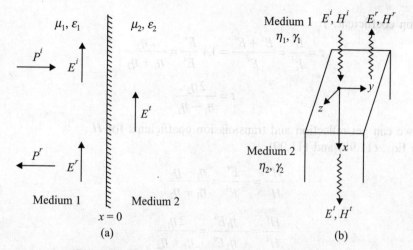

Figure 11.10 Incident, reflected and transmitted wave.

For first medium $\eta_1 = \sqrt{\dfrac{\mu_1}{\varepsilon_1}}$

For second medium $\eta_2 = \sqrt{\dfrac{\mu_2}{\varepsilon_2}}$

The relationship for electric and magnetic fields are

$$E^i = \eta_1 H^i \tag{11.98}$$

$$E^r = -\eta_1 H^r \tag{11.99}$$

(Here energy direction changed, so (–) negative sign)

$$E^t = \eta_2 H^t \tag{11.100}$$

The continuity of tangential components of E and H require that

$$H^i + H^r = H^t$$
$$E^i + E^r = E^t \tag{11.101}$$

we have from Eqs. (11.98) and (11.99)

$$H^i + H^r = \dfrac{1}{\eta_1}(E^i - E^r) = H^t = \dfrac{1}{\eta_2}(E^i + E^r)$$

or

$$\eta_2(E^i - E^r) = \eta_1(E^i + E^r) \tag{11.102}$$

we define reflection coefficient, Γ and transmission coefficient, τ from Eqs. (11.101) and (11.102).

Reflection coefficient Γ

$$\Gamma = \dfrac{E^r}{E^i} = \dfrac{\eta_2 - \eta_1}{\eta_2 + \eta_1} \tag{11.103}$$

Transmission coefficient, τ

$$\tau = \frac{E^t}{E^i} = \frac{E^i + E^r}{E^i} = 1 + \frac{E^r}{E^i} = \frac{2\eta_2}{\eta_1 + \eta_2}$$

or
$$\tau = \frac{2\eta_2}{\eta_1 + \eta_2} \tag{11.104}$$

Similarly, we can get reflection and transmission coefficients for H.

From Eqs. (11.98) and (11.99)

$$\frac{H^r}{H^i} = -\frac{E^r}{E^i} = \frac{\eta_1 - \eta_2}{\eta_1 + \eta_2} \tag{11.105}$$

and
$$\frac{H^t}{H^i} = -\frac{\eta_1 E^t}{\eta_2 E^i} = \frac{2\eta_1}{\eta_1 + \eta_2} \tag{11.106}$$

Equations (11.105) and (11.106) are valid for a boundary between any two materials.

In general, **E** and **H** are of form

$$\left.\begin{aligned} E^i(x,t) &= E_m^i e^{-\gamma_1 x} e^{j\omega t}\hat{j} \\ E^r(x,t) &= E_m^r e^{\gamma_1 x} e^{j\omega t}\hat{j} \\ E^t(x,t) &= E_m^t e^{-\gamma_2 x} e^{j\omega t}\hat{j} \\ H^i(x,t) &= H_m^i e^{-\gamma_1 x} e^{j\omega t}\hat{k} \\ H^r(x,t) &= H_m^i e^{\gamma_1 x} e^{j\omega t}\hat{k} \\ H^t(x,t) &= H_m^i e^{\gamma_2 x} e^{j\omega t}\hat{k} \end{aligned}\right\} \tag{11.107}$$

With normal incidence **E** and **H** are both tangential to the interface and thus are continuous. The boundary is at $x = 0$. The values of **E** and **H** in the Eqs. (11.98)–(11.106) represent E_m and H_m values.

The intrinsic impedances for various materials derived earlier are:

For free space, $\eta_0 = \sqrt{\dfrac{\mu_0}{\varepsilon_0}} = 120\,\pi\,\Omega$

For perfect dielectric, $\eta = \sqrt{\dfrac{\mu}{\varepsilon}}$

For partial conducting medium, $\eta = \sqrt{\dfrac{j\omega\mu}{\sigma + j\omega\varepsilon}}$

For conducting medium, $\eta = \sqrt{\dfrac{\omega\mu}{\sigma}}\angle 45°$

For *perfect dielectrics* $\mu_1 = \mu_2 = \mu_0$, Eqs. (11.103)–(11.106) become

$$\left. \begin{array}{l} \dfrac{E^r}{E^i} = \dfrac{\sqrt{\varepsilon_1} - \sqrt{\varepsilon_2}}{\sqrt{\varepsilon_1} + \sqrt{\varepsilon_2}} \\[2mm] \dfrac{E^t}{E^i} = \dfrac{2\sqrt{\varepsilon_1}}{\sqrt{\varepsilon_1} + \sqrt{\varepsilon_2}} \\[2mm] \dfrac{H^r}{H^i} = \dfrac{\sqrt{\varepsilon_2} - \sqrt{\varepsilon_1}}{\sqrt{\varepsilon_1} + \sqrt{\varepsilon_2}} \\[2mm] \dfrac{H^t}{H^i} = \dfrac{2\sqrt{\varepsilon_2}}{\sqrt{\varepsilon_1} + \sqrt{\varepsilon_2}} \end{array} \right\} \quad (11.108)$$

In Eqs. (11.103)–(11.104), the following observations can be made
(i) $1 + \Gamma = \tau$
(ii) Both Γ and τ are dimensionless and may be complex
(iii) $0 \leq |\Gamma| \leq 1$

If medium 2 is a perfect conductor, $\eta_2 = 0$, Eqs. (11.103) and (11.104) yield $\Gamma = -1$ and $\tau = 0$. Consequently $E^r = -E^i$ and $E^t = 0$. The incident wave will be totally reflected and a standing wave will be produced in medium 1. The standing wave will have zero and maximum values of the field at different points.

For lossless media, η_1 and η_2 are real. However, Γ can be positive or negative. Let us consider the following cases:

(i) $\Gamma > 0$ ($\eta_2 > \eta_1$): There will be a standing wave in medium 1 but there is also a transmitted wave in medium 2. However, the incident and reflected waves have amplitudes that are not equal in magnitude. Maximum and minimum values of $(E_1 = E^i + E^r)$ shall appear at different points.

(ii) $\Gamma < 0$ ($\eta_2 < \eta_1$): There will be a standing wave in medium 1 but the location of $(E_1 = E^i + E^r)$ maximum or minimum shall interchange with respect to the case $\Gamma > 0$.

The ratio of the maximum value to minimum value of the electric field intensity of a standing wave is called *standing wave ratio* (SWR), S.

$$S = \frac{|E_1|_{max}}{|E_1|_{min}} = \frac{|H_1|_{max}}{|H_1|_{min}} = \frac{1 + |\Gamma|}{1 - |\Gamma|} \quad (11.109)$$

An inverse relation of Eq. (11.109) is

$$|\Gamma| = \frac{S - 1}{S + 1} \quad (11.110)$$

while the value of Γ ranges from -1 to $+1$, the value of S will range from 1 to ∞. It is customary to express S on a logarithmic scale. The standing wave ratio in decibels (dB) is $20\log_{10}S$.

$|H_1|$ minimum occurs wherever there is $|E_1|$ maximum and vice versa.

The transmitted wave in medium 2 is purely a travelling wave and there is no maxima and minima in this region.

EXAMPLE 11.7 Determine the amplitudes of reflected and transmitted **E** and **H** at the interface between two regions. The characteristics of region 1 are $\varepsilon_{r_1} = 8$ and $\mu_{r_1} = 1$ and $\sigma_1 = 0$, and region 2 is free space. The incident E_0^i in region 1 is of 1.5 V/m. Assume normal incidence as shown in Figure 11.11. Also find the average power in two regions.

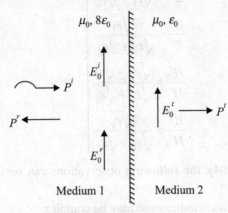

Figure 11.11

Solution
$$\eta_1 = \sqrt{\frac{\mu_1}{\varepsilon_1}} = \sqrt{\frac{4\pi \times 10^{-7}}{8 \times 8.854 \times 10^{-12}}} = 133.3\,\Omega$$

$$\eta_2 = \sqrt{\frac{\mu_0}{\varepsilon_0}} = 377\,\Omega$$

$$E_0^r = \frac{\eta_2 - \eta_1}{\eta_1 + \eta_2} E_0^i = 0.716 \text{ V/m}$$

$$E_0^t = \frac{2\eta_2}{\eta_1 + \eta_2} E_0^i = 2.216 \text{ V/m}$$

$$H_0^i = \frac{E_0^i}{\eta_1} = 1.13 \times 10^{-12} \text{ A/m}$$

$$H_0^r = \frac{\eta_1 - \eta_2}{\eta_1 + \eta_2} H_0^i = -5.4 \times 10^{-3} \text{ A/m}$$

$$H_0^t = \frac{2\eta_1}{\eta_1 + \eta_2} H_0^i = 5.90 \times 10^{-3} \text{ A/m}$$

The incident average power densities in two region:
$$P_1^i = \frac{1}{2} E_0^i H_0^i = 8.46 \times 10^{-3} \text{ W/m}^2$$

$$P_1^r = \frac{1}{2} E_0^r H_0^r = 1.93 \times 10^{-3} \text{ W/m}^2$$

$$P_1^t = P_2^i = \frac{1}{2}E_0^t H_0^t = 6.53 \times 10^{-3} \text{ W/m}^2$$

The energy is conserved

$$P_1^i = P_1^r + P_2^t$$

FORMULAE AT A GLANCE

11.1 Maxwell's equations in differential form.
 (a) $\nabla \cdot \mathbf{D} = \rho$ (b) $\nabla \cdot \mathbf{B} = 0$
 (c) $\nabla \times \mathbf{E} = -\dfrac{\partial \mathbf{B}}{\partial t}$ (d) $\nabla \times \mathbf{H} = \mathbf{J} + \dfrac{\partial \mathbf{D}}{dt}$

where $\mathbf{D}$ = electric displacement vector in C/m^2
 ρ = charge density in C/m^3
 $\mathbf{B}$ = magnetic induction in Wb/m^2 or Tesla
 $\mathbf{E}$ = electric field intensity in V/m or N/C
 $\mathbf{H}$ = magnetic field intensity A/m

11.2 Maxwell's equation in integral form
 (a) $\int_S \mathbf{D} \cdot d\mathbf{S} = \int_V \rho dV$ (b) $\int_S \mathbf{B} \cdot d\mathbf{S} = 0$
 (c) $\oint \mathbf{E} \cdot d\mathbf{l} = -\dfrac{\partial}{\partial t}\int_S \mathbf{B} \cdot d\mathbf{S}$ (d) $\oint \mathbf{H} \cdot d\mathbf{l} = \int_S \left(\mathbf{J} + \dfrac{\partial \mathbf{D}}{\partial t}\right) d\mathbf{S}$

11.3 Maxwell's equations in free space.
 (a) $\nabla \cdot \mathbf{D} = 0$ (b) $\nabla \cdot \mathbf{B} = 0$
 (c) $\nabla \times \mathbf{E} = -\dfrac{\partial \mathbf{B}}{\partial t}$ (d) $\nabla \times \mathbf{H} = \dfrac{\partial \mathbf{D}}{dt}$

With constitutive relations $\mathbf{D} = \varepsilon_0 \mathbf{E}$ and $\mathbf{B} = \mu_0 \mathbf{H}$.
Where ε_0 = absolute permittivity and μ_0 = absolute permeability.

11.4 Maxwell's equation in linear isotropic medium.
 (a) $\nabla \times \mathbf{E} = \dfrac{\rho}{\varepsilon}$ (b) $\nabla \cdot \mathbf{B} = 0$
 (c) $\nabla \times \mathbf{E} = \mu \dfrac{\partial \mathbf{H}}{\partial t}$ (d) $\nabla \times \mathbf{H} = \varepsilon \dfrac{\partial \mathbf{E}}{\partial t} + J$

11.5 Maxwell's equation for harmonically varying field.
 (a) $\nabla \cdot \mathbf{D} = \rho$ (b) $\nabla \cdot \mathbf{B} = 0$
 (c) $\nabla \times \mathbf{E} + i\omega \mathbf{B} = 0$ (d) $\nabla \times \mathbf{H} - i\omega \mathbf{D} = \mathbf{J}$

11.6 Energy stored

(a) In electric field $U_e = \dfrac{1}{2}\int_V \mathbf{E}\cdot d V$

(b) In magnetic field $U_m = \dfrac{1}{2}\int_V \mathbf{H}\cdot \mathbf{B}\, dV$

11.7 *Poynting theorem*

Electromagnetic energy theorem (conservation of electromagnetic energy).

This theorem states that "the rate of change of electromagnetic energy within a volume V is equal to the power radiated in the form of electromagnetic radiation through surface S enclosing volume V, plus rate of work done on the free charges in volume V".

$$-\dfrac{d}{dt}\int_V \dfrac{1}{2}(\mathbf{E}\cdot\mathbf{D}+\mathbf{B}\cdot\mathbf{H})\, dV = \int_S (\mathbf{E}\times\mathbf{H})\cdot d\mathbf{S} + \int_V (\mathbf{J}\cdot\mathbf{E})\, dV$$

or

$$\dfrac{\partial(U_{em}+U_m)}{\partial t} + \nabla\times\mathbf{S} = 0$$

11.8 Poynting vector

$$\mathbf{S} = \mathbf{E}\times\mathbf{H} = \dfrac{E^2}{Z_0}\hat{n}$$

Poynting vector $\mathbf{S}$ is defined as electromagnetic energy flowing per unit area per unit time $\left(\dfrac{\text{energy}}{\text{area.sec}}\right)$.

11.9 Velocity of plane electromagnet wave in free space

$$v = \dfrac{1}{\sqrt{\mu_0\varepsilon_0}} \cong 3\times 10^8\, \text{m/s} \cong c$$

= velocity of light in vacuum

$\mu_0 = 4\pi\times 10^{-7}$ Wb/Am

$\varepsilon_0 = 8.542\times 10^{-12}$ F/m.

11.10 Wave impedance of free space

$$Z_0 = \left|\dfrac{\mathbf{E}}{\mathbf{H}}\right| = \left|\dfrac{E_0}{H_0}\right| = \mu_0 c$$

$$= \sqrt{\dfrac{\mu_0}{\varepsilon_0}} \qquad \left[c = \dfrac{1}{\sqrt{\mu_0\varepsilon_0}}\right]$$

$$= 376.6\ \Omega$$

11.11 Average value of Poynting vector;

$$<\mathbf{S}> = \frac{1}{Z_0} E_{rms}^2 \,\hat{\mathbf{n}}$$

11.12 Wave propagation in lossy dielectric medium

(a) Attenuation constant $\alpha = \sqrt{\dfrac{\mu\varepsilon}{2}\left(\sqrt{1+\left(\dfrac{\sigma}{\omega\varepsilon}\right)^2} - 1\right)}$

(b) Phase constant $\beta = \sqrt{\dfrac{\mu\varepsilon}{2}\left(\sqrt{1+\left(\dfrac{\sigma}{\omega\varepsilon}\right)^2} + 1\right)}$

(c) Intrinsic impedance $\eta = \sqrt{\dfrac{i\omega\mu}{(\sigma + i\omega\varepsilon)}}$

11.13 Wave propagation in good dielectrics

(a) Attenuation constant $\alpha = \dfrac{\sigma}{2}\sqrt{\dfrac{\mu}{\varepsilon}}$

(b) Phase constant $\beta = \omega\sqrt{\mu\varepsilon}\left(1 + \dfrac{\sigma^2}{8\omega^2\varepsilon^2}\right)$

(c) Velocity of wave $v = \dfrac{\omega}{\beta} = \dfrac{1}{\sqrt{\mu\varepsilon}\left(1 + \dfrac{\sigma^2}{8\omega^2\varepsilon^2}\right)} = \dfrac{1}{\sqrt{\mu\varepsilon}}\left(1 - \dfrac{\sigma^2}{8\omega^2\varepsilon^2}\right)$

(d) Intrinsic impedance $\eta = \sqrt{\dfrac{\mu}{\varepsilon}}\left(1 + i\dfrac{\sigma}{2\omega\varepsilon}\right)$

11.14 Wave propagation in good conductor

(a) Attenuation constant $\alpha = \sqrt{\dfrac{\omega\mu\sigma}{2}} = \sqrt{\pi\nu\mu\sigma}$

(b) Phase constant $\beta = \alpha = \sqrt{\dfrac{\omega\mu\sigma}{2}} = \sqrt{\pi\nu\mu\sigma}$

(c) Velocity of wave $v = \dfrac{\omega}{\beta} = \sqrt{\dfrac{2\omega}{\mu\sigma}}$

(d) Intrinsic impedance $\eta = \sqrt{\dfrac{\omega\mu}{\sigma}} \angle 45°$

(e) Skin depth $\delta = \dfrac{1}{\alpha} = \sqrt{\dfrac{2}{\omega\mu\sigma}} = \dfrac{1}{\sqrt{\pi\nu\mu\sigma}}$

11.15 Reflection of uniform plane waves

If μ_1, ε_1 and μ_2, ε_2 are the permeability, permittivity of two media 1 and 2 respectively, then

(a) Intrinsic impedance of medium 1 is $\eta_1 = \sqrt{\dfrac{\mu_1}{\varepsilon_1}}$

(b) Intrinsic impedance of medium 2 is $\eta_2 = \sqrt{\dfrac{\mu_2}{\varepsilon_2}}$

(c) Reflection coefficient $\Gamma = \dfrac{\eta_2 - \eta_1}{\eta_2 + \eta_1}$

(d) Transmission coefficient $\tau = 1 + \dfrac{\eta_2 - \eta_1}{\eta_2 + \eta_1} = \dfrac{2\eta_2}{\eta_2 + \eta_1}$

(e) $\tau = 1 + \Gamma$

(f) Standing wave ratio (SWR) $S = \dfrac{1 + |\Gamma|}{1 - |\Gamma|}$

(g) $|\Gamma| = \dfrac{S - 1}{S + 1}$

SOLVED NUMERICAL PROBLEMS

PROBLEM 11.1 The electric field intensity of a uniform plane wave in air is 7500 V/m in the $\hat{j}$ direction. The wave is propagating in the $\hat{i}$ direction at a frequency of 2×10^9 rad/s. Find (a) the wavelength, (b) the frequency, (c) the time period, and (d) the amplitude of **H**.

Solution
$$E_y = 7500 \cos(2 \times 10^9 t - \beta x) \text{ in air}$$
$$\varepsilon_0 = 8.854 \times 10^{-12} \text{ F/m}, \quad \mu_0 = 4\pi \times 10^{-7} \text{ H/m}$$

(a) $\lambda = \dfrac{v}{\nu} = \dfrac{3 \times 10^8}{(2 \times 10^9 / 2\pi)} = 0.943$ m

(b) $\nu = \dfrac{\omega}{2\pi} = \dfrac{2 \times 10^9}{2\pi} = 318.3$ MHz

(c) $T = \dfrac{1}{\nu} = \dfrac{10^{-6}}{318.3} = 3.142$ ns

(d) $\dfrac{E}{H} = \eta = \sqrt{\dfrac{\mu_0}{\varepsilon_0}} \cong 377 \ \Omega$, E_y and H_z components shall exist.

$|H| = \dfrac{7500}{377} = 19.9$ A/m

$H_z = 19.9 \cos(2 \times 10^9 t - \beta x)$

PROBLEM 11.2 A lossless dielectric medium has $\sigma = 0$, $\mu_r = 1$ and $\varepsilon_r = 4$. An electromagnetic wave has magnetic field components expressed as

$$\mathbf{H} = -0.1\cos(\omega t - z)\hat{i} + 0.5\sin(\omega t - z)\hat{j} \text{ A/m}$$

Find
(a) the phase constant β,
(b) the angular velocity,
(c) the wave impedance, and
(d) the components of the electric field intensity of the wave.

Solution Given

$$\sigma = 0, \quad \mu_r = 1$$
$$\varepsilon_r = 4, \quad \alpha = 0 \quad \text{for lossless dielectric}$$

(a) Looking at the equations of **H**, we can find β directly. $\beta = 1$ rad/m
[as $H = H_m \cos(\omega t - \beta_z)$ type]

(b) $\beta = \omega\sqrt{\mu\varepsilon} = \omega\sqrt{\mu_0\varepsilon_0}\sqrt{\varepsilon_r\mu_r} = \dfrac{\omega}{c}\sqrt{4} = \dfrac{2\omega}{c}$

$$\omega = \frac{\beta c}{2} = \frac{1 \times 3 \times 10^8}{2} = 1.5 \times 10^8 \text{ rad/s}$$

(c) Wave impedance $\eta = \sqrt{\dfrac{\mu}{\varepsilon}} = \sqrt{\dfrac{\mu_0}{\varepsilon_0}}\sqrt{\dfrac{\mu_r}{\varepsilon_r}} = \dfrac{120\pi}{2} = 60\pi \ \Omega$

(d) $\mathbf{H} = -0.1\cos(\omega t - z)\hat{i} + 0.5\sin(\omega t - z)\hat{j}$ A/m

The wave is travelling in z-direction and has components of **H** in $\hat{i}$ and $\hat{j}$-directions, H_x and H_y, respectively, varying with respect to z. For finding **E**, we use Maxwell's equation for a lossless medium.

$$\nabla \times \mathbf{H} = \varepsilon\frac{\partial \mathbf{E}}{\partial t} \qquad [\because \sigma = 0]$$

$$\nabla \times \mathbf{H} = \begin{vmatrix} \hat{i} & \hat{j} & \hat{k} \\ \dfrac{\partial}{\partial x} & \dfrac{\partial}{\partial y} & \dfrac{\partial}{\partial z} \\ H_x & H_y & H_z \end{vmatrix} = -\frac{\partial H_y}{\partial z}\hat{i} + \frac{\partial H_x}{\partial z}\hat{j}$$

$$\nabla \times \mathbf{H} = 0.5\cos(\omega t - z)\hat{i} - 0.1\sin(\omega t - z)\hat{j}$$

$$E = \frac{1}{\varepsilon}\int(\nabla\times\mathbf{H})dt = \frac{0.5}{\varepsilon\omega}\sin(\omega t - z)\hat{i} + \frac{0.1}{\varepsilon\omega}\cos(\omega t - z)\hat{j}$$

or
$$E = 94.12\sin(\omega t - z)\hat{i} + 18.83\cos(\omega t - z)\hat{j} \text{ V/m}$$

PROBLEM 11.3 Find the skin depth δ of an electromagnetic wave in copper at $\nu = 60$ Hz. For copper $\sigma = 5.8 \times 10^7$ mho/m, $\mu_r = 1$ and $\mu_r = 1$.

Solution For copper at $\nu = 60$ Hz

$$\frac{\sigma}{\omega\varepsilon} = \frac{5.8 \times 10^7}{2\pi \times 60 \times 8.854 \times 10^{-12}}$$

$$= 1.74 \times 10^{14} \gg 1$$

Therefore, at $\eta = 60$ Hz, copper is a very good conductor. The depth of penetration

$$\delta = \frac{1}{\beta} = \sqrt{\frac{2}{\omega\mu\sigma}}$$

$$= \sqrt{\frac{2}{2\pi \times 60 \times 4\pi \times 10^{-7} \times 5.8 \times 10^7}}$$

$$= 8.53 \times 10^{-3} \text{ m}$$

PROBLEM 11.4 Starting from Maxwell's equations, establish Coulomb's law.

Solution From Maxwell's first equation,

$$\text{div } \mathbf{D} = \rho$$

Taking its volume integral over a sphere of radius r

We get,

$$\int_V \text{div } \mathbf{D} \, dV = \int_V \rho \, dV$$

As $\int_V \rho \, dV$ = net charge enclosed by sphere = q (say)

$\therefore$

$$\int_V \text{div } \mathbf{D} \, dV = q$$

Charging volume integral into surface integral, we get

$$\int_S \mathbf{D} \cdot d\mathbf{S} = q$$

But $\mathbf{D} = \varepsilon \mathbf{E}$

$\therefore$

$$\int_S \varepsilon \mathbf{E} \cdot d\mathbf{S} = q \quad \Rightarrow \quad \varepsilon \int \mathbf{E} \cdot d\mathbf{S} = q$$

$\Rightarrow$

$$\varepsilon E 4\pi r^2 = q \quad \Rightarrow \quad E = \frac{1}{4\pi\varepsilon} \frac{q}{r^2}$$

In vector form,

$$\mathbf{E} = \frac{1}{4\pi\varepsilon} \frac{q\hat{r}}{r^3}$$

Force on charge q_0 will be

$$F = q_0 \mathbf{E} = \frac{1}{4\pi\varepsilon} \frac{qq_0}{r^2} \hat{r}$$

This is Coulomb's law.

PROBLEM 11.5 Show that equation of continuity $\text{div } \mathbf{J} + \frac{\partial}{\partial t} = 0$ is contained in Maxwell's equations.

Or

Starting from Maxwell's equations, establish the equation of continuity.

Solution From Maxwell's 4th equation

$$\text{curl } \mathbf{H} = \mathbf{J} + \frac{\partial \mathbf{D}}{\partial t} \qquad \text{(i)}$$

Taking divergence of either sides of Eq. (i), we get

$$\text{div (curl } \mathbf{H}) = \text{div}\left(\mathbf{J} + \frac{\partial \mathbf{D}}{\partial t}\right) \qquad \text{(ii)}$$

But div (curl $\mathbf{H}$) = 0, since divergence of curl of any vector always vanishes, therefore, Eq. (ii) gives

$$\text{div}\left(\mathbf{J} + \frac{\partial \mathbf{D}}{\partial t}\right) = 0$$

or

$$\text{div } \mathbf{J} + \text{div} \frac{\partial \mathbf{D}}{\partial t} = 0$$

or

$$\text{div } \mathbf{J} + \frac{\partial \mathbf{D}}{\partial t}(\text{div } \mathbf{D}) = 0 \qquad \text{(iii)}$$

(Since space and time operations are interchangeable)

Also, from Maxwell's first equation div $\mathbf{D} = \rho$, therefore Eq. (iii) gives

$$\text{div } \mathbf{J} + \frac{\partial \rho}{\partial t} = 0 \qquad \text{(iv)}$$

This is required result.

PROBLEM 11.6 Starting from Maxwell's equations

$$\text{curl } \mathbf{E} = -\frac{\partial \mathbf{B}}{\partial t} \text{ and curl } \mathbf{H} = \mathbf{J} + \frac{\partial \mathbf{D}}{\partial t}$$

respectively, show that div $\mathbf{B} = 0$ and div $\mathbf{D} = \rho$.

Solution Given $\text{curl } \mathbf{E} = -\frac{\partial \mathbf{B}}{\partial t}$

Taking divergence of both sides

$$\text{div (curl } \mathbf{E}) = \text{div}\left(\frac{\partial \mathbf{D}}{\partial t}\right)$$

As divergence curl of any vector is zero and space and time operations are interchangeable

$$\text{div} \frac{\partial \mathbf{B}}{\partial t} = 0$$

$\Rightarrow$

$$\frac{\partial}{\partial t}(\text{div } \mathbf{B}) = 0$$

As isolated magnetic poles do not exist in nature, therefore for div **B** = 0.

$$\text{div } \mathbf{B} = \text{constant}$$

Given

$$\text{curl } \mathbf{H} = \mathbf{J} + \frac{\partial \mathbf{D}}{\partial t}$$

Taking divergence of both sides,

$$\text{div}(\text{curl } \mathbf{H}) = \text{div}\left(\mathbf{J} + \frac{\partial \mathbf{D}}{\partial t}\right)$$

As div (curl **H**) = 0

$$\Rightarrow \quad \text{div } \mathbf{J} + \text{div} \frac{\partial \mathbf{D}}{\partial t} = 0$$

$$\text{div } \mathbf{J} = -\text{div} \frac{\partial \mathbf{D}}{\partial t}$$

But

$$\text{div } \mathbf{J} + \frac{\partial \rho}{\partial t} = 0 \quad \text{(Equation of continuity)}$$

∴

$$-\text{div}\left(\frac{\partial \mathbf{D}}{\partial t}\right) + \frac{\partial \rho}{\partial t} = 0$$

$$\Rightarrow \quad \frac{\partial}{\partial t}(\text{div } \mathbf{D}) = \frac{\partial \rho}{\partial t}$$

On integrating div **D** = ρ.

PROBLEM 11.7 The relative permittivity of distilled water is 81. Calculate refractive index and velocity of light in it.

Solution If μ and ε are permeability and permittivity of a medium then the velocity of light in that medium is given by

$$v = \frac{1}{\sqrt{\mu \varepsilon}}$$

where $\mu = \mu_r \mu_0$
μ_r = relative permeability and
μ_0 = permeability of free space
$\varepsilon = \varepsilon_0 \varepsilon_r$
ε_0 = permittivity of free space and
ε_r = relative permittivity

∴

$$v = \frac{1}{\sqrt{\mu \varepsilon}} = \frac{1}{\sqrt{\mu_0 \varepsilon_0}} \times \frac{1}{\sqrt{\mu_r \varepsilon_r}} = \frac{1}{\sqrt{\mu_r \varepsilon_r}} \quad \text{(i)}$$

Here $\frac{1}{\sqrt{\mu_0 \varepsilon_0}} = c$, the velocity of light in vacuum

$$= 3 \times 10^8 \text{ m/s}$$

$\mu_r = 1$ [For non-magnetic material] and $\varepsilon_r = 81$ (given)

$$\therefore \quad v = \frac{3 \times 10^8}{\sqrt{81}} = 3.33 \times 10^7 \text{ m/s}$$

The refractive index of any medium is the ratio of the velocity of light in vacuum to the velocity of light in that medium.

$$\therefore \quad \text{Refractive index } (n) \frac{c}{v} = \sqrt{\mu_r \varepsilon_r}$$

$$= \sqrt{81} = 9.0 \qquad \text{[From Eq. (i)]}$$

PROBLEM 11.8 For silver $\sigma = 3.0$ MS/m. Calculate the frequency at which the depth of penetration is 1 mm. [Given $\mu_r = 1$]

Solution Since we know that $\delta = \sqrt{\dfrac{2}{\mu\sigma\omega}} = \sqrt{\dfrac{2}{\mu\sigma(2\pi v)}}$

$$\delta^2 = \frac{1}{\pi\mu\sigma v} \quad \text{or} \quad v = \frac{1}{\pi\mu\sigma\delta^2}$$

$$= \frac{1}{3.14 \times 3.0 \times 10^6 \times 1 \times 1 \times 10^6}$$

$$= 0.106 \text{ Hz} = 106 \text{ mHz}$$

PROBLEM 11.9 A plane electromagnetic wave travelling in the +ve z-direction in an unbounded lossless dielectric medium with relative permeability $\mu_r = r$ and relative permittivity $\varepsilon_r = 3$ has an electric field intensity $E = 6$ V/m. Find (i) speed of the electromagnetic waves in the given medium (ii) impedance of the medium.

Solution Given $\mu_r = 1$, $\varepsilon_r = 3$, $E_0 = 6$ V/m
but $\varepsilon = \varepsilon_r \varepsilon_0$ and $\mu = \mu_r \mu_0$ where ε and μ are the permittivity and permeability of the medium and ε_0 and μ_0 corresponding constant for free space.

(i) The speed of the electromagnetic waves in given medium

$$v = \frac{1}{\sqrt{\mu\varepsilon}} = \frac{c}{\sqrt{\mu_r \varepsilon_r}} = \frac{3 \times 10^8}{\sqrt{1 \times 3}} = \sqrt{3} \times 10^8 = 1.732 \times 10^8 \text{ m/s}$$

(ii) The impedance η of the medium is given by

$$\eta = \sqrt{\frac{\mu}{\varepsilon}} = \sqrt{\frac{\mu_r \mu_0}{\varepsilon_r \varepsilon_0}}$$

where $\mu_0 = 4\pi \times 10^{-7}$ H/m and $\varepsilon_0 = 8.86 \times 10^2$ F/m, we get

$$\eta = \sqrt{\frac{4\pi \times 10^{-7} \times 1}{8.86 \times 10^{-12} \times 3}} = 2.17 \times 10^2 \text{ } \Omega$$

PROBLEM 11.10 A medium like copper conductor which is characterized by the parameter $\sigma = 5.8 \times 10^{-7}$ mho/m, $\varepsilon_r = 1$, $\mu_r = 1$ supports a uniform plane wave of frequency 60 Hz. Find the attenuation constant, propagation constant, intrinsic impedance, wavelength and phase velocity of the wave.

Solution Let us obtain the ratio

$$\frac{\sigma}{\omega\varepsilon} = \frac{5.8 \times 10^7}{2\pi \times 60 \times 8.854 \times 10^{-12}} = 173 \times 10^{14}$$

This is very-very larger than 1, therefore, it is a very good conductor.

Attenuation constant $\alpha = \left(\dfrac{\omega\mu\sigma}{2}\right)^{1/2} = 117.2$ per m

Phase constant $\beta = \left(\dfrac{\omega\mu\sigma}{2}\right)^{1/2} = 117.2$ per m

Propagation constant $k = \alpha + j\beta = 117.2 + j117.2$ per m

Wavelength $\lambda = \dfrac{2\pi}{\beta} = \dfrac{2\pi}{117.2} = 0.0536$ m

Intrinsic impedance $\eta = \sqrt{\dfrac{j\omega\mu}{\sigma}} = (2.022 + j2.002) \times 10^{-6}\,\Omega$

Phase velocity of wave

$$v = \eta\nu = 0.0536 \times 60 = 3.216 \text{ m/s}$$

PROBLEM 11.11 Find the depth of penetration, δ of an EM wave in copper at $\nu = 60$ Hz and $\nu = 100$ Hz. For copper $\sigma = 5.8 \times 10^7$ mho/m, $\mu_0 = 1$, $\varepsilon_0 = 1$.

Solution For copper at $\nu = 60$ Hz.

$$\frac{\sigma}{\omega\varepsilon} = \frac{5.8 \times 10^7}{2\pi \times 60 \times 8.854 \times 10^{-12}}$$
$$= 174 \times 10^{14} \gg 1$$

Therefore at $\nu = 60$ Hz, copper is a very good conductor. The depth of penetration,

$$\delta = \frac{1}{\beta} = \sqrt{\frac{2}{\omega\mu\sigma}}$$

$$= \sqrt{\frac{2}{2\pi \times 60 \times 4\pi \times 10^{-7} \times 5.8 \times 10^7}}$$

or
$$\delta = 8.53 \times 10^{-3} \text{ m}$$

At $\nu = 100$ Hz.

$$\frac{\sigma}{\omega\varepsilon} = \frac{5.8 \times 10^7}{2\pi \times 8.854 \times 10^{-4}} = 0.10425 \times 10^{11}$$
$$= 10.425 \times 10^9 \gg 1$$

Copper is very good conductor at $v = 100$ MHz
The depth of penetration,

$$\delta = \frac{1}{\beta} = \sqrt{\frac{2}{\omega\mu\sigma}} = \sqrt{\frac{2}{2\times\pi\times 100\times 4\pi\times 10^{-7}\times 5.8\times 10^{7}}} = 6.608\times 10^{-6} \text{ m}$$

PROBLEM 11.12 The light is generally characterized by electric vector, although it also possesses the magnetic vector, why?

Solution Both electric and magnetic field vectors of the electromagnetic waves like light show their existence through the force exerted on the charged particles. The electric magnetic forces respectively are given by

$$\mathbf{F}_e = q\mathbf{E}, \ \mathbf{F}_m = q(\mathbf{v}\times\mathbf{B})$$

Their maximum magnitude are:

$$F_e = qE; \ F_m = qvB = qv\mu H$$

Here v is the velocity of the charged particles.

Now,
$$\frac{E}{H} = \sqrt{\frac{\mu}{\varepsilon}}$$

$\therefore$
$$F_m = \mu q v\sqrt{\frac{\varepsilon}{\mu}}E = q\frac{v}{c}E$$

$\therefore$
$$\frac{F_m}{F_e} = \frac{q(v/c)E}{qE} = \frac{v}{c}$$

Here c, the wave velocity which is very-very large as compared to the particle velocity v. Hence, F_m is very-very small as compared to F_e ($v \ll c$, $\therefore F_m \ll F_e$). Because of this reason, we characterize light through the electric vector.

PROBLEM 11.13 When the amplitude of magnetic field in a plane wave is 2A/m. (a) determine the magnitude of the electric field for the plane wave in free space. (b) determine the magnitude of electron field when the wave propagates in a medium which is characterized by $\sigma = 0$ and $\mu = \mu_0$ and $\varepsilon = 4\varepsilon_0$.

Solution (a) We have,

$$\frac{E}{H} = \eta_0\sqrt{\frac{\mu_0}{\varepsilon_0}} = 120 \ \pi\Omega \text{ for the free space}$$

But $H = 2$ A/m

$$E = \eta_0 H = 120\times 2 = 240 \ \pi \text{V/m}.$$

(b) $\sigma = 0$, $\varepsilon_0 = \mu_0 = 1$

$$\eta_0 = \sqrt{\frac{\mu}{\varepsilon}} = \sqrt{\frac{\mu_0}{\varepsilon_0}}$$

$$= \frac{1}{2}\sqrt{\frac{\mu_0}{\varepsilon_0}} = \frac{1}{2}\times 120 \ \pi\Omega = 60 \ \pi\Omega$$

$$E = \eta_0 H = 60\pi\times 2 = 120 \ \pi \text{ V/m}$$

PROBLEM 11.14 Assuming that all the energy from a 1000 W lamp is radiated uniformly, calculate the average values of intensities of electric and magnetic fields of radiations at a distance of 2 m from the lamp.

Solution If the total power P_0 is radiated uniformly in all directions, then the power or energy flux per unit area per second at a distance r from the point source (i.e., lamp) is

$$S_{av} = \frac{P_0}{4\pi r^2} = \frac{1000}{4\pi(2)^2} \text{ W/m}^2$$

From the definition of Poynting vector.

$$|\mathbf{S}| = (\mathbf{E} \times \mathbf{H}) = EH \sin 90°$$

$(\because$ **E** and **H** are perpendicular to each other$)$

$$EH = \frac{1000}{16\pi} (\Omega)^{-1} \quad \text{(i)}$$

But

$$\eta = \frac{E}{H} = \sqrt{\frac{\mu_0}{\varepsilon_0}} = \sqrt{\frac{4\pi \times 10^{-7}}{8.854 \times 10^{-12}}} = 376.72 \Omega \quad \text{(ii)}$$

Multiplying Eqs. (i) and (ii), we get

$$EH \cdot \frac{E}{H} = \frac{376.72 \times 1000}{16\pi}$$

or

$$E = \sqrt{\frac{376.72 \times 1000}{16\pi}} = 86.59 \text{ V/m}$$

From Eq. (i),

$$H = \frac{1000}{16\pi E} = \frac{1000}{16 \times 3.14 \times 86.59} = 0.23 \text{ A/m}$$

PROBLEM 11.15 Calculate the skin depth for an electromagnetic wave of frequency 1 MHz travelling through copper. Given $\sigma = 5.8 \times 10^7$ S/m and $\mu = 4\pi \times 10^{-7}$.

Solution We know that,

$$\delta = \sqrt{\frac{2}{\mu\sigma\omega}} = \sqrt{\frac{2}{\mu\sigma(2\pi\nu)}}$$

$$= \sqrt{\frac{2}{4\pi \times 10^{-7} \times 5.8 \times 10^7 \times 2 \times \pi \times 1 \times 10^6}}$$

$$= 0.0660 \times 10^{-3} \text{ m}$$

$$\delta = 66 \text{ μm}.$$

PROBLEM 11.16 Calculate the penetration depth for 2 MHz electromagnetic wave through copper. Given: $\sigma = 5.8 \times 10^7$ S/m, $\mu = 4\pi \times 10^{-7}$.

Solution We know that,

$$\delta = \sqrt{\frac{2}{\mu\sigma\omega}} = \sqrt{\frac{2}{\mu\sigma(2\pi v)}}$$

$$= \sqrt{\frac{2}{(4\pi \times 10^{-7} \times 5.8 \times 10^{7} \times 2 \times \pi \times 2 \times 10^{6})}}$$

$$= 0.0467 \times 10^{-3} = 46.7 \ \mu m$$

PROBLEM 11.17 Find the skin depth at a frequency 1.6 MHz in aluminium where σ = 38.2 MS/m and μ = 1.

Solution We know that,

$$\delta = \sqrt{\frac{2}{\mu\sigma\omega}} = \sqrt{\frac{2}{\mu\sigma(2\pi v)}}$$

$$= \sqrt{\frac{2}{1 \times 38.210^{6} \times 2 \times \pi \times 1.6 \times 10^{6}}}$$

$$= 0.0721 \times 10^{-6}$$

$$= 7.21 \times 10^{-8} \ m$$

CONCEPTUAL QUESTIONS

11.1 What do you understand by gradient of a scalar field? What is its significance?

Ans The gradient is a differential operator by means of which we can associate a vector field with a scalar field.

The gradient of a scalar field **S** is a vector whose magnitude at any point is equal to the maximum rate of increase of **S** at least point and whose direction is along the normal to the level surface at that point. This gives the physical significance of the gradient of scalar field.

11.2 What is divergence of a vector field? Write its physical significance.

Ans The divergence of a vector field at any point is defined as the amount of flux per unit volume diverging from that point.

The physical significance of divergence of a vector field is that at a point it gives the amount of flux per unit volume diverging from the point.

11.3 Define curl of a vector field.

Ans (i) The rate of change of a vector field is also called curl, which means circular rotation.
(ii) The curl of a vector field is defined as the maximum line integral of the vector per unit area.

11.4 Give statement of Gauss theorem.

Ans The electric flux (ϕ) through a closed surface is equal to $(1/\varepsilon_0)$ times the net charge (q) enclosed by the surface.

11.5 What do you mean by electric flux?

Ans The electric flux is defined as the number of lines of force that pass through a surface placed in the electric field. The electric flux ($d\phi$) through an elementary area dS is defined as the product of the area and the component of electric field strength normal to the area.

∴ The electric flux normal to the area $(dS) = d\phi = \mathbf{E}.d\mathbf{S}$.

11.6 Suppose that a Gaussian surface, encloses no net charge.
  (a) Does Gauss's law requires **E** equals zero for all points on the surface?
  (b) If the converse of this statement is true, i.e., if **E** equals zero everywhere on the surface, does Gauss's law require that there be no net charge inside?

Ans (a) When the Gaussian surface contains no net charge, Gauss's law becomes $\mathbf{E}.d\mathbf{S} = 0$, which does not require $\mathbf{E} = 0$. Here **E** and $d\mathbf{S}$ may be at right angles.
  (b) When $\mathbf{E} = 0$, everywhere on the surface, Gauss's law requires that there be no net charge inside.

11.7 What do you understand by displacement current?

Ans In one of the Maxwell's equations, the changing electric field term, is known as the displacement current **D**. This was an analogy with a dielectric material. If an electric material is placed in an electric field, the molecules are distorted, their positive charges moving slightly to the right, say, the negative charges slightly to the left. Now consider, what happens to dielectric in an increasing electric field. The positive charges will be displaced to the right by a continuous increasing distance, so, as long as the electric field is increasing and are moving the other way, adds to the effect of the positive charges motion.

11.8 What is absolute index of refraction?

Ans $n = \dfrac{c}{v} = \dfrac{1}{\sqrt{\varepsilon_0 \mu_0}} \sqrt{\varepsilon \mu} = \sqrt{\dfrac{\varepsilon \mu}{\varepsilon_0 \mu_0}} = \sqrt{\varepsilon_r \mu_r}$

11.9 Is gamma rays are electromagnetic radiations?

Ans Yes, these waves are electromagnetic radiations.

11.10 What is Faraday's laws of electromagnetism?

Ans Faraday's laws of electromagnetic induction are as below:

First law: Whenever the magnetic flux linked with a closed circuit changes, an induced e.m.f. is setup in the circuit whose magnitude, at any instant, is proportional to the rate of change of magnetic flux ϕ_B linked with the circuit.

i.e., $e \propto -\dfrac{d\phi_B}{dt}$

Second law (Lenz's law): The direction of induced e.m.f. or current in the circuit is such that it opposes the change in flux that produced it.

11.11 What is Ampere's circuital law?

Ans Ampere's law in magnetostatics is analogous to Gauss's law in electrostatics.

The line integral $\int \mathbf{B} \cdot d\mathbf{l}$ of magnetic induction **B** for a closed path is numerically equal to μ_0 times the current through the area bounded by the path. i.e., $\int \mathbf{B} \cdot d\mathbf{l} = \mu_0 i$, where μ_0 = permeability constant.

11.12 What do you understand by electromagnetic waves?

Ans Electromagnetic waves are coupled electric field and magnetic field vectors that move with the speed of light and exhibit typical wave behaviour i.e.,
   (i) they travel with speed of light
   (ii) they are transverse in nature
   (iii) the ratio of electric to magnetic field vectors in an electromagnetic wave equals to the speed of light
   (iv) they carry both energy and momentum

11.13 Write down Maxwell's equations.

Ans There are four fundamental equations of electromagnetism known as Maxwell's equations, which may be written in differential form as:
   (i) $\nabla \cdot \mathbf{D} = \rho$ (Differential form of Gauss's law in electrostatics)
   (ii) $\nabla \cdot \mathbf{B} = 0$ (Differential form of Gauss's law in magnetostatics)
   (iii) $\nabla \times \mathbf{E} = -\dfrac{\partial \mathbf{B}}{\partial t}$ (Differential form of Faraday's law in electromagnetic induction)
   (iv) $\nabla \times \mathbf{H} = \mathbf{J} + \dfrac{\partial \mathbf{D}}{\partial t}$ (Maxwell's modification in Ampere's law)

In above equations the notations have the following meanings :
$\mathbf{D}$ = electric displacement vector in C/m^2; ρ = charge density in C/m^3
$\mathbf{B}$ = magnetic induction in Wb/m^2 or tesla
$\mathbf{E}$ = electric field intensity in V/m or N/C
$\mathbf{H}$ = magnetic field intensity or strength in A/m

11.14 What do you understand by Poynting vector?

Ans Poynting vector is defined as the vector product of electric field and magnetic field vectors, which gives the time rate of flow of wave energy per unit area of the medium.

11.15 Why does not one see the portion other than visible one of electromagnetic spectrum?

Ans The retina of the eye is sensitive only to colours in the visible region i.e., wavelength lying between 3900 Å and 7800 Å. This region corresponds to visible part of the spectrum.

11.16 An electromagnetic wave carries momentum. What it signifies?

Ans An electromagnetic interaction between two electric charges means an exchange of energy and momentum between the charges.

11.17 A plane monochromatic wave, travelling in a homogeneous medium, meets a denser medium. What are the changes in (i) amplitude, (ii) frequency, (iii) speed of propagation, and (iv) phase of reflected and transmitted waves in comparison with the corresponding properties of the incident wave?

Ans (i) Amplitude decreases for both cases.
   (ii) Frequency remains same for both cases.
   (iii) Speed same for reflected wave and less for transmitted waves in comparison to reflected waves.
   (iv) Phase change by π for reflected wave. No phase change for transmitted wave.

11.18 What is difference between the propagation of electromagnetic waves in free space and the propagation of electromagnetic waves in conducting medium?

Ans

Propagation of electromagnetic waves in free space	Propagation of electromagnetic waves in conducting medium
Electromagnetic waves travel with speed of light in free space.	Electromagnetic waves in a conducting medium travel with a speed less than the speed of light.
In free space, the electric and magnetic vectors of electromagnetic waves are in phase.	In a conducting medium, the magnetic vector of electromagnetic wave lags in phase with respect to electric vector.
In free space, there is no attenuation in energy of electromagnetic waves.	In a conducting medium, the energy of electromagnetic waves is damped exponentially.
In free space, the electrostatic energy density of electromagnetic waves is equal to its magnetostatic energy density.	In a conducting medium, the magnetostatic energy density of electromagnetic waves is much greater than its electrostatic energy density.

EXERCISES

Theoretical Questions

11.1 Derive the wave equation from Maxwell's equations for free space and charge-free region.

11.2 Solve the wave equation for a uniform plane wave in an isotropic homogeneous lossy dielectric medium with no sources. Explain the terms propagation constant, attenuation constant and phase constant.

11.3 Show that for uniform plane waves in a perfect dielectric medium, **E** and **H** are normal to each other and the ratio of their magnitude is constant of the medium. What is the name of this constant? What is its significance?

11.4 What do you understand by skin effect? Define skin depth. Show that in case of a semiinfinite solid conductor, the skin depth (δ) is given by

$$\delta = \sqrt{\frac{2}{\omega\mu\sigma}}$$

Symbols have their usual meanings.

11.5 State and prove Poynting theorem. Explain the term instantaneous, average and complex Poynting vectors.

11.6 Using Maxwell's equations show that in free space, the propagation of an electromagnetic wave is given by

$$v = \frac{1}{\sqrt{\mu_0 \varepsilon_0}} = 3 \times 10^8 \text{ m/s}$$

11.7 Enumerate Maxwell's equations and show that they predict existence of electromagnetic waves.

11.8 From the electromagnetic point of view, discuss of the following:
   (a) Conductors
   (b) Insulators or dielectrics

11.9 Obtain the wave equation for a plane electromagnetic waves in a dielectric medium and show that its velocity of propagation is less than speed of light.

11.10 Obtain the equation of plane electromagnetic wave in a conducting medium.

11.11 Discuss the propagation of monochromatic plane electromagnetic wave in a conducting medium.

11.12 Define skin depth. Show that in case of good conductor, the skin depth is given by
$$\delta = \left(\frac{2}{\omega\sigma\mu}\right)^{1/2}.$$

11.13 Show that the electromagnetic waves are transverse in nature.

11.14 State and prove Poynting theorem. Give physical significance of Poynting vector.

11.15 Show that inside the conducting medium electromagnetic wave is damped and obtains an expression for skin depth.

11.16 Derive the wave equation using Maxwell's equations for a free space. Hence conclude that
   (i) The electric and magnetic vectors can be represented as an oscillatory wave with a given amplitude.
   (ii) The speed with which the electric and magnetic vectors are propagated in free space is equal to the speed of light.

11.17 Show that the velocity of the electromagnetic wave in a dielectric medium is always less than the velocity of free space.

Numerical Problems

11.1 Given $E(x, t) = 10^3 \sin(6 \times 10^8 t - \beta x)\hat{j}$ V/m in free space, sketch the wave at $t = 0$ and at time t_1, when it has travelled $\lambda/4$ along the x-axis. Find t_1, β and λ.
[**Ans** $t_1 = 2.62 \times 10^{-9}$s, $\beta = 2$ rad/m, $\lambda = \pi$]

11.2 A uniform plane wave in a medium having $\sigma = 10^{-3}$ S/m, $\varepsilon_r = 80 \varepsilon_0$ and $\mu_r = \mu_0$ is having a frequency 10 kHz. Calculate the different parameters of the wave.
[**Ans** $\alpha = 2\pi \times 10^{-3}$ Np/m, $\beta = 2\pi \times 10^{-3}$ rad/m, $\eta = 2\pi(1+j)\Omega$, $\lambda = 1000$ m. $v = 10^7$ m/s]

11.3 If the earth receives 2 cal/min/cm² solar energy, what are the amplitudes of the electric and magnetic fields of radiation?
[**Ans** $E_0 = E\sqrt{2} = 1024.3$ V/m, $H_0 = H\sqrt{2} = 2.717$ A/m]

11.4 The relative permittivity of distilled water is 81. Calculate the refractive index and the velocity of light in it.
[**Ans** $v = 3.33 \times 10^7$ m/s, $\mu = 9.0$]

11.5 A plane electromagnetic wave travelling in the positive z-direction in an unbounded lossless dielectric medium with relative permeability $\mu_r = 1$ and relative permittivity $\varepsilon_r = 3$ has an electric field intensity $E = 6$ V/m. Find

(i) the speed of the electromagnetic waves in the given medium, and
(ii) the impedance of the medium.

[**Ans** (i) $v = 1.732 \times 10^8$ m/s, (ii) $\eta = 2.17 \times 10^2 \Omega$]

11.6 Calculate the skin depth for an electromagnetic wave of frequency 1 MHz travelling through copper. Given $\sigma = 5.8 \times 10^7$ S/m and $\mu = 4\pi \times 10^{-7}$. [**Ans** $\delta = 66$ μm]

11.7 If the magnitude of **H** in a plane wave is 1.0 A/m. Find the magnitude of **E** for a plane wave in free space. [**Ans** $|E| = 377$ V/m]

11.8 Show that the energy stored in a magnetic field per unit volume in free space is $B^2/2\mu_0$.

11.9 Assuming that all the energy from a 1000 W lamp is radiated uniformly, calculate the average values of the intensities of electric and magnetic fields of radiation at a distance of 2 m from the lamp. [**Ans** $E = 48.87$ V/m, $H = 0.4$ A/m]

11.10 The ratio J/J_d (conduction current density to displacement current density) is very important at high frequencies. Calculate the ratio at 1 GHz for
(i) distilled water ($\mu = \mu_0$, $\varepsilon = 81\varepsilon_0$, $\sigma = 2 \times 10^{-3}$ S/m)
(ii) sea water ($\mu = \mu_0$, $\varepsilon = 81\varepsilon_0$, $\sigma = 25$ S/m)
(iii) lime-stone ($\mu = \mu_0$, $\varepsilon = 5\varepsilon_0$, $\sigma = 2 \times 10^{-4}$ S/m)

11.11 Assume that dry soil has $\sigma = 10^{-4}$ S/m, $\varepsilon = 3\varepsilon_0$. Determine the frequency at which the ratio of the magnitudes of the conduction current density and the displacement current density is unity. [**Ans** 600 kHz]

11.12 In a certain material
$\mu = \mu_0$, $\varepsilon = \varepsilon_0 k_e$ and $\sigma = 0$. If $H = 10 \sin(10^8 t - 2x)a_z$ A/m find J_d, E and ε_r.

11.13 In free space, $E = \dfrac{50}{\rho}\cos(10^8 t - kz)a_e$ V/m, find k, J_d and H.

[**Ans** 0.333, $\dfrac{-4.421 \times 10^{-2}}{\rho}\sin(10^8 t - kz)a_\phi$ A/m, and $\dfrac{2.5k}{2\pi\rho}\cos(10^8 t - kz)a_\phi$ A/m]

11.14 A certain material has $\sigma = 0$ and $\varepsilon_0 = 1$. If $H = 4 \sin(10^8 t - 0.01z)a_y$ A/m, make use of Maxwell's equations to find (a) μ_0 and (b) E.

11.15 Show that the skin depth in a poor conductor ($\sigma \ll \omega\varepsilon$) is $\dfrac{(2/\sigma)}{\sqrt{\varepsilon/\mu}}$ (independent frequency). Find the skin depth (in metres) for pure water.

11.16 Sea water at frequency $v = 4 \times 10^8$ Hz has permittivity $\varepsilon = 81\varepsilon_0$, permeability $\mu = \mu_0$ and resistivity $\rho = 0.23$ Ωm. What is the ratio of conduction current to displacement current?

11.17 If the average distance between the sun and earth is 1.5×10^{11} m, show that the average solar energy incident on the earth is 2 cal/cm^2/min (called solar constant).

11.18 A plane electromagnetic wave travelling in positive z-direction in an unbounded lossless dielectric medium with relative permeability $\mu_r = 1$ and relative permittivity $\varepsilon_r = 3$ has peak electric field intensity $E_0 = 6$ V/m. Find:
(i) the speed of the wave;
(ii) the intrinsic impedance of the medium;

(iii) the peak magnetic field intensity (H_0), and
(iv) the peak Poynting vector $S(z, t)$.

[**Ans** (i) 1.73×10^8 m/s; (ii) 217.6 Ω; (iii) 2.76×10^{-2} A/m and (iv) $S = 0.165$ W/m²]

11.19 A 1000 W radio station sends power in all directions from its antenna. Find E and B in its wave at a distance of 10 km from the antenna (B and E transport same amount of energy). [**Ans** $E = 0.024$ V/m, $B = 8.2 \times 10^{-11}$ T]

11.20 At what frequency may earth be considered a perfect dielectric, if $\sigma = 5 \times 10^{-3}$ S/m, $\mu = 1$, $\varepsilon = 8$? Calculate the value of attenuation constant at these frequencies. (take $\sigma/\omega\varepsilon \leq 0.01$) [**Ans** 1.13 GHz, 0.333 Np/m]

11.21 A radio station P is 100 km from Q. It radiates an average power of 50 kW. Assuming spherical radiation, find the value of S at Q. Calculate the values of E_0 and B_0 at an antenna located at Q.
[**Ans** $S = 0.4 \times 10^{-6}$ Wb/m², $E_0 = 1.7 \times 10^{-2}$ V/m and $B_0 = 5.7 \times 10^{-11}$ T]

11.22 Show that in the electromagnetic wave, the electrostatic energy is equal to the magnetic energy density. [**Ans** $U_e = U_m$]

11.23 The electric field intensity of a uniform plane wave in air is 8000 V/m in the y-axis. The wave is propagating along x-axis at the frequency 10^{10} rad/s. The compute:
(i) the wavelength λ,
(ii) the frequency ν,
(iii) the time period T, and
(iv) the amplitude of H

[**Ans** (i) $\lambda = 18.8 \times 10^{-2}$ m, (ii) $\nu = 1591.3$ MHz, (iii) $T = 0.63$ ns and (iv) 21.22 A/m]

11.24 Prove that
(i) $S_{av} = \mathbf{E}_{rms} \times \mathbf{H}_{rms}$
(ii) $I = S_{av} = \dfrac{1}{2} C \varepsilon_0 E_0^2 = \dfrac{1}{2} C \mu_0 H_0^2$

where I is the intensity of the electromagnetic wave.

11.25 Calculate the Poynting vector at the surface of the sun. Given that the energy radiated per second is 3.8×10^{26} J and radius of the sun is 0.7×10^9 m. Also calculate the amplitude of electric and magnetic field vectors on the surface of earth. The distance of the earth from the sun is 0.15×10^{12} m.
[**Ans** $S = 6.17 \times 10^7$ Jm⁻¹s⁻¹, $E_0 = 10^4$ V/m, $H_0 = 2.267$ A/m]

11.26 Show that the energy flux in a plane polarized electromagnetic wave in free space is the energy density times the velocity of the wave.

11.27 Find the energy stored in one metre length of a laser beam operating 1 mW.
[**Ans** $U = 3.3 \times 10^{-12}$ J]

11.28 Show that the energy density and Poynting vector of electromagnetic field are given by
$$U_e = \dfrac{1}{2}(\varepsilon_0 E^2 + \mu_0 H^2) \text{ and } \mathbf{S} = \mathbf{E} \times \mathbf{H}$$

11.29 Find the conduction and displacement current densities in a material having conductivity of 10^{-3} S/m and $\varepsilon_r = 2.5$, if the electric field in the material is

$$E = 5 \times 10^{-6} \sin 9.0 \times 10^9 \, t \text{ V/m}$$

[**Ans** $J_C = 5.0 \times 10^{-9} \sin 9.0 \times 10^9 \, t$ A/m^2 and $J_D = 1.00 \times 10^{-6} \cos 9.0 \times 10^9 \, t$ A/m^2]

11.30 A plane wave propagating in free space with a peak electric field of intensity 750 mV/m. Find average power through square area of 120 cm on a side perpendicular to the direction of propagation. [**Ans** $\rho_{av} = 10.7$ mW]

11.31 Show that in a conductor, the magnitude of the electric vector reduces to about 10% at a distance of 0.733 λ_c, where λ_c is the length of electromagnetic wave in the conductor.

11.32 The electric field in an electromagnetic wave is given by $E = E_0 \sin \omega(t - x/c)$, where $E_0 = 1000$ N/C. Find the energy contained in a cylinder of cross-section 10^{-3}m^3 and length 100 cm along the x-axis. [**Ans** 4.425×10^{-11} J]

11.33 A plane monochromatic linearly polarized light wave is travelling eastward. The wave is polarized with E directed vertically up and sown. Write expressions for E, H and B provided that $E_0 = 0.1$ V/m and frequency is 20 MHz.

[**Ans** $E_z = 0.1 \sin(4\pi \times 110^{-7} \, t - 0.419 \, x)$, $B_y = \dfrac{-E_z}{C}$, $H_y = \dfrac{B_y}{\mu_0}$]

11.34 The electric field in an electromagnetic wave is given by $E = 50 \sin \omega(t - x/c)$ N/C. Find the energy contained in a cylinder of cross-section 10 cm^2 and length 50 cm along the x-axis. [**Ans** 5.5×10^{-12} J]

11.35 Find the skin depth δ at a frequency of 3.0×10^6 Hz in aluminium where $\sigma = 38 \times 10^6$ S/m and $\mu = 1$. Also find out the propagation constant and wave velocity.

[**Ans** $\delta = 0.0416$ mm, $\gamma = 29.986 \times 10^3 \angle 45°$ m^{-1} and $v = 888.51$ m/s]

Multiple Choice Questions

11.1 The concept that a changing electric field in a conductor produces induced magnetic field was proposed by
(a) Faraday (b) Biot–Savart
(c) Maxwell (d) Oersted

11.2 The concept of displacement current was proposed by
(a) Maxwell (b) Faraday
(c) Ampere (d) Gauss

11.3 Maxwell's modified Ampere's law is valid
(a) only when electric field does not change with time
(b) only when electric field varies with time
(c) in the above situations
(d) none of the above

11.4 Which is the incorrect statement about the electromagnetic waves
(a) the electromagnetic field vectors **E** and **B** are mutually perpendicular and they are also perpendicular to the direction of propagation of the electromagnetic wave

(b) the field vectors **E** and **H** are in same phase
(c) the field vectors **E** and **H** are along the same direction
(d) electromagnetic waves are transverse in nature

11.5 Which one is the incorrect statement about the electromagnetic waves:
(a) the energy density associated with the electromagnetic wave in free space propagates with a speed less than the speed of light
(b) in free space, the electromagnetic waves travel with the speed of light
(c) the direction of flow of electromagnetic energy along the direction of propagation of wave
(d) the electrostatic energy density is equal to magnetic energy density

11.6 A monochromatic electromagnetic wave means that
(a) the field strength at a point varies with time according to sine and cosine function
(b) the wave always travels in the same direction
(c) electric field vector **E** lies in one direction only
(d) magnetic field vector **B** must be perpendicular to the direction of propagation

11.7 "The net power flowing out of a given volume V is equal to the time rate of decrease in energy stored within V minus the ohmic losses" is the statement of
(a) Gauss's theorem   (b) Stoke's theorem
(c) Poynting theorem   (d) none of these

11.8 The depth of penetration is the depth in which the electromagnetic wave has been attenuated to

(a) e of the original value   (b) $\dfrac{1}{e}$ of the original value

(c) 50% of the original value   (d) 100% of the original value

11.9 The skin depth for a good conductor is

(a) $\sqrt{\dfrac{2}{\omega\mu\sigma}}$ (b) $\sqrt{2\omega\mu\sigma}$

(c) $\sqrt{\dfrac{\omega\varepsilon\sigma}{2}}$ (d) $\dfrac{2}{\sqrt{\omega\mu\sigma}}$

11.10 The intrinsic impedance of good conductor is given by

(a) $\sqrt{\dfrac{\mu}{\varepsilon}\left(1+\dfrac{i\sigma}{2\omega\varepsilon}\right)}$ (b) $\sqrt{\dfrac{i\omega\mu}{\sigma}}$

(c) $\sqrt{\dfrac{i\omega\mu}{\sigma+i\omega\varepsilon}}$ (d) $\sqrt{i\omega\mu\sigma\left(1+i\dfrac{\omega\varepsilon}{\sigma}\right)}$

11.11 The displacement current can be represented as

(a) $i_d = \mu_0 \dfrac{d\phi_E}{dt}$ (b) $i_d = \varepsilon_0 \dfrac{d\phi_E}{dt}$

(c) $i_d = \mu_0 \dfrac{d\phi_B}{dt}$ (d) $i_d = \varepsilon_0 \dfrac{d\phi_B}{dt}$

11.12 Maxwell's modified Ampere's law is valid
(a) only when electric field does not change with time
(b) only when electric field varies with time
(c) in both of the above situations
(d) none of the above

11.13 The Maxwell's equation

$$\oint \mathbf{E} \cdot d\mathbf{l} = -\int \frac{\partial \mathbf{B}}{\partial t} \cdot d\mathbf{S}$$ is statement of

(a) Gauss's law (b) Ampere's law
(c) Faraday's law (d) Modified Ampere's law

11.14 The Maxwell's equation which interprets that isolated magnetic poles do not exist is
(a) $\nabla \cdot \mathbf{E} = \frac{\rho}{\varepsilon_0}$ (b) $\nabla \cdot \mathbf{B} = 0$
(c) $\nabla \times \mathbf{E} = -\frac{\partial \mathbf{B}}{\partial t}$ (d) $\nabla \times \mathbf{E} = -\mu_0 \mathbf{J} + \mu_0 \varepsilon_0 \frac{\partial \mathbf{E}}{\partial t}$

11.15 The Maxwell's equation which remains unchanged when a medium changes is
(a) $\nabla \cdot \mathbf{B} = 0$ (b) $\nabla \cdot \mathbf{E} = \frac{\rho}{\varepsilon_0}$
(c) $\nabla \times \mathbf{B} = -\mu_0 \mathbf{J} + \mu_0 \varepsilon_0 \frac{\partial \mathbf{E}}{\partial t}$ (d) none of these

11.16 In the following, which one is the consequence of the Maxwell's equations?
(a) $\mathbf{B} = \mu_0 \mathbf{H}$ (b) $\mathbf{D} = \varepsilon_0 \mathbf{E}$
(c) $\mathbf{D} = \varepsilon_0 \mathbf{H} + \mathbf{P}$ (d) $c = \frac{1}{\sqrt{\varepsilon_0 \mu_0}}$

11.17 A plane electromagnetic wave means that
(a) the field strength at a point varies with time according to sine and cosine function
(b) the wave always travels in the same direction
(c) electric field vector **E** lies in one direction only
(d) magnetic field vector **B** must be perpendicular to the direction of propagation

11.18 "The work done on the charges by the electromagnetic force is equal to the decrease in energy stored in the field, less the energy which flowed out through the surface" is the statement of
(a) Gauss's theorem (b) Stoke's theorem
(c) Pythagoras theorem (d) Poynting theorem

11.19 The energy per unit time, per unit area transported by the electromagnetic fields is expressed as
(a) $\mathbf{S} = \frac{1}{\mu_0}(\mathbf{E} \times \mathbf{B})$ (b) $\mathbf{S} = \mathbf{E} \times \mathbf{B}$
(c) $\mathbf{S} = \mu_0 (\mathbf{E} \times \mathbf{B})$ (d) $\mathbf{S} = \frac{1}{\varepsilon_0}(\mathbf{E} \times \mathbf{H})$

11.20 The total energy stored in electromagnetic fields is

(a) $\dfrac{\varepsilon_0}{2}\int E^2 dV$

(b) $\dfrac{1}{2\mu_0}\int B^2 dV$

(c) $\dfrac{1}{2}\int\left[\varepsilon_0 E^2 + \left(\dfrac{1}{\mu_0}\right)B^2\right]dV$

(d) $\dfrac{\varepsilon_0\mu_0}{2}\int E^2 B^2 dV$

11.21 A plane electromagnetic wave in a free space is represented by an equation

(a) $\nabla^2\mathbf{E} - \mu_0\varepsilon_0\dfrac{\partial^2 \mathbf{B}}{\partial t} = 0$

(b) $\nabla^2\mathbf{B} - \mu_0\varepsilon_0\dfrac{\partial^2 \mathbf{E}}{\partial t^2} = 0$

(c) $\nabla^2\mathbf{E} - \mu_0\varepsilon_0\dfrac{\partial^2 \mathbf{E}}{\partial t^2} = 0$

(d) $\nabla^2\mathbf{E} - \mu\sigma\dfrac{\partial^2 \mathbf{E}}{\partial t} + \mu\varepsilon\dfrac{\partial^2 \mathbf{E}}{\partial t^2} = 0$

11.22 For a static field in vacuum the correct form of the Maxwell's equation is

(a) $\nabla \times \mathbf{E} = 0$

(b) $\nabla \times \mathbf{E} = -\dfrac{\partial \mathbf{B}}{\partial t}$

(c) $\nabla \times \mathbf{E} = \mu_0\left[i + \varepsilon_0\left(\dfrac{\partial \mathbf{E}}{\partial t}\right)\right]$

(d) $\nabla \times \mathbf{E} = \mu_0\varepsilon_0\dfrac{\partial \mathbf{E}}{\partial t}$

11.23 Choose the incorrect Maxwell's equation for harmonically time varying fields.

(a) $\nabla \times \mathbf{E} = -j\omega\mathbf{B}$

(b) $\nabla \times \mathbf{H} = (\sigma + j\omega\varepsilon)\mathbf{E}$

(c) $\oint_S \mathbf{H}\cdot dl = (\sigma + j\omega\varepsilon)\int \mathbf{E}\cdot d\mathbf{S}$

(d) $\int_c \mathbf{E}\cdot dl = -\int_S \dfrac{\partial \mathbf{B}}{\partial t}\cdot d\mathbf{S}$

11.24 For a good conductor the skin depth is equal to

(a) $\sqrt{\dfrac{\omega\varepsilon\sigma}{2}}$

(b) $\sqrt{2\omega\mu\sigma}$

(c) $\sqrt{\dfrac{2}{\omega\mu\sigma}}$

(d) $\dfrac{2}{\sqrt{2\omega\mu\sigma}}$

11.25 The skin depth is
(a) directly proportional to attenuation constant, (α)
(b) inversely proportional to attenuation constant, (α)
(c) directly proportional to phase constant, (β)
(d) inversely proportional to phase constant, (β)

11.26 A good dielectric medium is one for which

(a) $\dfrac{\sigma}{\omega\varepsilon} \gg 1$

(b) $\dfrac{\sigma}{\omega\varepsilon} \ll 1$

(c) $\dfrac{\sigma}{\omega\varepsilon} = 1$

(d) $\dfrac{\sigma}{\omega\varepsilon} = 0$

11.27 A good conducting medium is one for which

(a) $\dfrac{\sigma}{\omega\varepsilon} \gg 1$ (b) $\dfrac{\sigma}{\omega\varepsilon} \ll 1$

(c) $\dfrac{\sigma}{\omega\varepsilon} = 1$ (d) $\dfrac{\sigma}{\omega\varepsilon} = 0$

11.28 A quasi-conducting medium is one for which

(a) $\dfrac{\sigma}{\omega\varepsilon} \gg 1$ (b) $\dfrac{\sigma}{\omega\varepsilon} \ll 1$

(c) $\dfrac{\sigma}{\omega\varepsilon} = 1$ (d) $\dfrac{\sigma}{\omega\varepsilon} = 0$

11.29 When an electromagnetic wave is propagated in good conductors, the velocity of wave propagation is reduced to

(a) $\left(\dfrac{2\omega}{\mu\sigma}\right)^{1/2}$ (b) $\dfrac{\sigma}{2}\sqrt{\dfrac{\mu}{\varepsilon}}$

(c) $\left(\dfrac{\omega\mu\sigma}{2}\right)^{1/2}$ (d) $\left(\dfrac{\omega\mu}{2}\right)^{1/2}$

Answers

11.1 (c) 11.2 (a) 11.3 (c) 11.4 (c) 11.5 (a) 11.6 (a) 11.7 (c) 11.8 (b)
11.9 (a) 11.10 (b) 11.11 (b) 11.12 (c) 11.13 (c) 11.14 (b) 11.15 (a) 11.16 (d)
11.17 (b) 11.18 (d) 11.19 (a) 11.20 (c) 11.21 (c) 11.22 (a) 11.23 (d) 11.24 (c)
11.25 (b) 11.26 (b) 11.27 (a) 11.28 (c) 11.29 (d)

Appendices

APPENDIX I

Refractive Index (μ) of substances at 15°C for *D*-line of sodium relative to Air

Substance (solid)	μ	Substance (liquid)	μ
Glass crown	1.50	Aniline	1.590
Glass flint	1.56	Benzene	1.504
Glass extra flint	1.65	Chloroform	1.530
Glass dense crown	1.62	Glycerin	1.449
Glass dense flint	1.62	Sulphuric acid	1.470
Diamond	2.417	Turpentine	1.430
Mica	1.56–1.69	Water	1.333
Sugar	1.56		
Quartz	1.544		

APPENDIX II

Specific Rotation of Important Optically Active Materials

Material	Solvent	Specific rotation (in degrees)
Cane sugar	Water	+66.5
Glucose	Water	+52.5
Fructose	Water	−91.5
Invert sugar	Water	−19.5
Camphor	Alcohol	+41.0
Turpentine	Pure	−37.0
Nicotine	Pure	−122.0

APPENDIX III

Wavelength of Important Spectral Lines (λ) [All in Å]

Hydrogen	Helium	Sodium	Mercury	Neon	Cadmium
6562–784	3889	5890–D_1 (Orange–yellow in colour)	4047 (Violet-1)	5765 (Yellow-1)	6438
4861–327	4026	5896–D_2 (Orange–yellow in colour)	4078 (Violet-2)	5853 (Yellow-2)	5085
4340–466	4471		4358 (Blue)	5882 (Orange)	4799
4101–736	5876		4916 (Bluish green)	6507	4678
			4960 (Green-1)		4662
			5461 (Green-2)		
			5770 (Yellow-1)		
			5791 (Yellow-2)		
			6152 (Red-1)		

APPENDIX IV

Dispersive Powers (ω) of Glasses (λ = 5893 Å)

Glass	ω
Crown	0.015
Dense crown	0.018
Flint	0.020
Dense flint	0.027
Extra dense flint	0.030
Very dense flint	0.033
Water	0.018

APPENDIX V

Accepted Values of Some Common Physical Constants

Physical constant	Value
Velocity of light in vacuum	$c = 3 \times 10^8$ ms^{-1}
Planck's constant	$h = 6.625 \times 10^{-34}$ Js
Boltzmann constant	$k_B = 1.381 \times 10^{-23}$ JK^{-1}
	$= 8.62 \times 10^{-5}$ eV K^{-1}
Universal gas constant	$N = 6.023 \times 10^{23}$ mol^{-1}
Gravitational constant	$G = 6.67 \times 10^{-11}$ Nm^2 kg^{-2}
Atomic mass unit	1 amu $= 1.66 \times 10^{-27}$ kg
Acceleration due to gravity (at sea level 45° altitude)	$g = 9.806$ ms^{-2}
Proton rest mass	$m_p = 1.672 \times 10^{-27}$ kg
Electron charge	$e = 1.602 \times 10^{-19}$ c
Electron rest mass	$m_e = 9.108 \times 10^{-31}$ kg
Radius of electron	$r_e = 2.82 \times 10^{-15}$ m
Permittivity of free space	$\varepsilon_0 = 8.854 \times 10^{-12}$ C^2/N–m^2
Permeability of free space	$\mu_0 = 4\pi \times 10^{-7}$ (H/m)
	$= 1.26 \times 10^{-6}$ N/A^2
Intrinsic impedance of free space	$\eta_0 = 120\pi \Omega$ or 377 Ω

Bibliography

Abhyankar, K.D. and Joshi, A.W., *An Overview of Basic Theoretical Physics*, Universities Press (India), Hyderabad, 2009.

Agarwal, B.K., *Thermal Physics*, Lokbharti Publications, Allahabad, 1988.

Agrawal, B.K. and Prakash, Hari, *Quantum Mechanics*, PHI Learning, Delhi, 2009.

Bekefi, G. and Barrett, A.H., *Electromagnetic Vibrations, Waves and Radiations*, MIT Press, Cambridge, MA, 1977.

Beynon, J., *Introductory University Optics*, PHI Learning, Delhi, 1998.

Cook, D.M., *The Theory of the Electromagnetic Field*, Prentice-Hall, New Jersey, 1975.

Deodhar, G.B., *Introduction to Optics*, The Indian Press Ltd., Allahabad, 1949.

Feyman, R.P., Leighton, R.B. and Sands, M., *The Feynman Lectures on Physics*, Vols. 1, 2 and 3, Narosa Publishing House, New Delhi, 2008.

French, A.P., *Vibrations and Waves*, Arnold-Heinemann, New Delhi, 1973.

Ghosh, K. and Manna, A., *Textbook of Physical Optics*, Macmillan, Delhi, 2007.

Giambattissa, A., Richardson, B.M., and Richardson, R.C., *Fundamentals of Physics*, Tata McGraw-Hill, New Delhi, 2009.

Goldstein, H., Poole, C.P. and Safko, J., *Classical Mechanics*, 3rd ed., Pearson Education, Delhi, 2014.

Goodman, J.W., *International Trends in Optics*, Academic Press, New York, 1991.

Griffiths, David J., *Introduction to Electrodynamics*, PHI Learning, Delhi, 2007.

Griffiths, David J., *Introduction to Quantum Mechanics*, Pearson Education, Inc., Noida, 2012.

Guru B.S. and Hiziroglu, H.R., *Electromagnetic Field Theory Fundamentals*, Cambridge University Press, New Delhi, 2007.

Hecht, E. and Ganeshan, A.R., *Optics*, Pearson Education, Singapore, 2009.

Hewitt, Paul G., *Conceptual Physics*, 10th ed., Pearson Education, Inc., Noida, 2015.

Jenkins, F.A. and White, H.E., *Fundamentals of Optics*, McGraw-Hill, Singapore, 1981.

Khandelwal, D.P., *Textbook of Optics and Atomic Physics*, Shiva Lal Agrawala & Co., Agra, 1979.

Kleppner, D. and Kalenkow, R.J., *Introduction to Mechanics*, McGraw-Hill, New York, 1973.

Longhurst, R.S., *Geometrical and Physical Optics*, Orient Longman, Hyderabad, 1999.

Pain, H.J., *The Physics of Vibrations and Waves*, 6th ed., Wiley, New Delhi, 2005.

Pedrotti, F.L., Pedrotti, L.S. and Pedrotti, L.M., *Introduction to Optics*, Pearson Education Inc., New Delhi, 2007.

Reif, F., *Fundamentals of Statistical and Thermal Physics*, McGraw-Hill, New York, 1965.

Sadiku, M.N.O., *Elements of Electrodynamics*, Oxford University Press, New Delhi, 2008.

Saha, M.N. and Srivastava, B.N., *A Treatise on Heat*, The Indian Press, Allahabad, 1969.

Saxena, A.K., *An Introduction to Thermodynamics and Statistical Mechanics*, Narosa Publishing House, New Delhi, 2010.

Sethi, N.K. and Raizada, S.B., *Textbook of Optics*, Premier Publishing Co., Delhi, 1953.

Singh, Devraj and Pandey, S.K., *Numerical Problems in Physics*, Vol. I, Narosa Publishing House, New Delhi, 2015.

Singh, Devraj, *Engineering Physics*, Vol. 1, 4th ed., Dhanpat Rai & Co., New Delhi, 2014.

Singh, Devraj, *Engineering Physics*, Vol. 2, 5th ed., Dhanpat Rai & Co., New Delhi, 2015.

Singh, Devraj, *Fundamentals of Engineering Physics*, Vol. 1, Dhanpat Rai & Co., New Delhi, 2009.

Singh, Devraj, *Fundamentals of Engineering Physics*, Vol. 2, 2nd ed., Dhanpat Rai & Co., New Delhi, 2010.

Singh, Devraj, Gautam, R.B., Shukla, A.K. and Mishra, P.K., *Applied Physics*, Second Edition, University Science Press, New Delhi, 2014.

Singh, Devraj, Mishra, G. and Yadav, R.R., *Thermal Physics: Kinetic Theory and Thermodynamics*, Narosa Publishing House, New Delhi, 2015.

Singh, Devraj, *Principles of Engineering Physics*, Vol. I and Vol. II, Dhanpat Rai & Co., New Delhi, 2012.

Singh, V.K., Singh, Devraj and Singh, D.P., *Mechanics and Wave Motion*, IK International Publishing House, New Delhi, 2013.

Sirohi, R.S., *Optics and Its Applications*, Orient Longman, Hyderabad, 2001.

Soni, V.S., *Mechanics and Relativity*, 2nd ed., PHI Learning, Delhi, 2011.

Upadhyay, J.C., *Mechanics*, Ram Prasad & Sons, Agra, 2004.

Wark, K., *Thermodynamics*, McGraw-Hill, New York, 1989.

Young, H.D., Freedman, R.A. and Ford, A.L., *Sears and Zermansky's University Physics*, 13th ed, Pearson Education, Delhi, 2014.

Index

Abbe's sine condition, 69, 76, 93
Aberrations, 69, 69
Absent spectra, 398
    missing order, 423
Acceleration amplitude, 229
Achromatic combination, 48, 96, 99, 102
Achromatic doublet, 86, 87, 88, 89, 94
Achromatic prisms, 49
Achromatism, 85, 86, 89, 100
Acoustic pressure, 211, 226
Adequate resolution, 105
Aircraft industry, 495
Air-wedge fringes, 296
Amplitude, 146, 201, 229, 230
Amplitude resonance, 167, 168
Analyzer, 453, 460, 465, 487
Anastigmat, 78
Angle of biprism, 267
Angle of deviation, 45
Angle of minimum deviation, 128
Angle of prism, 65, 128
Angular dispersion, 45, 46, 50, 59, 63
Angular frequency, 149
Angular fringe width, 259
Angular half width, 397
Angular half width of principal maxima, 423
Angular magnification, 42, 133, 134, 137, 139
Angular simple harmonic motion, 153
Anisotropic crystal, 493
Anisotropy, 492
Antenna, 523

Antinodes, 220, 227
Antireflection coatings, 321, 322, 343
Aperture, 69
Aplanatic points, 90, 91, 100, 101
Aplanatism, 90
Artificial rain, 66
Astigmatism, 69, 77, 78
Astronomical telescope, 112, 113, 114, 132, 135, 138, 139, 140
Average intensity, 251
Axial chromatic aberration, 82

Back focal length, 23
Ballistic galvanometer, 189
Bandwidth of resonance, 185
Bandwidth of the circuit, 190
Barrel-shaped distortion, 80
Beam splitter, 298
Biomechanics, 495
Biot's laws of optical activity, 481
    specific rotation, 481
Biprism, 268
    experiment, 347, 351
Biquartz plate, 489
Biquartz polarimeter, 489
Birefringence, 461, 492, 493
Blunt corners, 462
Brewster's angle, 456
Brewster's law, 456, 457, 458
Bright fringes, 257

581

582 Index

Bright ring, 292, 326
Bulk modulus, 208

Calcite, 464
    crystal, 462, 463, 470, 471, 475
    plate, 472
Calcium carbonate, 462
Camera filters, 478
Canada balsam, 464
Cardinal points, 23, 30, 36, 38, 40, 120, 123
Cartesian coordinates, 527
Cartesian sign convention, 114
Cassegrain reflecting telescope, 116, 132, 138
Cauchy's relation, 45, 59
    for a prism, 62
Chromatic aberration, 69, 81, 86, 94, 95, 99, 101, 102, 103, 117, 119, 69, 82
Circular aperture, 373
Circular disc, 372
Circular fringes, 298, 298
Circularly polarized light, 467, 470, 473, 474
Circularly polarized waves, 485
Circular ring, 301
Circular shape of fringe, 311
Coaxial lens system, 26
Coefficient of reflection, 271
Coherence, 247, 337
Coherent and incoherent sources, 248
Coherent sources, 248, 255
    of light, 337
Collimator, 126
Column matrix, 478
Coma, 69, 75, 76, 100
Comatic aberration, 76
Comatic error, 100
Combination of prisms, 47
Combination of two thin lenses, 26
Compared microscope, 111
Comparison of intensities, 252
Compensator, 318
Compound microscope, 105, 108, 110, 131, 134, 137, 140
Concave surface, 291, 292
Concave grating, 412, 413, 414
Concave lens, 25
Concentration, 491
    of sugar solution, 487
Concentric circles, 286

Condensation, 207, 210, 211, 226
Condenser, 187, 188
    lens, 130
Condition for
    brightness, 274, 275
    darkness, 274, 276
    good contrast, 264
    good visibility, 264
    interference of light waves, 263
    minimum chromatic aberration, 132
    spherical aberration, 74
    spherical aberration to be minimum, 103
    sustained interference, 263
Condition of achromatism, 103
Conservation of energy, 251
Constructive interference, 240, 247, 249, 250, 274, 275
Continuous medium, 203
Convergent lens, 24, 25
Convex lens, 24, 25, 367
Convex surfaces, 291, 292, 293
Critical angle, 464
Critically damped, 158, 161, 176
Cross-wires, 123
Crown glass, 47, 101
    prism, 49
Cryolite, 336
Crystallographic axis, 454
Curvature, 78
    of field, 69, 78
Cylindrical wavefront, 241

Damped
    and forced oscillations, 146
    harmonic oscillator, 157, 162, 163, 176, 186
    mechanical oscillator, 193
    oscillator, 181
    vibrations, 186
Damping
    coefficient, 155
    constant, 156, 188
    force, 155
    equation, 175
    motion, 155
Dark
    fringes, 257, 292
    ring, 292, 326

Index **583**

Decrement, 159, 176
Degree of contrast, 248
Depth of penetration, 539
Destructive interference, 240, 247, 249, 251, 274, 276
Detection of polarized light, 474
Deviation, 69
    angle, 62
    method, 266
    without dispersion, 47, 59, 63
Dextrorotary, 481, 484, 489
    substance, 484, 486
Dichroism, 476
Dielectrics, 537
Differential equation of motion, 225
Differential equation of wave motion, 202
Diffraction, 356, 357, 438, 453
    angle, 402
    grating, 391, 401, 407, 423, 432, 435
    and Huygen's principle, 357
    of light, 356, 437
    pattern, 398
Dilation, 207, 210, 211, 226
Direct vision spectroscope, 50, 51, 61, 63
Dispersion, 45, 54, 62, 315
    without deviation, 49, 63, 60
    of light, 44, 63, 65
    of white light, 44
Dispersive power, 45, 46, 48, 64, 90, 99, 420, 399
    of crown glass, 59
    of flint glass, 59
    of a grating, 399, 424
    of a prism, 62
Displacement, 146, 210, 225
    method, 265
Dissipative force, 155
Distortion, 69, 80
Divergent lens, 25
Division of
    amplitude, 245, 287
    wavefront, 245
Double
    concave lens, 22
    convex lens, 20
    diffraction pattern, 387
    image prisms, 466
    refraction, 461, 492
    slit, 330, 384

Driven, 164
    frequency, 195
Driver, 164

Eagle mounting, 416
Echelon grating, 418, 424
Effective focal length, 23
Electric field, 522, 536
Electromagnetic radiations, 4
Electromagnetic wave, 522, 523, 525
Electron gun, 130
Electron microscope, 105, 129, 130, 131, 138, 139
Elevator, 188
Ellipse, 469, 470
Elliptically polarized light, 467, 470, 474, 475
Energy density, 226
Energy of
    harmonic oscillator, 150
    plane progressive wave, 212, 226
    stationary wave, 221, 227
Epoch, 184
Equivalent focal length, 120, 123
Excess pressure, 208, 211, 226
Experimental arrangement, 286
    for Newton's rings, 286
Extended source, 278
External periodic force, 164
Extinction positions, 495
Extraordinary, 461
    image, 461
    ray, 461, 462, 465
    wavelets, 464
Eye lens, 120, 136
Eyepiece, 105, 109, 112, 117, 118, 137, 140

Fabry-Perot interferometer, 246, 307, 308, 313, 328, 342, 343, 344, 345, 346, 349, 353
Faraday's law, 522
Fermat's principle, 6, 7, 8, 11, 13, 14
    of extreme path, 5
    of extremum path, 13
    of least time, 4
Field lens, 120, 122
Finite distance, 365
First, 26
    focal point, 24

half-period zone, 360
principal plane, 24
principal point, 24
Flint glass, 47
Flint glass prism, 48, 49
Fluid media, 207, 208
Fluorescent screen, , 131
Focal length, 27, 41, 101
Focal points, 24, 28, 39, 121, 124
Focusing the
 collimator, 127
 eyepiece, 127
 telescope, 127
 collimator, 127
Forced
 harmonic oscillator, 163, 171, 173, 183
 (or driven) vibrations, 163
 oscillator, 177, 186
 vibrations, 185
Formation of Newton's rings, 286
Fraunhofer's class of diffraction, 358
Fraunhofer's diffraction, 357, 374, 377, 382, 384, 382, 384
 due to narrow slit, 425
 at two slits, 422
Fraunhofer type of diffraction due to single slit, 421
Free space, 532
Free vibrations, 185
Frequency, 146, 201, 229
Fresnel mirrors, 245
Fresnel's biprism, 245, 264, 265, 268, 344, 347, 351
Fresnel's biprism experiment, 351
Fresnel's class of diffraction, 357
Fresnel's diffraction, 357, 368, 372, 373
Fresnel's double mirror, 270
Fresnel's explanation, 358
Fresnel's explanation of optical rotation, 484
Fresnel's fringes, 267
Fresnel's half-period zone, 359, 420
Frictional force, 164
Fringe width, 255, 258, 283
Front focal length, 23

Gauss eyepiece, 125, 126, 140
Gauss points, 23
Geometrical optics, 4, 17, 38

Geometrical path difference, 287
Geometrical shadow, 356
Geometry of calcite crystal, 462
Glass tube, 487
Good conductor, 538
Governing factors of amplitude, 359
Graphical solution, 381
Grating, 399
 element, 402
 at oblique incidence, 400
 spectra, 396
 spectrum, 420

Haidinger fringes, 307, 344
Half fringe width, 311
Half-period zone, 359, 360, 420
Half shade device, 487, 488
Half shade polarimeter, 489
Half wave plate, 471
Half width, 169, 170
 of resonance curve, 178, 188
Halos, 57, 58
Harmonic motion, 146
Harmonic oscillator, 147, 175
 simple pendulum, 152
Hexagonal ice crystal, 58
High power microscope, 92
Horizontal plane, 78
Huygen's
 eyepiece, 119–121, 132, 136, 137, 139, 140
 principle, 243, 357
 theory, 240
 theory of double refraction, 464

Iceland spar, 462
Imple harmonic oscillator, 147
Incoherent sources, 248
Index of refraction, 493
Intensity, 214, 226
 distribution, 391
 distribution curve, 382
 variation in interference, 252
Interference, 240, 453, 320
 and diffraction, 437, 438, 358
 filter, 320, 343
 fringes, 255, 264, 267, 298, 267, 344
 pattern, 258, 285

refractometers, 316
  in thin films, 325
Internal reflection, 63
Intrinsic impedance, 531, 538, 539
Irrotational dispersion, 63
Isotropic crystal, 492

Jamin's refractometer, 317, 328, 342, 353
Jones matrix, 480
Just aperiodic, 161, 176

Kellner eyepiece, 140
Kinetic energy, 151, 152

Laevo-rotatory, 481, 489
Laser, 245, 329
Lateral chromatic aberration, 82
Laurent's half shade
  device, 471
  plate, 489
  polarimeter, 483, 486, 487, 488, 489
Laws of reflection, 6, 8
LCR circuit, 189
Left-handed or laevorotatory, 481
Lens equation, 17, 18, 33
Lens formula, 21
Lens maker's formula, 19, 20, 34, 38, 39
Levorotatory substance, 486
LG plate, 246, 343
Light vector, 478
Limit of resolution, 406
Linearly polarized light, 454
Linear vibration, 186
Lippich's polarimeter, 490
Liquid crystal display, 476
Lloyd's mirror, 269
Lloyd's single mirror, 269
Localized fringes, 282, 302
Logarithmic decrement, 159, 176, 180
Longitudinal, 453
  or axial chromatic aberration, 82
  chromatic aberration, 82, 83, 94, 103
  coherence, 247
  spherical aberration, 72
  wave motion, 200
Lossy dielectric medium, 534

Lummer–Gehrcke plate, 307, 313, 314, 328, 343, 346

Mach–Zehnder's refractometer, 319, 328, 342
Magnesium fluoride, 337
Magnetic field, 524, 529, 536
Magnetic (or electrostatic) lens, 130
Magnification, 33, 105
Magnifying glass, 133
Magnifying power, 106, 108, 113, 131, 138, 139, 141
Malus' law, 460
Marginal rays, 70
Material fringe value, 495
Matrix analysis, 34
Matrix method, 30
Maxwell's equations, 525
Maxwell's stress optic laws, 493
Mean deviation, 50
Mechanical and electrical oscillators, 185
Meridian plane, 78
Michelson interferometer, 246, 297, 303, 305, 306, 327, 335, 342, 343, 345, 349, 353
Microscope, 105
Minimum chromatic aberration, 122
Minimum deviation, 53, 54, 57, 128
  position, 128
Minimum spherical aberration, 73, 101, 122, 139
Missing orders, 398
  spectra, 398
Molecular rotation, 482
Monochromatic, 69
  aberration, 69, 100
  light, 255, 284, 285, 489
  plane polarized light, 471
Moon, 57
Motion of particles, 203
Mountings, 414
Multibeam interferometer, 321
Multiple beam interference, 246
Multiple beam interferometry, 306
Multiple lens eyepiece, 139
Multiple slit aperture, 391

Natural frequency, 187, 190
Nature of achromatism, 89
Negative crystal, 470, 471

Negative eyepiece, 120
Newtonian reflecting telescope, 116, 138
Newton's classic experiment, 44
Newton's corpuscular theory, 240
Newton's rings, 246, 285, 286, 291, 293, 294, 295, 296, 297, 326, 334, 338, 285, 340, 341, 344, 345, 348, 352, 353
  experiment, 293, 339, 349
  pattern, 339
  by reflected light, 288
  by transmitted light, 289
Nicol, 489
  prism, 464
  prism as an analyzer, 465
No chromatic aberration, 139
Nodal points, 25, 39, 121, 124
Nodes, 220, 227, 228
Non-linear vibration, 186
Normal adjustment, 114
Normal incidence, 401, 402
Normal spectrum, 400
N parallel slits, 390

Objective, 109, 112
  lens, 130
Oblique incidence, 400
Obliquity factor, 359, 361, 361
Oculars, 109, 117
Oil immersion, 92
  objective, 92
  objective of high power microscope, 92
Optical
  activity, 481, 482
  diagram of michelson's interferometer, 298
  flatness of surfaces, 284
  instrument, 137, 406
  microscope, 129, 130
  path, 4, 13, 14, 15, 493
  path difference, 493
  rotation, 481, 482
  separation, 39
  system, 23
  technology, 320
Optically active, 481
Optic axis, 462
Optics, 4
Ordinance, 495

Ordinary, 464
  image, 461
  ray, 461, 462
  wavelets, 464
Oscillatory motion, 146
Overdamped, 158, 160, 176

Parallel slits, 390
Paraxial
  optical system, 30
  rays, 70, 71
  zone, 70
Partially plane polarized, 459
Particle acceleration, 229
Particle velocity, 210, 225, 229
Path difference, 249, 324
Pencil-shaped, 59
Perfect blackness, 296
Perfect dielectric conditions, 526
Periodic motion, 146, 147, 198
Periodic time, 146
Petzval condition, 79
  for removal of curvature, 94
Phase, 146, 202
  constant, 149, 534
  difference, 249, 324, 494
  velocity, 204
Photoelasticity, 478, 491, 493, 495
Photoelastic material, 493
Photographic plate, 131
Physical optics, 4, 240
Pierre de Fermat, 5
Pile of plates, 458
Pin-cushion distortion, 80
Planatic lens, 77
Planconvex lenses, 122
Plane diffraction grating, 407, 408
Plane of polarization, 454
Plane of vibration, 454
Plane polarized, 453, 454, 459
  beam, 455
  light, 455, 467, 469, 470, 472–474, 476, 480, 487, 491
  progressive wave, 207
  progressive wave in media, 225
  transmission diffraction grating, 390
  transmission grating, 433, 439

Index **587**

wave, 382
    wavefront, 242, 358
Plane transmission grating, 409
Planoconvex lens, 73, 97, 285, 295
Polarimeter, 486, 490
Polarization, 453, 454, 486
    by double refraction, 461, 464
    by reflection, 455, 456, 458
    by scattering, 475, 476
    by selective absorption, 476, 477
    of light waves, 453
Polarized light, 486
Polarizer, 453, 460, 465, 475, 487
Polarizing angle, 456, 457
Polaroid, 476, 478
Polymers, 492
Position of
    crosswires, 120, 123
    focal points, 28
    principal points, 28
Positive crystal, 471
Potential energy, 148, 151, 152
Power absorption, 171, 179
Power dissipation, 162, 176
Power of the combination, 41
Poynting theorem, 542, 543
Poynting vector, 542, 544
Pressure wave, 231
Primary rainbow, 52, 53, 55, 56, 63
Principal maxima, 397
Principal plane of the crystal, 463
Principal points, 23, 28, 39, 121, 123
Principal section, 463
Principle of optical reversibility, 271
Principle of superposition, 244
Prism, 62, 410
    spectrum, 420
    table, 126
Production of
    coherent sources, 287
    electromagnetic waves, 523
    polarized light, 473
Progressive waves, 224, 227
Propagation constant, 534
Propagation of a plane wave, 524
Propagation of sound waves, 208

Quality factor, 163, 173, 177, 185, 187

Quantum optics, 4
Quarter wave plate, 470
Quartz crystal, 462, 471
Q-value, 186

Rack and pinion arrangement, 126
Rainbow, 51, 63
Raindrops, 52
Ramsden's eyepiece, 122, 123, 124, 133, 139, 140
Rayleigh criterion, 406
Rayleigh's refractometer, 318
Rectilinear propagation of light, 358
Reflectance, 216
Reflected light, 272, 288
Reflected wave, 218
Reflecting telescopes, 112, 116
Reflecting-type telescope, 117, 138
Reflection, 215, 226, 240
    electron microscope, 130
    matrix, 32, 34, 40
Reflectivity, 323
Refracted polarized light, 459
Refracting angle of the prism, 128
Refracting telescope, 139
Refraction in water drops, 63
Refraction matrix, 31, 32, 34
Refraction phenomenon, 13
Refraction through a double convex lens, 20
Refractive index, 15, 40, 41, 59, 62, 64, 65, 102, 48, 128, 291, 317, 327, 456
    by a spectrometer, 128
Refractive medium, 45
Reinforcement, 240
Relation between frequency and wavelength, 202
Relative strain optic coefficient, 493
Relaxation time, 155
Removal of
    astigmatism, 78
    coma, 76
    curvature, 79
    distortion, 81
Resolution of , 406
    spectral lines, 408
    spectral lines by a prism, 411
Resolving power, 129, 141, 406, 407, 410, 420
    of grating, 408, 409, 424
    of prism, 410, 424

Resonance, 166, 167, 184, 187
  curve, 169
Resonant frequency, 195
Restoring force, 146, 147, 157, 164, 184
Resultant amplitude of the wave, 361
Resultant mean deviation, 60
Retardation plates, 470
Rochon prism, 466
Rotatory dispersion, 491
Rotatory polarization, 481
Rowland circle, 414
Rowland (concave) grating, 436
Rowland mounting, 415, 417
Runge–Paschen mounting, 416, 417

Saccharimeter, 486
Sagittal, 78
  plane, 78
Sagittal/horizontal plane, 78
Scanning electron microscope, 130
Scanning tunnelling microscope, 130
Scattering particle, 476
Secondary bow, 57
Secondary maxima, 396
Secondary rainbow, 52, 56, 63
Second focal point, 25
Second half-period zone, 360
Second nodal planes, 26
Second principal plane, 23
Seidal aberrations, 70
Sequential refraction, 459
Setting the prism, 127
Shape of fringes, 327
Shape of the interference fringes, 259
Sharpness of fringes, 311
Sharpness of resonance, 169, 185, 195
Sign convention, 19
Simple harmonic, 192
  motion, 146, 147, 179, 192, 223
  oscillator, 175, 190, 192
Simple microscope, 105, 106, 108, 131, 137, 138, 139
Simple pendulum, 152, 175, 180, 188, 189
Single slit, 374, 377, 382
  pattern, 381
Sinusoidal electromagnetic waves, 532
Skin depth, 539

Slit width, 267, 387
Snell's law, 55, 92, 281
Sodium light, 283
Soleil's compensated biquartz polarimeter, 491
Soleil's polarimeter, 491
Solution of equation, 176
Sound waves, 438
Spatial coherence, 247, 248
Specific heats, 208
Specific rotation, 481, 482, 486, 487, 491
  of sugar solution, 488
Specific rotatory power, 482
Spectral lines, 407
Spectrometer, 105, 126
Spectrum, 44
Speed of light, 523
Spherical aberration, 69–74, 92, 94, 95, 99–102, 119
Spherical refracting surface, 91
Spherical surface, 71
Spherical wavefront, 241
Springs, 191
Spurious bows, 52
Standing wave, 219
Stationary wave, 222, 223, 224, 227, 231
Stokes' law, 271
  of reflection, 272
Straight edge, 368
  diffraction, 368
Stress optic law, 493
String, 215
Sugar solution, 482
Sundogs, 57, 58
Supernumerary, 52
Superposition of two plane polarized waves, 467
Superposition of waves, 324, 231
Superposition principle, 343

Tangential plane, 78
Telescopes, 105, 112, 127, 132, 138
Temporal coherence, 247, 248
Terrestrial telescope, 112, 115, 132, 137, 138
Theory of concave grating, 412
Theory of photoelasticity, 493
Thickness of thin film, 304
Thick lenses, 17
Thin lenses, 17

Thin films, 246, 272, 279
Thin transparent sheet, 260
Time-displacement curve, 160
Time period, 201, 225
Tint of passage, 491
Total internal reflection, 240, 465
Tourmaline, 476
    crystal, 454
Transient vibrations, 186
Translation matrix, 34
Transmission, 215, 226
Transmission electron microscope, 130
Transmittance, 216, 323
Transmitted beam, 454
Transmitted light, 274, 289
Transverse, 453
Transverse wave motion, 199
Tuning fork, 187
Two convex surfaces, 293
Two slits, 384

Undamped vibrations, 185
Underdamped, 158, 159, 176
    harmonic oscillator, 189
Uniaxial
    crystals, 464
    negative crystals, 462
    positive crystals, 462
Uniform plane waves, 528, 531
Unpolarized light, 453, 455, 458, 464
Uses of polaroids, 478

Velocity, 225, 228
Velocity amplitude, 229
Velocity of propagation, 522

Velocity resonance, 170, 178
Vertical plane, 78
VIBGYOR, 44
Vibration, 201, 223
Visibility of fringes, 252, 311, 311
Visual sensations, 4

Wave equation, 208, 225, 525
Wavefront, 240, 357, 438, 464
Wavelength, 201, 229, 258, 258, 303, 313, 401
    of incident light, 401
    of light, 303
    of monochromatic light, 290, 371
    separation, 303
Wavelets, 243
Wave motion, 198, 201, 232
Wave optics, 240
Wave propagation, 209, 537, 538
Wave velocity, 229
Wedge-shaped air film, 348
Wedge-shaped film, 280, 282
White light, 44, 258, 267, 258
Window panes of aeroplane, 478
Wind screens, 478
Wollaston prism, 466

Young's double slit experiment, 245, 246, 256, 258, 263, 324, 330, 338, 343, 344, 346, 347, 350
Young's experiment, 246
Young's experimental set up, 350
Young's two slit experiment, 261

Zero order fringe, 267
Zone plate, 363, 365, 367, 421